LIBRARY/NEW ENGLAND INST. OF TECHNOLOGY
3 0147 1002 7024 1

D1605409

Lockout/Tagout

The Process of Controlling Hazardous Energy

OCCUPATIONAL SAFETY AND HEALTH SERIES

The National Safety Council's OCCUPATIONAL SAFETY AND HEALTH SERIES is composed of five volumes and two study guides written to help readers establish and maintain safety and health programs. The latest information on establishing priorities, collecting and analyzing data to help identify problems, and developing methods and procedures to reduce or eliminate illness and accidents, thus mitigating injury and minimizing economic loss resulting from accidents, is contained in all volumes in the series:

ACCIDENT PREVENTION MANUAL FOR BUSINESS & INDUSTRY
(3-volume set)
Administration & Programs
Engineering & Technology
Environmental Management
STUDY GUIDE: ACCIDENT PREVENTION MANUAL FOR BUSINESS & INDUSTRY:
Administration & Programs, Engineering & Technology
OCCUPATIONAL HEALTH & SAFETY
FUNDAMENTALS OF INDUSTRIAL HYGIENE
STUDY GUIDE: FUNDAMENTALS OF INDUSTRIAL HYGIENE

Other safety and health references published by the Council include:

ACCIDENT FACTS (published annually)
LOCKOUT/TAGOUT: THE PROCESS OF CONTROLLING HAZARDOUS ENERGY
SUPERVISORS' SAFETY MANUAL
OUT IN FRONT: EFFECTIVE SUPERVISION IN THE WORKPLACE
PRODUCT SAFETY: MANAGEMENT GUIDELINES
OSHA BLOODBORNE PATHOGENS EXPOSURE CONTROL PLAN
(National Safety Council/CRC-Lewis Publication)
COMPLETE CONFINED SPACES HANDBOOK (National Safety Council/CRC-Lewis Publication)

Lockout/Tagout

The Process of Controlling Hazardous Energy

Edward V. Grund

National Safety Council
Itasca, Illinois

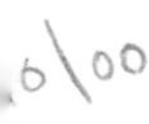

Project Editor: Patricia M. Laing
Associate Editor: Julie McIlvenny

Printed in the United States of America
10 9 8 7 6 5 4 3 2

Library of Congress Cataloging-in-Publication Data
Grund, Edward V.
Lockout/tagout : the process of controlling hazardous energy /
Edward V. Grund.
p. cm. — (Occupational safety and health series)
Includes bibliographical references and index.
ISBN 0-87912-189-0
1. Industrial safety. 2. Power (Mechanics) 3. Energy transfer.
I. Title. II. Series: Occupational safety and health series
(Chicago, Ill.)
[DNLM: 1. Accidents, Occupational—prevention & control—United
States. 2. Energy-Generating Resources. 3. Energy Transfer.
4. Safety Management—methods. WA 485 G889L 1995]
T55.G77 1995
363.11—dc20
DNLM/DLC
for Library of Congress
94-29822
5C695
Product Number: 12175-0100

Contents

Preface

Lockout/Tagout: The Process of Controlling Hazardous Energy examines the dimensions of society's challenge associated with the unexpected and sudden release of hazardous energy in the workplace. The author explores the nature and effect of these occurrences, the private sector's efforts to combat the problem, various governmental initiatives to develop standards and to regulate employers, and the evolution of control measures to prevent these incidents.

This text does not directly discuss hazardous energy release related to natural phenomena, off-the-job activities, ionizing radiation, or intentional releases of energy associated with explosives or controlled nuclear/chemical reactions. However, the concepts, principles, and techniques for energy isolation contained herein are generally applicable.

The purpose of *Lockout/Tagout: The Process of Controlling Hazardous Energy* is to begin the process of gathering and organizing the relevant information on the art and practice of controlling hazardous energy release. The work is viewed as a beginning, not a final treatise on the subject. Research found no comparable reference that was available to those with a need to protect workers at risk from energy release incidents. The book is intended to serve as a basic reference for all those with special interest or involvement—safety practitioners, regulators, researchers, manufacturers, designers, engineers, managers, craftsmen, workers, and academicians. It can be used as a guide to construct an effective hazardous-energy-control process.

Users of the text are encouraged to approach the subject of hazardous energy control from a systematic process perspective in order to improve their effectiveness. Far too frequently, existing lockout/tagout programs have not been well planned or failed to address all of the critical variables.

A model of a hazardous-energy-control system is proposed, consisting of six elements: (1) organizational culture, (2) management process-procedure, (3) human, (4) environment, (5) machine, and (6) design. The basic elements are further subdivided into 40 subelements that reflect the full range of considerations necessary for a coherent control approach.

In Chapter 1, Historical Perspectives, a summary of the evolution of energy use and basic machines and the dangers that came with them is presented. Early efforts at preventing or controlling energy transfer from ancient times through the industrial revolution are reviewed. The text provides the background for understanding the slowly evolving array of prevention initiatives and establishes reference points for cardinal events.

Chapter 2, Unexpected Energy Transfer, describes many of the most noteworthy studies conducted during the past 30 years. These morbidity/mortality studies define the nature and characteristics of energy-release incidents and reveal areas of prevention opportunity.

Chapter 3, Causation Analysis, contains a review of

accident/incident causation theory and its relationship to energy-release incidents. Only through a better understanding of the interconnectedness of root, contributory, and proximate causes and networking sequences can we expect to be more successful in our prevention efforts. The concepts of "design-induced error" and "error provocative" are introduced to offset the over-simplified views regarding human failure.

Chapters 4 and 5, United States and International Hazardous Energy Regulations, respectively, deal with the actions of unions, trade associations, public, standards organizations, and governments to establish guidance or requirements for preventing energy-release incidents. Although the general problem has persisted for centuries, formal attempts to constructively intervene have only occurred over the past 60 years. A broad overview of energy-release prevention tactics and practices is provided.

Chapter 6, The Process Approach, addresses the assessment and planning necessary to establish a complete hazardous-energy-control system. The process cycle involves five major steps: (1) design, (2) implement, (3) monitor, (4) evaluate, and (5) refine. Information is provided to enable the reader to organize the appropriate approach to establish a hazardous-energy-control system.

Chapter 7, System Elements, identifies all of the critical elements found in any well-developed lockout/tagout system. By addressing each subelement, those responsible for designing/developing hazardous-energy-control systems can ensure that all critical content has been included.

Chapter 8, Preventing Energy Transfer, describes the various methods, techniques, and hardware currently used to prevent unexpected or unintentional energy transfer. A 12-step action cycle for energy isolation is offered; it can be used to improve lockout/tagout performance. Alternative energy isolating practices are discussed as a means of identifying and effectively dealing with partially energized or energized task situations.

Chapter 9, Monitoring, Measuring, and Assessing, details what is necessary to ensure the continuing effectiveness of any hazardous-energy-control system. Without this dimension, lockout/tagout systems deteriorate and/or produce less than optimum results.

Chapter 10, Special Situations and Applications, covers what can be viewed as the unique or complex aspects of hazardous energy control. Group lockout/tagout, contractor issues, the role of machinery guarding/interlocks, automated systems/robotics, and high-voltage situations are thoroughly discussed.

Chapter 11, Beyond Compliance, reveals the ingredients necessary to move hazardous energy control from the incremental gain state to the quantum leap: (1) designing for safety, (2) managing change, and (3) continuous improvement of process and system. The road to performance excellence incorporates these ingredients with the six I's strategy, consisting of involvement, inquiry, innovation, intervention, intensity, and improvement.

The Appendixes contain the complete standard, *The Control of Hazardous Energy (Lockout/Tagout)*, U.S. 29 *CFR* 1910.147, Guidelines for Controlling Hazardous Energy During Maintenance and Servicing, NIOSH Pub. No. 83-125, a Diagram for Controlling Hazardous Energy During Maintenance and Servicing, NIOSH Pub. No. 83-125, a Sample Lockout/Tagout Policy and Procedure, Case Histories, and a glossary of energy-control-related terms to assist the user of this manual.

Throughout the text, considerable attention is directed at the safeguards and practices necessary to protect personnel when performing required activities such as servicing, repairing, adjusting, and inspecting machinery, equipment, and processes. In many respects, the safeguards and protective measures are necessary because of the absence of design features that would have reduced or eliminated worker risk during these activities. These situations where operators and maintenance personnel cope with inherent equipment limitations have been appropriately labeled "error provocative." Design improvements associated with access, location and provision of energy-isolating devices, remote lubrication, automatic safety blacks, self-cleaning mechanisms, jam-extraction systems, pressure-relief and venting features, and ergonomic enhancements should be viewed as a management priority.

Additionally, the frequency of worker exposure can be reduced by upgrading system reliability and performance. Fewer breakdowns, jams, product defects, translate into fewer opportunities for worker injury under conditions where energy has not been isolated or is required for the tasks being performed. A managerial mandate for "safety through design" offers the best long-term solution for eliminating the tragic consequences associated with hazardous-energy-release incidents.

There may be consequential energy-isolation information that may have been inadvertently overlooked during the preparation of this work. The author and the National Safety Council would welcome being informed of any material of this nature. We would also appreciate any suggestions or constructive criticism on how to improve the next edition.

Edward V. Grund, MS, CSP, PE, currently Director of Safety and Health, American National Can Company, Chicago, Illinois, has worked professionally for over 30 years in the metals, chemicals, glass, plastics, mining, and refractory industries. Since 1971, he has managed the worldwide safety, health, and loss prevention functions of Kaiser Aluminum and Chemical Corporation, American Metal Climax, Inc., and American National Can. His responsibilities have taken him to all corners of the United States and 20 other countries where he consulted with management groups, labor unions, government officials, and workers. The control of hazardous energy (lockout/tagout) has been of special interest to the author since his first safety assignment in the world's largest integrated steel plant.

As an original member of the ANSI Z244.1 Lockout/Tagout of Energy Sources Committee, he served as a Task Group Leader for drafting various sections of the standard. As current Chairman of the ANSI Z117.1 Committee on Safety Requirements for Confined Spaces, he has had opportunity to emphasize the recognition of the special risks associated with hazardous energy control during the standard's development and revision activity. During the past 10 years, he has conducted seminars and made numerous presentations on hazardous energy control involving thousands of participants.

Ed earned his undergraduate degree from the University of Maryland and master's degree in safety science from the University of Southern California. He has served on and chaired numerous governmental, association, and professional safety and health committees. He is a professional member of the American Scoeity of Safety Engineers and the American Industrial Hygiene Association. He currently serves on the National Safety Council's Industrial Division, Board of Directors and Foundation Board of Trustees. He was awarded the National Safety Council's Distinguished Service to Safety Award in 1992.

The author is deeply indebted to the National Safety Council Technical Publications team, Jodey Schonfeld, Pat Laing, and Julie McIlvenny, for their unrelenting support in all aspects of the completion of this work. Their contributions and encouragement allowed the author to eventually envision the "light at the end of the tunnel." My gratitude also to Sue Baugh, technical writer, whose special skills and assistance on chapters 1, 4, 5, 9, and 11 were invaluable.

Special thanks to the safety professionals of Bethlehem Steel and Kaiser Aluminum and Chemical, with whom the author had the pleasure and good fortune to work. Many of the lessons and ideas contained in this text were acquired from them during years of interaction, observation, and friendly debate. My appreciation and gratitude to my early safety mentors, Len Wozny and Homer K. Lambie, for their counsel and the various opportunities and challenges they placed before me.

To the men and women of General Motors and the United Auto Workers for their exemplary efforts during the past 20 years in pushing the hazardous-energy-control envelope to new dimensions. To Mike Taubitz, General Motors; Barrie Brooks, UAW; Mike Slyne, Nacanco; Fred Manuele, Hazards, Ltd.; Gary Fisher, TDC; Frank Hall, Esq.; and Frank Grimes, USWA my respect and appreciation for their technical contributions and sage advice. Thanks to the following NSC staff and volunteers who devoted their time and expertise during the technical review of the initial manuscript: J. David Amos, Ronald J. Koziol, Joseph Lasek, William McGill, and Fred Rine. To all those other safety practitioners, too numerous to mention individually, who in some say contributed to the final product, my thanks for your cooperation and special efforts.

My regards and appreciation to the original members of the American National Standards Institute Z244.1 Lockout/Tagout Committee, whose fiery debate and eventual consensus contributed immeasurably to the progress made on this critical subject. My appreciation to committee members Frank Rapp, UAW; Paul Shoop, IBEW (in memorium); Jim Geddings, Duke Power; John Henley, ATAA; and Paul Lang for sharing their expertise, experiences, and friendship.

To Gloria Steltman and Ken Kunz, my gratitude for your work processing talents and diligence during the long editorial journey. To my fiance, Regina, who tolerated the countless hours of research and writing in seclusion and assisted selflessly, my commitment and love.

A special tribute to all those laborers, farmers, seamen, soldiers, masons, miners, and industrial workers who fell victim to energy-release incidents over the ages. In your countless names, the quest continues for the ultimate prescription for prevention.

Edward V. Grund, MS, CSP, PE

Lockout/Tagout

The Process of Controlling Hazardous Energy

1

Historical Perspectives

Lockout/tagout refers to the placement of locks and/or tags on an energy-isolating device(s), in accordance with established procedure, to ensure that the machinery, equipment, or process cannot be started while anyone is exposed. The use of lockout/tagout methods for ensuring employee safety from hazardous-energy releases is a relatively recent development in workplaces, yet energy hazards have posed a threat since people first began making and using tools and machines thousands of years ago. Their ingenious accomplishments in avoiding backbreaking labor or increasing their own power or output always came with a cost in terms of human life. The death and casualty lists from hazardous-energy releases included laborers, masons, farmers, miners, soldiers, smelters, engineers, and experimenters, to name just a few.

Only during the past 100 years has the safety movement emerged as a potent force in protecting the lives of workers. Labor, government, and progressive employers have begun to address the age-old issue of safety, now magnified by the rapid evolution and growing complexity of modern industrial and information technology. Machine guarding, power transmission safeguards, electrical protective features, confined-space-entry methods, and hazardous-energy-control techniques are a few of the relevant prevention initiatives that are becoming common worldwide. Unfortunately, many of the lessons that led to these innovations were bought at the price of human pain and injuries. Because such injuries still occur, a complete program for hazardous energy control is presented in this book.

Recently, the safety movement has changed its perception of the events known as *accidents*. Few energy-release incidents are accurately described by that term (see Chapter 3, Causation Analysis, for a discussion of reasons why this is so). Therefore, throughout this text the term *incident,* rather than *accident,* is used. **An *energy-release incident* is any unwanted transfer of energy (electrical, mechanical, hydraulic, pneumatic, chemical, or thermal) that produces injury/damage to persons, property, or processes or otherwise interrupts, interferes with, or degrades the activity in progress.**

It may be easy to understand electric shock as the result of an unexpected energy release. But to view other incidents, such as worker injuries suffered while repairing equipment, in this way may require more explanation. To understand this perspective, one must have some idea of how various types of energy are converted to mechanical use and how energy can create hazards on the job.

This chapter provides a brief historical overview of the development and use of energy resources and the mechanical devices powered by energy. Of course, the risks of using new sources and forms of energy must be balanced against the benefit of increased productivity and efficiency (for example, nuclear power, ionizing radiation, lasers, and masers). As energy sources and machinery have become more sophisticated, so have the hazards; thus, there is a clear need for effective preventive measures to guard against future unwanted or abnormal energy releases.

A BRIEF HISTORY OF ENERGY USE

At its most basic level, *energy* can be defined as the capacity to do work. Bodies at rest are said to possess *potential energy,* whereas bodies in motion possess *kinetic energy* (Figure 1-1). Energy cannot be created or destroyed; it can just be transformed from one state to another. For example, coal can be transformed into heat and light, which can then be used to boil water to generate steam for engines. Types of energy include solar, water, wind, electrical, nuclear, ionizing/nonionizing radiation, chemical, thermal, and biological. Any one of these, under the wrong conditions, can cause injury and/or damage.

Energy should not be confused with force. Although these terms are often used interchangeably, they are quite different. *Force* is any push or pull, such as exerted by gravity, friction, or muscle power, that converts a body at rest into a body in motion, alters the course of a body in motion, or brings it to rest. Friction, for example, will gradually slow any body in motion until it stops. Four major types of force have been identified: gravity, electromagnetism, the strong nuclear force, and the weak nuclear force.

Work, in the scientific sense, is a measure of force and distance ($W = f \times d$). That is, work, measured in joules (J) or other units, reflects how much effort is required to move an object a particular distance. The greater the force needed and/or the longer the distance covered, the more work must be performed. By this definition, a worker pushing boxes up a 20-ft (6.1-m) ramp will perform twice as much work as a worker pushing the same number and weight of boxes up a

Energy is the capacity to do work. Energy cannot be created or destroyed; it can just be transformed from one state to another. Bodies at rest are said to possess *potential energy,* whereas bodies in motion possess *kinetic energy.* Work is done when potential energy is released as kinetic energy (for example, when a wound spring is released).

- *Chemical energy* is stored in chemical systems such as fuel cells, oil, gas, and so on. This type of energy can be released as heat, light, or electrical energy.
- *Electrical energy* is manifested as a stored electrical charge in a capacitor or as electrical charge flowing in a circuit as electrical current. It produces electrical heating and the electromagnetic and electrochemical effects of current. Electrical energy is measured in *ohms, watts, volts,* and *amperes.*
- *Mechanical energy* is a form of energy possessed by bodies or physical systems because of their position or motion—for example, vibrational, kinetic, potential, and rotational energy.
- *Nuclear energy (fission and fusion).* The former is produced by the fission of uranium-235 in a reactor, which releases heat and gamma radiation. The latter is produced by the fusion of hydrogen nuclei, which releases electromagnetic radiation.
- *Thermal energy* is heat possessed by a body or substance that manifests as vibrational energy.
- *Force* shows the presence of energy in a physical field. It can change the position or momentum of a body in space and enable work to be done. Force is measured in *newtons* (N). Examples of force include friction, gravity, heat, vibration, and magnetism.
- *Power* is a measure of the rate at which work is done or the rate at which energy is expended. Power is measured in *joules* (J) and in some instances, *horsepower* (HP).
- *Work* is done when energy interacts with a body or particle, causing it to move. Like power, work is measured in joules and is an expression of force times distance ($W = f \times d$).
- *Machines* are devices by means of which energy input at one point can produce work at another point. According to the law of conservation of energy, work output cannot exceed energy input. The *efficiency of a machine* is the ratio of its work output divided by the required energy input. The ratio has no units and is expressed as a percentage; its value is always less than 100% due to friction and other forces.

Newton's three laws of motion. (1) A body at rest will remain at rest, and a body in motion will continue in a straight line unless acted on by an outside force. (2) The rate and direction of change of a body's momentum is directly proportional to the strength and direction of the force exerted upon it. (3) Every action has an opposite and equal reaction.

Figure 1-1. Basic terms.

10-ft (3.5-m) ramp. *Power* is a measure of the rate at which work is done, either in a linear or rotary motion. Usually, power is measured in horsepower (HP) or watts (W).

These four concepts—energy, force, work, and power—form the basis for understanding how all tools and machines work, regardless of how simple or complex they may be. Our ancestors, although less technologically sophisticated, understood these concepts well.

Early Uses of Energy

The first humans had only to look to their environment to find sources of available energy: human and animal muscle, wind, water, fire, and gravity. While learning to harness them, however, humans also had to deal with their hazards and limitations.

Human and animal muscle. Excavations from the earliest settlements in the Middle East; Africa; North, Central, and South America; and Asia document the use of human and animal muscle as the primary motive power for hunting, gathering, and early forms of agriculture (Purcell, 1982). Because of their limited strength, humans had to find ways to multiply the power, force, and speed provided by their muscles. Even such primitive tools as digging sticks and spear throwers (which are types of levers) demonstrate the capacity to amplify muscle power. Over time, humans added other tools and weapons such as bows and arrows, hand axes, pottery wheels, bow-string drills, cutters, grinders, plows, and planters to gain an advantage over their environment. Many of these devices involved transforming potential energy (for example, pulling back a bow string) into kinetic energy (releasing an arrow from the bow).

The domestication of animals made available another form of motive power to push, pull, or carry loads; run threshing machines and various types of rotary wheels; and extend the range of human travel. While one man could exert 18 lb (8.2 kg) of force pulling a load, for example, a horse or an ox could exert 120 lb (54.5 kg) of force (Calder, 1968). Even with the greater individual power of draft animals, however, most of the larger work projects, such as the pyramids in Egypt and Central/South America, were built using masses of human workers, supplemented with horses and oxen.

As humans developed more sophisticated tools, machines, and forms of transportation, they also became more skilled at organizing human and animal labor to power these devices. Written records and paintings from early Egyptian and Babylonian times depict the use of draft animals and slave labor to work fields, build roads, construct buildings, and power ships and other forms of transport. Even with the development of waterwheels and treadmills, human and animal muscle often provided the motive power. During the Industrial Age, however, muscle power ceased to be the dominant source of energy for driving machinery.

Wind. Regions with large bodies of open water or level plains had an abundance of wind energy because there were few natural barriers to air flow. Likewise, people living in hilly regions and certain mountainous areas could also take advantage of wind currents, although they tended to be less reliable there.

Two applications for wind energy made it a valuable resource to early humans: lift and conversion of linear to rotary motion. Lift was used to propel ships at sea and later helped break the bonds of gravity with heavier-than-air craft. Fishermen and sailors discovered that their ships moved faster and farther with sails than on muscle power alone. Even though they believed that the wind was pushing their craft, it was actually pulling them through the water. The sails created an area of low pressure on the side away from the wind, toward which the ship moved. Skilled mariners could manipulate the sails to make ships travel at an angle or almost directly into the wind. The lift created by an aircraft wing is essentially the same as that created by a sail. The low-pressure area forms above the wing, thereby pulling the aircraft upward.

Conversion of linear to rotary motion was the principle behind the windmill, one of the earliest machines built in human settlements. The wind propelled blades or vanes that were connected to an axle or crankshaft. The axle, through a series of interlocking gears, turned a horizontal millstone that ground seed into flour.

The hazard of wind energy lay in the fact there was no way to control how hard, how often, or from which direction the wind blew. While the wind was blowing, the power, in effect, was always ON. This situation created hazards for sailors working in a ship's rigging and for millers, metalsmiths, and their employees working in the windmills.

Water. Early humans took advantage of the principles of water pressure, currents, and wave motion to accomplish work. For thousands of years, people dammed rivers, lakes, and streams to create irrigation systems and reservoirs for growing crops and ensuring a reliable water supply. Builders also found that by narrowing the channel of flowing water, they could increase its speed and power, thus increasing its ability to perform work. Running water was harnessed to turn waterwheels and transport by barge or ship heavy loads that would have otherwise required far more expenditure of animal or human muscle power. The energy of rising and falling tides was used to drive machinery and transport loads. More recently, the energy of running water has been used to generate electricity.

Hazards of water power involve tidal waves, flooding, death by drowning, and gradual weakening of machine parts or wooden structures directly exposed to standing or flowing water. Like wind power, water power may sometimes be impossible to control, posing a danger to workers who must maintain or repair machinery driven by this energy source.

Fire. Although it is not known how humans first discovered and used fire, it is certain that harnessing this form of energy radically altered human existence. Fire or combustion has been used in every industry from metalworking, construction, chemical, and transportation to space exploration. With it, humans have been able to create new alloys and chemical substances not found in nature, construct machines and buildings of these materials, heat permanent dwellings in colder latitudes, and develop increasingly sophisticated and powerful engines for work and transportation. Perhaps more than any other energy form, fire has given humans a unique status as the one species that continues to adapt nature to itself, rather than adapting itself to nature.

Hazards associated with fire include burns and the release of toxic fumes, explosions, and smoke. History offers many stories of towns and cities devastated by uncontrollable fires started by human errors or industrial accidents, such as the great Chicago fire that leveled most of the city in October 1871.

Gravity. Gravity, although a force and not a form of energy, was also harnessed by early humans to perform work. One of the simplest applications of gravity that they employed was to use it to convert the potential energy of objects into kinetic energy. For example, blocks of stone were guided by workers and allowed to slide, roll, or fall down an inclined ramp or slope.

In a more sophisticated application of this force, farmers devised gravity-fed irrigation systems, in which a water container was built higher than the irrigation channels so that gravity would pull the water into the channels and from there into the fields. Gravity was also used to turn overshot waterwheels; as water from an overhead sluice or spout filled up the separate segments or buckets of the wheel, their increasing weight pushed the wheel downward. The buckets then emptied as they neared the bottom of the cycle, keeping the wheel in motion.

Finally, gravity can be used to provide a sense of vertical and horizontal direction. Engineers and carpenters developed the plumb line, which took advantage of this principle to help them construct walls that were truly perpendicular to the earth. This method of construction prevented the walls from collapsing under their own weight or gradually settling until they toppled over. Plumb lines are still used by many bricklayers and construction workers today to find the true vertical and perpendicular angles of a wall.

Work hazards associated with gravity include falling from heights; being caught under or between falling objects, collapsing walls, or other structures; and bursting containers in which contents, such as water or other liquids, are under pressure due to their own weight. Again, because gravity cannot be turned off, these hazards are ever present in any work area in which objects are unstable or positioned at various heights or where workers must be elevated as part of their jobs. A shelf collapsing under too much weight, for example, is a graphic illustration of gravity converting potential energy into kinetic energy—with grave consequences for those directly below.

Five Basic Machines

Since prehistoric times, humans have used machines to take advantage of various forms of energy and types of forces to perform work and to alter their environment. Whether as unsophisticated as a hammer or as complex as a manufacturing robot, all machines serve one or more of five basic functions:

- *To transform energy. A* solar-powered battery transforms solar energy into chemical and electrical energy.
- *To transfer energy.* The crankshaft, drive shaft, and rear axle transfer the energy from a car's engine to the rear wheels.
- *To multiply a force.* A system of pulleys allows a worker to lift a heavy load by exerting a force that is smaller than the weight of the load. The machine is multiplying the force exerted by human muscle. In doing so, however, a machine generally sacrifices speed.
- *To multiply speed.* A machine can gain greater speed through the exertion of more force (for example, a bicycle moves faster when the rider increases his output of energy).
- *To change the direction of a force.* The track-switching mechanism on a railroad can change the direction of a freight train without affecting its speed.

Each of these five functions also carries the potential for an unwanted or abnormal release of energy. At any point in the process of transforming, transferring, multiplying, or changing the direction of energy or force, machines can injure those operating or working near them. The more powerful and complex the machine and the greater the force or energy used to run it, the higher the risk to workers.

The Greeks identified five simple machines that form the basis for every machine constructed: the lever, the inclined plane (and its cousin, the wedge), the screw, the pulley, and the wheel and axle.

The lever. The lever is perhaps the oldest machine in the world. Every lever is composed of three elements: a fixed point called the *fulcrum* or *pivot point,* a force or effort, and a resistance or weight. There are three classes of levers (O'Brien, 1968).

In the *first-class lever*, the fulcrum is located between the force and the weight. Seesaws, crowbars, oars in a rowboat, and pliers are good examples of this type of lever.

Second-class levers have the fulcrum at one end, a weight near the fulcrum, and a force applied at the other end. The wheelbarrow is the classic example of this lever. Both first- and second-class levers are generally used to overcome heavy weights with relatively small effort. In the wheelbarrow, for instance, 50 lb (22.7 kg) of effort applied to the handles can easily move 200 1b (90.7 kg) of weight.

In the *third-class lever,* the fulcrum is at one end, the weight is at the other end, and the force or effort is applied between the middle and the fulcrum end. The human arm, with the elbow joint serving as the fulcrum, is the perfect example of a third-class lever.

Levers come in countless forms from simple tools to heavy cranes, drill presses, and operating levers in machines. When Archimedes, the Greek philosopher and scientist, declared, "Give me a fulcrum on which to rest and I will move the earth," he was only slightly overstating the importance of the lever.

The inclined plane and the wedge. After the lever, the inclined plane and the wedge are the second-oldest mechanical inventions. Ramps, sloping treadmills, ax blades, and chisels are examples of these simple machines.

The inclined plane permits a worker to overcome large resistance or weight by applying a relatively small force through a distance that is longer than that through which the load is raised. The inclined plane splits the vertical force of gravity into two smaller forces, one perpendicular and one parallel to the plane (O'Brien, 1968). Only the parallel force needs to be counteracted. For example, if a worker must lift a 200-lb (90.7 kg) barrel straight up 3 ft (0.9 m) into a truck, the force required would be 200 lb (90.7 kg)—greater than most people could apply. However, by using a 6-ft (1.8-m) plank and rolling the barrel up the inclined plane, the worker needs to apply only 100 lb (45.3 kg) of effort to move the barrel into the truck.

The wedge is a special application of the inclined plane and can actually be viewed as two inclined planes joined at the base. By driving a wedge full-length into the material to be cut or split, the material is forced apart by a distance equal to the width of the broad end of the wedge. Long, slim wedges give a high mechanical advantage, particularly in situations where other simple machines will not work. Wedges also can be used to prevent movement by being forced under or between objects.

The screw. The screw can be regarded as a modified inclined plane wrapped around a central cylinder. Its operation depends on the twisting motion of a lever, either as a handle or as a driver. The screw can be used to raise weights or to press or fasten objects together. The mechanical advantage of the screw comes from the length of the lever that turns it and the distance between the threads (known as *pitch*). When a lever arm makes one full turn, the screw is raised a distance equal to its pitch.

The most famous screws of ancient times were the wooden screws of Archimedes used to raise water from one level to another in irrigation and to drag a three-masted ship fully loaded onto dry land (Usher, 1954). Screws are found in presses, worm gears, propellers, drill bits, adjustable wrenches, vises, bolts, and countless other applications.

The pulley. A pulley is basically composed of a wheel, a fixed support, and a rope or chain. The mechanical advantage is given by the number of ropes or chains that support a weight. A fixed pulley has a mechanical advantage of 1, because the rope gives the puller no help—it requires 50 lb (22.7 kg) of effort to move the same amount of weight. The single movable pulley, with two segments supporting the weight, has a mechanical advantage of 2 because it requires less effort to move the weight. The block and tackle has a mechanical advantage of 3, but that smaller effort must be exerted through a longer distance.

The main uses of pulleys are in moving or lifting objects; the most familiar pulleys include hoists and block and tackles. A complex series of pulleys can be used to lift massive weights in construction work, cargo loading, and salvage and mining operations.

The wheel and axle. The wheel and axle is a type of lever that rotates about its fulcrum a full 360°. The wheel and axle became one of the most important simple machines invented because it could be used not only to move heavy loads and perform work but also to give humans far more mobility over greater distances than ever before.

Like the lever, the amount of effort the wheel and axle magnifies depends on the ratio of the radius of the axle to the radius of the wheel. As a result, a 20-in. (50.8-cm) wheel on a 1-in. (2.5-cm) axle magnifies its force by 20. Furthermore, the mechanical advantage of the wheel and axle can be increased by the use of a series of gears, as in a bicycle or automobile transmission.

The wheel and axle is a remarkably versatile machine and can be applied in thousands of ways, from the simple mechanism of a doorknob to a windmill, a roller conveyor belt, or a 16-wheel trailer truck. Some of the earliest uses of the wheel and axle were the windlass, used to raise heavy weights, and the windmill or waterwheel, which was so important to early cultures (Sterland, 1967).

Early Energy-Release Incidents

Like the forms of energy and types of force harnessed to do work, each of these five simple machines poses risks as well as benefits to those who use it. As the following three scenarios suggest, workers and soldiers were not exempt from incidents in which hazardous-energy releases led to damaging consequences.

Pyramid construction. The construction of the pyramids of ancient Egypt remains one of the greatest engineering feats in history, both from a technical standpoint and from the sheer magnitude of the labor required to raise these massive structures (Williams, 1987). Some of the larger stone blocks in the Great

Pyramid of Giza, for example, weigh up to 100 tons (90,720 kg) each. Each block had to be cut and lifted from the quarry, transported down the Nile, hauled overland to the construction site, and then worked into its proper place in the pyramid. Without the benefit of heavy machinery, Egyptian engineers had to rely on the lever, the inclined plane and wedge, the pulley, the wheel and axle, and the screw to augment human and animal muscle power to maneuver these massive blocks.

The sheer weight and bulk of the stones made every step of the work process dangerous. Perhaps the greatest hazards were encountered during the arduous work of setting the stones in place as the pyramid was erected. Scholars believe that this was accomplished by building long inclined planes or ramps that encircled the pyramid as it rose higher (Williams, 1987). The stones may have been placed on wooden rollers and pulled by laborers up the ramps. Each time the laborers needed to stop and rest, other workers would place large wedges under the stone blocks and rollers to prevent them from sliding back down the ramp.

If the wedges were not properly sized or properly placed under the stones, however, the weight of the block could crush or dislodge the wedges, resulting in an unwanted energy release as the block suddenly moved downhill. Any workers caught in its path could be killed or maimed, while those hanging onto the ropes used to pull the stones upward could be dragged and trampled trying to get clear.

Siege machines. As early as 400 B.C., the Greeks took advantage of the forces of tension and torsion to build catapult machines to help in winning sieges of towns or fortifications (O'Brien, 1968). The Romans improved on the design and built catapults capable of hurling stones hundreds of yards.

The basic principle of the catapult was to put a lever under sufficient tension to create and store energy, like cocking back the hammer on a revolver, and then to release it in a powerful surge forward. Torsion machines were built so that the lever was pulled back against the force of tightly twisted ropes. Once released, the lever could throw a 50-lb (22.7-kg) stone some 1,200 ft (366 m). The catapult posed two main hazards: The machine tended to kick back once the lever was released, and the stones might be flung accidentally into one's own troops (an early type of friendly fire). This latter danger was particularly a risk in the heat of battle, when the restraining ropes might become frayed, the catapult components might need repair, or the release was premature. If the ropes holding the lever down slipped or broke, the stones could fall short of their enemy targets and strike supporting foot soldiers or cavalry advancing in front of the machine. The sudden energy release of the lever could also cause serious injuries to those in its immediate path (Figure 1-2).

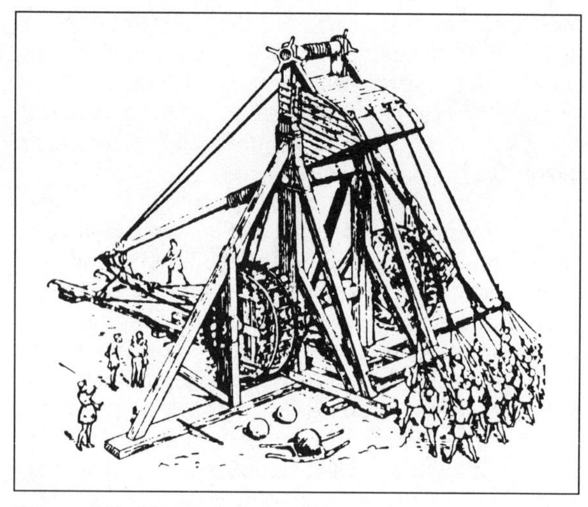

Figure 1-2. Mangonel.
Source: Reprinted by permission from *Webster's New Collegiate Dictionary.* © 1980 by Merriam-Webster, Inc., publisher of the Merriam-Webster® Dictionaries.

Windmills. As windmills became more sophisticated, particularly those developed in western Europe during the Middle Ages, they also became more dangerous for those who had to operate and maintain them. As shown in Figure 1-3, converting vertical mechanical energy to horizontal mechanical energy involved the use of a series of interlocking gear wheels to channel the energy from the drive shaft to the grinding stones (Reynolds, 1970). These gears, often constructed of wood and metal, had to be serviced constantly to prevent breakdown of the machinery.

To do this maintenance work, the worker had to stop the windmill from turning by securing the vanes with heavy ropes. The gears would then come to a halt, permitting the worker to make repairs or lubricate the mechanism safely. Should the ropes break or the vanes be improperly secured, however, the gears would begin moving again without warning. The worker could be caught in the mechanism with tragic results. Often, workers may have performed maintenance or servicing tasks while the mill machinery was still turning to save time and effort. It would take only a slight misstep or miscalculation to be caught in the moving gears. It is likely that over the centuries, a sizable number of millers, metalsmiths, and others—along with their laborers—were casualties of this type of incident.

THE EVOLUTION OF ENERGY USE

For several thousand years, humans were limited to only a few forms of energy, primarily from wind, water, fire, and human and animal muscle. Even with the invention of more complex machines during the centuries preceding the so-called Dark Ages and Middle Ages of Europe, the motive power was still supplied by these energy resources. Although the Romans

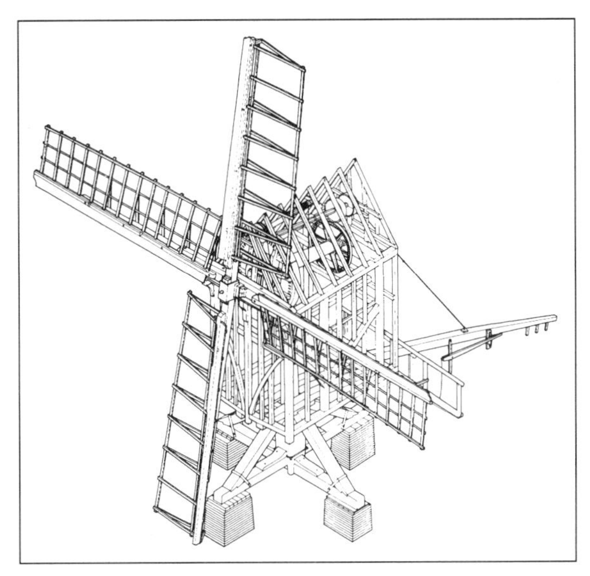

Figure 1-3. Bourn Mill, Cambridgeshire, 1636. An early open trestle post mill, much altered and repaired, but retaining the simple pitched roof of the medieval mill. The body has been extended at the rear to provide space for additional machinery, and the underframe raised on piers of brickwork.
Source: John Reynolds, *Windmills & Watermills* (New York: Praeger Publishers, 1970). Reprinted by permission of Henry Holt & Company.

built monumental machines, such as the Barbegal flour mills that could produce 28 tons (25,401.6 kg) of flour in one 10-hour day, their utility was limited by the water, wind, or muscle power that could be harnessed to drive them (O'Brien, 1968). And even though 15th-century Europeans were building quite sophisticated machines, they still had to rely on age-old methods of generating energy.

The Era of Steam and the Heat Engine

By the 13th century, waterwheels were powering machinery that sawed wood; processed cloth; and mined gold, zinc, lead, silver, copper, and iron from the earth (O'Brien, 1968). It was the thirst for metals and the advantages that came with them that sparked the next step in energy development.

The main problem in working mines was not simply digging out the ore but keeping the mine shafts and corridors dry so that work could continue. Conventional suction pumps could remove only a limited amount of water at any one time, and due to the mechanics of air and water pressure, they could lift the water no more than 33 ft (9.9 m).

In 1712, a young English ironmonger, Thomas Newcomen, invented the first steam engine, called an atmospheric engine, to be used to pump out a mine (Shepard, 1976). As shown in Figure 1-4, Newcomen's engine featured a massive wooden crossbeam; at one end hung a pump rod connected to a plunger in the mine below, and at the other end were

Figure 1-4. Newcomen engine, 1732.
Source: Abbott Payson Usher, *A History of Mechanical Inventions,* 2d ed. (Cambridge, MA: Harvard University Press, 1954; Dover Publications, Inc., 1988). Used with permission.

the piston and cylinder. The piston was driven by alternately applying steam and then condensing it with a spray of cold water, which rocked the beam up and down. The beam then moved the pump rod up and down, operating the pump. By 1769, many of these engines were being used in mines throughout the Northumbrian coal fields. But they had one drawback—they required large amounts of coal to run.

A young Scottish mechanic named James Watt asked himself why the engine had to use so much steam and fuel to produce so little power. To condense the steam, the main cylinder had to be heated and then cooled ad infinitum, which resulted in striking mechanical inefficiencies. In 1765, Watt hit on the solution—carry the steam to a separate condensing container that could be kept cold. The cylinder, in turn, could remain as hot as the steam through every stroke of the piston. This arrangement eliminated heat loss; with a more efficient arrangement of pistons and valves, the engine could produce far more power while using less fuel. Watt even devised a unit of measure—horsepower—to describe the engine's output.

The steam engine was fitted with a connecting rod and flywheel, along with a system of rotating gears, that converted its up-and-down, reciprocating motion into rotary motion. Later modifications to the engines alternately applied steam and vacuum to opposite sides of the piston, which doubled the cylinder's power and used higher pressures. Watt also devised two key refinements: a throttle instead of an ON or OFF control, and a centrifugal governor that automatically closed the throttle when the operating speed of the engine exceeded a set level. The governor was the first application of a feedback control to a heat engine. This was critical because a given volume of water could expand over a thousand times as it was converted to steam. With the contents of the steam engine continually under pressure, control and safety valves were vital. Even with these devices, however, steam engines were notorious for their tendency to explode and set their housings on fire. Tending these machines was regarded as a high-risk occupation.

After a few false starts, Watt built a successful model and launched the era of the steam engine. By the early 1800s, steam engines were making their mark not only in mines, paper and textile mills, and factories but also in transportation. Inventors such as William Hedley and Peter Cooper built the prototypes of the steam-powered locomotive, while John Fitch and Robert Fulton were among the first to design paddlewheel steamboats.

Eventually, the rate and severity of steam-related accidents became a matter of public concern and prompted the first efforts to develop safeguards and regulations governing the use of this volatile form of energy. On March 2, 1854, for example, an overheated boiler burst with a tremendous explosion that destroyed the boiler room and adjoining blacksmith shop of the Fales and Gray Car Works in Hartford, Connecticut (Weaver & McNulty, 1991). The blast killed 9 people outright; 12 more died later, over 50 people were seriously injured, and the property loss to the company amounted to more than $30,000. A coroner's jury investigating the tragedy not only found that the engineer had been careless in allowing an excessive accumulation of steam but also offered recommendations to help prevent future catastrophes:

- Regulations should be devised to prevent careless or irresponsible persons from being placed in charge of boilers.
- Regular safety inspections should be made by municipal or state authorities.
- Boilers should be placed outside the factories for which power is being provided.
- Employers using steam as power should pay close attention to the safety of workers.
- Some measure should be adopted to prevent steam boilers from being rated for carrying more steam pressure than would be consistent with safety (Weaver & McNulty, 1991).

Each of these proposals came to be adopted by public and private agencies that used steam power. In 1864, 10 years after the Fales and Gray disaster, the Connecticut State Legislature passed a boiler inspection law that authorized the governor to appoint one steam boiler inspector for each state congressional district and mandated yearly inspection of all boilers. Safe boilers would be issued a certificate, while unsafe boilers would be retired until all defects had been corrected.

Although a step in the right direction, such measures did not reduce the number of steam-related accidents. Despite the many thousands of boilers in use throughout the United States, there was an appalling lack of knowledge about the properties of steam or what caused boilers to explode. Even though one explosion occurred roughly every four days, most of these unwanted energy releases were regarded as acts of God rather than the result of predictable physical laws.

To address some of these safety issues, a group of young businessmen and industrialists organized the Polytechnic Club in 1857; membership was made up primarily of those who were associated with industries, businesses, or institutions that used steam power. Members of the club reasoned that research, improvements in boiler design and manufacture, and regular inspections could greatly improve the safety record of steam boilers. Unfortunately, manufacturers were generally not receptive to improvements in design if it meant adding to production costs, and users of steam were so accustomed to explosions that they regarded a certain loss of life and property as part of the inevitable cost of doing business.

However, some members of the Polytechnic Club discovered that agencies in England such as the Steam Boiler Assurance Company not only inspected boilers for safety violations but also insured them against the damage caused by explosions. The U.S. Civil War disrupted the club's work, and it took another spectacular explosion to push forward the development of boiler and pressure vessel standards and regulations. On April 27, 1865, the Mississippi River steamboat *Sultana* exploded in a fiery blast that killed 1,238 passengers (Weaver & McNulty, 1991). This disaster prompted the formation of insurance agencies to both inspect and insure steam boilers. One of the earliest of these agencies was the Hartford Steam Boiler Inspection and Insurance Company, officially chartered on October 6, 1866.

Inspection work was dangerous, and men hired as inspectors had few safeguards or protections. The only protective equipment was a kind of coverall that left the face exposed, and there was always the danger that someone would start the boiler while the inspector was inside. Nonetheless, these inspections helped to identify boilers unsafe for further use and to develop methods and standards for determining the condition and reliability of a steam boiler in a wide range of work settings.

Over the next few decades, there were marked improvements in boiler safety and design. J. M. Allen, one of the founders of the Hartford Insurance Agency, conducted scientific tests of boiler plates, seams, rivet sizes, and spacing to determine the optimal boiler design. His computations came to be known as the Hartford Standards and were adopted in 1889 by boilermakers under the name "Uniform Steam Boiler Specifications" at a meeting of the American Boiler Manufacturers Association. In 1914, the American Society of Mechanical Engineers (ASME) Boiler and Pressure Vessel Code was promulgated and adopted by a few states and insurance/inspection agencies. Most states, however, were slow to accept the standards set forth in the code. Anticipating this turn of events, a group of representatives from the boilermakers, boiler insurance companies, and steam users formed the American Uniform Boiler Law Society on July 28, 1915, later known as the Uniform Boiler and Pressure Vessel Laws Society, Inc., to promote boiler and pressure vessel safety. Today, all 50 states, the District of Columbia, Puerto Rico, Guam, the Panama Canal Zone, and Canada have adopted either a boiler law, a pressure vessel law, or both as their standard.

Since 1915, other organizations have been formed to help ensure the continued development of stronger safety guidelines, standards, and regulations regarding boilers and pressure vessels. These groups include the National Board of Boiler and Pressure Vessel Inspectors, the Factory Mutual Engineering Association, and Manufacturers of Boiler and Pressure Vessels. The National Board was organized

> for the purpose of promoting greater safety to life and property by securing concerted action and maintaining uniformity in the construction, installation, inspection and repair of boilers and other pressure vessels and their appurtenances; thereby assuring acceptance and interchangeability among jurisdictional authorities responsible for the administration and enforcement of the various sections of the ASME [American Society of Mechanical Engineers] Boiler and Pressure Vessel Code. (*National Board,* 1993)

Agencies such as the Hartford Steam Boiler Inspection and Insurance Company even developed some of the early educational and training programs such as their Correspondence Course for Firemen, which taught workers about the proper care and management of boilers and furnaces.

Overall, the frequency and severity of boiler explosions served perhaps as the precursor to interest in safety issues associated with hazardous-energy-release incidents. As in many safety areas, however, the focus was initially more on protecting property and the public than on protecting the health and lives of individual workers. Only in the past 50 years has the emphasis shifted to those who operate and maintain the machines.

Steam power enabled the United States to develop its transportation and manufacturing industries. In 1869, the first transcontinental railroad was completed, linking farmers and manufacturers in the West and Midwest to the markets in the East. In 1889 and 1890, Best & Holt built the first steam traction engines used in lumbering and farming. By 1892, the company, based in Stockton, California, was manufacturing 200 combines, 5 harvesters, and 10 steam traction engines a year (Shepard, 1976). These machines and similar inventions hastened the close of the frontier and helped to mark the United States as an emerging world power.

Hydraulic and Pneumatic Machines

Hydraulic and pneumatic machines also represented a significant step forward. The hydraulic ram pump, invented by J. M. Montgolfier in 1797, had no cylinder or piston, just two valves in a valve box and an air vessel on the delivery line (Burstall, 1965). The pump's action depended on establishing a flow of water through the valve box and interrupting the flow by suddenly closing a valve. The pressure waves that were set up continued to operate the valves indefinitely. The hydraulic press had numerous industrial applications, particularly where thousands of pounds of pressure were required to forge steel parts. Eventually, reciprocating hydraulic pumps increased the power delivered by earlier, simpler machines. Many, such as the 100-in. (254-cm) pump built for London's waterworks, were originally driven by steam engines.

Further developments led to the hydraulic jack and hydraulic air compressors, which were used in heavy industry and eventually in braking, transmission, and steering systems for trains and motor vehicles. Sir William Armstrong devised a hydraulic accumulator that enabled this form of power to be used on a large scale. The accumulator consisted of a cylinder containing a weighted piston whose size and weight was proportioned to equal the hydraulic pressure in the main. This device was used to operate such tools and machines as large cranes, lifts, and jacks. A hydraulic-pressure intensifier could be used to obtain pressures that were higher than the original supply pressure. Many of the same principles of hydraulics were used to develop pneumatic devices for heavy industry.

Chemical Energy

At the same time that steam was revolutionizing the face of commerce, the chemical industry was beginning to make its own contribution. Two major sources of stored or entrained chemical energy were developed in the form of explosives and petroleum products.

Black gunpowder, brought to Europe from China in the Middle Ages, had been used primarily in the development of weapons—the flintlock musket, the pistol, and the cannon. These weapons involved a controlled explosion contained in a metal cylinder that hurled a bullet or cannonball several hundred yards. In the 1700s, the Du Pont family saw commercial opportunities in this new form of power and began developing their own formulas for explosives, separating blasting powder from gunpowder used in firearms. In 1866, Alfred Nobel of Sweden, who had invented blasting caps, began producing dynamite, which proved to be safer and more practical industrially than black powder. Dynamite could be used in precisely controlled explosions to blast tunnels in mountains and create underground channels for sewer systems, extract minerals from the earth, and even sink new oil wells. The Du Ponts quickly began producing their own form of dynamite and used this key product to found an international chemical empire that helped usher in the era of fossil fuels (Colby, 1984).

In the mid-1800s, Nicolaus August Otto, a German engineer, began to wonder if the principle of contained explosions could be used to drive the pistons of a new kind of engine (O'Brien, 1968). Instead of gunpowder, Otto used a mixture of illuminating gas and air that, when ignited, became the working equivalent of a fluid, driving the piston in a cylinder just as the steam drove the pistons in Watt's engines. Over time, Otto perfected what came to be known as the four-stroke Otto cycle of the internal combustion engine: intake, compression, explosion, and exhaust. Otto incorporated his four-stroke principle into a horizontal engine, but it remained impractical until the invention of the carburetor in 1885 and the development of gasoline from petroleum.

By the 1880s, oil was becoming a vital source of heat, light, and power. The oil rush started in 1859 in Pennsylvania when Edwin Drake drilled a well 75 ft (22.9 m) deep and struck a pool of oil that gushed out at a rate of 23 barrels a day (a barrel holds 42.3 gal [160 l] of oil) (Potter, 1979). The petroleum industry provided a source of cheap, plentiful fuel that encouraged the development of heavy industry, mass transport, aviation, and international shipping. The industry's operations present a number of hazards and risks for workers. The threat of fire, explosion, toxic fumes, chemical spills, and hazardous wastes is present at every refinery, offshore drilling platform, and transport operation. As a result, oil companies developed and implemented some of the earliest health and safety measures for workers.

Electricity

The third new form of energy, electricity, completed the revolution begun in the 1700s. Electromagnetic energy had been known since the days of the Greeks and Romans, who had used magnetized lodestones as compasses. In the 1780s, Luigi Galvani and Alessandro Volta independently discovered that two dissimilar metals, combined with moisture, produced electricity when in contact with one another (Shepard, 1976). Volta constructed a type of primitive battery called a *voltaic pile* out of copper, zinc, and brine-soaked cardboard wafers that converted chemical energy into electrical energy. The metals served as electrodes, and the cardboard served as an electrolyte. The zinc reacted to the brine by releasing a stream of electrons that flowed through the cardboard to the copper electrode, completing the circuit through a wire attached to the zinc electrode.

Three more key discoveries were required before electrically powered machines became part of the industrial scene (Burstall, 1965). The first discovery was made by a Danish physicist, Hans Christian Oerstead, who found that wires carrying electrical current produced a force that acted like a magnet. He termed this force *electromagnetism.* Michael Faraday, a young English amateur scientist, found that the reverse was also true—magnets could produce electricity. He devised the dynamo in 1831, the first machine to generate electrical current. At about the same time, the Frenchman Hippolyte Pixii developed the generator, which produced a flow of alternating current. Finally, the U.S. inventor Joseph Henry created the forerunner of today's powerful electromagnets when he wrapped an electric current around a horseshoe bar of iron and magnetized it. Electrical motors were developed to drive mining equipment, manufacturing machines, steel drills, lathes, and short-haul trucks and trains.

The most prolific inventor of electrical devices, however, was Thomas A. Edison. His invention in 1879 of the tungsten lightbulb paved the way for the electrification of cities and rural areas. Huge steam- or water-driven turbines made it possible to generate extremely high-voltage current to run thousands of industrial and consumer electrical devices. High-tension power lines cut across city and rural skylines and brought the new energy into homes and workplaces all across the United States.

Perhaps the biggest impact of electricity on the workplace, however, was the advent of electrical motors that could drive single workstation machines and hand-held power tools in addition to the more massive machinery of factory production. Now even individual workers could multiply their own strength and productivity many times over. Unfortunately, the hazards of electrical energy were not well understood, and incidents involving shock, burns, electrocution, and property damage occurred in countless workplaces. Because such incidents could interrupt the power supply to large areas or interfere with production, companies were more motivated to institute safety measures for electrical power than they were for other types of energy.

By the 1940s, most modern workplaces used a combination of steam, chemical, hydraulic, and electrical energy to run machines and communication and transportation devices. The simple lever, wheel and axle, inclined plane and wedge, and screw had evolved enormously in terms of both increased speed, power, and work performed and the dangers they posed to workers.

ENERGY: RISKS VERSUS BENEFITS

Each time humans have discovered a new, more powerful form of energy to drive machines and provide services, users have had to weigh the risks of harnessing this resource against the benefits. *Risk* is a measure of the probability and severity of adverse effects, whereas *benefit* is the degree to which the end results of an action or device are judged desirable. *Safety* is the control of recognized hazards to attain an acceptable level of risk. These risks include not only the hazards and problems of containing and channeling the energy productively but also the risk of death and injury to workers and others who must use the machines or manipulate the energy. In most cases, workers have had to cope with risks associated with technological advancements for the good of society and the profit of those who were providing the venture capital. The problem, nevertheless, has been to find some way to determine what constitutes acceptable or reasonable risk.

Machine + Power = Work + Hazards

The original purposes of machines—to transform and transfer energy, multiply or change the direction of force, and multiply speed—were enhanced with each development of more powerful forms of energy. The multiplication of force, complexity, power, and output of machines created its own problems, as would using a very powerful horse to pull a milk cart. The horse can pull the heavy cart faster and farther than a human but also has the power to run away with the cart and possibly damage property or seriously injure the driver and others nearby. One can make a rough equation out of this by saying that the more power one adds to a machine, the more work can be done but also the greater the potential hazards involved: machine + power = work + hazards.

To add to the risk, new energy resources led to the design and construction of machines on a scale unprecedented in both size and number. Six inventions initiated the industrial revolution and helped to create the modern factory system: the flying shuttle, the spinning jenny, the water frame, the mule-spinner, the power loom, and the steam engine (Moore, 1961). By the time of the industrial revolution, thousands of men, women, and children no longer worked in their homes, family farms, or individual shops but in factories where they were expected to conform to machine-paced rates of productivity. These rates made no concessions to the fact that human workers were subject to boredom, illness, fatigue, and mistakes in actions and judgment. In addition, the emerging technologies held dangers not adequately recognized by designers, engineers, and owners. Thus, gains in productivity were offset by repeated explosions, fires, injuries, deaths, and damage to machinery and structures. The question arose, What is an acceptable or reasonable risk to assume for the benefits gained?

Determining Acceptable Risk

People generally expect that scientists, entrepreneurs, and government can and should solve some of the problems they create when developing new science and technology. Each new advance has its costs, and society must decide what it wants to pay for and whether it will follow through with the required commitment of resources. Because there never was (and never is likely to be) a golden age of risk-free living, the question of determining acceptable risk begins with four assumptions:

- Technology, although a mixed blessing, has enriched the human condition and will continue to be an important part of civilization.
- Many of society's problems are technological in origin and thus will have technological as well as political solutions.
- Human activity will always involve risks.
- To make the world safer, change must start with the way conditions are at the present moment. (Lowrance, 1976)

It is useful to divide the issue into two discrete activities: *measuring risk* and *judging the acceptability of that risk* (Lowrance, 1976). Measuring risk is an objective, empirical, and scientific activity that involves studying the various substances, devices, and procedures that pose a threat to human or environmental health and safety. This procedure can be used to establish standards for the quantities of emissions, waste products, electrical current, fumes, and so on that can be released or used at any one time. It can also help in the design of guards, barriers, and other protective devices to shield workers from harm and to reduce the risk of performing work tasks.

Judging the acceptability of risk, however, is a far less exact science. Safety, in its broadest sense, is a highly relative attribute that can change from worker to worker, from job to job, from one location to another, and from one era to the next. Also, an exposed individual's perception of risk may be quite different than that of the employer or even society. For example, conditions in most turn-of-the-century textile factories were considered dangerous by workers but safe by factory owners and most government inspectors. Today, those same conditions may be grounds for shutting down the plant and prosecuting the owners under various civil and criminal standards and laws.

In other instances, a tool or form of energy may be safe in one context but not in another, or it is not recognized as a hazard initially. For example, a power saw that is perfectly safe for an adult to handle would be potentially lethal in the hands of a child. Likewise, X rays were at first regarded as simply a novel, harmless form of photography; people sat in front of these machines for up to an hour or more, marveling at the sight of their own skeletal structures. The risks of X rays were not known until some years later.

Part of the problem is that in assessing acceptable risk, one must consider the matter of probabilities. Risk estimates can be used to assess the chance that an untoward event will occur, but they cannot predict a particular event at a specific time and place. It can be predicted that a certain number of steelworkers will be injured in the blast-furnace area of a steel mill during a 12-month period, but not exactly which steelworkers or exactly where and when the injuries will take place. On the other hand, probabilities can be used to indicate the low risk of a hazardous venture, such as the risk of nuclear versus nonnuclear facility incidents such as fire, explosion, toxic release, and vulnerability to natural disaster. Or risks can be assessed by their distribution over the population and over time, an evaluation that can be as important as the magnitude of the risk. For example, how many children five years of age and under are likely to be injured by portable electric fans? Research done in this area can help improve safety features on the fans, thereby eliminating more of the risk to specific groups of victims.

Using the term *acceptable risk* emphasizes that safety decisions are relative and not absolute. However, to whom and under what conditions are risks considered acceptable? Some guidelines for making these decisions include the following:

- *Reasonableness.* The Consumer Product Safety Act of 1972 directs producers and manufacturers to "reduce unreasonable risk of injury" associated with consumer goods. The concept of *reasonableness* is found in many economic analyses of hazard reduction and legal liability laws, yet the term is difficult to define. The National Commission on Product Safety has attempted to provide some guidance by stating that risks are *reasonable* when consumers understand the risks that exist, can appraise their probability and severity, know how to cope with them, and accept them to obtain the benefits of a product. Risks are *unreasonable* when consumers do not know they exist, are unable to estimate their frequency and severity, and do not know how to deal with them and when the risks are unnecessary because adequate substitutes for a product or less hazardous ways to produce it could be found (National Commission on Product Safety, 1970). The same principle could be applied to workers on the job.
- *Custom of usage.* If a substance, material, or thing has been in use for so long that it is "generally recognized as safe," then custom of usage can be cited as one way to determine an acceptable risk. Many food additives, for example, have been granted this status, although the Food and Drug Administration continues to review them.
- *Prevailing professional practice.* Initially established as a criterion for physicians' clinical practice, this principle is being extended to evaluate the risks involved in engineering and manufacturing. Such phrases as "common design," "normal intensities," or "prevailing local standards" are used to refer to machinery or products that have not been formally tested for safety but have not produced frequent injuries or serious side effects after long-term use. The assumption, as in custom of usage, is that if something has been in common use long enough, then it must represent an acceptable risk.
- *Best available practice, highest practicable protection, and lowest practicable exposure.* Many state and municipal occupational health and safety regulations have stipulated that manufacturers and owners must control workplace emissions, pollution, and other hazards by the "best available means," provide workers with "highest practicable protection," and ensure the "lowest practicable exposure" to hazardous substances and physical danger. These phrases are intended to give companies the opportunity to reduce risks at a cost that is not prohibitive.
- *Degree of necessity or benefit.* This guideline involves the principle that new procedures, machinery, or forms of energy should not be added until they have been evaluated on the basis of their risk to operators and workers as well as their benefit to the company.
- *The Delaney principle.* This principle was added to the Food and Drug Act introduced in 1958 by Congressman James J. Delaney and was meant to prevent carcinogenic substances from being added to foods. Although it has been controversial because of the extraordinary difficulty in determining carcinogenicity, its existence has been a strong deterrent to the indiscriminate use of risky additives in the public food supply. Some would like to see the principle extended to cover substances that may cause genetic mutations and birth defects as well.
- *"No detectable adverse effect" and "toxicologically insignificant levels."* These guidelines are meant to establish some parameters for items whose harm is difficult to measure or about which little is known as yet. However, their usefulness has been widely debated by scientific panels because of the lack of hard data to determine what "no detectable" and "toxicologically insignificant" mean in terms of public and worker health and safety.
- *Threshold principle.* This guideline states that if it can be proved there is a level of exposure below which no adverse effect occurs, then this level

might reasonably be considered safe. The problem with establishing such a threshold is finding a meaningful level that applies in all cases. What may not be harmful to one worker might be to another; what doesn't affect a healthy worker may adversely affect one who is ill. Some scientists believe that there are no safe threshold levels for some substances such as mercury, ionizing radiation, or certain chemicals used in manufacturing.

- *Exposure relative to natural background.* This guideline states that levels of a substance found in nature, such as natural background radiation, that are essentially irreducible can be used as a benchmark of safe exposure. However, one cannot assume that the levels found in nature are always beneficial to human life.
- *Occupational exposure precedent.* In setting standards, occupational records of worker exposure to various substances are usually consulted for some idea of the historical maximum exposure levels that workers have tolerated without experiencing ill effects. A generous safety factor is then applied to this figure to arrive at workplace health and safety standards. Depending on the accuracy of the records, this approach may or may not be useful in determining acceptable risk (Lowrance, 1976).

Evaluating workplace safety in terms of acceptable risk is one way to attack the issue of safeguarding workers and the environment from the hazards of new forms of energy and new manufacturing and production processes. Yet as can be seen by the guidelines above, this approach is often difficult to define and enforce because so much of the data on which guidelines are based tend to be relative and subject to various interpretations on the part of manufacturers, government, regulatory agencies, and workers. Questions that remain unanswered include: Who ends up paying the cost? Are those at risk the ones who benefit? Do those who benefit bear the risks?

Another approach that has gained more attention in recent years involves attacking the problem through a series of strategies that seek to prevent or neutralize any unwanted energy releases.

PREVENTING OR CONTROLLING ENERGY TRANSFERS

Haddon (1970) outlined 10 strategies to cope with potential or actual energy releases, listed in progressive order from preventing the release to treating victims after a release has occurred:

1. Prevent the marshaling or generating of the energy in the first place, whether it is thermal, kinetic, electrical, ionizing, explosive, biological, or of some other form. This strategy also recommends not elevating people or objects and not moving vehicles or operating machines. The sole objective is to prevent an occurrence.
2. Reduce the amount of energy marshaled or generated. If it is not possible to prevent the occurrence, the focus should be on minimizing the potential harm.
3. Prevent the release of energy. This strategy requires people to concentrate on anticipating how energy could be released accidentally and preventing it from doing harm. Such measures could be anything from putting a lock on firearms to installing a series of circuit breakers or fail-safe devices on heavy equipment to avoid unintentional start-ups.
4. Modify the rate or spatial distribution of an energy release. One could slow the burning rate of explosives, reduce the speed of machine operations, or limit access to high-voltage conductors.
5. Separate, in space and time, the energy being released from the structure or person that may be affected. For example, reroute traffic around downed power lines or evacuate personnel from areas contaminated by toxic fumes or radiation. The emphasis in this strategy is to prevent the intersection of energy and those who may be harmed by it.
6. Place material barriers between the energy and susceptible structures or people. These barriers include electrical and thermal insulation, safety glasses, steel-toed shoes, shields and guards on cutting or drilling machinery, gloves, fire doors, and radiation shields. It is important to note that some barriers may provide only temporary protection from certain forms of energy, such as gamma rays or toxic chemicals, and should be used as part of an evacuation procedure.
7. Modify surfaces that concentrate harmful contact. This strategy requires softening or rounding pointed or sharp edges and corners with which people come into contact. Physicians ever since Hippocrates (c. 400 B.C.) have noted that falls or blows onto or against hard, sharp surfaces produce more serious injuries than those occurring with flat or softer surfaces.
8. Strengthen the structure or person involved. This strategy can range from reinforcing the superstructure of a building to prevent earthquake damage to developing damage-resistant materials or conditioning workers so they are in better physical shape and more apt to resist harmful energy releases.
9. Reduce the losses produced by an energy release. This strategy includes rapid detection and evaluation of damage that occurs and responding quickly to control the hazard. Such a response requires an effective alarm system; devices such as sprinklers, automatic barriers, or other methods of isolating the damage; and emergency medical care.

10. Rehabilitate persons and objects. This strategy encompasses all the procedures taking place between the initial energy release and the final resolution of the emergency. The rehabilitation process might involve a return to preevent status or stabilizing the situation in an altered state (rebuilding a fire-damaged work area, for example).

In terms of lockout/tagout measures, strategy 3, preventing an unwanted energy release, is the top priority, followed by strategy 2, reducing the amount of energy released, and strategy 4, modifying the rate. It is impractical in most cases to prevent the marshaling or generation of energy because it is used to perform work. As a result, the larger the energy involved, the greater the need for control and for redundant, successive strategies and barriers. As discussed previously, many modern workplaces not only use elevated energy levels but also combine several forms of these energies—high-voltage electricity, acutely toxic chemicals, large stores of flammable materials, concentrated ionizing radiation sources, high temperatures and pressures, and so on.

Yet the systematic review of available strategies to reduce harm from energy releases has not been a customary feature of safety planning in many industries until quite recently. The likely reason for this lapse has its origins in the basic common-law principles that held that workers assumed responsibility for the risks of their jobs, that employers were not responsible for workers injured by the actions of other workers, and that most accidents could be attributed to worker negligence. These early employer defenses are no longer used because of the advent of various workers' compensation (no fault) laws.

ENERGY HAZARDS: A HISTORICAL PERSPECTIVE

Although occupational health and safety regulations and guidelines are found in every industry today, most of the truly effective measures for hazardous energy control have been developed only during the past 60 years. For many years, industry owners and managers, along with government officials, did little to ensure safe and healthful conditions for the average worker. Virtually no attention was paid to preventing or controlling hazardous energy releases even though the consequences in terms of injuries and fatalities were often serious. Nevertheless, although today's occupational health and safety laws are fairly recent, the health and safety movement itself is surprisingly old (Figure 1-5). Its written history begins in the ancient world (National Safety Council, 1992b). Around 1500 B.C., Ramses II, who ordered the construction of some of Egypt's largest temples, hired physicians to care for mine and quarry workers and others engaged in building public works.

The Industrial Revolution (1750 to 1900)

The use of new forms of energy and machinery in the factories and mills of Europe and the United States transformed the nature of work and the composition of the work force. In every urban center, men, women, and children labored for up to 16 hours, 6 or 7 days a week. Workers in mines, on railroads, in shipyards, and in other locations fared little better than those in factories and often worked under more hazardous conditions. The prevailing philosophy regarding safety was founded on three common-law doctrines that provided employers with a defense and gave workers little chance to obtain compensation for injury, illness, or death:

- *Assumption of risk.* Under this doctrine, workers, by virtue of being told the risks and dangers of a job, accepted responsibility for their own safety.
- *Contributory negligence.* An owner had merely to claim that workers, through their own error or negligence, contributed to the incident, and the owner was held blameless.
- *Fellow servant rule.* Employers were not liable for injuries to employees that resulted from the negligence of fellow workers.

Because employers could not be held liable for workplace injuries or fatalities, they had no need to change their operations, safeguard equipment, or train workers in safe operating procedures. Workers often received small increases in pay to entice them to accept greater hazards on the job without complaining. In addition, there was little effort to establish industry standards for emissions, sanitation, reliability of equipment, worker training, worksite pollution, equipment guarding or shielding, industrial housekeeping, or any other form of protection found in nearly every workplace today. The business of industry was production, not caretaking.

This attitude reflected the dominant economic and social belief at the time that those who invented and created industries should be allowed the freedom to set the conditions by which others participated in the commerce. In a free society, enacting laws that limited these entrepreneurs' freedom was seen as ultimately damaging to the freedom of the entire society. Workers were regarded as voluntary laborers, free to choose where they wished to work and for whom. If they found conditions unsafe in one factory, they were free to seek employment elsewhere. This philosophy was contradicted by reality, however. In most instances, workers were free only in theory to pursue work elsewhere; often there was only one factory in a town, or dozens of workers were competing for every available job.

Because working conditions by the early 1800s had become exceedingly grim in many areas, several countries attempted to improve the workers' lot. As early as

Inventions and Machines

Paleolithic Age to 20,000 B.C.: Use of wedge, lever, and inclined plane; hunting tools; awl and bone needle for sewing; fire, wind, water, muscle, and gravity harnessed as energy sources
3500 B.C.: Wheel and axle used for transport; spindle and loom used for textiles
2500 B.C.: Great Pyramids of Egypt constructed: use of seafaring sailing ships
2000 B.C.: First bronze implements; use of the plow
1500 B.C.: First evidence of pulley; glass-making developed
1000 B.C.: Development of iron implements
500 B.C.: Use of lathe for wood turning
400 B.C.: Invention of catapult by the Greeks
270 B.C.: Invention of force pump and water clock; principles of hydraulics discovered
250 B.C.: Archimedes devises water snail and endless screw
A.D. 60: Hero of Alexandria invents toy steam turbine and screw press; use of waterwheel spreads throughout Italy
600: Development of horizontal waterwheel in Persia
1185: Invention of horizontal-axle windmill in North Sea area
1300–25: Spinning wheel developed in Europe; cannon and first use of gunpowder
1440–50: Printing press; handgun
1642–60: First air pump and vacuum; frictionless electrical device (first machine to generate electricity)
1680–1740: Principle of internal combustion; first model of steam engine; first practical steam engine
1741-80: First electrical condenser (Leyden jar); spinning jenny; steam engine with separate condenser, double-action rotary feature, and speed governor; mechanized textile loom
1781–99: First successful power loom; first steamboat sails Delaware River; cotton gin
1800–29: Electric battery; carbon arc lamp; calculating machine, forerunner of computer
1830–40: Primitive electric motor; experimental electric generator; battery-powered electric motor; electric telegraph and steel plow; screw propeller
1841–50: First practical telegraph; first incandescent filament lamps
1851–60: Bessemer furnace for steelmaking; first oil well at Titusville, PA; petroleum products; chemical industry arises; first successful gas engine; experimental telephone
1861–80: Ring-dynamo for continuous flow of electricity; cathode-ray tube; gas engine; Bell's telephone; Edison's phonograph and incandescent lightbulb; dynamite; air brake
1881–99: First electric power system in New York City; compound steam turbine; first practical electrical transformer; gas-engine automobile; process of electrolysis; first electric streetcar system; AC motor; diesel engine; X rays discovered; hydroelectric power at Niagara Falls; radioactivity discovered
1900–20: First wireless signal sent across Atlantic; arc transmitter; Wright brothers pilot first airplane; first plastic; vacuum tube used in radio; Model T auto; atom split, thus achieving first atomic transmutation
1921–40: First commercial radio station opens in Pittsburgh; first motion picture with sound; first liquid-fuel rocket; transatlantic telephone service; first practical television system; cyclotron (atom smasher) built; deuterium, power source for hydrogen bomb; radar; first jet engine
1941–60: Atomic bomb and atomic energy made practical; mainframe computers; television; transistor; solar battery; maser; atomic power stations established; laser; integrated circuits invented
1961–80: Nuclear reactors built to generate electricity; first fully automated factory (Sara Lee); microprocessor chip by Intel; floppy disks; supersonic jet airliners; array technology for electronic equipment; LCD displays for commercial/industrial use; magnetic resonance imager; industrial robotics; start of the second industrial revolution—the computer age
1981–present: First space shuttle flights; laser disk storage system; minimills in steel industry; compact disk technology; mobile telephones and other hand-held communication devices; fiber optics; laptop computers; computer networks developed worldwide; multimedia and virtual reality technology

Health and Safety Measures—Hazardous Energy Control

3000 B.C.: *Ebers Papyrus* and *Edwin Smith Papyrus* discuss various medical treatments and ailments, including those of workers
2000 B.C.: Code of Hammurabi contains first known workers' compensation provisions

Figure 1-5. Time line of inventions and health and safety measures *(continued)*.

1500 B.C.: Ramses II adds physicians to work force to care for mine and quarry workers
500–200 B.C.: Nicander, Hippocrates, and Galen describe work-related illnesses of miners
98–55 B.C.: Lucretius produces *De Rerum Naturum* (*On the Nature of Things*), including description of ill effects of mining; Pliny the Elder, in *Natural History,* mentions use of pig bladders as respirators
1473: Ulrich Ullenbog discusses dangers of metalworking to artisans
1500s: Philippus Aureolus (Paracelsus) publishes *On the Miners' Sickness and Other Miners' Diseases*; Georgius Agricola writes *De Re Metallica* on miners' illnesses
1700s: Bernardino Ramazzini publishes *Discourse on the Diseases of Workers,* the first known text to offer medical model for treating worker diseases
1800s: Europe: Edwin Chadwick's *Report on the Sanitary Condition of the Laboring Population of 1842* published in England; England, Germany, and France pass laws for workshop and factory inspections; England establishes insurance for boiler accidents; United States: Steamboat Inspection Act (1852 and 1866); Polytechnic Club investigates boiler accidents (1860); Massachusetts factory inspections law (1867) and law on guarding machinery (1877); Hartford steam boiler insurance (1867); Illinois Steel establishes safety department (1892); Railroad Safety Appliance Act (1893)
1900–1950: United States: Workers' compensation laws passed in several states; Federal Workmen's Compensation Act (1908 and 1926); U.S. Steel Corporation establishes safety committee (1906); Haydon's *Synopsis of Accident Prevention* (1917); founding of National Council for Industrial Safety (1913), soon renamed the National Safety Council; first ASME boiler code (1914); machinery insurance initiated (electrical, flywheel, turbine); American Petroleum Institute establishes a Department of Accident Prevention (1931); U. S. Bureau of Labor Standards (1934); Walsh-Healy Public Contracts Act (1936); Association of Casualty and Surety Company's *Handbook of Industrial Safety Standards* (1937)
1951–present: United States: Coal Mine Safety Act (1952 and 1969); Federal Longshoremen's and Harbor Workers' Compensation Act amended to include rigid safety precautions (1958); Radiation Standards Act (1959); Natural Gas Pipeline Safety Act (1969); Occupational Health and Safety Act (1970); states pass similar OSHA-type safety and health codes; ANSI Z244.1 Lockout/tagout safety standards (1973 and 1982); NFPA *Electrical Requirements for Employee Workplaces* (1976); UAW lockout procedures (1979); NIOSH lockout systems (1980); 29 *CFR* 1910.147 lockout/tagout final rule (1989) and 29 *CFR* 1910.331–.335 *Electrical Safe Work Practices* (1990)
Europe and Canada: England, Sweden, France, Canada, and European Community (EC) take steps to adopt safety regulations and standards for hazardous energy control. British Standards Institute establishes standards that serve as guidelines for the EC. England develops "safe system at work" practices.

Figure 1-5. *(concluded)*

1833, England established a system for regular government inspection of factories. Both Germany and France also passed laws protecting worker health and safety and providing for regular inspections of factories.

In 1867, Massachusetts established the first system of factory inspections and 10 years later passed an employers' liability law that included a requirement to guard dangerous moving equipment. Companies in other states also began to take the first tentative steps toward addressing workplace hazards. Illinois Steel in Joliet, Illinois, created a safety department in 1892 to study ways of improving working conditions (Gersuny, 1981). This event probably marked the birth of the U.S. industrial accident prevention movement. These early measures, although a step in the right direction, left a great deal to be done.

The 20th Century

The mechanized era (1900 to 1950). One commentator described the lack of worker safety on railroads in early-20th-century United States by noting that "war is safe compared to railroading in this country." This observation applied to overall working conditions as well. The toll from worker injuries reached appalling rates in the years 1900–10:

- In 1904, 28 out of every 10,000 employees were killed in railroading accidents; by 1916, the injury rate was 1 in 10 employees. It was said that a man was killed for every mile of track that was laid. Equipment was often faulty, and some engineers were forced to operate trains for 36 hours at a time. By 1907, the annual death toll had reached 4,353.
- In 1906, the death toll in mining accidents in the United States was 48 out of every 10,000 workers.
- From 1900 to 1906, the average yearly death toll in the U.S. coal-mining industry was 35 out of every 10,000 workers, a figure that exceeded the rates of the other major coal producers: Great Britain, France, Belgium, and Prussia.
- Health hazards in cotton mills and textile factories,

including cotton and clay dust, heat, excessive noise and vibration, diseases, and exhaustion, accounted for high rates of worker illness, injury, and disability. (Gersuny, 1981)

One industry observer noted that "it is no wonder that accidents abound [in factories], considering the extent of machinery, the velocity of movement, the proximity of machines to each other, the loosely hanging gearing, and the . . . flying off of some parts of the machine" (Gersuny, 1981). Another observer, William Hard, asked in a 1906 article why society did not, "having invented machines which make business one long war, treat the enlisted men at least like enlisted men and, if they are incapacitated, assign them temporarily or permanently, to the rank and pay of pensioners of peace?" (Gersuny, 1981). The answer was simply that changes in the laws governing the treatment of workers injured or disabled on the job lagged far behind other changes in technology and in social relationships that accompanied the industrial revolution.

Unwanted energy releases were quite common in the turn-of-the-century work environment. Power was often supplied from a single source, and there were either inadequate or no protective guards for power transmission components, which resulted in many injuries and fatalities. Electrical wiring that was not adequately insulated or grounded made the work of electricians more hazardous than it had to be. And neither were there sufficient safeguards for switching machinery under repair on and off. In one instance in a carding mill, a loom fixer was repairing a machine when, according to the company report, "the girl tending the loom started it and . . . cut off the end of his left index finger" (Gersuny, 1981). Other workers suffered serious injuries when their hands, arms, or feet were caught in machinery gears and crushed or mutilated. Table 1-A shows an analysis of 1,000 accidents at the Pacific Mills in Lawrence, Massachusetts, from 1900 to 1905. Injury categories clearly show that machine-related accidents accounted for the largest number of injuries. Fifty-four percent of those injured were employed in the mill for less than one year. Table 1-B lists casualties from Lawrence Manufacturing Company from 1899 to 1905 and also shows that accidents involving machinery and unwanted energy releases dominated all other categories.

Even with such data available, companies were slow to respond. In fact, some industrialists and industry experts, such as Magnus Alexander, even criticized the request for protective devices, claiming it would make workers complaisant off the job: "Even if it were possible it would be unwise to so surround red-blooded human beings with safeguards as to convert them into such unthinking mollycoddles, while in the shop, that they would rush thoughtlessly into the very jaws of danger . . . the moment they step from the shop into the street" (Gersuny, 1981). Crystal Eastman, reporting from Pittsburgh in 1907, wrote that "most of the men in the community whose opinions count . . . believe that 95% of the accidents are due to carelessness of the men." (This view prevailed for another 50 years and can still be found in some workplaces today despite occupational health and safety legislation, education, and advances.)

Table1-A. Study of 1,000 Accidents at Pacific Mills in Lawrence, MA; August 1900–July 1905

Caught in operating machinery during the course of ordinary usage	320
Caught in machinery while cleaning contrary to orders	111
Careless handling of tools or implements of work	98
Handling machinery, merchandise, etc. in transportation	137
Slipped on floors, etc. and caught by machinery	61
Injured by trucks or wagons used in the work	36
Falls	47
Elevators	39
Injured by belts	32
Splinters	19
Flying shuttles	19
Other	81
	1,000

Source: Carl Gersuny, *Work Hazards and Industrial Conflict* (Hanover, NH: University Press of New England, 1981). Copyright 1981 by the Regents of the University of Rhode Island. Reprinted by permission of the University Press of New England.

Early workers' compensation laws passed in Massachusetts, New York, Wisconsin, New Jersey, and Washington were declared unconstitutional by the courts who claimed they deprived employers of property without due process of law. In 1916, however, five years after the tragic Triangle Fire in New York City's clothing district claimed 146 lives, the U.S. Supreme Court declared workers' compensation to be constitutional. After this ruling, many states passed compulsory workers' compensation laws.

Eventually, the growing pressure from citizens' groups, unions such as the United Mine Workers and American Federation of Labor, safety and health professionals, and industrialists and politicians of good conscience brought about reforms in industry work practices and workplace safety. The United Mine Workers of America, formed in 1890, developed a unique program in worker health care, establishing an industrywide trust fund jointly administered by the mine operators and the union. Union organizer John L. Lewis believed that the cost of maintaining workers in a safe and healthy condition should be

Table 1-B. Casualties Admitted to Lowell Hospital from Lawrence Manufacturing Company, 1899–1905, Lowell, MA

Age	Sex	Weekly wage ($)	Hospital days	Hospital cost ($)	Cause	Case description
15	M	3.94	27	10.61	machinery	scalp wound, lost part of ear
21	M	6.37	38	14.93	machinery	severed hand tendons
41	M	7.00	66	25.93	falling object	fractured leg
30	M	6.00	26	10.61	machinery	amputated finger
38	M	7.00	85	33.39	machinery	2 fingers amputated from each hand
40	M	9.86	2	.78	machinery	2 fingers amputated
28	M	11.50	2	.78	machinery	amputated finger
35	M	7.00	10	2.75	machinery	amputated finger
52	M	7.00	13	5.11	machinery	lacerated hands
42	M	7.00	12	4.71	machinery	fractured fingers, 1 amputated
16	F	4.90	19	4.75	machinery	fractured finger
19	M	2.94	6	2.36	sat on shears	punctured scrotum
56	M	9.80	16	6.28	fall	injured side
21	F	4.90	13	3.25	machinery	amputated finger
35	M	4.90	1	.39	machinery	severed arm; died 20 min. after admission
60	M	5.50	6	2.36	fall	sprained back
42	F	6.00	5	1.25	falling object	head wound
36	M	6.64	13	5.11	machinery	severed arm
17	F	4.90	8	2.00	machinery	fractured finger
45	M	6.64	7	2.75	fall	injured side
33	M	7.00	22	8.64	machinery	amputated arm
33	F	7.00	10	3.57	machinery	fractured fingers
63	M	8.70	12	4.71	machinery	lacerated hand; died from infection
25	M	7.00	17	6.68	elevator	crushed toe
25	M	7.00	13	5.11	splinter	infected foot
45	M	6.64	104	44.78	dye spill	scalded thigh; skin graft
26	M	7.00	43	16.89	machinery	3 fingers amputated
52	M	10.84	10	3.93	fall	fractured arm
21	F	4.50	8	2.34	falling object	knee injury
25	M	7.00	78	30.64	machinery	lacerated hand
22	M	6.64	62	24.36	machinery	fractured hand, 4 fingers amputated

Source: Carl Gersuny, *Work Hazards and Industrial Conflict* (Hanover, NH: University Press of New England, 1981). Copyright 1981 by the Regents of the University of Rhode Island. Reprinted by permission of the University Press of New England.

considered as much a part of production expenses as any other operational cost.

Among the first industries to take the initiative in the area of reducing work hazards were the electric power, railroad, steel, oil, and chemical companies, notably U.S. Steel, Du Pont, and Standard Oil. The hazardous nature of the processes and energy used and the need for a highly trained and educated work force contributed to the early steps in worker safety adopted by these industries. In 1906, U.S. Steel Corporation established its own safety department to investigate accidents and institute corrections, and its president, Judge Elbert Gray, wrote that the corporation "... expects its subsidiary companies to make every effort practicable to prevent injury to its employees." In 1931, the American Petroleum Institute established a department of accident prevention. In addition, insurance companies started to relate the cost of premiums for workers' compensation to the cost of accidents, and management began to understand the close connection between safe production and profitable production. As Admiral Ben Morell, president of Jones and Laughlin Steel Corporation, stated in 1948, "... I have often heard it stated that the cost of adequate health and safety measures would be prohibitive and that 'we can't afford it.' My answer to that is quite simple ... 'If we can't afford safety, we can't afford to be in business.'"

Although documentation is sparse, some voluntary hazardous-energy-control practices also were being developed and implemented by progressive employers during the post–World War I period through the end of World War II (1920–1945). These early efforts were related primarily to electrical power generation and transmission in the utility industry and electric power used by large industrial companies in the oil, steel, chemical, and mining businesses. Hazardous energy control during this period was most often the domain of the electricians or maintenance employees who used high-voltage lockout or tagout procedures to safeguard themselves and the equipment. Operating and servicing personnel were often prohibited from performing any energy isolation without assistance from electricians or supervisors or were otherwise generally discouraged from using this means of protection for their required tasks. Hydraulic, pneumatic, and gravitational energy control was rarely addressed with procedures unless some specific incident caused management to apply a particular control to a certain situation. Oil and chemical companies are credited with pioneering work on gas, oil, and liquid chemical energy isolation by developing techniques such as purging, blanking, bleeding, inerting, and so on.

Insurance organizations made early contributions to the control of hazardous energy by evaluating injury experience and publishing various guidelines or standards for prevention. For example, in 1917, George F. Haydon, engineer and underwriter of the National Underwriter Company, wrote *Haydon's Synopsis of Accident Prevention,* a pocket reference on industrial hazards and safety engineering (Haydon, 1917). It contained a variety of safety rules, practices, and machinery/equipment requirements related to a host of industrial operations and activities, including the following:

- Overhead cranes. "A separate [electrical] switch on top of crane for use of repairmen should be provided." (p. 61)
- *Electricity.* "Locking devices should be fitted on switches controlling lines on units upon which employees may be working and key should be kept in possession of such men." (p. 65)
- *Metalworking.* "It is most advisable to fit locking devices on all belt shifters to prevent machines starting unexpectedly." (p. 132)
- *Steel rolling mills.* "Safety switches should be installed in turntable pits to enable men going into pit to cut off power from table." (p. 165)
- *Wire drawing.* "Locking devices should be fitted to treadles . . . to prevent frames from starting unexpectedly." (p. 196)

In similar fashion, the Engineering Department of the American Mutual Liability Insurance Company of Boston published the *Handbook of Industrial Safety Standards* in 1924. Those standards relating to hazardous energy control are reproduced here to demonstrate the evolution of various preventive practices.

- *Safety switches.* "A switch in the main power circuit, capable of being locked when it is open, shall be placed above the crane cab where it can easily be reached from the bridge walk. . . ." (p. 16)
- *Boiler valve locks.* "When a person is working in any boiler of a battery, the steam, feed, and blow-off valves shall be closed and locked. When more than one boiler is connected into the same blow-off, feed, water, or steam pipe line, the valve nearest the boiler in each such connection shall be provided with a valve lock which will prevent the valve stem from being moved when the valve is locked." (p. 32)
- *Power transmission.* "Group drives do not eliminate transmission but they do reduce the danger of running on belts, making repairs and adjustments, etc. since the machines in the group can be shut down easily for the moment needed for such work." (p. 52)
- *Machines.* "All machines except grindstones or blowers should be provided with an effective starting and stopping device such as an individual belt shifter, clutch, switch, or valve which will effectively control the machine." (p. 55)
- *Electrical equipment.* "Voltage below 275 is rarely harmful to persons in good health [author's note: an obvious error; many deaths occur annually from 120 volts or less]. This section is therefore divided into two parts: protection needed at voltages above 275 and under 750, and additional protection needed for voltages above 750 volts." (p. 34)

The Association of Casualty and Surety Companies of New York (founded in 1926) also published its own *Handbook of Industrial Safety Standards.* In the seventh edition (1945), the following citation is found: "Steam Boilers—maintenance and inspection . . . When two or more boilers are operated in battery and are being prepared for an internal inspection, care should be taken to see that the stop valves are tight, and signs reading MAN IN BOILER should be hung on the main stop valve of the boiler that is being inspected and on the auxiliary valve." (p. 26)

Various states with progressive legislatures and labor departments developed standards addressing electrical/mechanical hazards and calling for lockout/tagout/blockout provisions or procedures on the job. California, Michigan, Pennsylvania, Washington, Maryland, New Jersey, and Ohio established standards and regulations related to hazardous energy control between 1930 and 1950. However, in most cases these standards were not comprehensive enough, focusing as they did primarily on maintenance work and the control of electrical hazards. In addition, most states were either too understaffed or underfinanced to make enforcement much of a priority.

In 1913, the National Council for Industrial Safety (later renamed the National Safety Council, or NSC) was founded to provide research, gather accident-related data, and develop safety standards and training programs for industry and the general public. The

NSC called a conference in 1919 to formulate uniform industry health and safety standards under the auspices and procedures of the American Engineering Standards Committee (AESC), which later became the American Standards Association (ASA) and, finally, the American National Standards Institute (ANSI). ASA handled the technical and equipment side of safety, while the National Safety Council focused on the worker side of prevention. In 1950, the NSC produced a data sheet entitled "Methods of Locking-Out Electric Switches."

Initiatives were being developed on the legislative front as well. The major legal reforms affecting work hazards and industrial conflicts passed before 1950 were the Workman's Compensation Acts passed in all states between 1911 and 1948; the National Labor Relations Act of 1935, which legalized unionization and collective bargaining; the Walsh-Healey Public Contracts Act of 1936, which established some of the first guidelines for wages, hours, and working conditions; the Fair Labor Standards Act of 1938, which outlawed child labor in most workplaces and established legal age limits and working hours for adolescent employees; and the National Management Labor Relations Act of 1947. These measures were not passed without a fight. Many employers, one observer reported, "regard safety work as a 'socialist fad'"; as a result, effective compulsion was exercised in only a few states (Gersuny, 1981). Nevertheless, workers' compensation laws ushered in the era of no-fault liability, in which workers were compensated for an injury regardless of the cause or origin of the incident.

The purpose of the workers' compensation laws was to cover part of the wages that employees lost and to defray the expenses of medical care and rehabilitation without regard to any determination of negligence. The National Labor Relations Act, along with various legislation enacted under Franklin D. Roosevelt's New Deal, established the right of workers to organize unions and required employers to bargain with them in good faith. Workers now had leverage to negotiate better working conditions and higher pay. The issue of safety, however, was not as clearly addressed in these laws and was left for later reforms.

The automation era (1950 to the present). The years following World War II witnessed the passage of major safety and health laws, the most notable being the Occupational Health and Safety Act, or OSHAct, which was passed in 1970. Prior to this law, the National Labor Relations Board (NLRB) had addressed the issue of worker safety in 1966 by declaring negotiation of safety rules to be a mandatory subject of collective bargaining. In a ruling arbitrating a dispute between Gulf Power Electrical Company and its unions, *Gulf Power and L.U.'s 1055 and 624 IBEW,* the NLRB stated that "safety provisions constitute an essential part of the employees' terms and conditions of employment, and, as such, are a mandatory part of bargaining" (Gersuny, 1981). Refusal to bargain on safety issues was declared to be an unfair labor practice. Companies could no longer simply claim they maintained safe operations; they now had to prove it to the unions and the NLRB.

OSHAct was the first comprehensive federal law to provide for the establishment and enforcement of uniform, nationwide safety and health standards. Previously, responsibility for such standards had been left up to individual states, which had differed widely in their legislation and enforcement. The only laws previously on the books had been poorly enforced federal legislation dealing with the health and safety of government workers and a few isolated laws covering hazards faced by miners, longshoremen, and atomic energy workers (Felton, 1993). Pressures leading to the passage of OSHAct had been the result of nearly 20 years of effort. Steadily rising industrial casualties during the 1960s and the growing public awareness of the threat to general health from industrial pollution spurred the belated congressional action.

In 1970, Labor Secretary Wirtz, in support of OSHAct, pointed out the need for uniform protection of workers, irrespective of the state in which they worked. The effects of work hazards were declared to be a burden upon and hindrance to interstate commerce and a threat to the general welfare. This gave Congress the authority to act. The OSHA legislation imposed a general duty on employers to "furnish each of [its] employees employment and a place of employment which are free from recognized hazards that are causing or are likely to cause death or serious physical harm . . . [and to] comply with occupational safety and health standards promulgated under this act" (Gersuny, 1981). Workers were now allowed to report dangerous or hazardous conditions to employers and OSHA inspectors and to request unannounced on-site inspections from the local OSHA agency.

OSHAct and its subsequent amendments established mandatory federal occupational health and safety standards, provided for safety research, mandated training of occupational and health medical professionals, and helped to establish cause-and-effect relationships between diseases and environmental work conditions. The law has been instrumental in transforming health and safety issues from low-priority to high-priority items on the bargaining and political agenda. It has given the entire field of safety planning and preventive action new importance and provided a counterargument to recalcitrant employer claims that safety devices and procedures are too costly to implement. Research under the auspices of OSHAct has enabled the safety movement to establish that putting safety high on the workplace agenda is far more cost-effective than paying for cleanup and compensation and the loss of public goodwill following accidents.

Other laws that have aided in the struggle to establish occupational health and safety regulations include the Coal Mine Health and Safety Act (CMHSAct) of 1969, which gave black-lung disease legal standing, and legislation passed under the umbrella of the Environmental Protection Act, including the Clean Air and Clean Water Acts; the Toxic Substances Control Act; the Resource Conservation and Recovery Act; the Comprehensive Environmental Response, Compensation, and Liability Act ("Superfund"); and the Superfund Amendments and Reauthorization Act (SARA).

From 1970 to 1982, various noteworthy voluntary initiatives also took place concerning the control of hazardous energy. The United Auto Workers lobbied, negotiated, and established guidelines for lockout procedures to help reduce the number of accidents occurring in the automotive industry. The American National Standards Institute published guidelines for lockout/tagout procedures and for what they termed "zero mechanical state" to ensure that machines could not be started inadvertently. Chapter 2, Unexpected Energy Transfer, and Chapter 4, United States Hazardous Energy Regulations, discuss other measures in more detail covering the period from 1983 to 1989 when the National Institute for Occupational Safety and Health (NIOSH) criteria documents, accident studies, and OSHA lockout/tagout standards were published.

NEW ENERGY: RISKS AND BENEFITS

Safety measures, including lockout/tagout procedures, are even more important today than in the past. Since World War II science has developed more forms of high-energy sources such as jet propulsion, liquid and solid rocket fuels, lasers and masers, new chemical fuels, and the most potentially hazardous source of them all, nuclear power. These have exponentially expanded the power and output of the ancient world's simple machines.

Yet the hazards have also been multiplied, not only those associated with handling the energy source itself but also those relating to the amount and intensity of light, sound, heat, radiation, and toxic wastes produced as by-products. Work accidents still destroy more than 9,900 lives a year, and some 1.7 million workers suffer disabling injuries, which cost industry more than $63.3 billion annually (National Safety Council, 1992a). Continuing efforts to improve health and safety at work are part of a coordinated effort by government agencies such as NIOSH; federal and state OSHA and EPA statutes; unions like the United Auto Workers, AFL-CIO, United Steel Workers of America, United Mine Workers, and Oil, Chemical, Atomic Workers; progressive companies including General Motors, Du Pont, U.S. Steel or USX, Exxon, Procter & Gamble, IBM, Dow Chemical, Kodak, General Electric, Johnson & Johnson, and Alcoa; and industry and safety associations such as the American Foundrymen's Society, Chemical Manufacturers Association, American National Standards Institute, American Society for Safety Engineers, American Petroleum Institute, Aluminum Association, and others.

The National Safety Council also supports research and education efforts to encourage adoption of lockout/tagout procedures in the workplace. Since the 1920s, the NSC has produced or assisted in the development of data sheets, manuals, safety brochures, pamphlets, and other publications that explain lockout/tagout procedures and illustrate the various tags and locks used to prevent unwanted energy releases. Through the 1930s and 1940s, the NSC's publication, *National Safety News,* also featured articles and safety tips on lockout/tagout measures.

As energy sources become more powerful and sophisticated, the knowledge required to handle them safely also becomes more complex. It is hoped that the procedures outlined in this book will contribute to the ongoing efforts of all concerned to create a safer work environment.

SUMMARY

- Humans have been developing increasingly complex and powerful machines and sources of energy for several thousand years. Yet only during the past 100 years has the safety movement emerged as a potent force in protecting the lives of workers from the associated hazards.
- Lockout/tagout methods for ensuring worker safety from hazardous energy release are part of the recent developments in the safety movement. These methods can be used to secure moving machinery or processes, lock out and tag out an energy source, or achieve zero mechanical or zero energy states.
- Energy, force, work, and power are the basis for understanding how all machines and tools work. Forms of energy and force used in ancient times included human and animal muscle, wind, water, fire, and gravity. Hazards associated with all forms of energy and force include unexpected movements, direct contact with an unshielded energy source, and being caught in a mechanism driven by a form of energy or force.
- Machines serve five purposes: to transfer energy, transform energy, multiply force, multiply speed, and change the direction of a force. Five basic machines developed in ancient times are the lever, the inclined plane and wedge, the screw, the pulley, and the wheel and axle. All machines, no matter how sophisticated, are combinations and variations on these five basic machines.
- The evolution of energy use has been a steady progression from simple to more powerful and dangerous sources such as steam and steam engines, fossil fuels and internal combustion engines, chemical energy, electricity, and radiant energy (lasers, microwave, nuclear, and solar).

- The dangers of new energy forms were slow to be recognized. For example, even though steam engines and boilers exploded frequently, it was not until the turn of the 20th century that improvements in engine and boiler design were made and safety regulations were passed.
- In general, the development of safety codes and regulations has been the result of the combined efforts of labor, enlightened employers, trade associations, safety organizations, state and federal government, and public opinion. In developing standards, the risks must be assessed against the benefits to determine what constitutes an acceptable or reasonable risk to workers, employers, and the public.
- Guidelines to evaluate the acceptability of risk include reasonableness, custom of usage, prevailing professional practice, best available practice, degree of necessity or benefit, the Delaney principle, "no detectable adverse effect," the threshold principle, exposure relative to natural background, and occupational exposure precedent.
- Haddon (1970) outlines 10 strategies to cope with potential or actual energy releases, listed in progressive order: prevent energy generation, reduce its amount, prevent its release, modify the rate or distribution of release, separate energy released from structures/persons, shield energy from structures or persons, modify surfaces, strengthen structure/person, reduce losses, and rehabilitate persons and objects.
- Lockout/tagout measures stress preventing unwanted energy release or reducing or modifying the amount released. Preventing the generation of energy (zero energy state) is nearly impossible in today's complex, high-technology work environments.
- For many years, little attention was paid to preventing or controlling hazardous energy releases. Common-law doctrines provided employers with three defenses—*assumption of risk, contributory negligence, and fellow servant rule*—that put the burden of safety on workers.
- By the mid-1800s, however, some safety measures were being taken. A few states passed workers' compensation laws; employers in the steel, chemical, transportation, and utilities industries established a few safety codes and standards; some insurance companies were publishing safety handbooks; safety organizations were developing consensus standards; and federal and state governments were beginning to act in support of these measures.
- Not until after World War II did the safety movement acquire any real legislative force. Several states with progressive legislatures passed lockout/tagout/blockout provisions or procedures. In 1970, the Occupational Health and Safety Act was passed, establishing for the first time a nationwide, uniform set of standards for worker health and safety. The act placed the main responsibility for compliance on the employers.
- Since 1970, various other initiatives have been taken to control hazardous energy releases. The federal government, unions, employers, and states have established detailed lockout/tagout provisions. As science and industry develop increasingly powerful and complex machinery, processes, and energy sources, there will be a greater need for stringent lockout/tagout systems.

REFERENCES

American Mutual Liability Insurance Company, Engineering Department. *Handbook of Industrial Safety Standards*. Boston: American Mutual Liberty Insurance Co., 1924.

Association of Casualty and Surety Companies. *Handbook of Industrial Safety Standards,* 7th ed. New York: Association of Casualty and Surety Companies, 1945.

Burstall AF. *A History of Mechanical Engineering*. Cambridge, MA: MIT Press, 1965.

Calder R. *Evolution of the Machine*. New York: American Heritage Publishing Co., 1968.

Colby G. *Du Pont Dynasty: Behind the Nylon Curtain*. Secaucus, NJ: Lyle Stuart, Inc., 1984.

Felton JS. History. In *Occupational Health and Safety*, 2d ed., ed. J LaDou. Itasca, IL. National Safety Council, 1993.

Gersuny C. *Work Hazards and Industrial Conflict*. Hanover, NH: Univ. Press of New England, 1981.

Haddon W, Jr. On the escape of tigers—An ecologic note. *Technology Review* 72, no. 7 (May 1970): 3–7.

Haydon GF. *Haydon's Synopsis of Accident Prevention*. Indianapolis, IN: The Rough Notes Co., 1917.

Johnson WG. Energy: For work or harm? In *MORT Safety Assurance Systems*. New York: Marcel Dekker, 1980.

Lowrance WW. *Of Acceptable Risk: Science and the Determination of Safety*. Los Altos, CA: William Kaufmann, Inc., 1976.

Moore FG. *Manufacturing Management,* 3d ed. Homewood, IL: Richard D. Irwin, Inc., 1961.

The National Board of Boiler and Pressure Inspectors. *Bulletin 48*. 1993.

National Commission on Product Safety. *Final Report of the National Commission on Product Safety*. Washington DC: U.S. GPO, June 1970.

National Safety Council. *Accident Facts*. Itasca, IL: National Safety Council, 1992a.

National Safety Council. Historical perspectives. In *Accident Prevention Manual for Business and Industry: Administration and Technology*, 10th ed. Itasca, IL: National Safety Council, 1992b.

National Safety Council. Safe practices for maintenance and repair men. Safe Practices Pamphlet No. 70. In *National Safety News*, National Safety Council XIII, no. 1 (Jan. 1926): 45–50.

O'Brien R. *Machines*. New York: Time-Life Books, 1968.

Potter N. *Oil*. Morristown, NJ: Silver Burdett Company, 1979.

Purcell J. *From Hand Ax to Laser*. New York: Vanguard Press, 1982.

Reynolds J. *Windmills & Watermills*. New York: Praeger Publishers, 1970.

Rickards T. *Barnes and Noble Thesaurus of Physics*. New York: Harper and Row, 1984.

Shepard ML, et al. *Introduction to Energy Technology*. Ann Arbor, MI: Ann Arbor Science Publishers, Inc., 1976.

Sterland EG. *Energy into Power: The Story of Man and Machine*. Garden City, NY: The Natural History Press, 1967.

Stephenson C, Asher R. *Life and Labor: Dimensions of American Working-Class History*. Albany, NY: State Univ. of New York Press, 1986.

U.S. Navy, Bureau of Naval Personnel Staff. *Basic Machines and How They Work*. Minneola, NY: Dover Publications, Inc., June 1971.

Usher AP. *A History of Mechanical Inventions*. Cambridge, MA: Harvard Univ. Press, 1954.

Weaver G, McNulty JB. *An Evolving Concern: Technology, Safety and the Hartford Steam Boiler Inspection and Insurance Company—*1866–1991. Hartford, CT: The Hartford Steam Boiler Inspection and Insurance Company, 1991.

Williams TI. *The History of Invention*. New York: Facts on File Publications, 1987.

2

Unexpected Energy Transfer

This chapter provides a basis for understanding, as completely as possible, the critical issues and facts of hazardous-energy-release incidents.

LIMITING FACTORS

To prevent incidents associated with the unexpected release of hazardous energy, the factors involved in their occurrence must be identified and clearly understood. Ideally, a sizable statistical data base providing details such as frequency of occurrence, primary and contributory factors, equipment involved, type of energy, nature of activity, extent of injury and/or damage, control measures, and so on would be available for assessment. Unfortunately, no satisfactory data collection system relating to the unexpected release of hazardous energy exists as of this writing.

State and federal governmental agencies do not maintain definitive information banks relating to energy-release incidents. Limited data are occasionally found when categorical breakdowns related to electrical and machinery incidents are available; however, too much inference negates this information's utility. Workers' compensation boards and commissions similarly do not collect and organize incident-specific case data. This problem was identified in a 1978 report by the President's Advisory Committee, National Safety Council, and remains an unresolved issue.

The report states:

> *Problems with the technical quality of current data.* These include the lack of detail, consistency, and relevance for the development of countermeasures in current occupational injury data. Existing data are very limited.
>
> Occupational injury data are often gathered primarily for the purpose of processing workers' compensation claims rather than for developing countermeasures. Most current information systems emphasize data about the injury, not the accident, and about the end results of the accident sequence, rather than the participating events and conditions leading to the injury event. Data of the latter type are often much more important for the development of countermeasures.
>
> The American National Standard *Method of Recording Basic Facts Relating to the Nature and Occurrence of Work Injuries* (ANSI Z16.2), or an adaptation of this coding method, is used almost exclusively among both the Federal and State Government agencies and private sector information systems. [In 1988 ANSI Z16.2 was withdrawn and is out of print.] The ANSI Z16.2 method is extremely limited and is designed primarily as a

> monitoring tool for grouping accidents by major types and for flagging a limited number of isolated factors about accidents so that the cases can be recalled when data analysis begins. The ANSI Z16.2 method is not designed as an analytical tool for identifying the patterns, the precipitating events and conditions in an accident sequence, or the relationships among contributing causal factors in an accident that are normally needed for countermeasure development. (1978)

Although large employers or trade associations should maintain injury/incident data bases that would be comprehensive enough to supply constructive details on energy-release incidents, such is not the case. In addition to those previously mentioned, another reason for this lack of data is probably the very nature of the incidents themselves. Injuries from the hazardous release of energy at a given facility are typically infrequent, leading to the conclusion that there is no pressing need to develop specific data collection and retrieval systems for energy-release incidents.

In spite of the absence of an ideal data base, a number of relevant studies will be examined.

OFFICE OF STATISTICAL STUDIES AND ANALYSES (1978)

The Occupational Safety and Health Administration (OSHA) Office of Statistical Studies and Analyses reviewed and analyzed 125 fatalities related to fixed-machinery incidents occurring between 1974 and 1976. Federal OSHA fatality/catastrophe investigating reports were used as the reference sources. The fatalities were analyzed, and the primary causal factors were classified as follows:

- *Operating procedures.* The employee failed to follow designated procedures, or there were none available. These cases involved production or maintenance activities that required the machine to be running—41 fatalities.
- *Accidental activation.* The deceased or another employee activated the equipment while the deceased was inside or in the way of moving parts of the machine—31 fatalities.
- *Machine deactivation.* Although the procedure for cleaning and maintenance required deactivation, employees attempted these activities with the machine in motion—23 fatalities.
- *Equipment failure.* Parts or components of the machine broke and flew off, or the machine "exploded," resulting in fatal injuries to the machine operator or others—21 fatalities.
- *Other.* These were incidents with no identifiable elements common to the other categories—9 fatalities.

Fatalities for all causal factors totaled 125. Each of these causal factors for fixed-machinery incidents will be discussed in more detail in the following paragraphs.

Operating procedure problems accounted for the largest number (41) of fatal incidents investigated. These incidents occurred in the following circumstances:

- Employees were inappropriately positioned relative to the machines.
- Employees were in locations that were inappropriate or prohibited during machine operation.
- Employees only partially followed or completely ignored safety procedures.
- Employees had received limited or no instruction in operating procedures.
- Employees had problems handling equipment.
- Safety procedures were not in effect, or safety devices were not operational.

Several of these factors may have contributed to the same incident. For example, an employee might have been in a prohibited position relative to an operating machine because he or she had not been adequately instructed how to operate the equipment.

Accidental activation occurred with all machine types, but apparently more frequently with mixers, conveyors, and presses. This type of incident occurred in several different ways:

- There was no intent to activate the machine, but either the victim or another employee had hit the starting button or switch.
- The intent was to activate the machine, but there was no awareness that a worker was inside or in the way of moving parts of the machine. This seemed to be the more frequent problem.

In many of these cases, a worker crawled inside a machine without following proper procedures or notifying other workers of his or her location. In other cases, no lockout system for machine control existed, or procedures for safety device use had not been established.

Machine deactivation failures before cleaning, repairing, or unjamming were a frequent factor in fatalities involving mixers. Workers caught their clothing and extremities in moving machine parts when they took unnecessary risks attempting to unjam or clean a running machine.

Other critical contributory factors and the problems causing them are identified in the study as follows:

- Illiteracy/language difficulties (problem: wrong switch activated)
- Housekeeping (problem: slips/falls)
- Loose clothing/long hair (problem: gets caught in machine)

- Multiple trades working together (problem: confusion with supervision)
- Machines in testing stage (problem: not enough precautions)
- Mills with series of machines (problem: not all parts of the system locked out)
- Worker alone (problem: ignores procedures)
- Start-up after break/lunch (problem: machine in reverse; location of other workers not known)
- Job time studied (problem: shortcuts in procedure to save time)
- Alcoholism (problem: unnecessary risks).

Machine and equipment operators and operational support personnel were found to be the most frequent victims in fixed-machinery incidents. Accidents characterized by the phrases *struck by* and *caught in* were predominant in the study.

Guarding and lockout/tagout procedures were the most frequently cited OSHA violations. Although these safety practices may help prevent some incidents, there were many cases where the employee removed guarding or did not use lockout/tagout procedures. The standards emphasize protecting the employee through machine alteration and do not seem to address the human factors contributing to these incidents.

In some of these cases, the incident was an unusual occurrence that would have been difficult to prevent. In other cases the employee was negligent, taking an unnecessary risk, or departing from usual practice for some unknown reason. Risk taking appeared to be more frequent among new workers who were probably not fully aware of hazardous situations. The mental lapses, on the other hand, seemed to occur among the experienced workers. Continued training and better supervision would likely address these problems.

Some general conclusions may be drawn from the analysis of fixed-machinery fatalities:

- Larger fixed machinery that can be entered or has exposed moving parts that can entangle employees are frequently involved in fatal accidents.
- As a machine type, conveyors account for a relatively high number of fatalities.
- Because the incident types occurred with all types of machines, countermeasures directed at the types of incidents may be more useful than focusing on a specific type of machine.
- Each of the incident types identified—operating procedures, accidental activation, machine deactivation, and equipment failure—involved behaviors that cause fatal incidents.
- Machine alteration alone will not adequately address the problems. Both human and machine factors contributed to the fatal incidents.

U.S. BUREAU OF LABOR STATISTICS STUDY (1981)

The U.S. Bureau of Labor Statistics (BLS) published a report in October 1981 entitled *Injuries Related to Servicing Equipment,* which summarizes a survey conducted by the Office of Occupational Safety and Health Statistics during the period from August to November 1980 in cooperation with 25 states.

State agencies reviewed about 500,000 injury reports, of which 1,285 were within the scope of the survey, which was limited to contact with machinery and electrical or piping systems that were energized or contained hazardous materials. The servicing activities included were unjamming, cleaning, repairing, maintenance, electrical work, setup work, installing, adjusting, inspecting, or testing. The survey covered injured workers in all occupations and industries except coal mining and metallic and nonmetallic mining. Cases were excluded if the injury resulted in a fatality or if more than 120 days had elapsed between the time of the injury and the beginning of the survey.

Sixty-five percent (833) of the workers selected as within the scope of the study responded to the mail questionnaire. Response was voluntary. Results of the survey were based upon the 833 questionnaires completed by the responding workers. The following tables were selected for inclusion because of their relevance to energy-release-incident causation: Table 2-A, Source of Injury; Table 2-B, Activity at Time of Accident; Table 2-C, Type of Power; Table 2-D, Extent of Power Shutdown; Table 2-E, Lockout Procedures and Instruction; and Table 2-F, Work Information.

The survey of workers injured while servicing equipment showed that most injuries were caused by contact with moving machine parts because of failure to turn off the equipment. Injuries occurring to workers who turned off equipment were most frequently caused by accidental reactivation. The survey covered workers who were injured while cleaning, repairing, unjamming, or performing similar tasks on nonoperating industrial equipment and electrical and piping systems.

The injuries studied occurred to workers in almost every industry group, although 74% were in manufacturing. The four industries showing the highest proportions of injuries were food and kindred products, 15%; paper and allied products, 7%; printing and publishing, 7%; and fabricated metal products, 6%.

About as many injuries occurred in establishments with 100 or more employees as in smaller firms. One-fifth of the injured workers were employed in firms with 500 or more workers. Slightly more than one-half of the injured workers reported that their establishments had a safety officer. Approximately one-fourth of the workers indicated that no safety officers were employed where they worked, while an equal number did not know if there were.

Table 2-A. Source of Injury: Injuries Related to Servicing Equipment, Selected States, August–November 1980

Source of injury	Workers	Percent
Total	833	100
Boilers, pressure vessels	5	1
Boxes, barrels, containers,	2	([1])
Chemicals, chemical compounds	11	1
Coal and petroleum products	3	([1])
Conveyors	65	8
Electric apparatus	45	5
Flame, fire, smoke	4	([1])
Hand tools, not powered	3	([1])
Heating equipment (nonelectric), not elsewhere classified	1	([1])
Liquids, not elsewhere classified	5	1
Machines	643	77
Machines, unspecified	30	4
Agitators, mixers	26	3
Agricultural machines, not elsewhere classified	6	1
Buffers, polishers, etc	15	2
Casting, forging, welding	11	1
Crushing, pulverizing	5	1
Drilling, boring	18	2
Highway construction	3	([1])
Office	15	2
Packaging, wrapping	63	8
Picking, carding, etc	4	([1])
Planers, shapers, molders	22	3
Presses (not printing)	44	5
Printing	78	9
Rolls	38	5
Saws	30	4
Screening, separating	3	([1])
Shears, slitters, slicers	27	3
Stitching, sewing	6	1
Weaving, knitting, spinning	3	([1])
Machines, not elsewhere classified	196	24
Mechanical power transmission apparatus	9	1
Metal items	13	2
Mineral items, nonmetallic, not elsewhere classified	1	([1])
Pumps and prime movers	2	([1])
Steam	5	1
Vehicles	1	([1])
Wood items	1	([1])
Working surfaces	2	([1])
Miscellaneous, not elsewhere classified	8	1
Nonclassifiable	4	([1])

[1] Less than 0.5 percent.

NOTE: Due to rounding, percentages may not add to 100. See appendix A for types of injuries included in the survey.

SOURCE: State workers' compensation reports.

Source: U.S. Department of Labor, Bureau of Labor Statistics, *Injuries Related to Servicing Equipment.* (Washington DC: GPO, 1981).

The predominant occupational class of workers injured was machine operators, who accounted for close to one-half of those included in the survey. Craft workers followed, representing slightly more than one-third. The injuries were widely dispersed within these two occupational classes. Mechanics and repairers represented 10% and printing press operators, 7%.

Lost workdays were reported by 686 respondents. Nearly one-half of the estimates of time lost exceeded 15 workdays. Moreover, the average was 24 workdays for those workers who lost time.

Three out of four injuries were to the hands and fingers. Although cuts were the most common injury, accounting for 1 out of 3 cases, fractures and contusions each occurred in 1 out of 7 cases and amputations in 1 out of 10 cases.

Nearly nine-tenths of the injuries resulted from contact with moving machine parts. Industrial equipment associated with the injury varied widely; printing presses, conveyors, and packaging machines were the more prevalent types, each accounting for about one-tenth of total injuries. Other sources of injuries included electrical apparatus, saws, agitators, and slicers. Workers' descriptions of the equipment indicated that rollers were a common hazard with many types of machinery. One-half of the workers indicated that no emergency shutoff was within their reach at the time of the accident.

Five percent of the injuries were caused by contact with electrical current. Such injuries usually occurred while doing electrical repairs or installations. Occasionally, workers performing other activities,

Table 2-B. Activity at Time of Accident: Injuries Related to Servicing Equipment, Selected States, August–November 1980

Item	Workers	Percent
Which of the following best describes what you were doing when the accident occurred?		
Total	833	100
Unjamming object(s) from equipment	250	30
Cleaning equipment	245	29
Repairing equipment	77	9
Performing maintenance (oiling, etc.)	34	4
Installing equipment	13	2
Adjusting equipment	99	12
Doing set-up work	57	7
Performing electrical work	29	3
Inspecting equipment	15	2
Testing material or equipment	12	1
Other activity	2	(1)
Was the task due to a breakdown during operation?		
Total	802	100
No	517	64
Yes	285	36
Had you done this type of task before?		
Total	798	100
No	81	10
Yes—on same or similar equipment	672	84
Yes—but on different equipment	45	6
How long would this task have taken to complete if you had not been injured?		
Total	814	100
Less than 2 minutes	328	40
2 to 15 minutes	253	31
15 minutes to 1 hour	124	15
1 to 8 hours	63	8
More than 8 hours	4	(1)
Don't know	42	5
Estimate how often you do this type of task.		
Total	768	100
Daily	429	56
About once a week	110	14
About once a month	106	14
About once a year	42	5
First time you did this type of work	81	11
How did your injury occur?		
Total	833	100
Injured by machine parts that were in motion	735	88
Injured by contact with electrical current	45	5
Injured by chemicals, hot liquids, or other hazardous material	29	3
Injured by falling machine parts	10	1
Other	14	2

[1] Less than 0.5 percent.

NOTE: Due to rounding, percentages may not add to 100. See appendix A for types of injuries included in the survey. Because incomplete questionnaires were used, the total number of responses may vary by question.

SOURCE: Survey questionnaire.

Source: U.S. Department of Labor, Bureau of Labor Statistics, *Injuries Related to Servicing Equipment.* (Washington DC: GPO, 1981).

Table 2-C. Type of Power: Injuries Related to Servicing Equipment, Selected States, August–November 1980

Item	Workers	Percent
Indicate all of the equipment's power sources.		
Total [1]	828	([1])
Electric	751	91
Hydraulic	112	14
Pneumatic (air)	155	19
Pressurized (as in pipeline)	32	4
Steam	16	2
Gravity	5	1
Spring action	21	3
Gas or diesel engine	25	3
Don't know power source	27	3
If electric:		
Indicate voltage.		
Total	622	100
120 volts	76	12
More than 120 volts	283	45
Don't know voltage	263	42
Indicate type of power.		
Total	474	100
Single phase	61	13
Multiphase	190	40
Don't know type of power	223	47
Was there an emergency shutoff within your reach at the time of the accident?		
Total	819	100
No	411	50
Yes	380	46
Don't know	28	3
What kind of warning system, if any, did the equipment have to indicate it was activated or about to be activated?		
Total [1]	779	([1])
None	596	77
Bells, alarms, or other audible warning system	54	7
Lights or other visual warning system	92	12
Don't know	43	6

[1] Because more than one response is possible, the sum of the responses and percentages may not equal the total. Percentages are calculated by dividing each response by the total number of persons who answered the question.

NOTE: Due to rounding, percentages may not add to 100. See appendix A for types of injuries included in the survey. Because incomplete questionnaires were used, the total number of responses may vary by question.

SOURCE: Survey questionnaire.

Source: U.S. Department of Labor, Bureau of Labor Statistics, *Injuries Related to Servicing Equipment.* (Washington DC: GPO, 1981).

such as cleaning, received shocks because of faulty or exposed wiring.

Workers described the type of service work they were performing at the time of the injury and how familiar the work was to them. The tasks that led to the greatest number of injuries were unjamming and cleaning activities, each accounting for about 30% of the accidents. Twelve percent occurred while workers were making adjustments to the equipment. Maintenance and repair work was involved in 13% of the injuries, and set-up work accounted for 7%. Less frequent activities were electrical work, installing, inspecting, and testing equipment.

Workers' lack of experience or familiarity with the

Table 2-D. Extent of Power Shutdown: Injuries Related to Servicing Equipment, Selected States, August–November 1980

Item	Workers	Percent
Was the equipment turned off before doing the task?		
Total	833	100
No [1]	653	78
Yes	180	22
If equipment was not turned off, indicate reason(s).		
Total [2]	592	([2])
Felt it would slow down production or take too long	112	19
Not required by company procedures	69	12
Didn't know how to	8	1
Didn't think it was necessary	209	35
Could not do task with equipment off	209	35
Did not realize power was on	62	10
Other reason	61	10
If equipment was turned off:		
a. Indicate what happened at the time of the injury.		
Total	176	100
You accidentally turned equipment or system back on	20	11
Co-worker accidentally turned equipment or system back on	15	9
Co-worker turned equipment or system back on, not knowing you were working on it	56	32
Equipment or material moved when jam-up was cleared	9	5
Parts were still in motion (coasting)	30	17
Other reason	46	26
b. Were any additional steps taken to shut down equipment before the accident?		
Total [2] [3]	160	([2])
No—felt it would slow down production or take too long	8	5
No—not required by company procedures	23	14
No—didn't have supplies or tools to do this	4	2
No—didn't think it was necessary	49	31
No—other reason	20	13
No—reason not given	37	23
Disconnected main power	14	9
Tagged equipment power, valves, etc	6	4
Locked out equipment power, valves, etc	2	1
Removed fuse	-	-
Disconnected electrical line or broke circuits	5	3
Removed section of pipe	2	1
Drained pressure or hazardous materials from system	9	6
Installed blank flange	1	1
Restrained parts that could move, fall, or slide with blocks, chains, clamps, etc.	4	2
Other	4	2

[1] Of these, 61 reported accidental activation of auxiliary controls (e.g., foot pedals) and 19 reported using jog or inch controls.

[2] Because more than one response is possible, the sum of the responses and percentages may not equal the total. Percentages are calculated by dividing each response by the total number of persons who answered the question.

[3] Includes 33 respondents who took additional steps to shut down equipment and 127 who took no further action.

NOTE: Dashes indicate that no data were reported. Due to rounding, percentages may not add to 100. See appendix A for types of injuries included in the survey. Because incomplete questionnaires were used, the total number of responses may vary by question.

SOURCE: Survey questionnaire.

Source: U.S. Department of Labor, Bureau of Labor Statistics, *Injuries Related to Servicing Equipment.* (Washington DC: GPO, 1981).

Table 2-E. Lockout Procedures and Instructions: Injuries Related to Servicing Equipment, Selected States, August–November 1980 *(continued)*

Item	Workers	Percent
Have you ever padlocked power controls or valves in "off" position before servicing or repairing equipment?		
Total [1]	655	([1])
No	383	58
Yes—on equipment involved in accident	101	15
Yes—on other equipment	112	17
Yes—at another place of work	47	7
Don't know	45	7
Are the equipment controls designed for padlocking main power source or valve in "off" position?		
Total	694	100
No	246	35
Yes	273	39
Don't know	175	25
If equipment controls are designed for a lockout:		
a. Are the lockout controls within reaching distance or within sight of this equipment?		
Total	264	100
No	43	16
Yes	221	84
b. Would the lockout procedure for the equipment involved in the accident require any of the following?		
Total [1]	259	([1])
A written permit	2	1
Supervisor's authorization or participation	34	13
Protective equipment (such as gloves)	11	4
Special tools other than locks (such as ladder, blocks, etc.)	6	2
Special skills (such as strength, electrical knowledge, etc.)	17	7
Help from co-workers	19	7
None of the above	172	66
Don't know	26	10
c. How long would it take to lock out this equipment?		
Total	262	100
2 minutes or less	225	86
2 to 15 minutes	22	8
15 minutes to one hour	6	2
One hour or more	-	-
Don't know	9	3
d. Would the lockout and restart be supervised?		
Total	264	100
No	183	69
Yes—by foreman or other supervisor	60	23
Yes—by safety officer	2	1
Don't know	19	7

See footnotes at end of table.

Table 2-E. (*concluded*)

Item	Workers	Percent
What type of policy, if any, does your employer have for locking out equipment before doing service or repair work?		
Total	653	100
Single lockout requirement covering all equipment	64	10
Specific lockout requirements for each type of equipment	107	16
No policy	210	32
Don't know	272	42
Were you provided any instructions on how to do a lockout of equipment power before servicing?		
Total [1]	554	([1])
Provided printed instructions	25	5
Lockout procedures posted on equipment	37	7
Given instructions as part of on-the-job training	176	32
Given formal training at meetings, etc	28	5
Other	7	1
No instructions on lockout provided	340	61
If lockout instructions were provided:		
a. What did they include?		
Total [1]	183	([1])
When to lock out	160	87
Where to place locks on equipment	91	50
Tagging in addition to locking out	48	26
Restraining parts that could move, fall, or slide with blocks, chains, clamps, etc	30	16
Clearing the area of personnel	16	9
Testing lockout to be sure power is off	52	28
Procedures for storing keys and removing locks	32	17
Controlling access to locks and keys	25	14
Lockout procedures covering change in work shifts	21	11
Group lockouts	18	10
b. When were the lockout instructions given to you?		
Total [1]	186	([1])
After the accident	15	8
One to six months before the accident	36	19
Six months to a year before the accident	28	15
Upon hiring	84	45
Over a year ago	60	32

[1] Because more than one response is possible, the sum of the responses and percentages may not equal the total. Percentages are calculated by dividing each response by the total number of persons who answered the question.

NOTE: Dashes indicate that no data were reported. Due to rounding, percentages may not add to 100. See appendix A for types of injuries included in the survey. Because incomplete questionnaires were used, the total number of responses may vary by question.

SOURCE: Survey questionnaire.

Source: U.S. Department of Labor, Bureau of Labor Statistics, *Injuries Related to Servicing Equipment.* (Washington DC: GPO, 1981).

job did not appear to be a contributing factor in most cases. Eighty-four percent of the respondents had done the task before on the same or similar equipment. The vast majority of the workers performed this type of work daily or weekly, and most had more than one year's experience.

Nearly 8 out of 10 workers surveyed failed to turn off the equipment before performing the service work that resulted in injury. It should be noted that some machines were equipped with activating controls, such as foot pedals or jog buttons, in addition to ON/OFF switches. Sixty-one respondents commented that their accidents were caused by accidental activation of these auxiliary controls.

The reasons most frequently given for not turning off the equipment were that workers thought it was

Table 2-F. Work Information: Injuries Related to Servicing Equipment, Selected States, August–November 1980

Item	Workers	Percent
What are your regular job duties?		
Total [1]	814	([1])
Operating equipment	532	65
Unjamming equipment	294	36
Making minor adjustments to equipment	330	41
Set-up work	296	36
Servicing equipment or systems	220	27
Electrical work	87	11
Plumbing or pipefitting work	61	7
Supervising other workers	136	17
Other	109	13
How long have you had these job duties at the place where you work?		
Total	801	100
Less than 6 months	174	22
6 months to a year	132	16
1 to 3 years	202	25
3 to 5 years	104	13
5 years or more	189	24
Are you paid on an incentive basis (piecework, production bonus, or profit sharing)?		
Total	768	100
No	659	86
Yes	109	14
How many people are employed at the place where you work?		
Total	794	100
1 to 19	159	20
20 to 49	123	15
50 to 99	120	15
100 to 499	234	29
500 or more	158	20
Does the place where you work have a safety officer or safety representative?		
Total	791	100
No	206	26
Yes	401	51
Don't know	184	23

[1] Because more than one response is possible, the sum of the responses and percentages may not equal the total. Percentages are calculated by dividing each response by the total number of persons who answered the question.

NOTE: Due to rounding, percentages may not add to 100. See appendix A for types of injuries included in the survey. Because incomplete questionnaires were used, the total number of responses may vary by question.

SOURCE: Survey questionnaire.

Source: U.S. Department of Labor, Bureau of Labor Statistics, *Injuries Related to Servicing Equipment.* (Washington DC: GPO, 1981).

unnecessary at the time or that the task was not possible with the power off. The latter explanation, however, often reflected the difficulty rather than the impossibility of doing the work with the equipment shut down. For example, many workers injured while cleaning rollers remarked that wiping across rotating rolls was the most efficient way to do the job. One out of eight workers claimed that the company did not require them to turn off the equipment during the activity performed at the time of injury. Pressure to keep production on schedule was mentioned by one out of five workers, some of whom noted that deactivating a machine, such as a conveyor, would shut down an entire production system. In addition, 1 out of 10 reported that they did not realize the power was on. Comments by these workers often reflected a lack of knowledge about such features as automatic cycling systems or multiple power sources.

About one-half of the workers who turned off the power were injured by accidental reactivation of equipment, most frequently by a co-worker who was unaware that the equipment was being serviced. More than one-fifth of the workers who turned off the power were injured by residual energy when either the moving parts continued to coast or the machinery moved when a jam-up was cleared. Most of the remaining injuries to workers who had turned off the power were the result of faulty power switches or valves that did not work properly.

Among those who were injured after turning off the power were nearly one-fifth who took additional steps such as disconnecting the main power source, breaking the circuit, or tagging equipment. (*Tagging* refers to the attachment of tags on equipment's power sources to advise co-workers not to turn on power.) These extra precautionary measures were sometimes carried out for reasons other than safety. For example, in cases involving pipe systems, draining the system after closing the valve was necessary to accomplish the service work. Only two workers had attempted to fully lock out the equipment to prevent accidental reactivation or contact with electricity. (*Lockout* was defined as "disconnecting or shutting down and locking equipment controls in the OFF position.") These lockouts, however, were apparently not tested before servicing the equipment. In one case, the lockout had been done on the wrong power line, and in the other, a second power line had been spliced into the wiring beyond the point of lockout.

The workers surveyed generally indicated little experience with lockout procedures. About two-thirds noted that they had never done a lockout, and a nearly equal proportion had received no training on lockout procedures. Those most likely to have experience in lockouts were electricians and mechanics. Only one-fourth of the workers were aware of any policy their employers had regarding lockouts. About two-fifths did not know of any policy, and one-third reported there was no lockout policy.

OFFICE OF STATISTICAL STUDIES AND ANALYSES (1982)

The OSHA Office of Statistical Studies and Analyses reviewed 83 fatalities related to lockout/tagout problems that had occurred between 1974 and 1980. Federal OSHA fatality/catastrophe investigation reports were used as the reference sources. The study was divided into three major sections: Machines and Conveyors (60 fatalities), Vehicles and Equipment (14 fatalities), and Electrical (9 fatalities). Each incident was further assigned to one of five categories of factors most likely responsible for causing the accident even though several factors may have been present. The categories were (1) operating procedures, (2) accidental activation of machinery, (3) failure to deactivate machinery, (4) equipment failure, and (5) other. Ninety-four percent (78) of the incidents were classified in categories 1 through 4, described as follows:

- Operating procedure incidents occurred when the employee or employer failed to follow the designated work or safety procedures or when there were no procedures to follow. These incidents included entering unauthorized areas, taking risky work positions, or failing to follow completely all steps in the lockout/tagout procedures. They do not include incidents where the lockout/tagout procedures were ignored or where there were none to follow.
- Accidental activation of machinery incidents occurred when the victim or another employee accidentally activated the machinery. These incidents included cases where the machinery was knowingly or intentionally turned on but the employee was unaware of the consequences; where the machine was accidentally turned on by striking against it, pushing the wrong button, and so on; and where there was no explanation of how the activation occurred.
- Failure to deactivate machinery incidents occurred when the employee attempted to clean parts of the machine, clear out clogged material, do maintenance work, and the like while the machine was turned on or still in motion. There may or may not have been a procedure for such activities. In some incidents, lack of proper machine guarding and interlocking devices contributed to the problem.
- Equipment failure incidents occurred when the machine or parts of it malfunctioned, resulting in fatal injuries. There may or may not have been lockout/tagout of the machine or equipment.
- Other incidents were those with no identifiable elements common to the above categories.

Machines and Conveyors

These fatal incidents occurred primarily in components (frequently conveyors and augers) of large fixed machinery. Sixty fatalities (72%) were attributable to this grouping, 27 of which (45%) were related to accidental activation of machinery. The next largest category was "failure to deactivate machinery," which involved 18 of the 60 incidents (30%). In practically all cases, the worker died from being caught in moving parts of the machinery. Crushing injuries were compounded by others such as lacerations, fractures, contusions, and so on. Fifty of the fatalities occurred while workers were performing normal job activities, which included cleaning, repairing, unclogging, and otherwise servicing machinery. There were no

Table 2-G. Types of Lockout/Tagout-Related Fatal Incidents: Machines and Conveyors

Type of Incident	Fatalities
A. Operating procedure	11
Workers entered unauthorized and hazardous machine areas.	4
Lockout procedures were carelessly or incompletely followed.	2
Workers performed job from improper location or position on machine.	2
Worker was not authorized to operate machinery.	1
Incorrect placement of lockout system.	1
Dangerous improvised work procedures were used.	1
B. Accidental activation	27
Worker could not see the deceased when activating machine.	9
The machine was unintentionally activated by another worker, e.g., bumped against controls, pushed wrong buttons, etc.	5
The machine started for reasons unknown or understood.	5
The deceased unintentionally activated the machine.	3
The deceased pushed the wrong button.	2
Another worker activated the wrong controls	2
Another worker, following instructions from foreman, turned machine on.	1
C. Failure to deactivate	18
Machine was cleaned or washed while it was running.	8
Machine was unjammed or unclogged while it was running.	3
Machinery was adjusted or repaired while it was running.	4
Materials, e.g., dough, was removed from the machine while it was running.	2
Worker was cleaning around machine and fell in while it was running.	1
D. Equipment failure	4
The interlock malfunctioned when the worker attempted to clear the machine.	1
The hydraulic pressure failed, lowering roll.	1
The belt holding a safety bar failed.	1
The relay activity switches were not working properly.	1
TOTALS	60

Source: U.S. Occupational Safety and Health Administration, Office of Statistical Studies and Analyses. *Selected Occupational Fatalities Related to Lockout/Tagout Problems as Found in Reports of OSHA Fatality/Catastrophe Investigations.* PB83-125724 (Washington DC: GPO, 1982).

incidents where the worker was operating the machinery during routine production (Tables 2-G and 2-H).

Vehicles and Equipment

These fatal incidents occurred when vehicles rolled into or over workers, parts of equipment dropped down on them, and so on. Fourteen fatalities (17%) were attributable to this grouping. Nine of the 14 (64%) were related to operating procedure problems in which the worker failed to set parking brakes, chock wheels, block overhead components of equipment, and the like. The worker was performing normal job activities in 11 of the 14 cases (79%, Tables 2-I and 2-J).

Electrical

These fatal incidents occurred when the deceased came in contact with an energized conductor and was electrocuted. Nine fatalities (11%) were attributable to this grouping. Failure to deactivate accounted for one-third of the incidents. In eight of the incidents, the worker was performing a normal job activity (Tables 2-K and 2-L).

Table 2-H. Employee Activity at Time of Lockout/Tagout-Related Fatal Incidents: Machines and Conveyors

Activity	Fatalities
A. Performing normal job	50
Cleaning machinery.	13
Repairing machinery or other maintenance work.	12
Unclogging material from machinery.	10
Adjusting machinery.	7
Oiling or greasing machinery.	2
Observing or inspecting operations.	2
Removing product material from machine.	2
Applying silicon to machine.	1
Cleaning up adjacent area fell into machinery.	1
B. Performing other than normal job	10
Worker in area where he should not have been.	4
Operating machinery that was not his duty.	1
Deceased, found in tank, reason unknown.	1
Entered machine to retrieve hammer.	1
Attempted to fill drinking water container.	1
Left assigned area to do another task.	1
TOTALS	60

Source: U.S. Occupational Safety and Health Administration, Office of Statistical Studies and Analyses. *Selected Occupational Fatalities Related to Lockout/Tagout Problems as Found in Reports of OSHA Fatality/Catastrophe Investigations.* PB83-125724 (Washington DC: GPO, 1982).

A review of the available data shows that slightly over 60% of all deaths were associated with operational occupations—for example, machine operator, helper, laborer, baker, and so on. Seventy-five percent of the incidents involved workers being caught in or between parts of machinery, and over 80% of those killed were performing their normal job duties.

In summary, a number of problem areas were identified that had special significance for each major grouping:

- **Machines and conveyors**
  - Workers entered unauthorized and hazardous areas while machinery was operating or performed jobs from an improper location or position on machine.
  - Lockout procedures were carelessly or incompletely followed.
  - When activating machines, workers were out of sight of the injured.
  - Machines were unintentionally activated by the injured or other workers when they bumped against controls, pushed wrong buttons, or made similar accidental moves.
  - Workers attempted to clean, unjam, adjust, repair, remove material, and the like while machines were operating or had not come to a complete stop.
  - Workers were fatally injured when crucial parts of the machines malfunctioned, for example, hydraulic pressure failure.
- **Vehicles and equipment**
  - Workers failed to set parking brakes and chock wheels of stationary vehicles.
  - Workers failed to lower fully or block overhead equipment components.
  - Workers failed to shut down equipment completely before leaving the operating position.
- **Electrical**
  - Workers did not follow directions when shutting off electrical power.
  - Workers failed to deactivate main switches before making repairs.
  - Workers were electrocuted when it was assumed that power had been disconnected or there were failures in the switches.

Table 2-I. Types of Lockout/Tagout-Related Fatal Incidents: Vehicles and Equipment

Type of Incident	Fatalities
A. Operating procedure	9
Worker failed to set parking brakes and chock wheels.	3
Worker failed to set brakes and place in neutral.	1
Brakes were not maintained in operable condition, and treads were not chocked.	1
Worker failed to shut down equipment completely (as instructed) before leaving operator's position.	1
Dump body, not blocked, on dump truck.	1
Worker failed to fully lower or block bucket on front-end loader.	1
Worker failed to block crane boom.	1
B. Accidental activation	1
A shop-made tie press was accidently (reason unknown) activated while being moved.	1
C. Equipment failure	1
A safety latch did not hold on the blade of a bulldozer.	1
D. Other	3
The worker placed his hand through the opening in an elevator shaft door as elevator descended.	1
The worker was in an elevator pit with the floor of the elevator above him; he did not use the safety switch.	1
A skip bucket tripped out, crushing the deceased; it could not be determined how it was activated.	1
TOTALS	14

Source: U.S. Occupational Safety and Health Administration, Office of Statistical Studies and Analyses. *Selected Occupational Fatalities Related to Lockout/Tagout Problems as Found in Reports of OSHA Fatality/Catastrophe Investigations.* PB83-125724 (Washington DC: GPO, 1982).

Table 2-J. Employee Activity at Time of Lockout/Tagout-Related Fatal Incidents: Vehicles and Equipment

Activity	Fatalities
A. Performing normal job	11
Repairing or other maintenance work.	3
Moving equipment or vehicles.	3
Cleaning equipment.	2
Dismantling equipment.	1
Unloading material.	1
Drilling rocks for road construction.	1
B. Performing other than normal job	3
Attempted to stop movement of vehicles.	1
Walked in front of moving vehicle.	1
Signaling for elevator.	1
TOTALS	14

Source: U.S. Occupational Safety and Health Administration, Office of Statistical Studies and Analyses. *Selected Occupational Fatalities Related to Lockout/Tagout Problems as Found in Reports of OSHA Fatality/Catastrophe Investigations.* PB83-125724 (Washington DC: GPO, 1982).

Table 2-K. Types of Lockout/Tagout-Related Fatal Incidents: Electrically Related (1)

Type of incident	Fatalities
A. Operating procedure	1
The deceased did not follow directions in turning off the power.	1
B. Accidental activation	1
Another worker mistakenly followed a signal and turned the power on.	1
C. Failed to deactivate	3
Worker failed to deactivate main switch before repairs.	1
Worker failed to turn off power at the worksite.	1
Worker failed to tell another worker to turn off power.	1
D. Equipment failure	2
The main power switch was turned off but failed to de-energize the bridge runway.	1
There was an error in splicing of the junction box.	1
E. Other	2
Power had been disconnected but two wires remained alive.	1
Conductors that were being relocated were energized when they were supposedly dead.	1
TOTALS	9

Source: U.S. Occupational Safety and Health Administration, Office of Statistical Studies and Analyses. *Selected Occupational Fatalities Related to Lockout/Tagout Problems as Found in Reports of OSHA Fatality/Catastrophe Investigations.* PB83-125724 (Washington DC: GPO, 1982).

Several secondary factors that contributed to the incidents and appear to cut across all categories of lockout/tagout-related fatalities were also identified. These include the following:

- communication problems
- unnecessary personal risks
- dangerous work procedures
- incorrect work positions
- short time period on the job
- not being authorized to operate machine
- inability to read or write English.

A review of the types of incidents and secondary factors makes clear that occupational fatalities are complex events. Multiple points of attack are needed to address the human, mechanical, and environmental interactions that result in fatal incidents. Preventive measures include the following:

- development of strict adherence to lockout/tagout procedures
- establishment and enforcement of safe working procedures
- training
- supervision
- strengthening and modifying existing standards and developing new ones where the need is clearly indicated.

Table 2-L. Types of Lockout/Tagout-Related Fatal Incidents: Electrically Related (2)

Activity	Fatalities
A. Performing normal job	8
Installing or removing equipment.	3
Repairing or other maintenance work.	3
Checking connections and removing dust.	1
Climbing out of crane.	1
B. Performing other than normal job	1
Superintendent touched live wires for unknown reasons.	1
TOTALS	9

Source: U.S. Occupational Safety and Health Administration, Office of Statistical Studies and Analyses. *Selected Occupational Fatalities Related to Lockout/Tagout Problems as Found in Reports of OSHA Fatality/Catastrophe Investigations.* PB83-125724 (Washington DC: GPO, 1982).

U.S. NATIONAL INSTITUTE FOR OCCUPATIONAL SAFETY AND HEALTH/BOEING STUDY (1983)

The Division of Safety of the U.S. National Institute for Occupational Safety and Health (NIOSH) and Boeing Aerospace (under contract) prepared a document entitled *Guidelines for Controlling Hazardous Energy During Maintenance and Servicing.* Its data did not include enough specific information to identify factors causing or contributing to accidents that occur during maintenance as a result of inadequate energy control measures. However, out of 300 accident scenarios that could be attributed to this cause, 59 cases were selected for closer study. Due to limited detail, exact cause determination was not possible. However, the scenarios did provide a general indication of primary or contributing causes as well as insight regarding the types of accidents that can occur when hazardous levels of energy are not controlled. NIOSH, after review and analysis, determined the following:

- initiated activity without attempting to lockout/tagout power sources—27 accidents
- energy isolation attempted but inadequate—6 accidents
- residual energy not dissipated—1 accident
- accidental activation of energy—25 accidents
- all causes—59 accidents.

Over 80% of the accidents involved mechanical motion, either linear translation or rotation, whereas

11% resulted from direct live electrical conductor contact. Production machinery accounted for over 30% of the incidents in which employees were caught between moving parts or between a moving part and a stationary object. Approximately 25% of the incidents were associated with process or support equipment (such as mixers, blenders, shakers, dryers, and washers). Conveyors were involved with 15% of the occurrences.

A sizable proportion (35%) of accidents in the study were characterized by someone other than the injured worker activating the machine or equipment. In some cases, a co-worker cycled the equipment or closed the circuit knowingly in an attempt to accomplish some task without realizing that another employee was vulnerable. In others, an employee activated equipment without knowing work was in progress. These incidents demonstrate the criticality of effective communication and proper de-energizing procedures.

A large number of accidents (again, 35%) involved cleaning or clearing jams/obstructions. These tasks were invariably performed by operating or production personnel. One could surmise that the injured workers assumed that the activity would be accomplished quickly with little risk of accident.

The energy-isolating or control method was noted to be an electrical disconnect in 50 of the 59 accidents. Blocking/pneumatic hardware, blinds or blanks, and valves accounted for the balance of energy control devices reported in the case histories.

Furthermore, a review of the detail found in the study's Appendix B shows the following: Approximately 50% of those who were injured/died worked in maintenance classifications, while operational personnel accounted for 40% of the casualties. Thirty-five of the accidents resulted in fatalities, and seven involved amputations. The severity of all of the cases was very high.

A review of the literature did not produce statistical evidence of the effectiveness of one specific energy-control method over another, and neither did it identify accident-causative factors leading to injuries. No values could be developed to differentiate accidents and injuries occurring during maintenance from aggregate U.S. injury statistics—that is, data to provide indexes on the probability of injury occurrence, the magnitude of the injuries, and the degree of exposure to the hazard (the number of workers at risk) that correlate with the identified primary hazard causes.

Industrial accident and injury data available in the United States describe the injuries sustained by the workers in greater detail than the causes of the accidents that inflicted the injuries. The hazard causes found in the analyzed accident reports are categorized by fire, explosion, impact, fall, being caught in or between machinery, and so on. The hazard causes identified by this study are different. Consequently, the published statistics (aggregates of accident/injury reports) cannot be broken down into accident causes identified as specific to maintenance and servicing activities.

The study identified the following hazard causes:

- Maintenance activities were initiated without attempting to de-energize the equipment or system or to control the hazard with energy present.
- Energy blockage or isolation was attempted but was inadequate.
- Residual energy was not dissipated.
- Energy was accidentally activated.

The worker population at risk could not be defined because employers who implemented procedural controls may choose to make the procedures applicable to (1) all the personnel; (2) personnel engaged in any one or any combination of activities such as construction, operations, manufacturing, and testing; or (3) maintenance personnel only. Thus, the effectiveness of one method of control with respect to another cannot be determined. However, the implementation of any method that increases worker awareness of potentially hazardous energy sources is better than no method at all.

Existing federal and state government safety regulations (California and Michigan excepted) of energy control during maintenance are inconsistent, fragmented, and only applicable to certain equipment, processes, and industries (for example, telecommunication and construction). Most of these existing regulations use the concept of "power off" to prevent injuries and do not provide guidance on how to discern when to apply locks, tags, or a combination thereof. They also do not allow for the performance of maintenance with power on, even under normal, everyday conditions or when such maintenance cannot be performed with power off.

STUDY OF HAZARDOUS-ENERGY-RELEASE INJURIES IN OHIO (1983)

The Division of Safety Research, NIOSH, prepared a draft report entitled *Study of Hazardous Release of Energy Injuries in Ohio in 1983.* Three hundred thirty-nine accidents were selected because (1) they fell into likely categories of industry, occupation, type of accident, source of injury, and diagnosis of injury; (2) the workers' compensation claim narrative suggested applicability; and (3) questionnaire responses by plant officials positively identified the injuries as resulting from an unexpected energy release during equipment repair, servicing, or maintenance. The study indicated that 70% of the reported injuries occurred to production workers as a result of servicing that took place during normal production operations. However, it was not possible to determine from the data whether the tagout procedures that were in use by the firms had actually been applied. The study attempted to focus on the issue of locks versus tags while not considering other elements of the lockout/tagout programs that were in place.

The following tabulation was made of the task being performed at the time of the accident:

- unjamming object—84 (25%)
- cleaning equipment—75 (22%)
- repairing equipment—41 (12%)
- adjusting equipment—41 (12%)
- doing set-up work—27 (8%)
- inspecting equipment—11 (3%)
- testing equipment—9 (3%)
- installing equipment—9 (3%)
- electrical activity—8 (2%)
- other tasks—34 (10%)
- total—339 (100%).

The NIOSH *Ohio Study,* although submitted to OSHA for review, remains unpublished.

NATIONAL SAFETY COUNCIL STUDY (1984)

The Electronic and Electrical Equipment Section of the National Safety Council's Industrial Division surveyed the division from 1981 to 1982 regarding accidents where contact with live electrical conductors either directly caused injury or was the underlying cause of an accident. A total of 828 employee injuries were reported and analyzed for several variables. The principal occupational distribution was as follows: electrician (26.7%), lineman (8.3%), maintenance worker (8.0%), supervisor (5.4%), service repair workers (5.2%), and production worker (5.1%).

Sixty percent of the injured employees had received training in handling electricity. The distribution for employee electrical experience was less than 1 year (19.6%), 1 to 5 years (19.3%), 5 to 10 years (18.3%), 10 to 15 years (16.5%), 15 to 30 years (20.2%), and over 30 years (5.1%). Fifty-seven percent of the injuries involved contact with 480 volts or less. Nine percent of the total cases involved fatal injury.

The following contributing factors were identified most frequently during the analysis of the injury incidents (the totals exceed 100% because multiple factors are identified as contributing to a single injury event):

- failure to follow safety rules—48.3%
- deliberately working with hot (energized) machinery—20.8%
- unknowingly working with hot machinery—16.8%
- lack of personal protective equipment—16.0%
- crowded work space—13.8%
- inadequate lockout—12.3%.

With regard to the age of the workers associated with these contributing factors, the following was reported:

- Workers 20 to 29 years of age accounted for 30.7% of the accidents caused by failure to follow safety rules.
- Workers 40 to 49 years of age accounted for 35.8% of the accidents caused by deliberately working with energized machinery.
- Workers 40 to 49 years of age also accounted for 26.1% of the accidents caused by lack of personal protective equipment.
- Workers 50 to 59 years of age accounted for 20.4% of the accidents caused by unknowingly working with energized machinery.

Trained workers were involved in 61% of the incidents involving permanent disability and 67% of the fatalities.

U.K. HEALTH AND SAFETY EXECUTIVE STUDY (1985)

From 1980 to 1982, fatal accidents reported to the Agricultural, Factory, and Mines and Quarries Inspectorates were studied by the U.K. Health and Safety Executive Accident Prevention Advisory Unit. The study focused on "maintenance activities," which were defined as "the repair, restoration, or cleaning of process machinery, plant, vehicles, buildings, structures and roads."

A general report entitled *Deadly Maintenance—A Study of Fatal Accidents at Work* was published. After further research, two additional reports were prepared because of the high number of deaths associated with roof work and plant/machinery maintenance. *Deadly Maintenance—Roofs* and *Deadly Maintenance—Plant and Machinery* outline various prevention methods but make no attempt to produce a definitive safe system of work.

The initial study revealed that 326 people died performing maintenance activities. This represents over 21% of the total number of fatal accidents. However, because fatalities represent only the tip of the iceberg, broader concern needs to be focused on approximately 10,000 accidents involving major injury that were estimated to have occurred during the same period. Fatalities were distributed in the main employment sectors as follows: manufacturing (99), construction (117), services (58), agriculture (22), and mines and quarries (30).

Plant and machinery maintenance, with 106 deaths, was the largest category. Conveyors, elevators, and overhead traveling cranes figured prominently in the types of machinery involved. Thirty-four people died while working on vehicles, mostly while making mechanical repairs. Twenty-five were killed during the maintenance of electrical, gas, and water services.

The relationships between accident type and maintenance activity is shown in Table 2-M. Eighty people were crushed or entangled in machinery and vehicles—almost a quarter of the total. The majority involved powered motion, but 26 were attributable to such things as gravity, fall of machinery, vehicles falling off supports or rolling forward while on sloping ground (Figure 2-1). In four incidents, sufficient energy had been retained to cause movement even after the isolation of the power supply.

An analysis of the causative factors associated with the 326 fatalities is shown in Table 2-N. Failure to solve people-related problems accounted for almost 48% of the causative factors, which primarily consisted of the lack of appropriate safe systems of work, management failures, and inadequate communications. Elementary failure to provide adequate physical safeguards produced 35% of the causes. These related to unsafe facilities, equipment, working platforms, or safety equipment. Human error was responsible for only 9% of the accident causes mentioned in inspectors' reports.

Management was found to be primarily responsible in 54% of the cases and shared responsibility with the work force, not necessarily the deceased, in over 15%. Workers were deemed to be primarily responsible in 16% of the accidents.

The main conclusions drawn from the study are as follows:

- Most (83%) of the accidents could have been eliminated by taking reasonably practical precautions.
- In nearly 70% of the cases, positive management action could have saved lives.
- The causes of injury most commonly identified were lack of adequate safe systems of work; failure to provide proper physical safeguards; poor management organization; and inadequate information, instruction, and training.
- Seven major maintenance activities featured in the accidents:
  - plant and machinery maintenance—106 deaths
  - roofwork—63 deaths
  - vehicle repairs—34 deaths
  - painting—33 deaths
  - electrical, gas, and water systems maintenance—25 deaths
  - building maintenance—21 deaths
  - window or industrial cleaning—20 deaths
- Falls caused 49% of lives lost, and nearly 25% of

Table 2-M. The Relationship Between Accident Type and Maintenance Activities

	Accident type							
Maintenance activity	**Crushed or entangled**	**Falls**	**Burns**	**Asphyxiation, gassing, or drowning**	**Electrocution**	**Impact**	**Struck by falling object**	**Total**
Plant or machinery maintenance	50	21	10	14	6	—	5	106
Roofwork	—	61	1	—	1	—	—	63
Vehicle repairs	26	—	1	—	5	2	—	34
Painting or decorating	—	29	1	1	1	—	1	33
Electrical, gas, and water system maintenance	1	3	5	6	9	1	—	25
Building and structure maintenance	—	20	—	—	—	—	1	21
Window and industrial cleaning	—	18	—	1	1	—	—	20
Road or river works	3	1	—	1	—	8	—	13
Scaffold work	—	6	—	1	—	—	—	7
Dockyard maintenance	—	1	—	3	—	—	—	4
Total	80	160	18	27	23	11	7	326

Source: Health and Safety Executive Accident Prevention Advisory Unit, *Deadly Maintenance—A Study of Fatal Accidents at Work* (London: HMSO, 1985). British Crown copyright. Reproduced with the permission of the Controller of Her Majesty's Stationery Office.

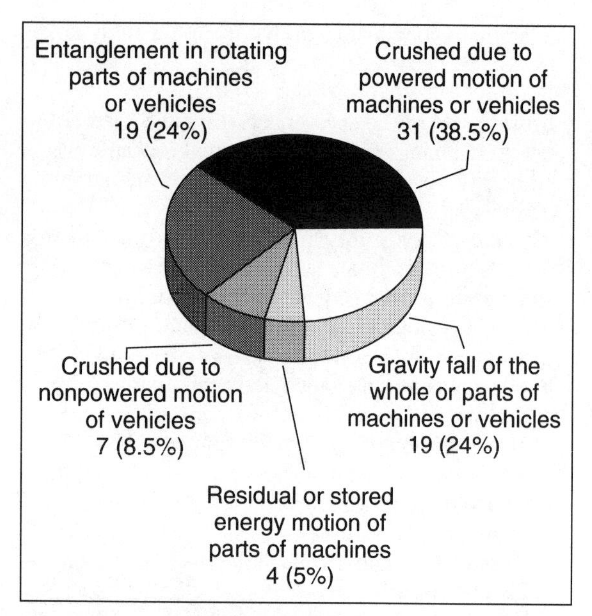

Figure 2-1. Machinery- and vehicle-related accidents. Source: Health and Safety Executive Accident Prevention Advisory Unit, *Deadly Maintenance—A Study of Fatal Accidents at Work* (London: HMSO, 1985). British Crown copyright. Reproduced with the permission of the Controller of Her Majesty's Stationery Office.

the victims were crushed or entangled in machinery, facilities, or vehicles.

- In the construction sector, maintenance work was responsible for almost a third of the annual death toll and in some manufacturing industries accounted for as much as 40% of yearly fatalities.

In summary, most people from a wide range of trades and industries will be engaged in maintenance activities and be exposed to its special risks. During the three-year period of this study, maintenance was responsible for 326 deaths, the vast majority of which were foreseeable and preventable.

The impression that the authors of this study have formed in compiling this report is that management personnel fail to recognize or choose to ignore the dangers and that maintenance work does not receive an appropriate share of the available management resources. Unless serious thought and attention are given to planning maintenance work, this high death toll can only continue. Safe systems of work form the core of safe maintenance.

UNION STUDY (1987)

The International Union, United Automobile, Aerospace, and Agricultural Implement Workers of America (UAW) petitioned OSHA on May 17, 1979, to establish an Emergency Temporary Standard (ETS) for locking out machinery and equipment. The petition stated that a need existed to recognize the complexities of modern industrial equipment that uses sources of energy other than electricity. It discussed the increasing need for locking out equipment to prevent that equipment from cycling without warning while work was in progress. It also emphasized the importance of applying lockout procedures to systems using hydraulic or pneumatic power, to energy stored in springs and electrical capacitors, and to potential energy from suspended parts. Abstracts of case studies for fatalities involving 22 UAW members that were attributed to lockout-related causes since 1974 were submitted with the petition.

In 1987, the Health and Safety Department of the UAW released a report entitled *Occupational Fatalities Among UAW Members: A Fourteen Year Study.* The report analyzed 252 deaths that occurred between 1973 and 1986. The analysis attacked the conventional wisdom about injury prevention that OSHA and many companies used to set safety policy.

A striking concentration of fatalities among experienced skilled trades workers was apparent when the job titles of victims were reviewed. The risk ratio was about 3.6 times greater for skilled trades compared to all other workers (Figure 2-2). Within the skilled trades, the greatest number of fatalities was among electricians (24.5%) and millwrights (22.7%) who are engaged in maintenance and repair operations (Figure 2-3). The average age of fatality victims was 45 years, with average seniority of 15 years, where such data were available. Thus, the high-risk group appeared to be highly trained, experienced workers who were involved in maintenance, repair, and service operations. Materials handling and production service were also high-risk occupations. Populations in the job categories of materials handling and production service were not readily available, so risk ratios were not calculated. The data contradict the idea that unskilled younger workers are at greater risk for injuries.

In addition, prior experience in the industry had suggested that failure to de-energize and lock out power from equipment being set up, serviced, repaired, or cleaned was a major cause of fatal accidents. Review of case reports determined that lockout was a cause of 60 out of 252, or 24%, of all reported fatalities. This supports the suggestion that complex, automatic activated equipment poses a hazard in itself, rather than inherently protecting workers from hazards.

The concentration of fatalities among maintenance and materials-handling workers suggests a reason for lack of progress in reducing fatalities over the last decade. No OSHA standard for lockout existed, and standards in general did not apply well to maintenance operations. In contrast to machine guarding, unsafe maintenance situations are brief in duration and therefore are less likely to be observed during routine facility inspections by OSHA or other enforcement agencies or even by internal company safety personnel.

Table 2-N. The Relationship Between Accident Causes and Maintenance Activities

Accident Causes		Maintenance activity: Plant/machinery	Vehicles	Electrical, gas, and water systems	Dockyards	Roofwork	Painting/Decorating	Window/Industrial cleaning	Building/Structure	Road/River works	Scaffold work	Totals: Subtotal	Total	Percentage
System of work	None or Unsafe	38	16	8	–	23	9	6	5	1	1	107	178	24
	Inadequate	14	3	6	–	9	4	3	5	2	–	46		
	Not Maintained or Used	15	5	2	–	1	–	1	–	–	1	25		
Unsafe	Plant or Equipment	29	11	6	1	6	5	4	5	1	1	69	157	21
	Working Platform	7	–	1	–	4	10	4	6	–	1	33		
	Workplace or Access	3	–	2	–	25	12	5	7	1	–	55		
Management organization and/or supervison	None	2	2	–	–	1	1	1	–	–	–	7	100	13.5
	Inadequate	39	5	10	–	19	9	4	4	2	1	93		
Safety equipment or guards	None	9	1	3	–	20	2	8	3	–	1	47	95	13.0
	Inadequate	10	2	–	–	13	2	–	1	–	1	29		
	Not used	4	–	2	1	6	2	3	1	1	–	19		
Information instruction or training	None	11	1	5	–	2	4	1	2	1	2	29	68	9
	Inadequate	22	4	6	–	4	1	1	–	1	–	39		
Human error		23	13	5	–	6	6	4	3	7	1	–	68	9
Unforeseeable event or not known		11	1	3	–	3	3	2	1	5	3	–	32	4.5
Unauthorised activity		13	–	1	3	4	1	2	1	–	–	–	25	3.5
Communication failures		7	1	1	–	–	1	1	1	–	–	–	12	1
Defective design of plant or equipment (Section 6 HSW Act)		6	–	2	–	–	–	–	2	–	–	–	10	1
Adverse weather			–	1	–	2	–	–	–	–	–	–	3	0.5
Total number of accident-associated causes													748	

Source: Health and Safety Executive Accident Prevention Advisory Unit, *Deadly Maintenance—A Study of Fatal Accidents at Work* (London: HMSO, 1985). British Crown copyright. Reproduced with the permission of the Controller of Her Majesty's Stationery Office.

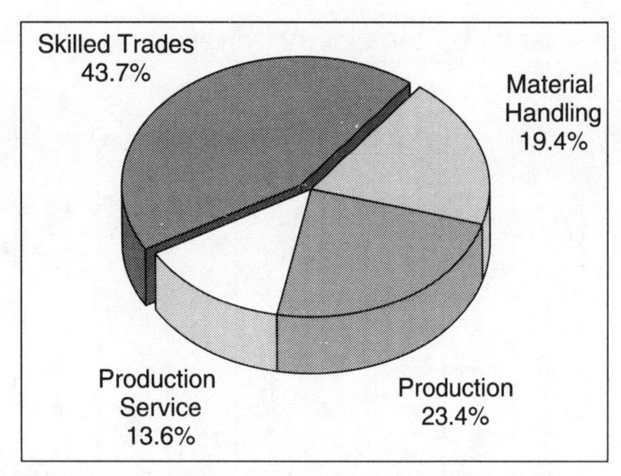

Figure 2-2. UAW occupational fatalities, 1973–1986.
Source: UAW, Health and Safety Department, *Occupational Fatalities among UAW Members: A Fourteen Year Study* (Detroit: UAW, March 1987).

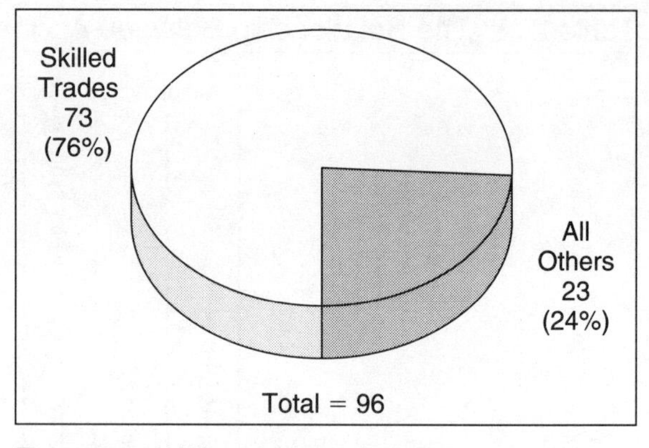

Figure 2-4. UAW no-lockout, January 1973–December 1993.
Source: UAW, Health and Safety Department, *Occupational Fatalities among UAW Members* (Detroit: UAW, 1993).

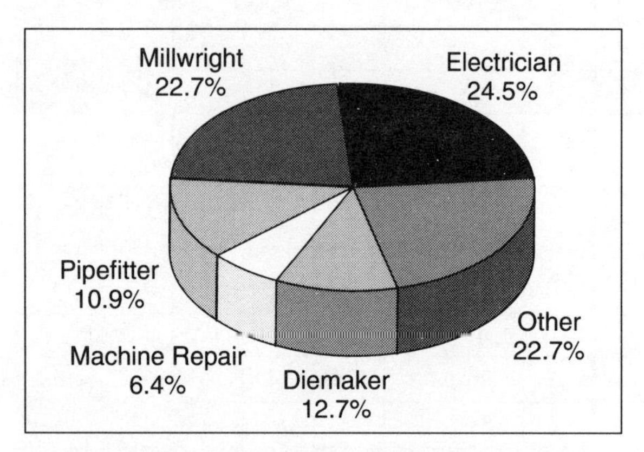

Figure 2-3. UAW skilled trades fatalities
Source: UAW, Health and Safety Department, *Occupational Fatalities among UAW Members: A Fourteen Year Study* (Detroit: UAW, March 1987).

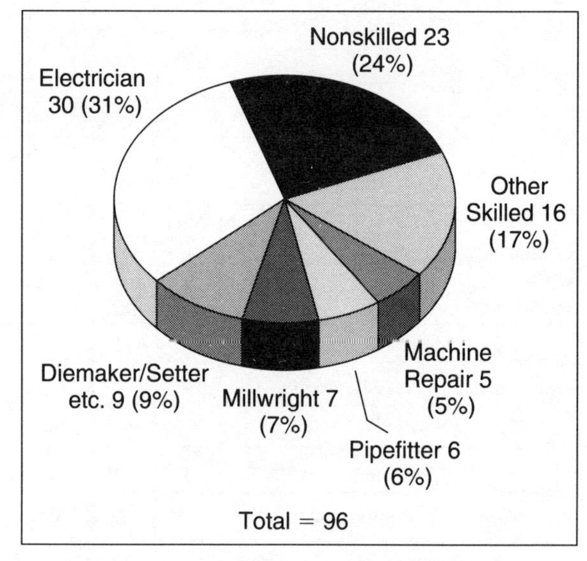

Figure 2-5. UAW no-lockout fatalities distributed by victim's trade, January 1973–December 1993.
Source: UAW, Health and Safety Department, *Occupational Fatalities among UAW Members* (Detroit: UAW, 1993).

During October 1988, at a public hearing conducted by OSHA on the proposed lockout/tagout standard, the UAW testified that there had been 72 "lockout fatalities" among its members between 1973 and 1988. Of these cases, there had been "inadequate training" in 49 cases (68%); "inadequate procedures" in 50 cases (69%); and "adequate, but unenforced procedures" in 19 cases (26%).

Data indicated that 26% of all reported UAW fatalities were attributable to "no lockout." Skilled trades (electrician, die maker/setter, pipe fitter, machine repairer, millwright, and so on) accounted for 77% of the deaths. Repair activities ranked first (65%) in terms of lockout-related fatal accident occurrence, servicing maintenance ranked second (16%), and operation ranked third (10%). The primary causes of death were crushing injuries (54%) and electrocution (22%).

Mechanical energy was a factor in 36% of the reported lockout fatalities. Electrical energy was a close second, with 32%. Pneumatic, gravitational, chemical, and hydraulic energy types were involved in 16% of the incidents, and multiple energies accounted for the balance. Stored or residual energy was involved in 16% of all lockout fatalities. Minor repairs, adjustment, operation, and cleanup of machinery and equipment accounted for 36% of the lockout incidents.

The UAW has continued to develop its occupational fatality data base as part of its ongoing commitment to aggressively represent its membership's safety interests. The "lockout" component of these mortality data from 1973 to 1993 continues to provide insight with regard to the occurrence of "no lockout" fatality incidents. Ninety-six (27%) of no-lockout cases were reported, representing the UAW's single largest fatality cause. Crushing injuries or electrocution were the causes of death in 79% of the incidents.

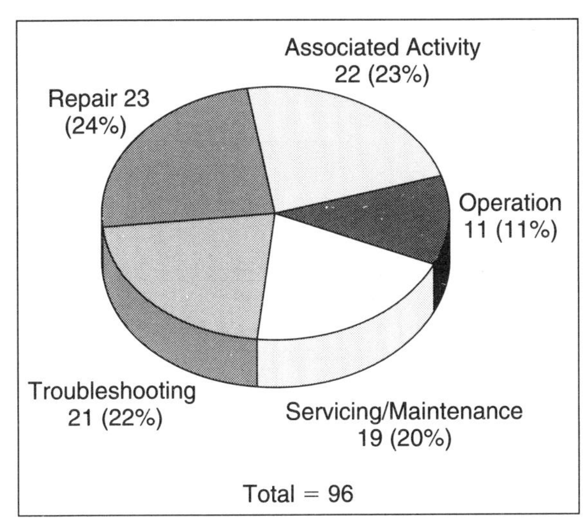

Figure 2-6. UAW no-lockout fatalities distributed by work activities, January 1973–December 1993.
Source: UAW, Health and Safety Department, *Occupational Fatalities among UAW Members* (Detroit: UAW 1993).

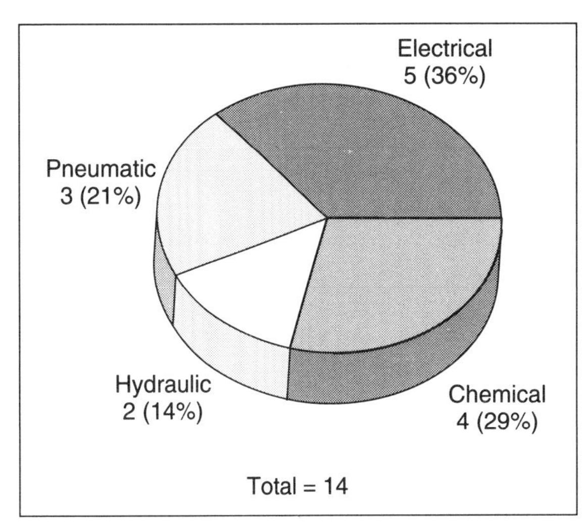

Figure 2-8. UAW no-lockout fatalities distributed by type of stored energy, January 1973–December 1993.
Source: UAW, Based on reports to the Health and Safety Department.

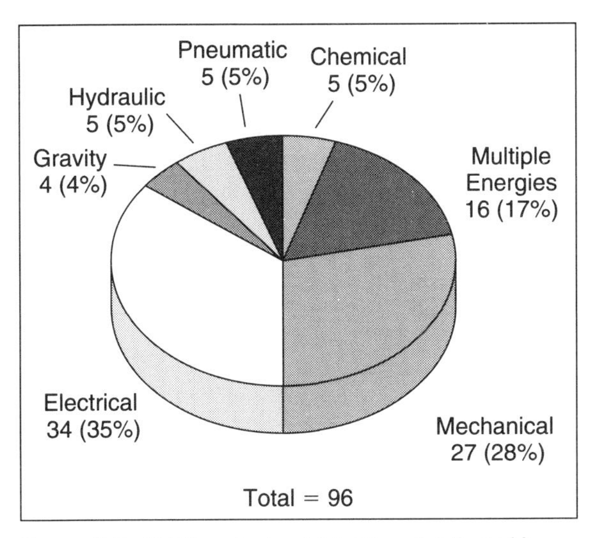

Figure 2-7. UAW no-lockout fatalities distributed by types of energy involved, January 1973–December 1993.
Source: UAW, Health and Safety Department, *Occupational Fatalities among UAW Members* (Detroit: UAW, 1993).

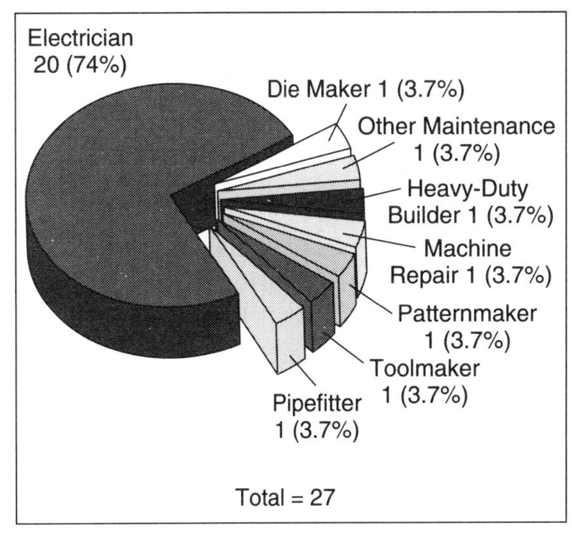

Figure 2-9. Skilled trades fatalities caused by failure to lock out electrical energy, January 1973–December 1993.
Source: Includes plug and cord–type equipment.
UAW, Based on reports to the Health and Safety Department.

Skilled trades personnel sustained a significant percentage (76%) of the fatal no-lockout injuries (Figure 2-4). Electricians were the most represented in these incidents (Figure 2-5).

Repair work was involved 24% of the time in no-lockout incidents (Figure 2-6). Electrical or mechanical energy was the primary source of injury in 63% of the cases reported (Figure 2-7). Stored or trapped energy was involved in 14 of the incidents (Figure 2-8).

Twenty-seven cases involved deaths that occurred due to failure to lock out electrical energy (Figure 2-9). One-half of these incidents involved plug and cord equipment.

QUEBEC RESEARCH INSTITUTE STUDY (1989)

A study entitled "Accidents Related to Lockout in Quebec Sawmills" was published by the Quebec Research Institute in 1989. From January 1, 1986, to December 10, 1987, 1,164 incident reports from eight Quebec sawmills employing 1,342 workers were reviewed. Machine incidents numbering 691 (59.4%) were isolated for analysis. These represented only stationary machines operating in sawing and planing shops. Furthermore, 51 (7.4%) of the machine cases were separated out where there was an obvious lack or nonuse of padlocking or blocking measures; these were called padlock cases.

The padlock cases were classified into the following three categories according to worker activity at the time of the accident or incident:

- unjamming, extraction, and unjamming/extraction assistance activities—40 (78.4%)
- maintenance, repair, and maintenance/repair assistance activities—10 (19.6%)
- production or production assistance activities—1 (2.0%).

The padlock cases were further studied and classified according to four types of events that triggered the accidents:

- motor not stopping—28 (54.9%)
- motor start-up—10 (19.6%)
- abnormal movement (linear mechanical movement [not caused by the revolution of a motor] considered as abnormal to the production process; for example, wood ejected from an edger or machine breakage)—7 (13.7%)
- unwanted movement (linear mechanical movement [not caused by the revolution of a motor], usually normal, but that occurred without the accident victim wanting it)—6 (11.8%).

The following recommendations were formulated to improve the protection of workers carrying out interventions on Quebec sawmill machines.

1. Protection relating to unjamming intervention:

- Specific devices (safety stops) to protect workers against untimely start-ups or to stop motors safely should be implemented in addition to normal devices for padlocking isolating switches.
- Existing safety stop devices should be improved, particularly in relation to their effectiveness in stopping and preventing all loads (electrical, pneumatic, or hydraulic) from restarting.
- Safety stop devices should meet the following criteria:
  - Activation will stop all loads and prevent their being restarted until the operator has made a positive movement to repeat the command—for example, using a restart push button that is physically different from the padlockable blocking switch. The activation of these switches will also neutralize the automatic start-up functions by the action of photoelectric cells or other means.
  - All loads (electrical, pneumatic, and other) controlled by a console or panel can be maintained in a safe stop position from this panel by one or several safety stop devices.
  - Depending on the operation's requirements, for one control console or panel, certain loads could be regrouped so that the entire machine and its auxiliary equipment do not have to be stopped during an unjamming intervention. However, this regrouping must not introduce new risks.
  - The loads controlled by any of the safety stop devices will be clearly identified. Several means are available: regrouping control push buttons, displayed list, color coding, and separation lines.
  - A standard padlock can lock each safety stop device in the activated position.
  - The safety stop device will be chosen mainly for its safety, installation possibilities, ease of use, and compatibility with the workplace habits in each sawmill.
  - Padlocking procedures in sawmills should specifically mention what means of protection, including safety stop, should be established in any case of intervention—not only for padlocking during repair. Particularly, guidelines should be established in each sawmill to indicate what operations must be stopped, kept in a safe stop state, or have their power circuit padlocked for wood unjamming or regular maintenance operations.
  - Complete padlocking of electric-isolating switches must remain in effect according to regulations, and it must be specifically stated that for each tool change, the machine must be isolated electrically and its isolating device must be padlocked. Application of this principle can lead to modifications in existing equipment, particularly through the installation of additional padlockable circuit breakers.

2. Elimination of risks at the source:

- Improve processes and machines in order to reduce operational anomalies that require workers to make various corrections and to effect incident recovery, repositioning, unblocking, unjamming, and so on.
- Improve machines so that maintenance is easier, reducing the number of maintenance interventions and increasing the safety of those interventions.

3. Machine improvement:

- Each sawmill should devote a sufficient period of time to the evaluation of its specific needs for machine and process improvement in order to reduce the number of worker interventions on machines and in their vicinity.
- Evaluations of each machine's needs should be based on the number of interventions required, their duration, and their type (for example, unjamming, unblocking, repositioning, and maintenance) as well as on previous accidents and incidents.
- A systematic intervention survey system should be implemented.
- The existing log and board unscramblers should be analyzed and improved in their two functions of extraction and separation.
- The material feed in sawmills should be the subject

of in-depth improvement studies, particularly in cases where moving materials change directions.
- Machine protectors should be studied, improved, or added according to the following main factors: the proximity of movable parts or components hazardous to the physical well-being of workers, frequency of intervention in the danger zone or near it, the type of intervention to be carried out, difficulty of access, the risk of being struck by moving materials, and so on.

FATAL ACCIDENT CIRCUMSTANCES AND EPIDEMIOLOGY PROJECT (1993)

NIOSH investigates selected workplace fatalities through the Fatal Accident Circumstances and Epidemiology (FACE) project. This project is designed to (1) collect descriptive data on selected fatalities, using epidemiological methods; (2) identify potential risk factors for work-related deaths; (3) develop recommended intervention strategies; and (4) disseminate findings that increase employer and employee awareness of the danger of fatal workplace injuries.

From November 1982 through December 1990, 201 electrocution incidents, resulting in 217 deaths, were investigated as part of FACE, whose protocol defines *electrocutions* as deaths occurring as a result of contact with an electrical agent or a vehicle carrying electricity.

The program is designed to collect data on selected work-related deaths using an epidemiological model that evaluates the workplace environment, the worker, tasks performed, tools used, energy exchange resulting in fatal injury, and management's role in the interaction of these factors. FACE aims to prevent worker fatalities by identifying potential risk factors and developing intervention strategies that are disseminated in an attempt to reduce the risk of fatal workplace injuries.

Occupational fatalities associated with electrocutions are a significant and ongoing problem. Data from NIOSH's National Traumatic Occupational Fatality (NTOF) data base indicate that nearly 6,500 traumatic work-related deaths occur each year in the United States; an estimated 7% of these fatalities are electrocutions. The National Safety Council reported that electrocutions were the fourth leading cause of work-related traumatic death in 1990, whereas the Bureau of Labor Statistics, in a study entitled *Occupational Injuries and Illness in the United States by Industry*, reported that electrocutions accounted for 9% of private sector fatalities during 1989 and were the fifth leading cause of death on the job (U.S. NIOSH, 1983).

Electrical hazards represent a prevalent occupational danger. Most members of the work force are exposed to electrical energy while performing daily duties, and electrocutions routinely occur to workers in various job categories. Many workers are unaware of potential electrical hazards in the work environment, which increases their vulnerability to the danger of electrocution.

NIOSH has identified five categories that describe the 201 work-related electrocution incidents that occurred between November 1982 and December 1990 and resulted in 217 fatalities:

- direct worker contact with an energized power line—58 incidents, 60 fatalities
- boomed vehicle contact with an energized power line—37 incidents, 40 fatalities
- conductive equipment contact with energized equipment—32 incidents, 43 fatalities
- direct worker contact with energized equipment—38 incidents, 38 fatalities
- improperly installed or damaged equipment—36 incidents, 36 fatalities.

In 71 incidents (35%), no safety program or written safe work procedures existed. In an additional 23 incidents (11%), it could not be determined if a safety program or written safe work procedures existed.

Figure 2-10 shows the 10 job classifications with the highest number of fatalities. Laborers, the leading classification, typically received minimal training in electrical safety. Linemen and electricians (in industries other than electric utilities) had the next highest number of fatalities; however, they generally received extensive training in electrical safety. Of the fatalities among linemen, 20 (67%) were due to failure to use required personal protective equipment (gloves, sleeves, mats, blankets, and the like).

Common factors in these incidents were lack of enforcement of existing employer policies concerning personal-protective-equipment use and lack of supervisory intervention when existing safety policies were violated. Supervision was present at the site in 106 incidents (53%), and 37 victims were supervisors.

Of the 217 victims, 180 (83%) had received some electrical safety training. Eighty-eight victims received on-the-job training, whereas 36 victims received no training. Ninety-three victims (43%) had been on the job for less than one year.

Alternating current (AC) was involved in 198 incidents (99%), one incident involved direct current (DC), and two incidents involved AC arcs. Of the 198 AC electrocutions, 66 (33%) involved low voltage (less than or equal to 600 volts, as established by NEC); 132 (67%) involved high voltage (greater than 600 volts). Figure 2-11 shows the number of electrocutions at different voltage levels. Thirty-eight (58%) of the low-voltage electrocutions involved household currents of 110 to 240 volts. Manufacturing companies accounted for 33 (50%) of low-voltage incidents.

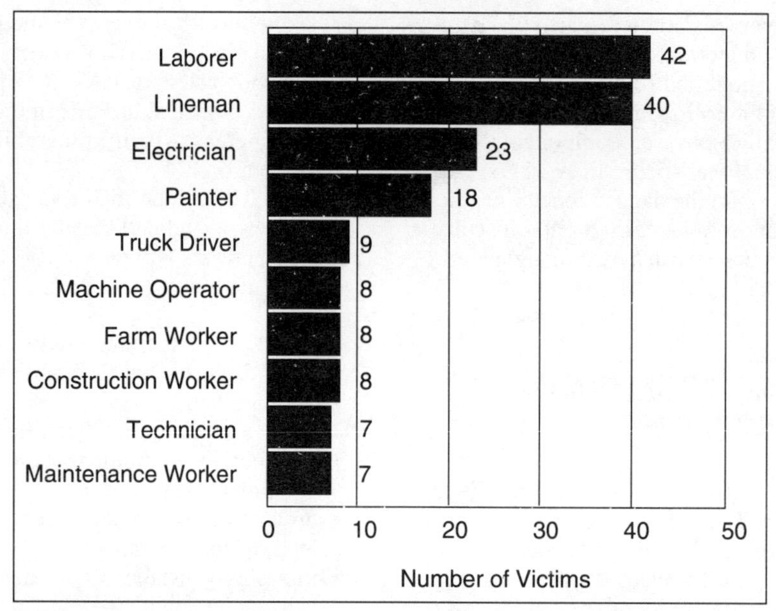

Figure 2–10. FACE electrocution data, 1982–1990—Top 10 job classifications (n = 217).
Source: V. Casini; (NIOSH, Division of Safety Research), Occupational electrocutions: Investigation and prevention. *Professional Safety* 38, no. 1 (Jan. 1993):34–39.

Of the 132 high-voltage incidents, 101 (77%) involved distribution voltages (7,200–13,800 volts), and 17 incidents involved transmission voltages (above 20,000 volts). Of those incidents involving at least 7,200 volts, 37 (31%) resulted from contacting an energized power line with a boomed vehicle. Thirty-two incidents occurred when conductive equipment, such as an aluminum ladder or scaffold, contacted an energized power line. This equipment's weight often required more than one worker to move or position it, which resulted in multiple fatalities. Thirteen deaths occurred in six separate incidents when workers erected or moved scaffolds that contacted energized overhead power lines. Electric power line mechanics were victims in 42 (35%) of the incidents involving transmission and distribution voltages.

During interviews, witnesses stated that 81 victims (37%) of both low- and high-voltage incidents knew they were working on energized circuits. Twenty-three fatalities (60%) caused by direct contact with energized equipment or conductors involved work on circuitry that co-workers or witnesses stated was known (by the victim) to be energized.

The FACE surveillance system is passive and limited to those fatalities reported by NIOSH; therefore, caution should be exercised when interpreting these data. However, project data can provide valuable information regarding occupationally related deaths, which can be used to target intervention efforts and identify further studies needed to evaluate prevention strategies for occupational electrocutions.

At least one of the following causal factors was involved in each of the 201 incidents evaluated: (1) established safe work procedures were either not implemented or not followed; (2) adequate or required personal protective equipment was not provided or worn; (3) established lockout/tagout procedures were either not implemented or not followed; (4) existing OSHA, NEC, and NESC regulations were not complied with; and (5) worker and supervisor training in electrical safety was not adequate. Although a safety program existed in 107 incidents (53%), it was not always thoroughly administered and in some cases was inadequate to ensure worker safety.

Many employer representatives and other personnel with fatality-reporting functions did not realize that contact with 110–120 volts could cause death. It is possible that because of this misconception, the number of low-voltage electrocutions that occur each year are greater than the number actually reported.

Of particular concern is the high proportion (53%) of low-voltage electrocutions that occurred in manufacturing companies. Electrical equipment used in manufacturing settings has various safeguards incorporated into its design (that is, controlled environment, electrical grounding, interlocked covers, and so on). However, many of these cases involved inadequate work practices, such as performing maintenance on energized equipment, a lack of equipment knowledge or familiarity on the part of employees working with these voltages, or improper installation of equipment grounding.

This demonstrates the need to educate and train all workers about hazards associated with electrical en-

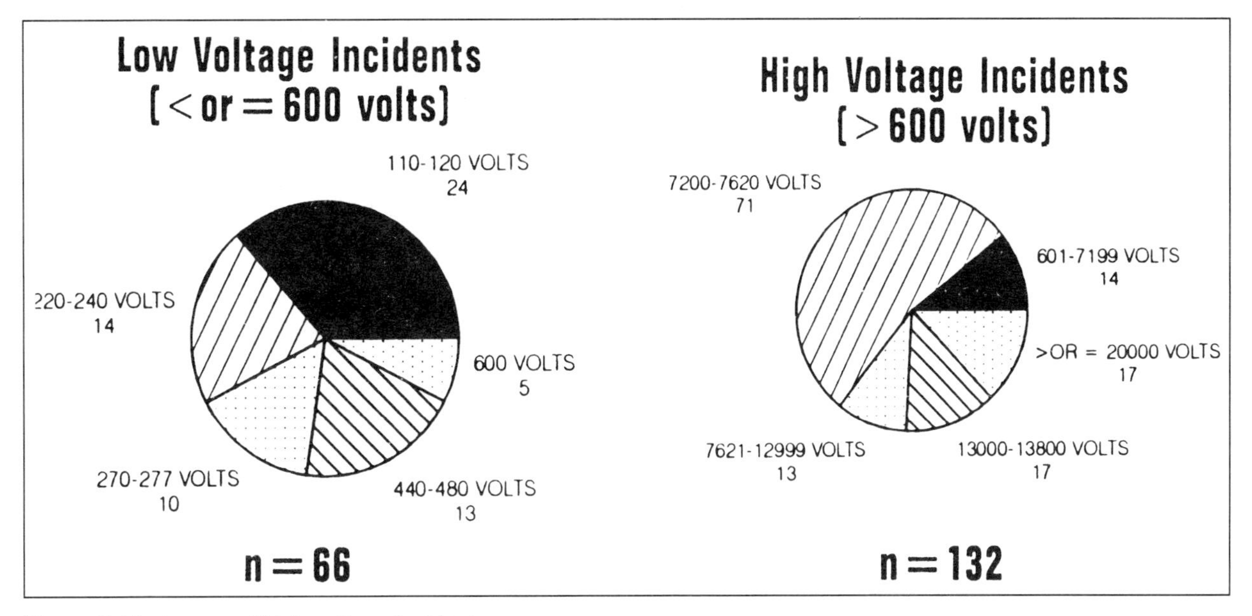

Figure 2-11. Low- and high-voltage incidents.
Source: V. Casini; (NIOSH, Division of Safety Research), Occupational electrocutions: Investigation and prevention. *Professional Safety* 38, no. 1 (Jan. 1993):34–39.

ergy and ensure supervisory enforcement of established safe work practices and procedures as well as personal-protective-equipment use. Because 93 victims in these incidents were employed for less than one year, training should be initiated upon hire.

Most of the 201 occupational electrocution incidents could have been prevented through compliance with existing OSHA, NEC, and NESC regulations; by following established safe work procedures; and/or by using adequate personal protective equipment. All workers should receive hazard awareness training so that they can identify existing and potential hazards in the workplace and understand fully the potential seriousness of injuries associated with each hazard. Once hazards are identified, control measures should be developed.

Based on analysis of these data, employers should take the following steps:

- Develop and implement a comprehensive safety program and, when necessary, revise existing programs to thoroughly address electrical safety
- Ensure compliance with existing regulations (OSHA, NEC, NESC) and establish safe work procedures
- Provide workers with adequate training in identifying and controlling hazards associated with electrical energy in the workplace
- Ensure that proper personal protective equipment is available and worn by workers.

Data indicated that although many companies had comprehensive safety programs, they were often not completely implemented. The data further underscore the need for increased understanding, awareness, and ability on the part of management and workers to identify hazards associated with working on or in proximity to electrical energy.

The study concluded that management must provide a safe workplace and develop and implement a comprehensive safety program. This can necessitate developing additional worker training and/or evaluating and restructuring existing safety programs. Management should also provide adequate training in electrical safety to all workers and strictly enforce adherence to established safe work procedures and policies. Additionally, adequate personal protective equipment should be available where appropriate. A strong commitment to safety by both management and labor is essential in preventing occupational injuries and death.

SUMMARY

The limitations of the existing data collection and analysis systems for hazardous-energy-release incidents prevent us from obtaining a complete picture of the problem. However, we can visualize and construct an operative model of the finished work.

- All types of energy (electrical, hydraulic, pneumatic, gravitational, and so on) are involved in hazardous energy-release incidents. A sizable number of incidents associated with the simplest energy type, gravity, have taken place with surprising regularity.
- The frequency of energy-release incidents is deemed to be low when compared with other groupings of injuries/incidents that preoccupy management's time. Therefore, as a class of injuries/

incidents, they receive a low priority during quantitative assessments. However, these occurrences invariably produce high-severity outcomes with correspondingly high levels of emotional response. Because of this, it appears that both management's and workers' mind-sets generate an approach to hazardous energy control that is cyclical and fragmented. The infrequency of energy-release injuries convinces employers and employees to believe that all is well in their domain. Nothing could be further from the truth. As with incidents associated with elevated work, confined space entry, and the like, which are characterized by low frequency and high severity, one has to look deeper into the problem of causation.

- Conventional wisdom would suggest that repair and maintenance personnel are at particular risk for hazardous-energy-release incidents. Not true. Operating and operational support workers are injured as often or more frequently than trade and craft employees. It is also generally believed that this problem especially impacts blue-collar workers. However, the casualty rates for supervisory/management personnel are relatively high when their percentage of total exposure hours under lockout/tagout circumstances is taken into consideration.
- The data suggest that extensive work experience is no shelter from the hazards of installing, repairing, adjusting, inspecting, cleaning, operating, or maintaining the equipment/process. Workers with long histories of service and respectable qualifications are often severely injured in energy release incidents, although inexperienced employees also account for a sizable share of total morbidity/mortality. Jobs or tasks in which injury occurs are frequently ones that have been done many times without adequate precautions or safeguards. The data support the view that deficient conditions, procedures, and work practices are routinely present, but often circumstances are not suitable to trigger the negative consequences.
- Large numbers of employees sustain injuries when they are caught in or between objects. This might occur in a machine's point of operation, in-running rolls, or power transmission gear; between a fixed object and a moving object, such as a rotating loader assembly and a column; between objects believed to be fixed, such as a vehicle under repair and the ground; or between two moving objects, such as bridge cranes on a common runway. Direct or indirect electrical contact with energized conductors is also responsible for a significant number of cases in the data bases.
- A recurring theme is that of an employee being seriously injured because of the direct activity of a coworker. Often shared responsibility exists for the failure(s) that precipitated the incident's occurrence. Collective experience strongly indicates that crew activity places additional pressure and demands on all participants to adhere to well-designed procedures, use available safeguards, and communicate effectively.
- Cleaning and unjamming equipment, which is most often associated with production personnel, alone may account for one-third of total energy-release casualties. This activity is often viewed as nonstandard and therefore subject to a free-style approach. Management may believe that employees doing the work are best suited to determine how it should be done. The data suggest that this neglected area should be evaluated with much more care. Employees may assume significant risk where the trade-off is perceived perks such as (1) ease of performance, (2) greater convenience, (3) faster return to production, (4) better results, and so on. Management may knowingly or unknowingly send messages that contribute to the belief that cleaning and unjamming are but momentary distractions to the real work and don't deserve any detailed attention.
- The frequency of cleaning and unjamming activity as well as other breakdown maintenance requirements is also a critical factor in generating energy-release injury statistics. The reliability and functioning effectiveness of machinery, equipment, and processes have a direct bearing on the work force's opportunity for risk exposure. Improving machine reliability/performance or minimizing/eliminating cleanup is an ideal way to reduce the opportunity for injury.
- Other contributing factors in hazardous-energy-release incidents were identified, such as complex systems, emerging technology, production incentive rates, language barriers, part-time employees, multiple-employer work sites, obsolete equipment, and the like.
- Common preventive denominators are found in the available literature. Employers need to emphasize developing and applying hazardous-energy-control procedures, effectively training the work force, providing resources and support for the energy-control process, monitoring the application of lockout/tagout, auditing the energy-control system, and enforcing all requirements with consistency.
- To break the patterns of the past, employers/management need a thorough knowledge of the problem and the strength of conviction to initiate change based on their new understanding.

REFERENCES

International Union, United Automobile, Aerospace and Agricultural Implement Workers of America. Health and Safety Department. *Occupational*

Fatalities Among UAW Members: A Fourteen Year Study. Detroit, MI.: UAW, March 1987, updated 1993.

National Safety Council. President's Advisory Committee. *National Program to Improve Occupational Injury Information.* Chicago: National Safety Council, 1978.

National Safety Council, Industrial Division, Electronic and Electrical Equipment Section. *Study of Electrically Related Industrial Accidents.* Itasca, IL: National Safety Council, 1984.

Paques JJ, Masse S, Belanger R. Accidents related to lockout in Quebec sawmills. *Professional Safety* 34, no. 1 (Sept. 1989):17–20.

Rapp FJ. *Testimony on OSHA's Proposed Standard for Lockout.* Houston, TX: UAW International Union, Oct. 13, 1988.

U.K. Health and Safety Executive Accident Prevention Advisory Unit. *Deadly Maintenance—A Study of Fatal Accidents at Work.* London, England: HMSO, 1985.

U.S. Department of Labor, Bureau of Labor Statistics. *Injuries Related to Servicing Equipment.* Bulletin 2115. Washington DC: U.S. GPO, Oct. 1981.

U.S. National Institute for Occupational Safety and Health. *Guidelines for Controlling Hazardous Energy During Maintenance and Servicing.* Publication No. 83–125. Morgantown, WV: National Institute for Occupational Safety and Health, 1983.

U.S. National Institute for Occupational Safety and Health, Division of Safety Research. Occupational electrocutions: Investigation and prevention. Article prepared by Casini VJ. *Professional Safety* 38, no. 1 (Jan. 1993):34–39.

U.S. Occupational Safety and Health Administration. Final rule: Control of hazardous energy source (lockout/tagout); *Federal Register.* Washington DC: GPO, Sept. 1, 1989. pp. 36648–36651.

U.S. Occupational Safety and Health Administration, Office of Statistical Studies and Analyses. *Occupational Fatalities Related to Fixed Machinery as Found in Reports of OSHA Fatality/Catastrophe Investigations.* Washington DC: GPO, May 1978.

U.S. Occupational Safety and Health Administration, Office of Statistical Studies and Analyses. *Selected Occupational Fatalities Related to Lockout/Tagout Problems as Found in Reports of OSHA Fatality/Catastrophe Investigations.* PB 83–125724. Washington DC: GPO, Aug. 1982.

3 Causation Analysis

After reviewing the benchmark studies presented in the preceding chapter, the impatient professional safety practitioner or manager might be tempted to leap directly into improvement activities such as worker safety awareness and training, procedure development, rules promulgation, and so on. However, such a decision would be premature. Although these initiatives might be warranted and of possible general benefit, they might not have any real bearing on preventing future hazardous-energy-release incidents. A more systematic and deliberate approach is necessary to ensure that all of the influences and causative factors related to hazardous energy releases are identified and understood so that a more coherent and effective control strategy can then be developed.

INCIDENT VERSUS ACCIDENT

Use of the term *accident* presents a number of problems regarding prevention in general and hazardous energy releases in particular. *Accidents* have been classically defined as unplanned or unwanted events that may result in personal injury or illness and/or damage to property. Many energy-release incidents certainly might fit this definition at first glance, but they may not upon further reflection. A more in-depth look at this term may expose its operative limitations.

Over time, employers and management have become comfortable with the idea that *injuries* and *accidents* are somewhat interchangeable terms. Their focus and response to accidents has been driven principally by the outcome of the accidents. Even in this age of enlightened management, the size and seriousness of prevention campaigns are often directly proportional to the severity of the accidents being targeted. Unfortunately, this preoccupation with results does very little for the problem of causation. Even if management's thinking has advanced to the stage where accident results are considered to include damage, interruption, and near misses, there is no guarantee that any constructive effort will be directed further upstream in the causation chain.

In a presentation on human factors and accident prevention, B. Lunsford, president of Performance Partners International, stated that "when an employee leaves the workplace as a result of an injury/illness, we need to understand the sequential nature of the incident and the accident." He offered the following definitions:

- An *incident* is any deviation from the established acceptable standard.
- An *accident* is an incident that results in damage to person or property.

It follows then that all accidents are incidents, but not all incidents result in accidents. Lunsford went on to say that once an incident occurs, the resulting damage and its extent are influenced by chance and circumstance. The following ramifications of the incident/accident relationship were offered:

- One or more incidents are in existence when an accident occurs.
- Multiple incidents may exist at the same time, any one or combination of which could allow the accident to occur.
- An accident does not occur every time an incident is in existence, but there is always that potential.
- Some incidents have a greater potential for accident than do others.
- Some incidents are capable of producing devastating damage.
- All incidents have a human factor involved in them somewhere.
- The individual who perpetrates the incident is not necessarily the one who will suffer the damage when the accident occurs.
- An incident can exist momentarily or continue indefinitely.
- An incident can be the result of the action of an individual, a physical condition, or a combination of both. . . .
- Because incidents are a deviation from the acceptable standard, they can be identified and eliminated or controlled.
- Intervention measures can be set up anywhere in the sequence of events, but they are more effective at some points than at others.
- Whenever an incident exists, something within the organization is at risk.
- Incidents that occur without damage lead to behavior that ignores their potential threat (Lunsford, 1990).

Bird and Germain, in their text *Practical Loss Control Leadership,* define *accident* and *incident* somewhat differently. According to them,

> an accident is an undesired event that results in harm to people, damage to property or loss to process. It is usually the result of contact with a substance or a source of energy (chemical, thermal, acoustical, mechanical, electrical, etc.) above the threshold—limit of the body or structure. There are three important aspects of this definition. First, it doesn't limit the human results to "injury" but says "harm to people." This includes both injury and illness, as well as adverse mental, neurological or systemic effects resulting from an exposure or circumstances encountered in the course of employment. Second, this definition does not confuse "injury" with "accident." They are not the same. Injuries and illnesses result from accidents. But not all accidents result in injuries or illnesses. This distinction is critical to significant progress in safety and health. The occurrence of the accident itself is controllable. The severity of an injury that results from an accident is often a matter of chance. . . . Third, if the event results in property damage or process loss alone, and no injury, it is still an accident. Often, of course, accidents result in harm to people, property, and process. However, there are many more property damage accidents than injury accidents (Bird & Germain, 1990).

They go on to say that "an incident is an undesired event which, under slightly different circumstances, could have resulted in harm to people, damage to property or loss to process as well as an undesired event which could or does result in a loss."

William G. Johnson, author of *MORT Safety Assurance Systems,* provides a more complex definition of *accident* that is presented here in a somewhat simplified form. According to Johnson, an accident is

1. An unwanted transfer of energy that occurs
2. Because of a lack of barriers and/or controls,
3. Producing injury to persons, property, or processes
4. Preceded by sequences of planning and operational errors that
    a. Failed to adjust to changes in physical or human factors and
    b. Produced unsafe conditions and/or unsafe acts,
5. Arising out of the risk in an activity, and thereby
6. Interrupting or degrading that activity.

He further stated that an *incident* is similar to an accident without injury or damage. The near-miss incident does have great import for safety. An incident with high potential for harm (HIPO) should be investigated as thoroughly as an accident (Johnson, 1980).

The National Safety Council, in its *Supervisors' Safety Manual,* eighth edition, defines *accident* as "an unplanned, undesired event, not necessarily injuring or damaging, that disrupts the completion of an activity."

Merriam-Webster's Collegiate Dictionary, 10th edition (1993), defines accident as "an unforeseen and unplanned event or circumstance; an unfortunate event resulting especially from carelessness or ignorance."

Robert E. McClay, in his article "Towards a More Universal Model of Loss Incident Causation," used the operative term *loss incident* and defined it as "any event resulting from uncontrolled hazards, capable of producing adverse, immediate or long-term effects in the form of injury, illness, disability, death, property damage or the like" (McClay, 1989).

We could continue this semantical journey around the proverbial mulberry bush without reaching any constructive consensus. The fact remains that the term *accident,* whether used in common conversation or within the context of managerial dialogue, is an apologetic avoidance of responsibility. The expression "It wasn't my fault, it was an accident" is engraved in the thought process of the general public. Continuing to use this term sends out the same old signals:

- The occurrence could *not* be anticipated or predicted.
- It resulted from some uncontrollable defect in the person proximate to the occurrence.
- The sequence of events was cosmically influenced.
- The outcome was a real surprise.
- Something that is beyond our capacity to understand is at work.

Therefore, as previously stated in Chapter 1, any unwanted transfer of energy (electrical, mechanical, hydraulic, pneumatic, chemical, thermal, gravitational, and so on) that produces injury/damage to persons, property, or processes or otherwise interrupts, interferes with, or degrades the activity in progress is an *energy-release incident.* This definition does not include those causal events that may have preceded the incident in time, place, intensity, and interconnection, even though they may have collectively contributed to or influenced the outcome.

One of the practical problems relating to the connection of all causal events with a given incident is their strength—in other words, the degree, if any, to which they singly or in combination contributed to the outcome—and their distance in time. For example, a decision made in an engineering department to limit the number of intermediate electrical energy-isolating devices on certain installations to reduce costs may play a role in an energy-release incident years later. That causal event may or may not be identifiable as the sequential trail cools with the passage of time. We may be able to conclude, for example, that an intermediate disconnect was missing or required, but we may not be able to precisely identify the deficiency or decision-making flaw that precipitated the situation. However, Fred Manuele, in his recent text *On the Practice of Safety,* helps with this dilemma by suggesting that an appropriate causation model for hazard-related incidents is one that views the incident as a process and requires a determination of when it began and ended. This type of thinking is necessary in order to depart from the usual obsession with outcome and trace all sequencing paths back in time toward the causal genesis.

CAUSATION THEORY AND CONCEPTS

H. W. Heinrich, in his classic work *Industrial Accident Prevention,* mentions 10 axioms of industrial safety. Several of these have had a dramatic influence on how accidents and their causes have been viewed. The following axioms are of special relevance:

1. The occurrence of an injury results from a sequence ending in the accident itself. The accident is caused by the unsafe act of a person and/or a mechanical or physical hazard.

   Five factors are involved in the accident sequence. These are illustrated by comparing them with a row of dominoes placed on end; the fall of the first precipitates the fall of the rest:

   $A > B > C > D > E$

   where the

   A factor is ancestry and social environment
   B factor is the fault of a person
   C factor is an unsafe act and/or physical hazard
   D factor is the accident
   E factor is the resulting injury

   In other words, he asserts that the injury (E) is invariably caused by an accident (D) and that the accident is produced by the factor that immediately precedes it (C). He postulates further that the central factor (C), the unsafe act and/or physical hazard, is the critical one and that its removal makes the preceding factors inconsequential.
2. Unsafe acts of persons are responsible for most accidents (88%).
3. The severity of an injury is largely a matter of chance.
4. Unsafe acts of persons are generally caused by
   - improper attitude
   - lack of knowledge or skill
   - physical unsuitability
   - improper physical environment

   These factors apply principally to the persons who commit the unsafe acts and are proximate or direct in nature. Underlying or indirect reasons why the act was committed may also exist. For example, an employee may fail to de-energize equipment before servicing it because a co-worker convinces him that it is a waste of time.

Heinrich then concludes that remedial action can be classified into four groupings:

- engineering changes or improvements
- persuasion and appeal (including training)
- personnel adjustments (such as reassignment, personal counseling, and the like)
- discipline (last resort) (Heinrich, 1959).

After more than 50 years, Heinrich's work should not be judged too harshly by current practitioners. Unfortunately, little was done by the professional safety movement to advance his concepts until the 1970s, and management groups worked hard at distorting his original intent. I recall that during the early 1960s I frequently heard managers of progressive companies pontificating about what they called the "80/20 rule": 80% of accidents are caused by unsafe acts of employees, and 20% are caused by unsafe physical conditions. I came to realize that they based their premise on Heinrich's original figure of 88%. However, their real interest was to allocate blame or responsibility using the 80/20 proportion, that is, employees are responsible for unsafe acts and management is responsible for unsafe conditions. This is a convenient division for management, but hardly representative of the real roots of causation. Investigation of accidents and their causes was therefore invariably oriented toward identifying only the proximate or direct causes, with bias toward the unsafe act. If an employee failed to isolate energy during a repair procedure, it was considered enough to warn, counsel, or reprimand the individual without examining what precipitated the failure. No effort was usually made to determine if there were shortcomings in the design of the system, if the procedure was deficient, if training was inadequate, if energy-isolating devices were conveniently located and identified, or if management consistently enforced its rules and requirements.

Heinrich's linear model may now appear too simplistic and limited, but anything more elaborate and definitive may have been premature for its time and certainly would have been too much for managements to digest and accept.

Dan Petersen, in his article "Other Voices—Safety's Paradigm Shift," addresses some of the changes that have caused us to question our past rules, beliefs, and patterns of thinking related to safety. In particular, he presents Heinrich's axioms as not being axioms at all because they are not self-evident truths supported by good research.

Petersen outlines 10 new principles of safety management that I have somewhat condensed here for ease of review:

1. Unsafe acts/conditions are symptoms of something wrong in the management system.
2. Severity is predictable under certain sets of circumstances:
   - unusual, nonroutine, nonproductive activities
   - high-energy sources
   - certain construction activities
3. Safety should be managed like any other company function.
4. Safety performance is achieved by management procedures that fix accountability.
5. The function of safety is to locate and define operational errors that permit accidents to occur.
6. Causes of unsafe behavior are identifiable and controllable.
7. Management's job is to change the environment that produces unsafe behavior.
8. Physical, managerial, and behavioral subsystems must be addressed.
9. Organizational culture and safety system should fit.
10. An effective safety system:
    - forces supervisory performance
    - involves middle management
    - reflects top management commitment
    - enables employee participation
    - is flexible and perceived positively (Petersen, 1991).

Principles 1, 2, 5, and 7 are particularly important for anyone addressing hazardous-energy-release incidents and their prevention. Principle 2, dealing with severity, perfectly fits the idea that energy-release incidents are characterized by low frequency and elevated severity of outcome, which is a standard feature of these incidents, particularly when examining injury occurrence. This would also be true of activity associated with elevated work, confined space entry, explosive demolition, excavation, and so on.

Much of the work done under conditions with the potential for hazardous energy release occurs in unusual, nonroutine, or nonproductive situations. This is borne out by the data presented in Chapter 2 as well as by the nature of maintenance activity associated with machine/equipment/process breakdown.

Principles 5 and 7 are central to the dilemma encountered by those attempting to prevent the occurrence of energy-release incidents. A significant part of the operational error/unsafe behavior that drives energy release–related loss statistics is not addressed by current conventional prevention policies and practices. Furthermore, compliance with regulatory standards alone is unlikely to provide the improvements desired in future performance.

In adapting to these evolving new paradigms, a degree of discomfort may be experienced as our "security blanket" is slowly removed and the cool, crisp air of change chills our temperate mind-set. However, we have certainly gone as far as we can go with post–World War II prevention concepts, and change is in the wind.

With regard to causation, Johnson (1980) suggests that accidents result from a long series of events and related causal factors in the following arrangement:

ENERGIES > BARRIERS > ERRORS >
CHANGES > RISKS

This arrangement of causative elements is likely to be amenable to improvement and control. Johnson states:

> Energy is performance oriented, that is, it does work. Barriers direct energy into useful channels and prevent unwanted energy transfers. Analysis of energy transfer is objective and likely to produce a realistic initial understanding of error-producing situations. . . .

> Errors are deviations from some standard. They occur all up and down the chain of control of work. Error-provocative situations can be improved. . . .
>
> Both energies and errors have substantial effects on costs, performance, and profitability, as well as safety.
>
> Change is often a cause of trouble and is increasing exponentially. Even well-motivated change may have unwanted side effects unless analysis is adequate. Sensitivity to change is a critical management survival skill. Perceptive change analysis methods are one of the most effective of the new techniques. . . .
>
> Lengthy sequences of events and causal factors erode control and lead up to troubles. There are many opportunities to interrupt the sequences. The length and complexity of the sequences show vividly that "it is the system that fails." . . .
>
> Risk is ever present. It should be quantified and analyzed. Risk assessment and decision analysis methods can contribute to accurate evaluation. Properly analyzed risk usually leads to risk reduction. . . . (Johnson, 1980, p. 17)

Manuele (1993), in searching for a more effective causation model, suggests that we may need to create new words to convey precise and understood meanings, for example, *hazop* (Hazard and Operability Study). He focuses on hazards and defines them as the potential for harm or damage to people, property, or the environment. They include the characteristics of things and the actions or inactions of people. If a hazard is not avoided, eliminated, or controlled, the potential will be realized.

He proposes that all hazard-related incidents be referred to by the term *hazrin,* which will by definition de-emphasize outcome and focus on the potential for harm or damage, in other words, the incidents that influence outcome. A hazrin is an unplanned, unexpected process deriving from the realization of an uncontrolled hazard or hazards, which could be an unwanted flow of energy or an unwanted release of hazardous materials, that is likely to result in adverse consequences.

Hazrin, as defined by Manuele, appears to parallel my general definition of energy-release incident from the preceding section, with the exception of my excluding certain causal events that may have contributed to the occurrence. The inclusion of "uncontrolled hazard(s)" in Manuele's definition does appear to encompass "actions or inactions of people," which represent a significant portion of the contributory causal events that eventually lead to energy-release incidents. For example, a plant manager might direct a purchasing agent to reduce spending by 10% next year, resulting in a unilateral procurement decision to buy nonlockable pneumatic valves. That order and response may be the beginning of a causation sequence that is many years removed from a specific energy-release incident. The actions of the manager and purchasing agent could be considered causal factors (hazards) and the purchase and installation of the valves a hazrin (incident) with future potential.

Manuele provides more food for thought when he adds the following suggestions to complement and extend his proposal:

- Consider Haddon's energy-release theory as fundamental in dealing with causal factors.
- Extend Haddon's energy-release concepts to include the release of hazardous materials.
- Treat a hazrin as a complex process with an identifiable beginning and end.
- Recognize that multiple events, occurring sequentially or in parallel in time and influencing each other, may precede the incident that results in, or could have resulted in, injury or damage.
- Realize that all events deriving from things or from the actions or inactions of people that contribute to hazrins are causal factors.
- Establish the significance of an organization's culture in relation to hazard-related incidents.
- Give design and engineering considerations distinct and primary status, with an emphasis on ergonomics.
- Unwanted releases of energy and hazardous materials are the fundamental causal factors for hazard-related occurrences.

McClay (1989), in pursuing a conceptual model of loss-incident causation, provides further insight to sequence and causal events. He indicates that prior to the point of irreversibility in the loss-incident sequence, three types of factors exist:

1. proximal causal factors
2. distal causal factors
3. factors unrelated to the occurrence.

Proximal causal factors exist or occur within the same specific time frame or location as the loss incident, whereas distal causal factors do not. If there is no evidence or suspicion that any relationship exists between some occurrence and the loss incident, we can say it is unrelated.

He suggests that the point of irreversibility is that point in the sequence beyond which a loss incident of some sort can no longer be prevented. Conversely, until this point is reached, the opportunity for hazard control still exists, and circumstances can be returned to normal.

He describes *hazards* as anything that directly contributes to the occurrence of a loss incident. Further, all proximal causes are to be viewed as either states (the way that things are at any point in time) or events (the occurrences that produce change in the states). States as a distinct proximal cause or hazard type are identified as a physical condition. Events are divided into two groups: human actions and exceeded functional limitations.

McClay carefully expands upon the phrase "exceeded functional limitations," because it is not as easily understood. It refers to situations in which the limitations of the elements composing the physical condition are surpassed. A potentially harmful instability is produced and causes some change in the physical condition. As he states:

> Anything that exists can be identified as a system, and each system is made up of elements—functional building blocks which combine efforts to produce the overall output from the system. Each element performs one or more functions, and when we speak of limitations, it is in the ability to perform these functions that the limitations are seen. So, if one identifies the function of a thing, then he or she has also found where the limitations will exist.

As an example, if the blade of a knife has a function to cut and it becomes dull, then its ability to cut could be exceeded by any particular cutting task undertaken. If a metal bar is to hold a particular load and the load is excessive, we say that the (functional) strength of the bar has been exceeded. If a tire is designed to contain air under pressure, and the tire is punctured by a sharp object, its ability to contain pressure will be exceeded by whatever pressure is present. (McClay, 1989)

He also points out that human functional limitations can be exceeded, causing us to do things that are unexpected or inappropriate. Human actions are also capable of producing a change in physical conditions, and because we react to the conditions around us, it is true that physical conditions can give rise to human actions as well. Therefore, each of the three hazard types, physical conditions, human actions, and exceeded functional limitations, can cause either of the other two types.

McClay notes that with regard to human performance in the operation of systems, there are many important limitations, such as endurance, dexterity, reaction speed, motivation, visual acuity, memory recall, attention span, knowledge, force capability, and so on, that must be considered.

His universal model, Figure 3-1, graphically presents all of these elements. It shows management policies as distal causes, three types of hazards and their interaction as proximal causes, and the sphere of control where hazard prevention efforts must be exerted. If prevention efforts are not successful, an adverse, unexpected event can occur, and the point of irreversibility can be passed. If this happens, a release or transformation of mass or energy will occur, triggering a loss incident. The sequence may stop, or aggravating factors may develop producing further incident(s).

Without attempting to challenge or dislodge any of the aforementioned causation concepts and models, I offer yet one more. Through the process of examining and re-examining the elements associated with human activity and the phenomena that disrupt its progress, we can greatly improve our understanding of causation and increase prevention effectiveness. The model presented in Figure 3-2 has some elements that are identical or closely resemble certain features of other concepts or theories. It is intended to stimulate additional perspectives on the issue of incident causation and although it portrays a business or enterprise situation, it could be relabeled for other applications.

The outermost area of Figure 3-2, the business dimension, symbolizes the set or backdrop upon which pertinent events unfold—for example, a multinational company with nationalistic influences and indigenous social mores that form the existing organizational culture, which is represented by the next layer. This could be comprised of a set of coexisting cultures or a single monolithic structure. Things can change seemingly overnight in today's business world—what was once U.S. Steel is now USX, American Can is now Primerica, Inc., or American National Can, and International Harvester is now J. I. Case or Navistar. The cultures of a significant number of companies are in transition, merging, or homogenizing without sharply defined beliefs, principles, or vision to provide stability and guidance. This can present serious challenges for safety prevention initiatives, particularly when the original organization and its culture is slowly being swallowed and digested by the newly dominant one. Overcoming resistance and managing change is critical!

Within the business, various groups/units perform work according to the the organization's mission and objectives. Thus, Figure 3-2 focuses on a single, discrete job/task activity that is deemed necessary. To move toward job/task accomplishment, management employs various process-procedure techniques (shown in the circle) to plan, organize, and control the basic human/machine/environment system, which is used to produce products, deliver services, meet requirements, and so on. The human component includes individual, group, and functional entities; the machine component represents mechanical contrivances, equipment, materials, auxiliary hardware, and utilities; and the environment includes both internal and external factors such as lighting, noise, vibration, heat, cold, radiation, weather pressure, and so on. The role of safe design is represented by the intersection of these three components. Engineering design provides management's best opportunity to eliminate/control hazards with special emphasis on the unwanted release of energy/hazardous materials and design of work. (See also Chapter 11, Beyond Compliance.)

The potential for hazards (unacceptable conditions/actions), shown as a triangle, is an intervening or obstructing element to job/task accomplishment. Con-

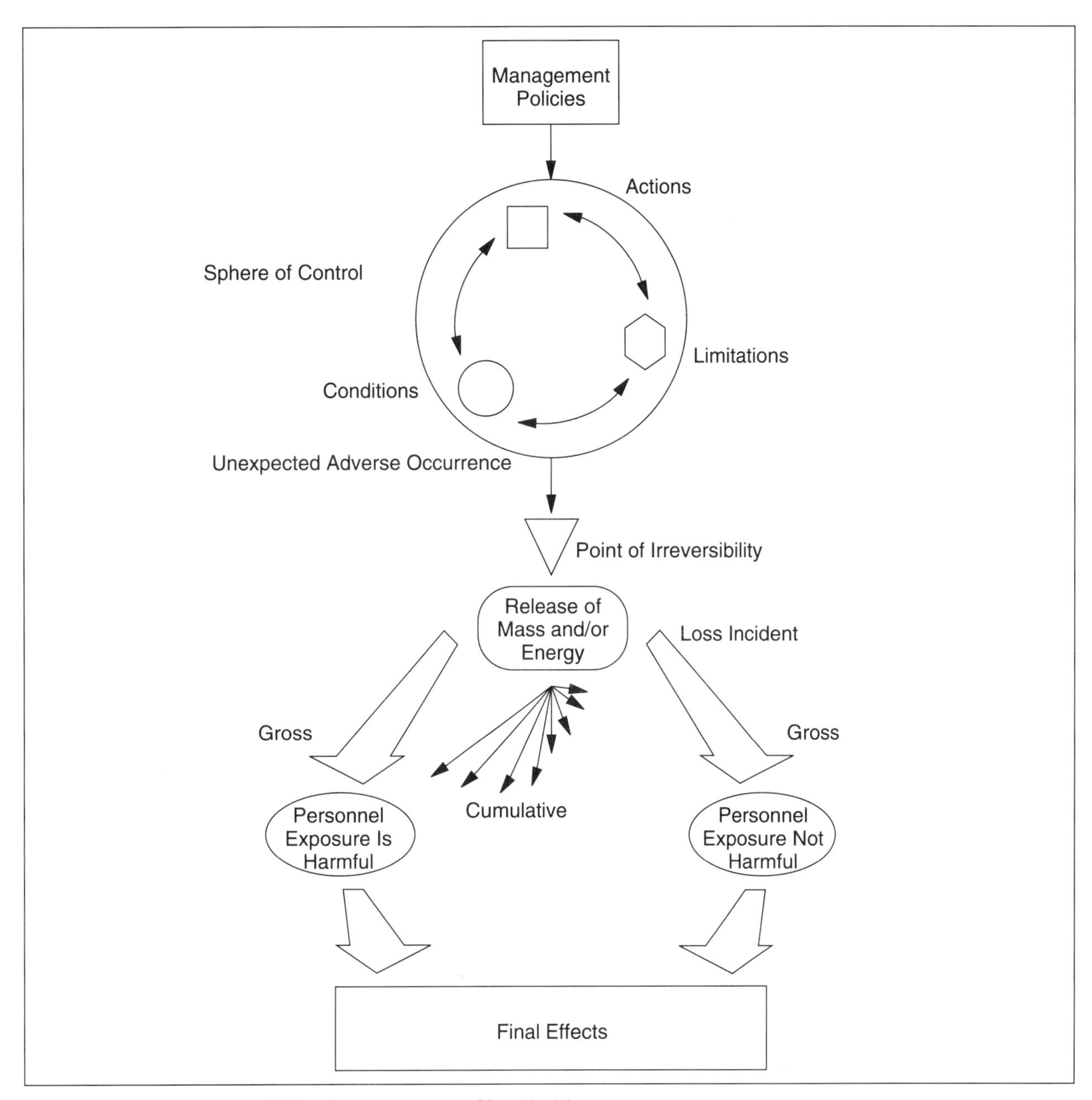

Figure 3-1. A universal model for the occurrence of loss incidents.
Source: R. E. McClay, Toward a more universal model of loss incident causation. *Professional Safety* 34, no. 1 (1989):20. Used with permission.

ditions are physical agents/states that are nonhuman functions of the environment and machine components of the system. Actions are those things done or not done by individuals, groups, or organizations that are functions of the human component of the system. The unacceptable conditions/actions may also be considered deviations from the specified requirements, norms, or standards. No implied intent exists to measure or assign a degree of fault/culpability or comparative negligence in the causal analysis of human actions. The conditions/actions that are unacceptable carry a higher relative risk. *Risk* is defined as the measure of the probability and severity of adverse effects. Currently, most prevention work directed at hazards attempts to reduce risk by means of various intervention tactics or energy management/barrier strategies promoted by Haddon (1970) or Johnson (1980). However, risk is best reduced by inherent safe design or redesign.

The causes of hazards are shown inside the triangle as being either root, contributory, or proximate in nature. Their position in the triangle is relative to their spatial distance from the incident that they precipitated. Proximate causes are directly connected or closest to the occurrence of the incident; they can be considered triggering in nature. Contributing causes are events that connect the proximate and root factors and whose existence allows the sequence to extend toward incident occurrence. Root causes are defined

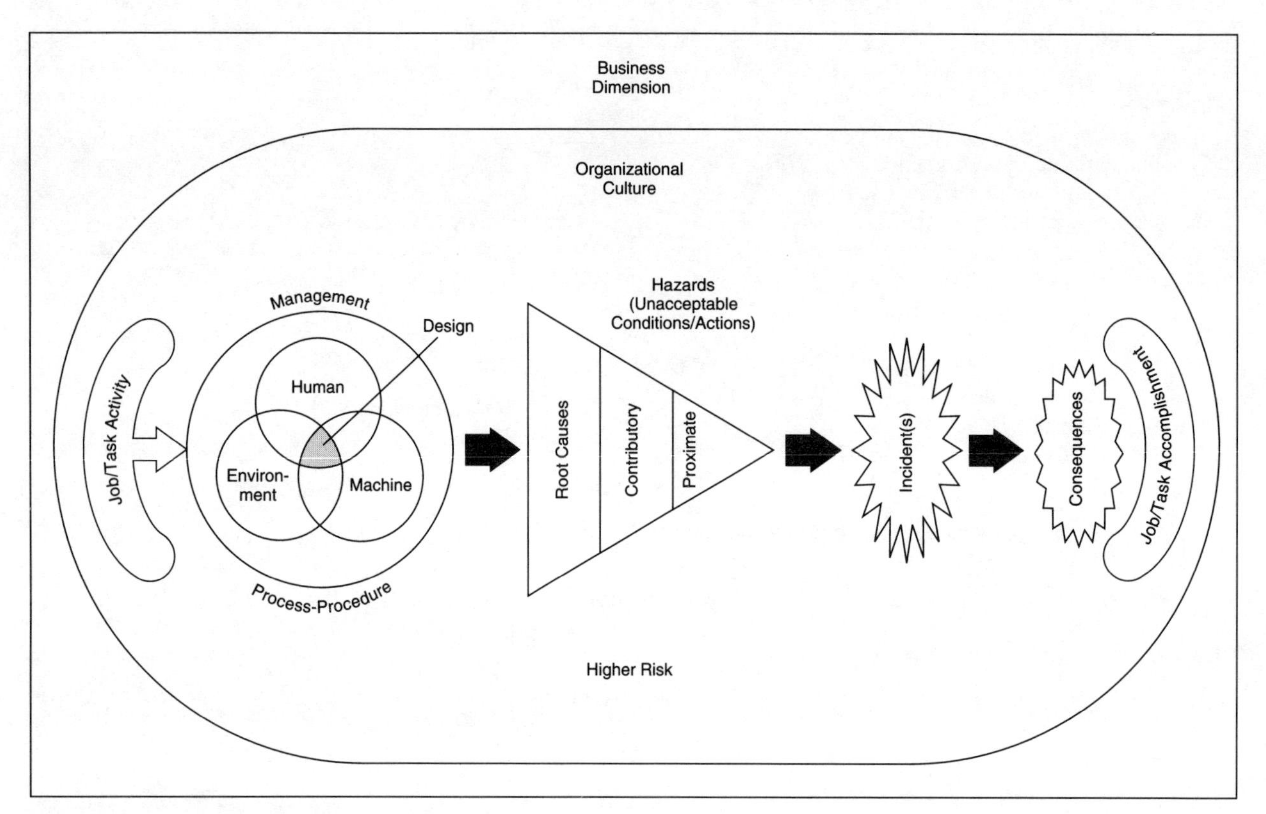

Figure 3-2. Incident model (1).
Source: Developed by E. Grund. © National Safety Council.

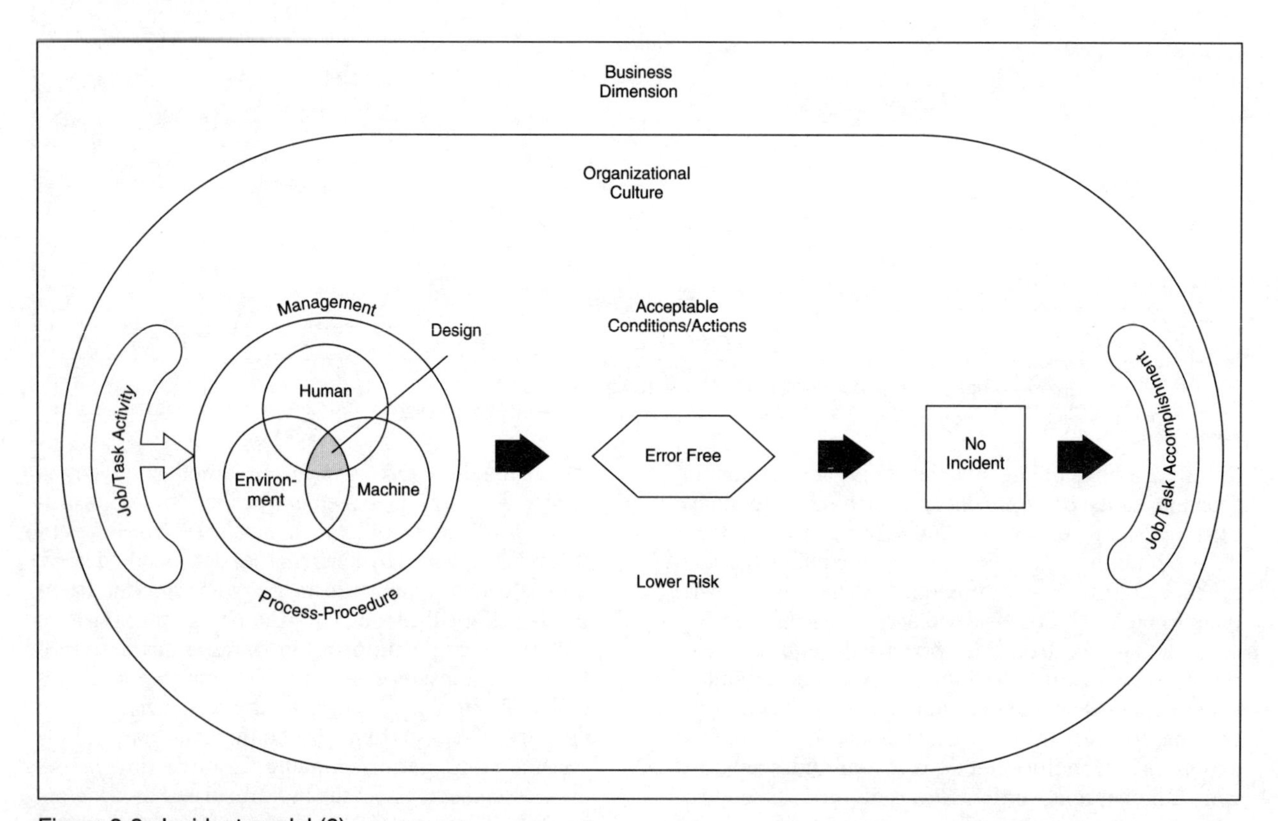

Figure 3-3. Incident model (2).
Source: Developed by E. Grund. © National Safety Council.

as those events that are the sequential genesis or source of the existence of the unacceptable conditions/actions.

The incident is triggered by the confluence of hazards (unacceptable conditions/actions) and time, circumstance, and opportunity. It may result in consequences that interrupt, interfere with, degrade, or produce injury/damage to persons, property, or processes. Of course, the occurrence of the incident may well have some negative impact on the completion or accomplishment of the job/task.

Figure 3-3 depicts a model without unacceptable conditions/actions and so reflects lower risk (safer circumstances) and orderly progression without incident to job/task accomplishment.

HUMAN ERROR AND ITS SIGNIFICANCE

The expression "To err is human, to forgive divine" suggests that the more you err, the more human you are and also provides an absolution; the error is not only accepted or tolerated, but the individual is also granted a pardon with respect to any personal responsibility. This attitude sets the stage for much of our endemic problems with the performance of human systems. However, we need to examine the real basis and limitations of human performance to understand how error can be minimized.

Peters, in his article "Human Error: Analysis and Control," defines *error* as "any significant deviation from a previously established or expected standard of human performance that results in unwanted delay, difficulty, problem, trouble, incident, accident, malfunction or failure" (Peters, 1966).

Rigby (1970) suggests that "human error is any member of a set of human actions that exceeds some limit of acceptability." This view is consistent with Lunsford's definition of *incident* (any deviation from the established acceptable standard). It is possible, therefore, to use *incident* and *error* interchangeably in the Lunsford proposal. Kjellan (1987) expands on this idea by establishing four criteria for assessing compliance with the norm:

- standard rule or regulation
- appropriate or acceptable
- normal or usual
- planned or intended.

Salvendy (1987) introduces a slightly different perspective by stating that human error is an out-of-tolerance standard action, where the limits of acceptable performance are defined by the system.

McCormick and Sanders (1987) provide additional depth by defining *human error* as an inappropriate or undesirable human decision or behavior that reduces, or has the potential for reducing, effectiveness, safety, or system performance.

Individuals are frequently singled out as the cause of error-induced mishaps; however, doing so fails to consider the total system and all of its influences and combinations. Johnson (1980) emphasizes that errors are symptoms of a failed system. People operating a system can't perform as expected because training, supervision, procedures, and/or specifications are less than adequate. Furthermore, error reduction must be accomplished by management identifying and eliminating error-provocative situations. Employee failures are often preordained by inherent deficiencies in the system in which they work.

Swain and Gutherman contend that

> a means of increasing occupational safety is one which recognizes that most human initiated accidents are due to the features in a work situation which define what the worker must do and how he must do it. The situation approach emphasizes structuring or restructuring the work situation to prevent accidents from occurring. Use of this approach requires that management recognize its responsibility to (1) provide the worker with a safety-prone work situation, and (2) forego the temptation to place the burden of accident prevention on the individual worker (Swain & Gutherman, 1983).

Various subsets for cataloging human error have been proposed. Swain and Gutherman suggest the following:

- Errors of omission: Failure to execute an assigned task
- Errors of commission: Failure to perform properly
- Sequence error: Failure to perform a task or step in the proper order
- Timing error: Failure to execute an assigned task at the proper time
- Extraneous acts: Introduction of a task or step that should not have been performed (Swain & Gutherman, 1983).

Keller and Stern provide additional categories for differentiating human error as follows:

- Operator-induced error: Incorrect conclusion, decision, or action by someone with knowledge, experience, and skill to make the correct response
- System-induced error: Errors induced from integration of a total system that can lead to injury or create a catastrophic event
- Design-induced error: Poor equipment design, arrangement, or fabrication
- Input error: Associated with information processing, error detecting, coding, and classifying information
- Output error: Failure to execute procedures in proper sequence

- Low-stress error: An action planned or intended but not executed to plan or as intended
- High-stress error: Wrong decision or action under life-threatening/emergency situations (Keller & Stern, 1991).

J. W. Altmann, in "Behavior and Accidents," proposes five error classes:

- sensing, detecting, identifying, coding, and classifying
- chaining or sequencing
- estimating and tracking
- decision making
- problem solving (Altmann, 1970).

DeJoy proposes a human-factors model for workplace accident causation that features three key elements. These general causal-factor categories, (1) person-machine communication, (2) environment, and (3) decision making, link human error and various subsets and their individual contributions or influences. This segment of the overall model is as follows:

A. Human error
    1. Person-machine communication
        a. Words/symbols
        b. Displays
        c. Controls
    2. Environment
        a. Anthropometry/biomechanics
        b. Microtask environment
        c. Macrotask environment
        d. Ambient physical
    3. Decision making
        a. Predisposing factors
        b. Enabling factors
        c. Reinforcing factors.

DeJoy states that a "human error is presented as the focal point of concern in the prevention of accidents, and use of the model involves systematically identifying error-provocative situations, and then tracing the causal factors involved. The causal factors categories include traditional human factors considerations as well as those related to worker risk appraisal and self-protective behavior" (DeJoy, 1990).

The aforementioned categories or types of human error suggest a complex web of opportunity that is too often assigned simplistic labels such as "carelessness" or "unsafe act." Thompson (1989) expands this framework by stating that often human error is inappropriately labeled a "careless act." He contends that personal limitations involving physical and mental incompatibilities cause employees to be out of sync with their immediate task. Physical incompatibilities such as anatomical dimensions, dexterity, congenital defects, strength, visual acuity, and the like, coupled with mental incompatibilities such as attention span, alertness, attitude, motivation, stress, and so on, influence task performance. In addition, chemical impairment, associated with prescription drugs, alcohol, and illegal drugs, impacts the physical and mental state by deteriorating sensory response, muscle control, communication, reasoning, and so on. Human error, from an incident-causation standpoint, is multifactorial and demands detailed examination to determine what prompted or was responsible for the failure in question.

Johnson elaborates on various error-reduction techniques, such as:

1. Knowing human characteristics and designing tools/equipment to fit capacities and limitations
2. Redesign, automation, or "goof-proofing"
3. System improvement by redesign, which is better than what can be achieved by selection, training, supervision, or motivational campaigns
4. Having employees participate in job safety improvements, which increases the likelihood that they will accept any necessary requirements
5. Designing or redesigning to deal with situations that
    - provide too little or too much information
    - provide inadequate facilities
    - violate predictable patterns, responses, or expectations
    - require performance beyond capacity
    - are unnecessarily difficult, onerous, or dangerous.

According to DeJoy, all work has error rates, that is, errors ÷ opportunities. A simple calculation can determine what error rates can be tolerated.

Errors/Opportunities × Consequences =
Judgment of Importance
(DeJoy, 1990).

A great deal of effort should be expended in determining the criteria for what is acceptable and unacceptable performance. The importance attached to the consequences will dictate what is tolerable. Rigby has developed a hierarchy of tolerance limits in order of effectiveness:

1. Physical: Barriers prevent or limit unacceptable performance (stops, dividers, guards, etc.).
2. Fixed: Limits are clearly and permanently established (aisle lines, red lines on gauges, etc.).
3. Measurable: Limits are checked by measurement or sampling during or after performance (ladder angle, lockout/tagouts, etc.).
4. Descriptive: Reference limits can be compared with work (step-by-step procedures, crane signals, etc.).
5. Cautions: Limits are reinforced by warnings, signs, or other indications (not always available at the time of action).

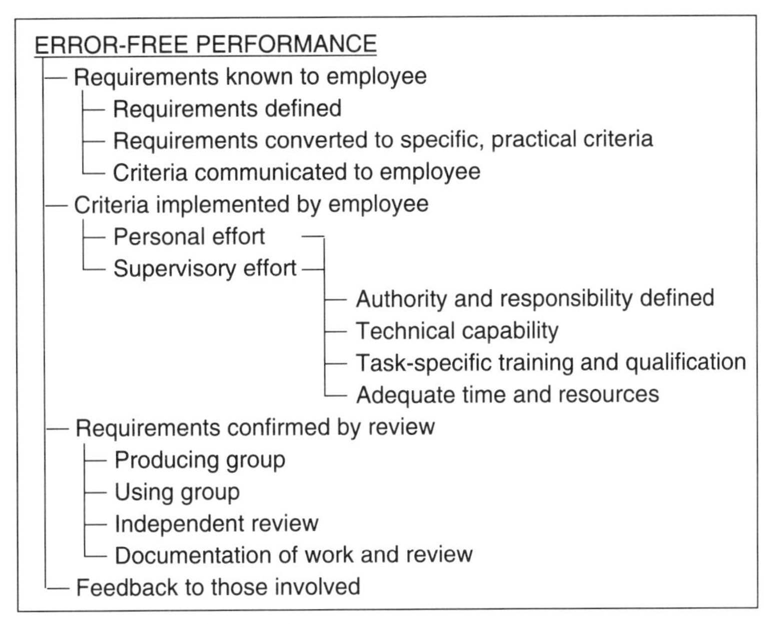

Figure 3-4. Error-free performance tree.
Source: W. G. Johnson, *MORT Safety Assurance Systems.* (New York: Marcel Dekker, Inc., 1980).

6. Custom: Conventional limits are instilled by training/practice (not habitual unless reinforced).
7. Debatable: Limits are subject to argument after the fact by hearing or other consensus (Rigby, 1971).

Rigby's list is in order of decreasing effectiveness but also, unfortunately, of increasing frequency of application. Errors are fewer when employees understand and can work within the relevant established tolerance limits.

The error-free performance tree shown in Figure 3-4, an approach first developed by Nertney, is the opposite of the pattern of error-provocative situations where apparent blameful failure occurs. Failure to provide requirements is a leading source of safety deficiency.

BEHAVIORAL PERSPECTIVE

Over the years, investigators have had a virtual field day identifying the unsafe act, human error, or unsafe behavior as the prime culprits in accident causation. Heinrich's domino theory, which describes key elements such as "fault of person" and "unsafe act," was easily distorted by those who failed to understand his ideas. Behavioral failures were easy to identify because of their commonly proximate connection to the accident. As soon as the employee responsible for the incident was located, management was all too ready to close the books and move on to the next issue or crisis. Very little interest or motivation existed to analyze the behavior and determine *why* it had occurred. Conclusions such as "mental lapse, carelessness, preoccupation, disregard, lack of attention, etc." were prevalent, and more thorough examinations of the total causative sequence were not often undertaken.

During the 1980s, safety researchers and practitioners, behavioral scientists, and progressive managers began to reexamine the behavioral component of accident causation. Human-factors specialists working in aerospace and the military had for some time looked at error, human performance, and behavior much more objectively than those outside their domain. More recently, significant attention has been directed at safe behavior and its development. Leaders in this field such as Krause, Hidley, Topf, Sloat, Petersen, and others have written and lectured extensively on the subject. This seems to be a time during which new ideas and perspectives thrive and accelerate toward acceptance, and this certainly appears to be the case in the behavioral movement.

Behavior Theories

Krause, Hidley, and Hodson, in their text *The Behavior-Based Safety Process* (1990), describe a multiphase methodology. In the assessment phase, behavioral methods such as Antecedent-Behavior-Consequence (ABC) Analysis are used to evaluate the strengths and weaknesses of existing safety measures/culture and to target opportunities for improvement. In the implementation phase, principles of work-pattern search and operational definitions are used to develop an inventory of behaviors critical to safety. Sampling of these critical behaviors establishes a baseline expressed as percent safe behavior within the area being assessed. In the feedback phase, verbal and written information is used to modify and improve the facility's safety-related behaviors. Four principal functions of the behavior-based safety process are identified: planning, facilitating meetings, observation, and training.

The authors indicate that discovering and addressing the root causes of accidents is a critical function of safety management. They note that any safety initiatives that work, whether or not explicitly behavioral in orientation, are effective because they ultimately influence behavior. Therefore, behavioral analysis provides invaluable assistance to organizations in evaluating the factors that are really responsible for their safety progress. ABC Analysis is suggested as the basic tool and foundation of behavioral change. In it, *antecedents* are seen as events that trigger observable behavior and consequences.

For example, if the phone rings (antecedent), we answer it (behavior) to find out who is calling (consequence). Even though the antecedent is a powerful stimulus to behavior, B. F. Skinner has demonstrated that consequences are stronger determinants.

Krause et al. summarize ABC Analysis by stating that consequences control behavior powerfully and directly, whereas antecedents control behavior indirectly, primarily serving to predict consequences.

They suggest that many safety programs are ineffective because they are antecedent-oriented, for example, safety rules, meetings, procedures, and so on. Too often these antecedents have no powerful consequences backing them up. Three features determine which consequences are stronger than others:

- Timing: A consequence that follows soon after a behavior . . . controls behavior more effectively than a consequence that occurs later.
- Consistency: A consequence that is certain to follow a behavior . . . controls more powerfully than an unpredictable or uncertain consequence.
- Significance: A positive consequence controls behavior more powerfully than a negative consequence (Krause et al., 1990).

These rules mean that those consequences with the most power to influence behavior are simultaneously soon, certain, and positive. By contrast, the weakest consequences are those that are late, uncertain, and negative. Table 3-A shows the range of consequences and their relative strength.

Krause et al. observe that the most common factor in favor of unsafe behavior is simply that most unsafe behavior is not even observed, let alone noted and addressed. In this instance, not being observed is an inducement for unsafe actions. The worker knows (certain) that unsafe behavior goes unnoticed (non-negative = positive) continually (soon). For example, when an employee fails to use an available energy-isolating device for his or her protection, the perception of the employee might be that the device is inconviently located and that its use will take too much time and effort relative to the task required. The employee's past experience indicates that it is highly unlikely the behavior will be detected, much less acted upon by supervision.

Table 3-A. Range of Consequences/Strengths

Strength	Timing	Consistency	Significance
Strongest	Soon	Certain	Positive
	Late	Certain	Positive
Stronger	Soon	Uncertain	Positive
	Soon	Certain	Negative
	Late	Uncertain	Positive
Weaker	Soon	Uncertain	Negative
	Late	Certain	Negative
Weakest	Late	Uncertain	Negative

Source: Reprinted with permission from T. R. Krause, J. H. Hidley, and S. J. Hudson, *The Behavior-Based Safety Process* (New York: Van Nostrand Reinhold, 1990). © 1990 by Van Nostrand Reinhold.

Saving time, effort, and inconvenience is soon-certain-positive, whereas detection, reprimand, and injury is likely to be late-uncertain-negative! The consequences favoring the unsafe behavior are appreciably stronger than those favoring the safe behavior.

When conducting ABC Analysis, the following three steps are taken:

1. Analyze the unsafe behavior (an undesired higher-risk, present employee action).
2. Analyze the safe behavior (a desired, lower-risk, future employee action).
3. Develop an action plan (define roles, responsibilities, what must be done, and when).

Krause et al. state that

> employees are justifiably resentful in cases where their unsafe behavior has been either encouraged or allowed until an injury occurs and then they are not only injured but subjected to disciplinary action. It is the responsibility of management to establish and maintain systems that produce safe behavior and discourage unsafe behavior (Krause et al., 1990).

An example of ABC Analysis as it applies to energy-isolation behavior is illustrated in Figures 3-5 and 3-6. Failure to de-energize a machine before cleaning or unjamming is a very common occurrence in manufacturing. Too often the knee-jerk reaction is for reinstruction

A Antecedents (Triggers)	B Unsafe Behavior	C Consequences	Values S/L	C/U	+/–
1. Isolation point inconvenient	Failure to de-energize machine before clearing jams	1. Save effort	S	C	+
2. Co-worker influence		2. Peer approval	S	C	+
3. Supervisor indifference		3. Reprimand	L	U	–
4. Successful experience		4. Injury	L	U	–
5. Frequent task		5. Saves time	S	C	+
6. Production quota		6. Less downtime	S	U	+
7. Complicated process		7. Make quota	S	U	+

Figure 3-5. ABC analysis of unsafe behavior.
Source: Developed by E. Grund, using the technique outlined in T. R. Krause, J. H. Hidley, and S. J. Hudson, *The Behavior-Based Safety Process* (New York: Van Nostrand Reinhold, 1990). © National Safety Council.

A Antecedents (Triggers)	B Safe Behavior	C Consequences	Values S/L	C/U	+/-
1. Relocate breaker	De-energize machine before clearing jams	1. Less effort	S	C	+
2. Supervisor modeling		2. Positive reinforcement	S	C	+
3. Review relevant injury data		3. Improved awareness	S	U	+
4. Repair root cause (jams)		4. Observation-feedback	S	C	+
5. Simplify isolation process		5. Improved machine reliability	S	U	+
6. Practice involvement by employees		6. Quicker isolation	S	C	+
7. Skills training		7. Pride-ownership	S	C	+

Figure 3-6. ABC analysis of safe behavior.
Source: Developed by E. Grund, using the technique outlined in T. R. Krause, J. H. Hidley, and S. J. Hudson, *The Behavior-Based Safety Process* (New York: Van Nostrand Reinhold, 1990). © National Safety Council.

or some form of discipline. This usually does not address the real issues or the consequences and their antecedents and is therefore ineffective or at best temporarily beneficial. In Figure 3-5, we see the unsafe behavior analyzed and the strength of the consequences evaluated. It is apparent that reprimand and injury are the weakest forms of consequence (Table 3-A) and therefore not effective in altering the existing behavior pattern. The other consequences are stronger and support the continuation of the present unsafe behavior. In Figure 3-6, by contrast, existing antecedents/consequences are assessed and dealt with in order to create the necessary circumstances for the desired safe behavior to occur—in other words, "de-energize machine before cleaning jams." The consequences are now strongly in favor of the new, lower-risk behavior pattern. The application of ABC Analysis can provide a sound basis for dealing with behavioral issues in a way that focuses on system deficiencies as a root cause rather than on personal faults or attitudes.

M. Topf and R. Preston of the Topf Organization are also promoting and encouraging new perspectives on the behavioral component of incident causation. They recommend a multidimensional behavior modification approach that involves the following strategies:

- Cognitive behavior modification: Understanding how beliefs, values, thoughts, and other aspects of people's mental processes contribute to both "automatic" and "calculated" behaviors that are unsafe.
- Reality behavior modification: Dealing with current behaviors and their consequences to achieve a shift to personal responsibility and self-management.
- Management by commitment: Once individuals realize that they and management are committed to safety, they are motivated by the perceived benefits.

The interaction of these elements—thoughts, action, and commitment—help individuals and organizations to increase their awareness of the variables that cause people to place themselves and others at risk as a result of their behavior. This "awareness approach" addresses fundamental human mechanisms, attitudes, and behaviors. The following awareness-action model describes the key steps to safe action:

Awareness → Focus → Distinction →
Responsibility → Action → Result

The following human mechanisms often cause people to act without regard for safety or take unnecessary risks, according to Topf and Preston (1991):

- Going away: Daydreaming, inattention, preoccupation, and losing focus as a result of boredom, repetitive tasks, etc.
- Awareness: Not being alert or vigilant for hazards or danger
- Focus: Lack of concentration on tasks and attention to detail
- Distinction: Inability to sense (see, hear, smell, etc.) potential hazards and take appropriate action
- Responsibility: Not responding when safety demands immediate action due to attitudinal influences
- Changing behavior: Resistance to certain behaviors and safety feedback from others
- Early warning: Ignoring the mental alarm system that provides alertness signals
- Agreements: Ignoring or not communicating and discussing safety rules and regulations.

Their approach for optimizing employee safeguards is accomplished by providing employees and management with the skills to maintain or improve safety performance. This process is tailored to supplement existing prevention activities by focusing on the following areas:

- Safety awareness: Learning new skills of observation for safety behavior and the identification and response to early warning signs
- Modifying risk behavior: Utilizing vital steps of the awareness-action model and discovering how change actually occurs
- Safety-thinking process: Learning to manage automatic and premeditated responses and essential attitudes for appropriate behaviors
- Personal responsibility: Increasing the level of personal ownership or responsibility
- Leadership: Enabling others to manage behaviors by providing empowerment skills
- Safety and personal commitment: Realizing what is at risk and how to make the right choices.

Petersen, in *Safety Management,* suggests that "the worker who regularly performs unsafe acts when he 'knows better' is possibly following Skinner's law of operant conditioning. The unsafe act has been learned and is maintained because it has been (and continues to be) reinforced by satisfying events."

He continues by listing a number of reasons why workers perform in a way that management would describe as unsafe:

- The benefits gained by the worker (in his view) outweigh the disadvantages.
- The behavior makes legitimate sense to the employee after being reinforced over time.
- The behavior provides personal satisfaction, for example, peer approval, thrill seeking, attention, individuality, and so on.
- The behavior may be perceived as having job-related advantages such as saving time, being less fatiguing, increasing piece rate bonus, and so on.

Petersen suggests that from a strategy standpoint it is more productive to bolster the probabilities for safe behavior than to attempt to eliminate unsafe acts through discipline and punishment. Basically, there are two main avenues to follow (1) increase the satisfactions for working safely and (2) eliminate the obstacles that inhibit working safely (Petersen, 1975).

While serving as deputy director of safety for the U.S. Army, F. S. McGlade examined the relevancy of the concept of adjustiveness in behavior to proficiency in driving performance. He contends that consistency in performing successfully is the basis of adjustive behavior. *Adjustiveness* was defined as the capacity to mesh the abilities, skills, and tasks used in behavior into a meaningful whole that brings about successful performance of an activity in virtually all situations and under almost all conditions.

He identifies and describes the mechanics of the three forms of adjustive behavior:

- Unguided adjustive behaviors: More or less habituated, reflexive, or preset responses to stimuli that lead to successful results (instinctive responses)
- Directly guided adjustive behaviors: Require sensory feedback from the environment; tied to specific situations and are employed in connection with those situations; the individual learns that a given behavior brings success in a given situation and therefore uses it each time environmental cues antecedent to the particular situation are perceived (dependent and largely involuntary).
- Apperceptively guided behaviors: The means for combining or integrating independent, voluntary acts and systems of dependent, involuntary acts into adjustive patterns of behavior; apperceptive guidance has significance for safe performance—it connotes thinking, interpretation, anticipation, and appropriate use of information and prior experiences as a basis for action.

The adjustiveness of behavior—or more specifically, apperceptive guidance—is important for dealing with the substantial number of interacting variables associated with controlling hazardous energy releases.

McIntire and White (1975) refer to B. F. Skinner's work in behavior modification. They state that "the foundation of behavior modification is based on the principle that people will act by a set of rules (even safety rules) if they are 'paid' (reinforced) in a direct, immediate and consistent manner. The application of behavior modification starts with an objective analysis of behavior that emphasizes a person's actions and the results of these actions. References to 'inside events' such as feelings, intellect or attitudes are generally avoided."

McIntire and White provide behavior modification guidelines that contain essential criteria for promoting and reinforcing safe behavior. These guidelines are presented here in outline form:

A. Consequences of behavior
  1. Reinforcement: Through appropriate use of reinforcers, behaviors related to safe performance can be increased and unsafe performance decreased.
  2. Punishment: Punishment given for unsafe behavior will tend to suppress the behavior; other undesirable effects may occur.
  3. Avoidance: Employees avoid undesirable consequences; therefore, organizations should communicate the consequences of unsafe behavior.
  4. Response cost: Increasing the amount of effort required for unsafe behavior decreases its occurrence and conversely reducing the effort necessary for safe behavior increases its occurrence.
  5. Social reinforcement (supervisory): Appropriate reactions from supervisors can increase the frequency of safe behavior and decrease the frequency of unsafe behavior.
  6. Social reinforcement (peers): Appropriate response from one's peers can increase the frequency of safe behavior and decrease the frequency of unsafe behavior.
  7. Modeling: The safe behavior of an individual can be altered by observation of the reinforcers received by others.

B. Scheduling of consequences
  1. Immediacy: Reinforcers given immediately after the behavior are more effective than delayed reinforcers.
  2. Consistency: The environment must ensure that reinforcers are consistently paired with certain behaviors over repeated occurrences of the behavior.
  3. Shaping: Reinforcers must be provided frequently when complex sequences of safe behavior are involved, at least initially for close approximations; over time tighter performance can be required for the reinforcers.
  4. Partial reinforcement: After behavior has been established, reinforcement frequency can be reduced.
  5. Extinction: In order to eliminate or reduce the incidence of unsafe behavior, the reinforcers should be identified and neutralized/removed.
  6. Compensatory reinforcement: If certain reinforcers are removed from the work environment, others should be added to prevent deterioration of performance.
  7. Stimulus control: Vary the time and circumstances under which reinforcement is provided to prevent situational behavioral response.

C. Pitfalls in scheduling consequences
  1. Noncontingent reinforcement: If reinforcers are not linked to specific behaviors according to an explicit/predictable pattern, they cannot be expected to increase the desired safe behavior (randomness).
  2. Noncontingent punishment: Punishment for no apparent reason may produce a variety of effects, including suppression of safe behavior (unpredictable).
  3. Reinforcement opportunity: Workers who engage in legitimate or safe behavior should have sufficient reward opportunity.

D. Implementation—General consequences
  1. Practice: To be effective in modifying behavior, practice should involve not only repetition of desired behavior but repetition of the behavior-consequence pairings.
  2. Record keeping: Records can serve as both effective consequences of behavior and as a means for monitoring changes in safe behavior.

During 1990, E. V. Grund and B. Lunsford conducted a safety seminar for line supervisors in Manchester, England. A sizable portion of the presentation dealt with supervisors and their responsibility for and influence on safe behavior. Several graphic representations were used to communicate the criticality of the supervisory role. The feedback after the seminar indicated the supervisors had gained new insight regarding the impact of their everyday actions or inaction. They were better able to visualize certain aspects of human behavior that they had thought were too complex for them to affect in any way. They did understand, all too well, the idea of punishment as a deterrent of unsafe behavior.

Most had never participated in any discussion of what produces behavior and what can be done about it.

The hourglass-shaped time-behavior graphic (Figure 3-7) was used to explain the development of behavior over time and what supervisors must know regarding their influence and opportunity for positive intervention. They were instructed that employees develop behavior through observation, risk perception, experiencing results, assessing the incentive, experiential learning, and training.

Employees observe what is happening around them and respond in ways that are shaped by the surrounding variables. They watch co-workers follow or not follow procedures, supervisors detect and correct or fail to detect unsafe practices, and so on. Workers develop a perception of risk based on their personal attitudes toward danger or chance and their intuitive assessment of severity potential. They also absorb the outcome or results of various behavioral cycles—for example, a job was accomplished without incident, there is no flack from the boss, co-workers respect for them increased and so on. The employee has some developed idea or feeling of what value the incentive has or means to them personally. This could translate into saving time or effort, avoiding unpleasant conditions, and the like. Learning from experience can be positive or negative. Too often, experiential learning occurs by trial and error and eventually boils down to "what works for me." In addition, experience was shown to include the effects of OJT (on-the-job training), where co-workers directly influence the behavior of those learning for better or worse. Finally, the impact of training was addressed as a factor in future behavior. The content, quality, methodology, and relevance of the acquisition of knowledge and skill was also covered.

In the hourglass analogy, behavior development was likened to grains and layers of sand piling on top of each other, cycle after cycle, over time. Unlike the hourglass, however, employees cannot simply be turned over and reverted to their original, clear behavioral states.

The average participant in our sessions seemed to be well equipped to deal with things—machinery, processes, and regulations—in the workplace but inadequately prepared with respect to the human element. Feelings of frustration, resignation, and indifference were evident. We proceeded to discuss the relationships between software (procedures, policies, etc.), hardware (machines, equipment, etc.), and humanware (the employee). Insufficient understanding and effort surrounded the element of human behavior and its determinants.

In order to keep our discussions from becoming too theoretical, we again attempted to demonstrate visually how the supervisor could increase his/her effectiveness in developing safe behavior. Attitudes were pitted against behavior on a flip chart in an attempt to dislodge the supervisors' preoccupation with their workers' mental state. The chart was presented as follows:

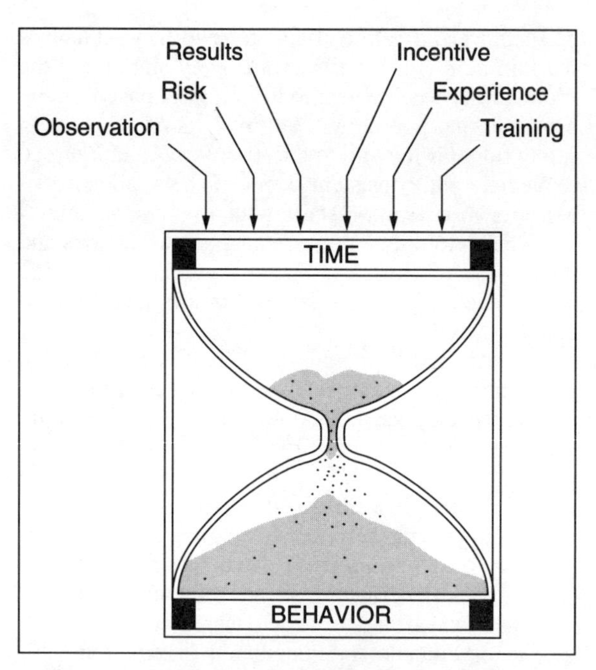

Figure 3-7. Development of behavior over time.
Source: Developed by E. Grund. © National Safety Council.

Attitudes (internalized) => emotion

Behavior (externalized) => motion

Supervisors were advised to concentrate on observable behaviors as the culprits behind unacceptable performance and not on the mind-set or attitude of each employee. We then altered the flip chart:

Attitudes (internalized) => emotion

Emotion minus e => motion (behavior)

By eliminating the "psych 'e'" or mind state in the attitudinal approach—the quicksand of emotions, impressions, feelings, and so on—the supervisors can focus on motion, which is connected with and observable as behavior. Although supervisors may not be equipped to deal with employees on an attitudinal level, they are capable of operating effectively on a behavioral basis.

When determining what is safe or acceptable behavior, one needs to focus on relative risk. *Risk* as a concept was illustrated in our seminar by the graphic reproduced here as Figure 3-8. Supervisors were instructed that any given situation or behavior carries some degree of risk. The spectrum ranges from zero (a state simply shown for the purpose of example) to infinite (defined as certain occurrence). Risk is quantified by these factors: (1) frequency of exposure, (2) severity of consequences, and (3) probability of occurrence.

The supervisors' first obligation is to promote, establish, or reinforce safe (lower-risk) behavior in the workplace. A secondary obligation is to convert or change unsafe (higher-risk) behavior to behavior exhibiting a lower degree of risk. This risk reduction ap-

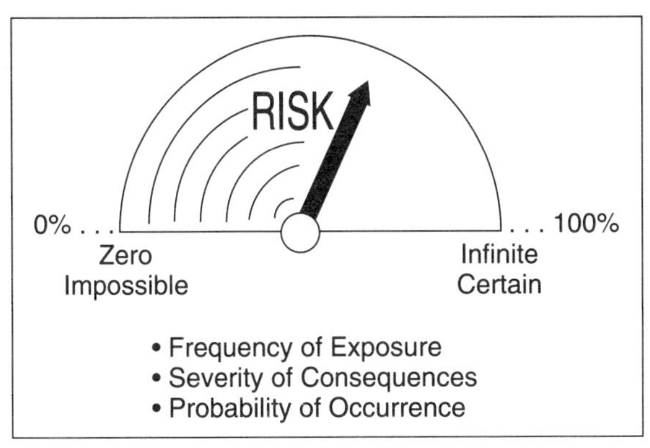

Figure 3-8. The spectrum of risk.
Source: Developed by E. Grund. © National Safety Council.

proach provides supervisors with more options for improvement than the absolute perspective of safe versus unsafe. The participants were provided a risk evaluation and control tool (Figure 3-9) to assist them in orienting their thinking along these lines. Hazards can be evaluated as a function of risk, and control priorities can be established according to the ratings determined by the techniques.

Considerable attention was directed to the issue of converting or changing unsafe behavior. A four-step approach was proposed:

1. establishing or identifying the safe behaviors
2. detecting or identifying the unsafe behaviors
3. intervening or acting
4. taking remedial or corrective action.

Because steps 1 and 2 are simple enough to accomplish, attention was placed on step 3 as a formidable obstacle. Two critical factors why this is so and the reasons for their existence were addressed:

1. Supervisory reluctance to act
   - time consuming
   - confrontation avoidance
   - uncertainty (questionable basis for acting)
   - poor-performance atmosphere
   - assumption (skill + long service = acceptable behavior)
   - fear of reprisal (slowdown, cooperation, etc.)
   - conflicting signals from management
   - lack of knowledge (of safe behavior standards)
   - past behavior pattern
   - poor time-management skills
   - no management emphasis/priority
   - no feedback from superiors
2. Employee resistance to change
   - supervisory inconsistency
   - peer viewpoint/approval
   - failure to understand true consequences
   - lack of hazard knowledge/skill
   - confusion over management intent
   - resentment over double standards
   - personal relationship issues
   - supervisory approach and reaction
   - incentives for status quo
   - macho needs/style/reputation
   - opposition to regimentation
   - prior lack of attention
   - local politics (informal/formal).

Risk Evaluation and Control

Guidelines to assess and prioritize detected hazards:

1. Hazard (Risk) Evaluation

a. Injury/damage severity potential:
High (6 points); Moderate (4 points); Low (2 points)

b. Frequency of employee exposure:
High (3 points); Moderate (2 points); Low (1 point)

c. Probability of accident occurrence:
High (3 points); Moderate (2 points); Low (1 point)

d. Violation of safety codes, standards, rules, etc.: Yes (1 point); No (0 Points)

NOTE: Evaluate each of the above factors before assigning priority for hazard control action.

2. Hazard Control Priority Rating:

Priority Rating	Risk Rating
E—emergency	10 – 13 points
A—today	8 – 10 points
B—one week	6 – 8 points
C—one month	5 – 6 points
D—three months	4 points

This priority rating scale is provided as a supervisory education tool and should not be considered as an absolute measurement system.

Figure 3-9. Risk evaluation and control.
Source: Developed by E. Grund. © National Safety Council.

Many of these reasons need to be addressed so that the workplace can become one in which the supervisor is free to act and the employee is free to change. Obviously, the root causes of behavior associated with the overall human-machine-environment system also have to be treated as mentioned earlier in this chapter. If design factors are implicated and viewed as error provocative, they will also have to be addressed to achieve any meaningful preventive change.

A "light" postulate, introduced as "Lunsford's Law of Behavior," actually had more serious ramifications. It states: "My behavior in a given situation is not based on common sense as to what might happen but

Figure 3-10. Behavioral cycle.
Source: Developed by E. Grund. © National Safety Council.

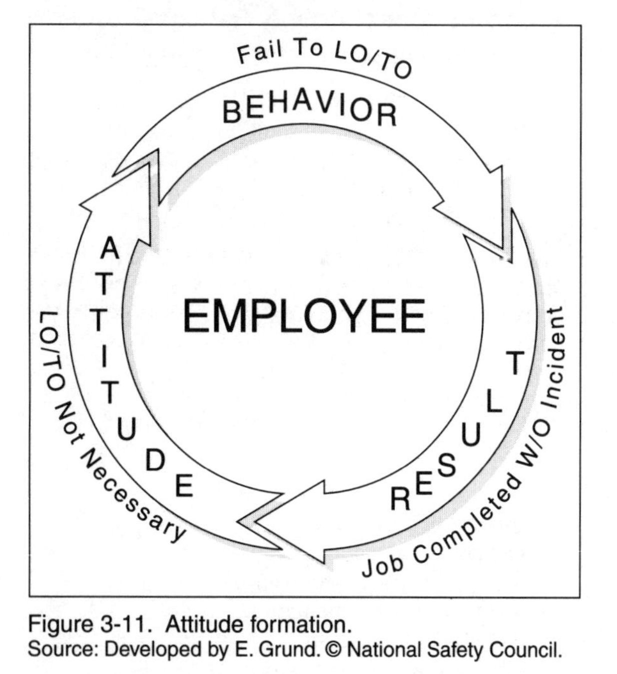

Figure 3-11. Attitude formation.
Source: Developed by E. Grund. © National Safety Council.

rather on what did happen in the past, when I was in a situation that I perceive to have been essentially the same as this one." This was used to set the stage for discussion of terminal or conditioned behavior. A behavioral cycle graphic (Figure 3-10) was used in our seminar to illustrate the process of conditioning or reinforcement and how intervention could be used to change the pattern. We presented a behavior (not stopping at a traffic stop sign) that was followed by a result (uninterrupted passage through the intersection) and then by an attitude (deciding that the stop sign had been improperly placed or that there was little chance of getting a ticket), which supported more of the same

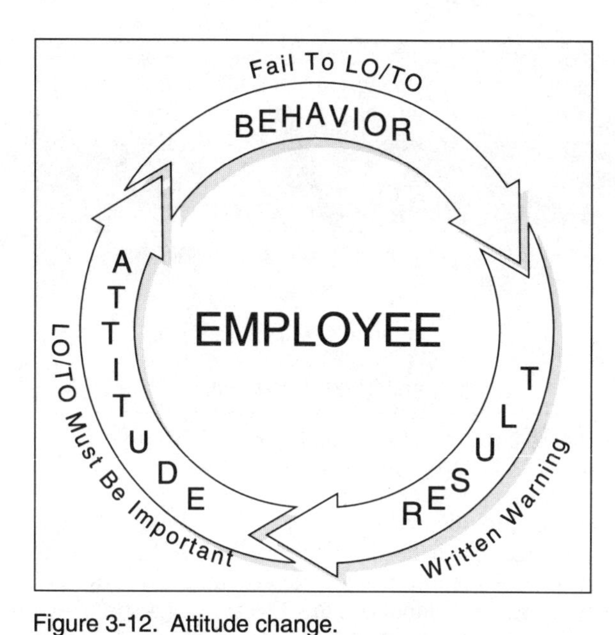

Figure 3-12. Attitude change.
Source: Developed by E. Grund. © National Safety Council.

behavior in the future. Each repetition reinforced the depth or the commitment to the behavior. Eventually, some result might occur—for example, a traffic citation or a collision—that would likely produce a different attitude/behavior.

We can see this cycle further depicted in Figures 3-11 and 3-12 as it might apply to a situation where an employee has failed to lock out/tag out an energy-isolating device. In Figure 3-11, the job is completed without incident, thus forming the attitude that lockout/tagout really isn't necessary. As the behavioral cycle is repeated, the pattern is strengthened if the results continue to be positive from the employee's viewpoint.

In Figure 3-12, the supervisor observes the failure and intervenes. In this case, counseling and a written warning serve as discipline to cause a change in attitude and future behavior. No attempt was made to suggest that this is the ideal or preferred tactic; it was presented to show what would happen over time when a supervisor was oblivious or indifferent to the occurrence of unsafe behavior. It also revealed how supervisory detection and correction of existing unsafe behavior is critical to improved safety performance. Our emphasis remained with the more productive process of defining and supporting safe behavior.

The intervention scenario we discussed involved (1) seizing the first opportunity to stop the behavior from proceeding and drawing attention and emphasis to its unacceptability; (2) awareness raising to uncover why the behavior is occurring and what is wrong with it; (3) helping the employee to understand the consequences/risk as well as the rules, standards, requirements; (4) providing knowledge about acceptable behavior and instructing the employee on the specifics of what's right; (5) revealing the expectation for the employee's future behavior and what the consequences for persisting with the unacceptable behavior might be; (6) providing reinforcement during follow-up that supports the new acceptable or safe behavior.

Tasks with Transfer-of-Energy Potential

Various tasks or activities are performed each day that have energy-transfer or energy-release potential. In some cases, energy isolation is not currently possible because of considerations such as the need to evaluate (troubleshoot) the machine/equipment or intermittent requirements to cycle or test performance or function. These POWER ON or ENERGY PRESENT situations require alternative protective measures or safeguards that provide acceptable levels of employee safety short of achieving a zero energy state. However, a significant percentage of the following tasks/activities may be accomplished with isolation of the hazardous energy and minimal risk to the worker:

- *Installing:* new equipment placement; additional supporting accessory component installation; erection activity, etc.
- *Operating:* control activity, monitoring, cycling machine/equipment/process, unjamming/unplugging, re-setting, etc.
- *Inspecting:* tolerance checks, gauging, performance evaluation, etc.
- *Setting up:* die change/adjustment; product change; process transition, etc.
- *Adjusting:* alignment, dimensional preparation, trial runs, verifying balance or chemical condition, etc.
- *Testing:* debugging, troubleshooting, specification checks, functional performance, etc.
- *Modifying:* redesigning, altering, expanding, replacing, etc.
- *Cleaning:* cosmetic treatment, decontamination, removal of buildup, sanitizing, etc.
- *Servicing:* fueling, preventive maintenance, oiling/lubricating, supplying, painting, etc.
- *Repairing:* welding, fabricating, rigging, removal/replacement, restoration, fixing wear and tear, etc.

Accomplishment of each of these tasks/activities is affected by various influences that determine the proper approach to energy isolation. Every job that requires the achievement of a zero energy state has an accompanying backdrop of influences or variables that tend to make it unique—in other words, as the set of circumstantial influences varies, so does the job's demands and employee response to them. The approach to the job of replacing backup rolls on a continuous hot-strip mill during a planned outage is often different than the proper approach under breakdown circumstances. These influences impact both management's and employees' judgment, composure, sense of urgency, and tolerance; they are very real and often present. They are not to be considered excuses but rather part of the total environment in which the work is done. Ideally, smooth and safe workplace function will not be affected by the surrounding circumstances, but employees under pressure in complex situations may be more likely to make errors or fail to behave as expected.

In addition to the normal requirements for energy isolation when inspecting, servicing, repairing, and so on as well as those for adjusting to unusual conditions, employees must be aware of and prepare for various external influences. Consider the following partial listing:

- *Status:* routine/nonroutine, breakdown/planned, emergency, continuous/intermittent isolation, etc.
- *Personnel:* operational, maintenance, mixed, contractors, few/many, etc.
- *Energy state:* energized, de-energized, cycling, partial power, etc.
- *Criticality:* machine/process output, batch/continuous, cost impact, sequential impact, etc.
- *Frequency:* rare/high rate of occurrence, machine/process reliability, etc.
- *Severity:* low/high risk, damage potential, etc.
- *Technology:* low-tech/high-tech, networked, complex process, advanced human skills, etc.
- *Urgency:* normal/tight schedule, incentive clauses, overtime, etc.
- *Hardware:* common isolation devices, custom appliances, unique arrangements, support equipment, etc.
- *Isolation complexity:* simple systems, multiple energy types, numbers/locations of isolating devices, pyramid control schemes, computerized systems, advanced machine systems, robotics, etc.
- *Perception:* easy/difficult, normal/high pressure, safe/dangerous, reasonable/unreasonable, clean/dirty, etc.

SYSTEMS APPROACH—CAUSAL FACTORS

When assessing the potential for the occurrence of energy-release-incidents, it is beneficial to use a systems approach to identify the causal factors or failure modes that may possibly be involved. The author's incident model, described in Figures 3-2 and 3-3, shows a human/machine/environment system within an overall management structure of process/procedure. The human component includes individual, group, and functional entities. The machine component represents mechanical contrivances, equipment, materials, auxiliary hardware, utilities, and so on. The environment encompasses both internal and external factors such as lighting, noise, vibration, heat, cold, radiation, weather, pressure, and the like. Management process/procedure relates to activities such as planning, training, controlling, and so on.

Figure 3-13, a partial listing of potential causal factors for energy-release incidents, is provided to stimulate the development of a hazardous-energy-control system to eliminate or control such deficiencies. The list is also a valuable tool for pursuing and rectifying the root causes of incidents while de-emphasizing simplistic conclusions related to individual unsafe behavior.

Many possible causes (root, contributory, or proximate) exist that could interact to produce an energy-release incident. A mind-set for *sequencing and networking* should be established so that in-depth analysis (before the fact) or investigation (after the fact) can occur. Petersen, in his discussion of the limitations or interpretations of Heinrich's domino theory, suggests that the theory of multiple causation, which describes how many contributing factors, causes, and subcauses combine in random fashion, is a more appropriate concept.

An illustration of the multiple causation that reflects factor interrelationships (sequencing and networking) is provided in Figure 3-14. The interconnected ellipses represent the relationships between the human and physical causative factors in perhaps one influencing sequence. Under certain circumstances, the factors could be individually inclusive or exclusive of each other. As shown in Figure 3-14, the human side, action/inaction/reaction, represents acts of omission, commission, response, oversight, or induced error. The physical side, condition, represents that aspect of the machine, equipment, process, materials, or environment that is unacceptable or fails to meet requirements.

It would not be unusual if the maintenance mechanic was injured (item 6) as a result of his/her reaction, identified as having committed an unsafe act, and possibly admonished for not doing something else. This example also reveals how complex the sequences and networking may become. One could obviously continue to expand it in all directions and dimensions with additional sequences/causes. A U.S. Department of Energy investigation of the Elk Hills accident (October 25, 1977), for instance, revealed 22 direct sequence events, 65 contributing events and conditions, and 24 systemic inadequacies.

CASE HISTORIES—CAUSAL FACTORS

The following energy-release incident case histories provide insight into the mechanisms of causation and reveal how the proximate cause is often the predictable product of many other factors in the sequence of events. One example for each of the major elements of the human-machine-environment system is detailed. The categorization is subjectively based on the principal or critical cause among all those associated with each incident. Only a condensed version of each case is provided to illustrate the need for investigative depth.

Case A—Machine (Mechanical Repairman—Fatality)

A mechanical repairman was assigned the task of troubleshooting an overhead bridge crane's main hoisting mechanism. He ascended a set of stairs leading to the runway that provided access to cranes in several building bays. He traveled along the runway and proceeded to investigate the operational problem with the hoisting trolley by climbing over the end trucks and getting on top of the bridge girders. Apparently while moving either to or from the crane bridge to the runway, he was trapped and rolled against a column by movement of the crane.

Initial viewpoint. A failure to lock out the electrical breaker for power to collector rails occurred (unsafe behavior); a facility lockout/tagout program had been formally in place since start-up ten years earlier.

Further investigation. The following consequential factors were identified as having critically influenced the proximate action of the repairman: (1) No specific written procedure for this job existed; (2) no sign was placed in the crane cab at the controls to indicate work was in progress; (3) the crane operator was not told that work would take place (he was working elsewhere prior to incident); and (4) access to the cab was via a vertical ladder, so the operator did not see the repairman on the bridge.

Root/primary cause. The electrical disconnect had been relocated from floor level to 12 ft (3.66 m) up a building column, thus requiring a ladder for access; maintenance supervision had breakers relocated because of frequent physical damage/contamination from surrounding work activity; ladders were kept in a maintenance toolroom 200 yards (183 m) from area and available as 18-ft (5.5-m) extension types only; employees were unlikely to acquire a ladder to cycle a breaker.

Case B—Human (Truck Serviceman—Fatality)

Two forktruck service company employees were assigned under an annual contract to maintain and repair a fleet of forktrucks on the host's premises. They had experience in excess of 10 years performing this type of work. They were working on the lift cylinder controlling a dual set of forks. The rack frame was raised about 12 ft (3.66 m) off the floor, and a safety chain was tied from the lift rack to the frame that encloses the driver. A 0.375-inch (0.9-cm) carriage bolt was inserted through the chain links. One serviceman was on the top of the driver cage of the forklift assisting in lining the cylinder up. The deceased was kneeling on the floor under the lift rack, either assisting in lining up the cylinder or adjusting a floor jack located under the bottom of the cylinder. The bolt in the chain sheared in half and the forks dropped, striking the deceased on the left side of his head and face.

I. MANAGEMENT PROCESS/PROCEDURE
  A. Culture (absent or inadequate)
    - Safety/values, vision, beliefs
    - Objectives/goals
    - Interest, promotion, motivation
    - Participative process/environment
    - Emphasis, priority, commitment
  B. Policy
    - Absence of hazardous energy control (lockout/tagout) policy
    - Flawed hazardous energy control policy
    - Incomplete hazardous energy control policy
    - Policy developed unilaterally by management
  C. Resources
    - Inadequate funds to accomplish energy control objectives
    - Absence of or inadequate personnel support (quantitative)
    - Inappropriate personnel used (qualitative)
    - No benchmarking funds
    - Insufficient time allocation for energy system
  D. Planning
    - No general energy control plan
    - Absence of appropriate schedule
    - Inappropriate representation during planning
    - Incomplete or flawed plan
    - Responsibilities not defined
  E. Engineering
    - Absence of or inadequate "safety through design" approach
    - Limitations in energy control specifications
    - Absence of feedback on energy isolation/control specifications
    - No review of incident data to improve design
    - Cost-driven installation mind-set
    - No pre/post start-up review
    - Inadequate drawings/prints energy supplies
  F. Training
    - Absence of or inadequate training plan/schedule
    - Insufficient time commitment
    - Poor quality of training materials/methods
    - Inadequate delivery system
    - No training effectiveness measurement
    - Awareness training only
    - Failure to provide specific procedure training

II. PEOPLE
  A. Managers
    - Failure to define energy control expectation
    - Failure to communicate energy control plan
    - Inadequate assignment of responsibility
    - No accountability system in place
    - Absence of continuous improvement posture
    - Delegation of total responsibility to staff group
    - Emphasis on standards compliance not prevention
    - Inconsistent emphasis—wrong signals
    - Insufficient determination for enforcement
  B. Supervisors
    - Failure to set example
    - Failure to enforce lockout/tagout requirements
    - No positive reinforcement for safe behavior
    - Insufficient technical knowledge
    - Inconsistency supervisor to supervisor
    - No effort to detect and correct deficiencies
    - Absence of or inadequate coaching on procedures
    - Failure to audit application of energy isolation
  C. Workers
    - Improper peer influence
    - Failure to communicate with co-workers
    - Confused communication for energy isolation
    - Lack of knowledge/skill for lockout/tagout
    - General disregard for requirements/risk taking
    - Improper perception of risk
    - Failure to communicate energy isolation deficiencies
    - Language limitations
    - Substance abuse
    - Preoccupation or inattention
    - Personal protective equipment discomfort/obstruction

III. MACHINES
    - Poor layout/arrangement—safe access
    - Performance—breakdown, jam, overheat, etc.
    - Absence of or inadequate energy-isolating device identification
    - Poor energy-isolating device accessibility
    - Deficient design re: human response
    - Deficiencies for servicing—oiling, lubrication, cleaning
    - Confusing isolating device arrangement
    - Absence of or inadequate lockout appliances
    - Ergonomic deficiencies
    - Absence of or inadequate energy isolating devices
    - No system for relieving stored energy
    - Poor component reliability
    - Energy-isolating device without lockout feature

Figure 3-13. Potential causal factors for energy release incidents. *(continued)*
Source: Developed by E. Grund. © National Safety Council.

IV. ENVIRONMENT
- Poor or inadequate lighting
- Interfering or excessive noise
- Limiting layout or spatial arrangement
- Obscured visibility—dust, smoke, etc.
- Congestion—cramped quarters, confining, etc.
- Poor air quality—respiratory protection
- Excessive heat—process, weather
- Cold conditions—frozen valves, etc.
- Distracting weather conditions
- Dirty/unsanitary conditions
- Spills, overflows, etc.—hazardous/ nonhazardous

NOTE: Chapter 11, MORT User's Manual, of W. G. Johnson, *MORT-Safety Assurance Systems,* provides an excellent resource for expanding the search for all causal factors connected with energy release incidents.

Figure 3-13. (*concluded*)

MULTIPLE CAUSATION

1. ACTION
2. INACTION
4. CONDITION
5.
CONDITION 3.
REACTION 6.

1. Engineering Manager, to reduce costs, eliminates intermediate disconnects and breakers from installation plan.
2. Operations Superintendent fails to check prints. Supervisors fail to check energy-isolation capability before start-up.
3. Machinery system A is in operation without intermediate electrical disconnect.
4. Motor control center can isolate machinery system A, but electrical supply also services system B.
5. Supervisor instructs (Action) Maintenance Mechanic not to use motor control center to isolate system A because system B will be affected.
6. Maintenance Mechanic uses control circuitry for isolation of system A.

Potential: Maintenance Mechanic may be injured when someone else actuates controls while he/she is out of sight repairing system A.

Figure 3-14. Sequencing and networking phenomena.
Source: Developed by E. Grund. © National Safety Council.

Initial viewpoint. Defective bolt failure and/or improper rigging were responsible; no energy isolation concepts/procedures were available or contemplated in regard to gravity.

Further investigation. The following consequential factors were identified as having contributed to the outcome: (1) this procedure had been done many times, but normally 0.5-inch (1.3-cm) bolts on single

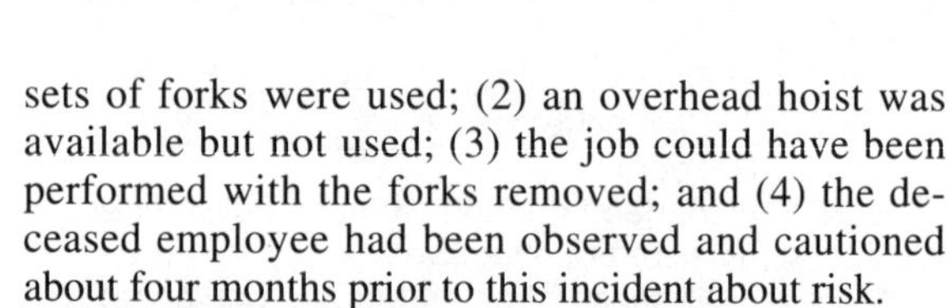

sets of forks were used; (2) an overhead hoist was available but not used; (3) the job could have been performed with the forks removed; and (4) the deceased employee had been observed and cautioned about four months prior to this incident about risk.

Root/primary cause. The deceased failed to use hoist, blocking, or fork rack removal as safer alternatives; the practice of using the chain was established over time, and no on-site service company supervision to identify and correct this behavior was available. The co-worker did not object to the method of work that led to the fatality.

Case C—Environment (Machine Operator—Amputation Incident)

A machine operator/attendant was scheduled on his normal day off to assist in light maintenance work as part of general process shutdown. The operation was a continuous (seven-day/24-hour) type of metal-forming manufacturing. He had previously done this work as part of his involvement with periodic maintenance days. He was assigned to change filters and adjust the air-conveying blower fan belts. He had completed one installation and had moved to the second. The fan units were 10 ft (3 m) off the floor on a platform and were accessible by a scissor lift. He changed the filter and reached to test the tension on the belt with his hand. His left hand was pulled into the nip point between the pulley and the V-belt.

Initial viewpoint. The employee failed to use the electrical disconnect within 6 ft (1.8 m) of the belt position on the blower platform; specific lockout/tagout procedures were posted on most machines but not on small equipment like air-conveying system blowers.

Further investigation. The following consequential factors were identified as having critically influenced the proximate action of the machine operator: (1) The employee had previously changed filters but not checked belt tension; (2) the supervisor making assignments did not know of the employee's work competence in this area of the facility; (3) the belt-pulley guard had been taken off and not replaced; and (4) no energy isolation review had been provided before work proceeded.

Perceptions Survey Form

	Rating
• Lack of specific LO/TO procedures	( )
• Inadequate specific LO/TO procedures	( )
• Lack of LO/TO training	( )
• Inadequate LO/TO training	( )
• Negative supervisory role model	( )
• Negative co-worker role model	( )
• Weak management LO/TO emphasis	( )
• Employee blatant disregard	( )
• No survey to locate isolating devices	( )
• Absence of energy-isolating devices	( )
• Poor location of isolating devices	( )
• Lack of management enforcement	( )
• No compliance auditing	( )
• No identification of isolating devices	( )
• Bypass of tags	( )
• Negative environmental influences	( )
• Other	( )

Causative Rating
A—Major; B—Average; C—Minor
NOTE: Circle top 5 causative factors

Figure 3-15. Perceptions survey form: Energy release incident causative factors.
Source: Developed by E. Grund. © National Safety Council.

Root/primary cause. The employee failed to use the disconnect based on his assumption that the system blower was not running like the prior similar unit he had worked on; failure to verify its power status was influenced by masking general noise levels, marginal platform lighting, and restricted elevated access; the belt and pulley were not directly visible from the position he had assumed.

PERCEPTIONS—ENERGY-RELEASE INCIDENT CAUSATION

Between January 1990 and March 1994, the author surveyed over 500 safety professionals, supervisors, operating personnel, maintenance craftspeople, engineers, and so on regarding their perceptions about the causation of energy-release incidents. The data were reviewed and discussed with an additional 300–350 management personnel from various disciplines, organizational backgrounds, and positions. During the reviews, very little difference of opinion surfaced regarding the critical issues presented. Factors other than those listed were brought up but with limited general support. The survey was not intended to be a statistically refined sampling tool but rather a method of obtaining a general sense of what those involved believed to be true. Those who completed the survey form did belong to a cross-sectional group of individuals directly or indirectly connected with prevention efforts related to energy-release incidents.

Summary

Rank	Factors
1	Lack of management enforcement; negative supervisory role model
2	Blatant disregard by employees
3	Weak management lockout/tagout emphasis
4	Inadequate lockout/tagout training Lack of specific lockout/tagout procedures
5	Lack of lockout/tagout training

Figure 3-16. Perceptions about energy release incidents: Causative factors.
Source: Developed by E. Grund. © National Safety Council.

The perception survey form (Figure 3-15) was completed by rating each factor as either *A* (major), *B* (average), or *C* (minor) based on their contribution to incident causation. After this was accomplished, participants were asked to circle what they believed were the top five factors. In most cases, these were selected from the factors marked *A*.

The results were tabulated and are presented in Figure 3-16. "Lack of management enforcement" and "negative supervisory role model" shared the top position. This suggests that a weak commitment to safety on the part of management and its interface position with the hourly work force is an important deficiency in the control of hazardous energy.

"Blatant disregard by employees" ranked second. This seems to suggest a chicken-or-the-egg quandary—did disregard generally cause management to lose interest in enforcement or did weak enforcement, over time produce complete disregard? The first-ranked factor sends a strong signal regarding management's intensity and commitment to putting in place a process that works. "Weak management lockout/tagout emphasis" ranked third, which underscores even more the strength of these perceptions.

The fourth and fifth factors related to training deficiencies and the absences of specific energy-control (lockout/tagout) procedures. These were included by OSHA in their top five causal factors in the preamble to the lockout/tagout final rule (29 *CFR* 1910.147). The other three causal factors identified by OSHA related to employees failing to ensure that energy was isolated. This proximate behavioral issue is the result of many diverse contributing factors or error-provocative situations and is largely, in my opinion, a by-product of factors 1–3 from Figure 3-16.

SUMMARY

- Hazardous-energy-release incidents continue to occur because of failure to adequately examine the chain of events that are responsible. Using the term *accident* tends to impede the search for real causes. The causes of energy-release incidents are known, repetitive, and quite predictable. For this reason, *incident* is used as a descriptor, not *accident.*
- Johnson (1980) and Haddon (1970) have elaborated on the concept of unwanted transfers of energy and the control methodology of barriers with emphasis on isolation or separation of energy and the exposed individual(s). For the purposes of this text, I have defined an energy-release incident as any unwanted transfer of energy (electrical, mechanical, hydraulic, pneumatic, chemical, thermal, gravitational, and so on) that produces injury/damage to persons, property, or processes or otherwise interrupts, interferes, or degrades the activity in progress.
- Heinrich's domino theory is no longer sufficient to adequately describe the current understanding of the dynamics of incident occurrence. Petersen (1991) expands our thinking with his new principles of safety that emphasize deficiencies in the physical, managerial, and behavioral subsystems that need to be addressed. He indicates that unsafe acts/conditions are symptoms of something wrong in the management system and that severity, in many cases, is predictable. His observation on severity is supported by the author's experience, particularly with respect to energy-release incidents.
- Manuele (1993) introduces the term *hazrin* to encompass all hazard-related incidents. It is intended to de-emphasize outcome and focus on the potential for harm or damage, in other words, the incidents that influence outcome. Manuele also cites Haddon's energy-release theory as being fundamental to dealing with incident causation. He adds the release of hazardous materials to the Haddon approach and gives design and engineering special significance with regard to control of incidents at the source.
- McClay (1989) provides a universal model for the occurrence of loss incidents (Figure 3-1) that explains the relationships between management policies (distal causes) and physical conditions, human actions, and exceeded functional limitations (proximal causes). If control efforts are not successful, an adverse, unexpected event can occur and the point of irreversibility can be passed. A release or transformation of mass or energy will then occur, producing a loss incident.
- Grund's incident model shows the influence of the business dimension and organizational culture on management process and procedure. This, in turn, impacts the human-machine-environment system, which has the potential to generate hazards of higher risk that can produce incidents if not eliminated or controlled. Hazards are categorized as being either the proximate, contributory, or root causes of incidents.
- Human error and its significance to incident causation is explored. Rigby (1970) defines human error as "any member of a set of human actions that exceeds some limit of acceptability." Kjellan (1987) adds four criteria for assessing compliance with the norm:
  —standard rule or regulation
  —appropriate or acceptable
  —normal or usual
  —planned or intended
- Swain and Gutherman (1983) state that "most human initiated accidents are due to the features in a work situation which define what the worker must do and how he must do it." Therefore, management must (1) provide the worker with a safety-prone work situation and (2) avoid the temptation to place the burden of prevention on the worker. Swain and Gutherman identify various subsets of human error: errors of omission and commission; sequence and timing errors; extraneous acts; operator- and system-induced errors; input and output errors; low- and high-stress errors; and design-induced errors.
- DeJoy (1990) proposes a human-factors model for causation that links human error and three causal subsets: person-machine communication, the environment, and decision making. He emphasizes the need for identifying error-provocative situations and tracing the causal factors involved.
- Rigby proposes a hierarchy of tolerance limits that relate to acceptable or unacceptable performance. He lists them in descending order of effectiveness as physical, fixed, measurable, descriptive, cautions, custom and debatable. He finds, however, that the least effective measures are used most often.
- Johnson, working from ideas developed by Nertney, presents an error-free Performance Tree that identifies failure to provide requirements as a leading source of deficiency.
- A behavioral perspective has been provided by Krause, Hidley, and Hodson (1990) in which behavior is examined in terms of antecedent-behavior-consequence relationships. They suggest that most organizations have what amount to very strong incentives in favor of unsafe behaviors. They introduce ABC Analysis as an analytical tool for changing unsafe behavior to safe behavior.
- Topf and Preston (1991) recommend a multidimensional behavior modification approach that provides intervention initiatives as follow: (1) safety awareness, (2) modifying risk behavior, (3) safety-thinking process, (4) personal responsibility, (5) leadership, and (6) safety and personal commitment.
- Petersen (1975) adds that from the standpoint of strategy, it is more productive to bolster the probabilities for safe behavior than to attempt to eliminate unsafe acts through discipline and punishment. Basically, there are two main avenues to follow: (1) increase the satisfactions for working safely and (2) eliminate the obstacles that inhibit working safely.
- McIntire and White (1975), referring to B. F. Skin-

ner's work in behavior modification, state that "the foundation of behavior modification is based on the principle that people will act by a set of rules (even safety rules) if they are 'paid' (reinforced) in a direct, immediate, and consistent manner." They provide guidelines in the following general areas: (1) consequences of behavior, (2) scheduling of consequences, (3) pitfalls in scheduling, and (4) implementation.

- Grund and Lunsford, in training supervisors, have illustrated the relationships of attitudes and behavior with regard to incident causation. Risk and risk reduction were discussed in terms of hazard evaluation and control. A method of simply quantifying risk was presented to established priority for control. Problems associated with detecting and controlling unsafe behavior were categorized into (1) supervisory reluctance to act and (2) employee resistance to change. A behavioral cycle graphic showed how behavior, result, and attitude interconnect either to reinforce unsafe behavior or to halt its progress and begin the process of converting to safe behavior. Intervention and its criticality were discussed with regard to early detection and control of unsafe practices.
- Tasks with energy-release potential were described as well as the influences surrounding them that may compromise their safe completion. A systems approach was proposed in which human, machine, and environmental elements were examined to reveal the variety and scope of possible or potential causes of energy-release incidents. Multiple causation, as described by Petersen (1991), was elaborated upon by the author. A graphic illustration was provided that showed the sequential and multidimensional nature of causation. Action, inaction, and reaction were depicted in a hypothetical energy-isolation scenario as a chain of events that influenced other causative factors. Case histories were provided that contrasted the superficial or apparent causes of energy-release incidents with those uncovered after deeper probing.
- The results of a survey involving over 500 respondents that examined common perceptions of the causes of energy-release incidents were reviewed. Weaknesses in management enforcement, training, and the development of specific energy-control procedures were identified as principal causative factors.

REFERENCES

Altmann JW. Behavior and accidents. *Journal of Safety Research* 2 (1970):110–20.

Bird FE, Germain GL. *Practical Loss Control Leadership.* Loganville, GA: Institute Publishing, 1990.

DeJoy DM. Toward a comprehensive human factors model of workplace accident causation. *Professional Safety* 35, no. 5 (May 1990):11–16.

Haddon W, Jr. On the escape of tigers: An ecologic note. Insurance Institute for Highway Safety. *MIT Technology Review* 72, no. 7 (1970).

Heinrich HW. *Industrial Accident Prevention,* 4th ed. New York: McGraw-Hill, 1959.

Johnson WG. *MORT Safety Assurance Systems.* New York: Marcel Dekker, Inc., 1980.

Keller RR, Stern A. Human error and equipment design in the chemical industry. *Professional Safety* 36, no. 5 (May 1991):37–41.

Kjellan V. Deviations and the feedback controls of accidents. In *New Technology and Human Error.* New York: John Wiley & Sons, Inc., 1987.

Krause TR, Hidley JH, Hodson SJ. *The Behavior-Based Safety Process.* New York: Van Nostrand Reinhold, 1990.

Lunsford B. Human Factors and Accident Prevention. Unpublished paper, 1990.

Manuele FA. *On the Practice of Safety.* New York: Van Nostrand Reinhold, 1993.

McIntire RW, White J. Behavior modification. In BL Margolis and WH Kroes (eds), *The Human Side of Accident Prevention: Psychological Concepts and Principles Which Bear on Industrial Safety.* Springfield, IL: Charles C. Thomas, 1975.

McClay RE. Toward a more universal model of loss incident causation. *Professional Safety* 34, no. 1 (Jan. 1989):15–20.

McCormick EJ, Sanders MS. *Human Factors in Engineering and Design,* 6th ed. New York: McGraw-Hill, 1987.

McGlade FS. *Adjustive Behavior and Safe Performance.* Springfield, IL: Charles C. Thomas, 1970.

Merriam-Webster's Collegiate Dictionary, 10th ed. Springfield, MA: Merriam-Webster, Inc., 1993.

National Safety Council. *Supervisors' Safety Manual.* Itasca, IL: 1993.

Peters GA. Human error: Analysis and control. *ASSE Journal* XI, no. 1 (Jan. 1966):9–16.

Petersen D. Other voices—Safety's paradigm shift. *Professional Safety* 36, no. 8 (Aug. 1991):47–49.

Petersen D. *Safety Management.* Englewood Cliffs, NJ: Aloray, Inc., 1975.

Rigby LV. "Nature of Error." Sandia, Alberquerque, 1970.

Rigby LV. "The Nature of Human Error." *Chemtech.* 1971.

Salvendy G. *Handbook of Human Factors.* New York: John Wiley & Sons, 1987.

Skinner BF. *Science and Human Behavior.* New York: Macmillan, 1963.

Swain A, Gutherman H. *Handbook of Human Reliability Analysis with Emphasis on Nuclear Power Plant Applications.* Washington DC: Nuclear Regulatory Commission, 1983.

Thompson CI. Investigating the careless act. *Professional Safety* 34, no. 12 (Dec. 1989):39–41.

Topf MD, Preston RT. Behavior modification can heighten safety awareness and control accidents. *Occupational Health and Safety* 60, no. 2 (Feb. 1991):43–49.

4

United States Hazardous Energy Regulations

The development of lockout/tagout safety guidelines, standards, and regulations to prevent the unwanted transfer or release of energy has been a slow, often painstaking process of consensus and refinement. Today, the emergence of international business and manufacturing is prompting industries, unions, governments, and trade associations to join forces in establishing adequate, up-to-date lockout/tagout standards and guidelines to protect those who work on or near machinery and power sources.

This chapter looks at the development of major voluntary initiatives and state and federal standards that address hazardous energy control.

VOLUNTARISM OR REGULATION

Under common law, as it slowly evolved in Great Britain and Europe, an employer was obligated to provide employees with the following:

- A safe workplace
- Safe tools to perform the work
- Knowledge of any hazards that might be encountered in performing the work
- Competent fellow workers and managers
- Rules establishing safe work practices and a means to enforce these rules.

However, because European factory law and trade unionism were in their embryonic states, little relief was actually provided for the early industrial worker coping with energy hazards on a daily basis. There is no substantial evidence of the existence of any equipment design features, established safe practices or techniques, or employee-training efforts directed at controlling the unexpected release of energy from factory machinery prior to 1850.

It is important to remember, however, that in 1850 animal and human muscle still supplied 75% of the total energy supply. The invention and development of steam engines in the late 1700s unleashed a new and frightening hazard for industrial workers: large engines, or "prime movers" as they were then known, driving long overhead shafts that supplied power via connecting rods and belts to machine after machine. This cost-effective way to transmit power from one source (the engine) to multiple workstations or machines brought with it danger of unmeasurable proportion. Thousands of factory workers in Europe and the United States were maimed or killed while attempting to operate or service their machines amidst swirling belts, gears, and shafts.

If safety techniques that were within the industry's developmental capacity had been used, such as machine guarding, isolation features, remote lubrication, warning methods, and protective devices, much of this tragedy could have been prevented. Employee safety would have to wait 100 more years until the factory system steamed into the 20th century.

The Voluntary Movement

For the first few decades of the industrial revolution, however, determining what *safe* meant, what workplace standards and guidelines should be established, and how they should be enforced were matters left up to individual employers. Local, state, and federal governments were reluctant to intervene in the private sector. As a result, workers were exposed to numerous hazards on the job when they tried to repair machinery, remove jammed material, replace parts, or work on machines, while they were still energized. No formal methods were used for cutting off power, blocking out moving parts, and ensuring that machinery was not accidentally started or controls triggered while employees were still exposed. At first, workplace reforms were brought about by the efforts of trade unions such as the International Workers of the World and the United Mine Workers as well as by private citizens and charity groups. Eventually, the general public also began pressuring private industry and government for legislation to correct some of the worst excesses in factories, offices, transportation systems, mines, and other workplaces.

By the early 1900s, the rate of employee fatalities and injuries caused by being caught or crushed in machinery or shocked or burned by an energy source was high enough that just about everyone recognized that something had to be done to ensure workers' safety. Over time, progressive companies and some legislatures identified major safety issues regarding maintenance, repair, and the operation of machinery and developed early standards and guidelines to prevent unwanted energy transfers or releases and established safe practices for those who worked on or near electrical and mechanical equipment.

Codes of Safe Practice: State of the Art

Lockout/tagout codes of safe practice must deal with two distinct protective issues:

- Activities on machines, equipment, or processes that must occur under energized conditions with guards temporarily removed and other safety devices bypassed
- Activities on de-energized or energized machines, equipment, or processes in which some part of the worker's body may be in the path of potential machine movement or flow of materials or exposed to electrical current.

The issue of the necessity of working on energized machines, equipment, or power sources deserves further comment. Currently, a legitimate need exists for certain work to be performed under live or hot conditions. However, a great deal of energized work that occurs is actually the result of deficiencies or limitations in the original *design* of the machine, equipment, or system. Employees are left to cope with inherent flaws on older equipment and omissions/oversights on newer installations. Prevailing operational realities create an environment in which the worker's risk is increased because of the need for frequent servicing and maintenance of equipment or machines that do not have the requisite safety design features.

The need for all energized-status work should be evaluated and determined to be legitimate with regard to safety. Furthermore, a continuous process of either eliminating or minimizing those needs should be in place. Finally, alternative protective measures or safeguards should be developed and implemented where zero energy conditions are not feasible.

Lockout/tagout procedures should include all forms and sources of energy such as electrical power, hydraulic fluids under pressure, compressed air, energy stored in springs, potential energy from suspended parts, and other sources that might cause unwanted mechanical movement. The practice of locking out electrical energy only will not necessarily achieve maximum protection for machine operators and maintenance and service people.

Because of the complexity of today's computerized equipment, robotics, and automated machinery systems, lockout/tagout codes and standards must reflect state-of-the-art methods and procedures that will truly render a machine or power source safe for those who must work on or around it. These methods and procedures include but are not limited to:

- Communicating to all concerned that repair or maintenance work will be done on the machinery, equipment, or process and ensuring that nothing can be started or set in motion
- Neutralizing all energy sources, thus placing the machine in a zero mechanical state. This approach provides the greatest protection to workers against unexpected energy transfers either in the power supply or through mechanical movement. Electrical energy is locked out, and all kinetic and potential energy is isolated, blocked, supported, restrained, or controlled. (ANSI standard Z241.1-1989, American Foundrymen's Society, explains the concept of zero mechanical state.)
- Locking out and blocking all energy sources and all kinetic and potential energy:
  - bleed down steam, air, or hydraulic cylinders
  - block gears, dies, other mechanisms
  - release coiled springs, spring-loaded devices and securing cams
  - place blocks under any equipment or parts that may descend, slide, or fall
  - use blocks or special stands under raised vehicles, carts, or other equipment to prevent failure or slippage
- Placing personal padlock(s) and tag(s) on energy-isolating devices. Tagging energy-isolating devices to indicate the reason for isolation.
- Verifying the isolation effectiveness by trying the controls or switches to determine if the machine, equipment, or process is truly isolated or de-energized
- Setting controls and switches to the OFF or NEUTRAL position to prevent accidental start-up when power is restored
- When work is completed, ensuring the removal of all locks and tags, the removal of all blocks and guards used to immobilize the equipment, and clearance of all tools and equipment
- Communicating to all concerned that maintenance or repair work is finished and re-energizing is about to be initiated
- Returning the machine, equipment, or process to the control of the responsible individual(s).

Maintaining state-of-the-art codes of safe practice requires periodic review of current standards and regulations to determine how they relate to changes in machine design, technology, and/or energy sources. A procedure that may be adequate for a simple milling machine may not be sufficient for an automated multistation welding machine. Or procedures that apply to one industry—such as steel—may not be completely adequate for another industry—such as petroleum refining. The types of energy-isolating devices in use must also be updated to make sure that they can eliminate the risk of accidental energy release while machines, equipment, or processes are being repaired or serviced.

Three Approaches to Regulation

Ideally, all of the parties involved in developing safer working methods directed at hazardous-energy-release issues would support a universal standard that applies across all industries and all worksites. Unfortunately, the special requirements and problems of industries and individual businesses are too varied at present for any universal standard. However, employers, workers, unions, trade associations, and city, state, and federal governments have made steady progress toward that goal, particularly over the past 50 years. International efforts to create uniform standards are also under way as we approach the 21st century.

From a regulatory perspective, three principal ways to address the problem of developing lockout/tagout safety standards exist: mandates, performance criteria, and specifications.

Mandates. These establish a general requirement to provide safe and healthful working conditions. They are expressed in regulations as "general duty" clauses for employers, employees, manufacturers, and so on. The General Duty Clause 5(a) in the U.S. Occupational Health and Safety Act (1970) and Part I, Sections 2–9 of the British Health and Safety at Work Act (1974) reflect this broad approach. The following excerpts illustrate the nature and scope of the governing language:

> United States: Sec.5 (a) Each employer—
>
> (1) shall furnish to each of his employees employment and a place of employment which are free from recognized hazards that are causing or are likely to cause death or serious physical harm to his employees. . . .

> United Kingdom: Section **7.** It shall be the duty of every employee while at work—
>
> (*a*) to take reasonable care for the health and safety of himself and of other persons who may be affected by his acts or missions at work; and
>
> (*b*) as regards any duty or requirement imposed on his employer or any other person by or under any of the relevant statutory provisions, to co-operate with him so far as is necessary to enable that duty or requirement to be performed or complied with.

Mandates tell employers or workers in a general way *what* they must do but not *how* they must do it.

Performance criteria. These define more clearly the safety problem that needs to be addressed (for example, dangers of shock while repairing electrical equipment) and *what* employers or workers must do to establish safe working conditions (for example, lock out power supply). However, like mandates, performance criteria typically do not provide details about how to accomplish the job. For example, 29 *CFR* 1910.147, Control of Hazardous Energy Sources (Lockout/Tagout), lists the general procedures for lockout of energy sources but does not specify the precise type of lock, tag, or securing appliance to be used.

Specifications. These are the most detailed and explicit of the three approaches. They tell employers and workers *what* to do and exactly *how* to do it. Consider this excerpt from OSHA in 29 *CFR*, 1910.27, *Fixed Ladders* (c) Clearance (1) Climbing side: "On fixed ladders, the perpendicular distance from the centerline of the rungs to the nearest permanent object on the climbing side of the ladder shall be 36 in. (0.9 m) for a pitch of 76 degrees, and 30 in. (0.7 m) for a pitch of 90 degrees . . . with minimum clearances for intermediate pitches between these two limits in proportion to the slope, except as provided in subparagraphs (3) and (5) of this paragraph."

Of these three approaches, performance criteria are considered the most effective at establishing uniform guidelines. They not only identify safety problems but also allow employers some flexibility to devise their own procedures or equipment to comply with the intent of the standard. Mandates are generally too broad and lack sufficient detail to be enforceable in any meaningful way. However, they set the social objectives and alert those being regulated of their obligations. Specifications, on the other hand, can be too restrictive and may inhibit employers or workers from changing safety measures either to improve them or to adapt to changing technology or conditions. Figure 4-1 shows the hierarchy of standards and the relative strength of their attributes.

Evolution Toward a Universal Standard

The development of ever more stringent lockout/tagout standards and regulations has not been a neat, orderly progression from mandate to specification. In many instances, specifications came first, followed by the passage of performance standards and/or mandates. In addition, the past 20 years in particular have seen the development of a kind of hybrid regulation that combines mandates and performance standards (such as Great Britain's Safe Systems of Work) or performance standards and specifications (29 *CFR* 1910.331-335, covering prevention of electrical hazards on the job).

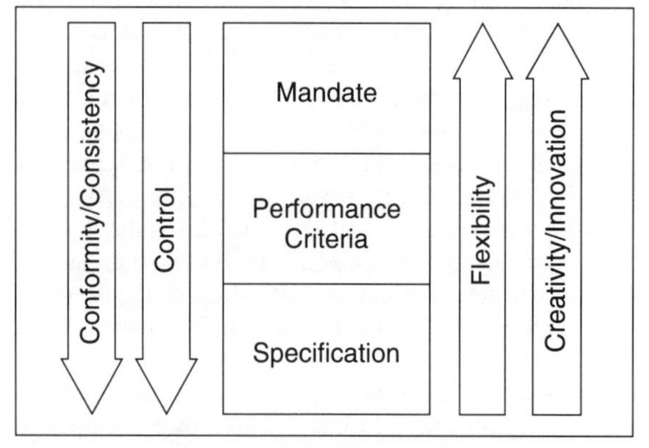

Figure 4-1. Standards hierarchy and attribute strength.
Source: Developed by E. Grund. © National Safety Council.

Unions generally appear to be in favor of the creation of a strong central regulation that covers lockout/tagout issues in all industries with tough enforcement by the government. That, they feel, will go far toward ending the problem of injuries and fatalities related to hazardous-energy-release incidents. Companies and industries, on the other hand, would prefer more flexibility and voluntary self-regulation. They feel that too much interference by government will stifle private safety initiatives. It is interesting to note, however, that only a small percentage of all employers actually develop and/or pursue innovative or progressive approaches to hazardous energy control. Insurance companies, trade associations, and other concerned groups appear to want a compromise between the two positions. All groups would like to see more uniformity in both regulations and enforcement.

Over the past 50 years, the safety movement has used the mandate, performance criteria, and specification approaches to produce various legal guidelines, standards, codes, and regulations covering lockout/tagout issues. These measures have the dual purpose of safeguarding workers and establishing reasonable levels of care for employers and industries. Because the terms *guidelines*, *standards*, *codes*, and *regulations* are often confused or used interchangeably, the next section discusses definitions of these key terms.

Definitions of Safety Measures

Because the legal definitions of *guideline, standard, code, regulations,* and *safe practice* are still being developed, the dictionary definitions of these terms are used in this section (*Merriam-Webster's Collegiate Dictionary*, 10th edition, 1993).

Guideline. An indication or outline [as by a government] of policy or conduct. [Guidelines are less formal than either standards or codes, such as guidelines for storing nonhazardous chemicals, and compliance is generally voluntary.]

Standard. Something established by authority, custom, or general consent as a model or example. [Standards, such as the ANSI standard for fixed ladders, ANSI A14.3-1992, are generally accepted by industry, but compliance is not mandatory unless standards are further used in either codes or regulations.]

Code. A systematic statement of a body of law . . . a system of principles or rules. [Compliance with codes is generally mandatory, and many government agencies have established safety and health codes for workplaces. Violation or noncompliance can be punished by fines or other penalties.]

Regulation. An authoritative rule dealing with details or procedure; a rule or order issued by an executive authority or regulatory agency of a government and having the force of law. [Federal regulations for air quality or worker exposure to lead are examples of this type of safety measure. Compliance with regulations is compulsory, and violation or noncompliance is punishable by fines, criminal prosecution, and other penalties.]

Limitations of Safety Measures

Although these dictionary definitions seem straightforward, the actual use of these terms in safety regulation is less clear. Once a hazard or problem has been recognized, such as the need to guard a machine's point of operation or lock out a power source, the next problem is how to develop an adequate measure that truly controls the hazard or can prevent an incident. These safety measures can have significant limitations that weaken their effectiveness. Simply developing and complying with a code, regulation, or standard does not automatically ensure that a hazard is completely controlled or that risk has been eliminated.

The term *standard*, for example, has been defined by the American Society for Testing & Materials (ASTM) in their *ASTM Glossary* as

1. A reference used as a basis for comparison or calibration.
2. A concept that has been established by authority, custom, or agreement to serve as a model or rule in the measurement of quantity or the establishment of a practice or a procedure.

However, there is no standard definition of a standard, either by consensus or mandate. And neither, at present, is there any recognized general standard for the preparation of standards. The closest guideline may be the ANSI *Guide for Consumer Product Standards—ANSI Consumer Council Publication No. 1*, which provides a list of factors to consider in preparing consumer product standards. This publication can be used as a reference to assist those who must develop lockout/tagout standards for their own industries or companies.

As a result, most standards, including those related to lockout/tagout, have inherent weaknesses that need to be recognized by employers, unions, trade associations, government agencies, and workers. Among the more common limitations of present safety standards are the following:

- *Lack of consistency across an industry, among industries, from state to state, or across national borders.* Standards are often developed on an as-needed situational basis; frequently little effort is exerted to make them consistent from one location to another. Indeed, local conditions may prevent such an effort.
- *Lack of time and money to ensure compliance or to maintain compliance.* Standards that require considerable time and expense to implement are not likely to be followed closely unless sanctions for noncompliance are strong or the risks are perceived as significant.
- *Unreasonableness either in design of the standard or in expected level of compliance.* Some standards may dictate that employers must meet conditions that are not reasonable requirements for workplace safety, such as requiring two exits for every occupied space when spaces vary in size, occupancy, degree of risk, or layout.
- *Too much reliance on company or industry self-regulation to ensure compliance with the standard.* If standards are voluntary, compliance can be unpredictable or lax.
- *Major hazards not addressed or focus is on minor problems rather than on more important safety issues.* Some standards are the result of industry initiatives, or negotiation or may be drafted by a group not familiar with specific on-the-job conditions. The standard may focus on issues of concern to the industry but not to the workers.
- *Conflicts or variation between or among standards or between standards and other codes, guidelines, or regulations.* State standards, for example, may be more stringent than federal standards. Agency overlap creates competition to regulate the same areas or imposes extra demands to establish preeminence.
- *Minimal safeguards accepted as the result of a compromise or consensus procedure.* For example, some ANSI safety standards state the following on the title page overleaf: "The existence of an American National Standard does not in any respect preclude anyone, whether he has approved the standard or not, from manufacturing, marketing, purchasing, or using products, processes, or procedures not conforming to the standard . . . (ANSI Z88.2-1992). Such standards are called "negotiated," "maximum acceptance," or "voluntary" and provide only minimum safeguards.
- *Hazards reduced rather than eliminated.* This is desirable but not acceptable if state-of-the-art methods exist that are technically and economically feasible to eliminate the hazard.
- *Standards fail to incorporate sections on generic hazards and safeguards.* Most standards are designed to address specific procedures or hazards but do not approach the problem from a more general viewpoint, such as incorporating principles and concepts like Haddon's energy-management strategies to prevent injuries or fatalities.
- *Lack of updating and supplementing standards.* Standards that once reflected the latest in machine

safeguarding and energy control are no longer adequate to protect workers using new technology. Thus, even if these outdated standards are followed to the letter, workers may still not be protected.

- *Lack of cross referencing or linking standards with other scientific, medical, or safety and health documents.* Important information about exposure levels, medical treatment techniques, proper work procedures, and other information is not readily available to managers and workers. Absence of consensus for any subject area is often influenced by national agenda or politics.
- *No distinction between point-of-operation hazards and maintenance hazards.* Guards and blocks that protect machine operators may not protect maintenance workers who must work on machinery from different vantage points and under different conditions that entail higher risks.
- *Human error not accounted for in standard.* The common mistakes that operators or maintenance personnel may make should be taken into consideration when developing a standard, and the proper cautionary language should be incorporated into the step-by-step procedures.
- *Failure to incorporate safe design requirements.* When developing standards for machinery, equipment, or processes, guidance is necessary to eliminate hazards that will confront workers performing servicing, maintenance, or repair activities. Inherent design limitations will often lead to error-provocative situations that cannot be addressed effectively by safe procedures.
- *Dissenting opinions not disclosed.* Some standards represent the best collective efforts of associations, groups, or committees—and certain provisions may be controversial. There may be important alternative approaches or minority views regarding work methods, levels of exposure, use of protective equipment, energy isolation techniques, and other issues that managers and workers should know about in order to choose the best safe practices for each situation.

Hidden Hazards of Compliance

Simply because a procedure is in technical compliance with safety standards/codes does not mean that all risk has been eliminated. Companies must guard against the danger of complacency and misunderstanding on the part of their managers and workers. The mere presence of sophisticated lockout/tagout procedures may tempt employees to feel that there is no danger that the procedure or system cannot neutralize. Each situation is unique, and generic standards/procedures may be adequate with regard to *intent* but deficient with regard to *content*. Compliance with prevailing standards should be a starting point for control and not the end.

The need for workers to comprehend fully what they are doing and to follow and apply lockout/tagout procedures and systems systematically cannot be overemphasized. The general purpose of safety standards, regulations, and codes is to establish the minimum threshold for safety while addressing the central issues of hazard control for a given subject. Compliance can be viewed as a subset of prevention, whereas the reverse is not true.

The following sections examine the contributions of various players in the safety movement toward the development of more stringent, comprehensive lockout/tagout safety measures. The early efforts toward a consensus on lockout/tagout standards began primarily with voluntary efforts on the part of employers, trade associations, and other groups such as insurance companies and safety organizations. Gradually, more specific procedures were formalized and appeared as data sheets, safety guidelines, and legal regulations. Table 4-A presents a timeline for the more significant of these developments.

EMPLOYER INITIATIVES

Progressive companies, particularly those in the chemical, petroleum, steel, automotive, and electrical utility industries, have realized since the early 20th century that energy-release incidents were costly in terms of property damage, lost production, and workers' compensation.

As a result, the first safety requirements developed in these industries tended to be more like specifications or performance standards. These internal directives identified the problem of hazardous energy release and emphasized protection of property and worker safety to maintain productivity levels and to reduce workers' compensation costs. They resulted from the hard lessons gained by experience and were perfected by trial and error. Between 1900 and 1925, pioneering work regarding hazardous energy control emerged in the nation's principal industrial sectors. Safety during maintenance or repair work was emphasized, and the main responsibility for compliance was placed on the workers and their immediate supervisors, generally without involving upper management.

A good example of such an early directive is the Republic Steel Corporation's *General Safety Order No. 19: High Voltage Rules*, published May 1, 1934. The corporation used the general order like a performance standard and recommended that "every Plant should develop safety rules governing local hazards, in connection with High Voltage." The safety order stated who was covered by the order and who was responsible for compliance, and it outlined procedures for opening breakers and disconnects, grounding, clearing the line, and restoring power as well as precautions for exposure to high-voltage circuits.

Table 4-A. Benchmark Events in U.S. Lockout/Tagout Standards Evolution (*continued*)

Date	Organization	Event
1950	National Safety Council	*Methods of Locking Out.* D-GEN-41; revised 1971, 1978.
1961	Manufacturing Chemists Association	*Electrical Switch Lockout Procedure,* SG-8.
11/71	National Safety Council	*Guidelines for a Lockout Program.* Review draft.
3/73	American National Standards Institute	Z244.1 Lockout/tagout safety standard committee organized—New York (National Safety Council, secretariat).
3–11/73	American Foundrymen's Society (Frank B. Hall and Robert Mitchell)	Zero mechanical state (ZMS) concept proposed.
12/75	National Safety Council (Charles Price)	Zero energy state (ZES) concept proposed (Price, 1975).
1/7/76	National Fire Protection Association, 70E Committee	*Electrical Safety Requirements for Employee Workplaces:* Part I, Installation (NEC); Part II, Work Practices; Part III, Maintenance Requirements.
1/7/77	Occupational Safety and Health Administration	*Federal Register,* Machinery and machine guarding (subpart O) technical issues—update of machine guarding standards and lockout/tagout.
1978	Occupational Safety and Health Administration	*Fatality/Catastrophe Investigations,* occupational fatalities related to fixed machinery; 125 cases (1974–76).
5/17/79	United Auto Workers, AFL/CIO, Allied Industrial Workers, United Steel Workers of America	Petition for emergency temporary standard regarding lockout of machinery and equipment.
6/79	United Auto Workers	*Machinery Lockout Procedures* development.
6/80	National Institute for Occupational Safety and Health Administration	*Federal Register.* Lockout and Interlock Systems/Devices, request for information.
6/17/80	Occupational Safety and Health Administration	Advanced Notice of Proposed Rulemaking (ANPR) for lockout/tagout standard.
1981	American National Standards Institute	ANSI Z241.1-1975 *Sand Preparation, Molding, and Coremaking in the Sand Foundry Industry,* revised (American Foundrymen's Society—secretariat); ZMS concepts included.
10/81	U.S. Bureau of Labor Statistics	Bulletin 2115: *Injuries Related to Servicing Equipment.*
3/8/82	American National Standards Institute	ANSI Z244.1 for personal protection, lockout/tagout of energy sources, minimum safety requirements (National Safety Council—secretariat).
8/82	Occupational Safety and Health Administration	*Fatality/Catastrophe Investigations*—Selected occupational fatalities related to lockout/tagout problems: 83 cases (1974–80).
7/7/83	Occupational Safety and Health Administration	*Control of Hazardous Energy Sources (Lockout/Tagout)* preproposal draft developed. Draft distributed to key parties; 80 comments received.
10/83	National Institute for Occupational Safety and Health	*Guidelines for Controlling Hazardous Energy During Maintenance and Servicing.*
1985	United Auto Workers, Aerospace and Agricultural Implement Workers of America (UAW) and General Motors	*Lockout Training* publication, GM/UAW National Joint Health and Safety Committee.
11/30/87	Occupational Safety and Health Administration	*Federal Register,* 29 *CFR* 1910, "Electrical Safety Related Work Practices," proposed rule 1910.331-335.
4/29/88	Occupational Safety and Health Administration	*Federal Register,* 29 *CFR* 1910, "The Control of Hazardous Energy Sources (Lockout/Tagout)," proposed rule 1910.147.
8/9/89	Occupational Safety and Health Administration	Public Hearing Notice—lockout/tagout Washington, DC—9/22–23/89; Houston—10/12–13/89.

Table 4-A. *(concluded)*

9–10/89	International Union-UAW, American Petroleum Institute, Chocolate Manufacturers Association DOW Chemical Company, National Association of Manufacturers, National Confectioners Assoc.	Lockout/tagout standard court challenge—U.S. Court of Appeals, District of Columbia Circuit
8/6/90	Occupational Safety and Health Administration	*Federal Register,* 29 *CFR,* 1910, Electrical Safety-Related Work Practices, final rule 1910.331–335 (effective 12-9-90)
7/12/91	U.S. Court of Appeals, District of Columbia	Lockout/Tagout standard (1910.147) remanded to OSHA for further explanation of three issues: 1. criteria for setting standards under section 3(8) of OSHAct 2. preference for lockout over tagout 3. application to all general industry.
3/30/93	Occupational Safety and Health Administration	Response to remand issues raised by U.S. Court of Appeals, Washington DC.
5/27/93	National Association of Manufacturers	Request to U.S. Court of Appeals to suspend enforcement or vacate the Lockout/Tagout Standard.

Source: Developed by B. Grund. © National Safety Council.

The typical objective in such employer safety efforts was threefold: (1) to state the problem (for example, accidental contact with high-voltage circuits or other hazardous energy must be prevented), (2) to establish the uniform application of general performance standards throughout the corporation, and (3) to allow each facility or subsidiary freedom to develop its own specifications.

In the post–World War II period, many companies such as Bethlehem, Republic Steel, and U.S. Steel; Du Pont, DOW, Allied & Monsanto Chemical; Standard Oil of California, New Jersey, and Indiana; General Motors and Ford; ALCOA, ASARCO, and Goodyear developed internal hazardous-energy-control standards and directives. These benchmarks were often shared with subsidiaries and second-tier companies that benefited from their acquisition. Sharing of acquired knowledge continued through associations, professional societies, personal and professional relationships, and technical publications. However, prior to 1970, the focus or emphasis for lockout/tagout was still on maintenance personnel performing electrical tasks under energized or de-energized conditions. Furthermore, most medium and small employers were grossly uneducated and poorly informed about hazardous-energy-control practices and techniques.

Only the most progressive companies prior to 1950 exhibited a sound understanding of a system of safety management and the critical nature of the role and commitment of upper management. In spite of the general duty of employers to provide safe working conditions, as found in common law, workers' compensation statutes, or state orders, safety in the workplace was usually left to the workers and their supervisors. This was also true for hazardous energy control and the application of lockout/tagout.

Until the 1970s, employers had little in the way of resources to draw upon for the control of hazardous energy. What was available had either been developed by employers themselves; by insurance leaders such as Liberty Mutual, Hartford Steam Boiler, and so on; or other trade or safety organizations. State or federal regulation was either minimal or nonexistent, and no uniform performance standard was available to create the pattern for prevention.

UNION ACTIVISM

The United Automobile, Aerospace, and Agricultural Implement Workers of America (UAW), International Brotherhood of Electrical Workers (IBEW), and United Steelworkers of America (USWA) have acted as advocates and helped to pioneer hazardous-energy-control requirements. The booklet published by the UAW, *Machinery Lockout Procedures*, for example, presents case studies illustrating energy hazards, policies for management, recommended lockout procedures, testing and releasing of equipment, and methods of lockout. The booklet establishes minimum guidelines for the neutralization of all energy sources that could cause injury or death. Such sources may be electrical, mechanical, hydraulic, pneumatic, chemical, and so on. It also cites procedures to achieve a zero energy state for workers' safety.

However, the unions approached the problem of hazard control from their own perspective. For example, although OSHA ultimately developed a standard embracing much of the content of the UAW booklet, it did not require lockout procedures only. However, OSHA did invoke strict criteria for companies using a tagout approach. The UAW booklet claims that tags cannot substitute for locks, can be removed too easily, and can be left on machinery accidentally.

United Automobile, Aerospace and Agricultural Implement Workers of America (UAW)

The UAW challenged federal standards on lockout/tagout as defined in 29 *CFR* 1910.147 and restated in 1910.331–.335. It questioned the criteria that OSHA used to develop the tagout portion of the regulations, the definitions used, and other parts of the requirements. The union wanted to exert some influence over what it saw as a major issue for workers and to give workers more freedom to participate in the development of appropriate lockout measures.

The UAW also played an active role in the development of the ANSI Z244.1 lockout/tagout standard and began developing a data base on member deaths in lockout/tagout incidents during the 1970s. This activism led to a petition for an OSHA "emergency temporary standard" for lockout/tagout in 1979.

International Brotherhood of Electrical Workers

The International Brotherhood of Electrical Workers (IBEW) predominantly represents workers in the electric utility industry. The union participated actively in the original ANSI Z244.1 lockout/tagout consensus standard process. The IBEW has collaborated with the Edison Electric Institute in developing a draft-negotiated standard for improving safety requirements to be used during the operation and maintenance of electric power generation, transmission, and distribution facilities. This collaboration led to OSHA's rules 29 *CFR* 1910.269, Electric Power Generation, Transmission, and Distribution, and 29 *CFR* 1910.137, Electrical Protective Equipment, proposed on January 31, 1989.

The IBEW has also maintained a data base of electrically related fatal and nonfatal incidents involving its membership for over 20 years; the data base includes information relating to lockout/tagout incidents. However, much of the data are associated with specialized activity involving work on overhead power transmission lines, substations, power generation equipment, and the like.

United Steelworkers of America

The United Steelworkers of America (USWA) played a major role in lobbying for the passage of the 1970 OSHAct and has been very active in its ongoing development and administration. The steel industry began using lockout/tagout procedures as early as the 1920s with involvement of local steelworker unions. A great deal of significant pioneering energy isolation work was accomplished as a result of its investigation of energy-release incidents and collaboration of management and organized labor in the development of control measures.

In the early 1970s, the USWA published its booklet *USWA Safety and Health Program*, which provided recommended language for negotiating safety/health provisions in agreements. Section 3, "Plant Inspections," refers to conducting monthly inspections by union and company representatives and to observing and evaluating such things as switch lockout procedures. In a subsequent revision of the booklet, the inspection section was relabeled 2B, and the term *lockout procedure* was inserted. In addition, the following note was added: "A lockout procedure should be mandatory for all places of employment. This is one of the most vital safety procedures to prevent injury from unintentional machine or process operation."

In 1979, the USWA also petitioned OSHA for an emergency temporary standard for lockout of machinery and equipment. The union proposed rule-making activity on lockout/tagout, participated in hearings, and provided testimony.

In addition, the USWA has developed a data base concerning fatal accidents involving its members from various industrial sectors such as steel, aluminum, copper, and so on. Since 1980, 26 fatal energy-release incidents have been recorded in which lockout system failures were the primary cause.

CONTRIBUTIONS OF TRADE ASSOCIATIONS

Trade associations such as the Chemical Manufacturers Association, American Foundrymen's Society and American Petroleum Institute attempted to develop broader, more uniform safety guidelines and standards regarding lockout/tagout issues with a specific industry perspective. They recognized that as industrial machinery, equipment, and processes became more complex, so would the procedures needed for energy isolation.

Also, ongoing technological changes in manufacturing processes made old lockout/tagout methods obsolete or dangerously ineffective. By the 1970s, trade associations were actively involved in research and development of safety measures to address the problem of unwanted energy releases. Four key issues were identified that must be considered for machines, equipment, and processes:

- Protecting the operator at the point of operation
- Protecting workers from contact at places other than the point of operation while motion or movement is occurring
- Protecting service, maintenance, and repair people while guards are temporarily removed and other safety devices bypassed
- Protecting workers when their jobs involve placing some part of their bodies in the path of a machine, equipment, process movement, flow, or cycle.

These four issues moved industry focus beyond the concept of safety for maintenance and repair work only. Instead, attention was expanded to consider normal operational activity, service tasks, interruption responses, and incidental work where exposure existed.

Chemical Manufacturers Association

Some associations, such as the Chemical Manufacturers Association (CMA), emphasized building lockout/tagout procedures into the initial planning of any manufacturing process or handling of products. CMA also attempted to involve top management in safety. Its Safety Guide SG-8, *Recommended Safe Practices and Procedures: Electrical Switch Lockout Procedure*, adopted in 1961, stated as its final recommendation for developing standards: "Do not put the standard in effect unless and until top management has accepted it in both spirit and in wording." There was a growing recognition that without executive support, safety would not be regarded as a high priority within the organization.

American Foundrymen's Society

The concept of zero mechanical state (ZMS) was developed by the American Foundrymen's Society to develop stronger, more appropriate safety measures for modern machines that had many moving parts and often relied on a combination of energy sources. Merely interrupting an energy source by pulling a plug or flipping a switch was not enough to ensure worker safety or prevent an accidental energy release. In addition, not everyone had the same mental checklist of steps to take or things to watch out for in shutting down or starting up machines. Too many times, accomplishing these steps safely depended not on following prescribed standards but more on individual experience, dexterity, and skill. For example, not every worker would know to check that pneumatic lines were safely vented or would understand that air needed to be vented on the machine side port of a three-way valve. Clearly, a new approach was needed in the age of growing technological complexity.

Zero mechanical state was first conceived by Frank Hall and Robert Mitchell, representing the Foundry Equipment Manufacturers Association, who served on the D10 Safety Committee of the American Foundrymen's Society. ZMS received formal recognition as part of the American National Standard Z241.1-1975, *Safety Requirements for Sand Preparation, Molding, and Coremaking in the Sand Foundry Industry*, sponsored by the American Foundrymen's Society. In brief, *ZMS* is defined as "that state of the machine in which the possibility of an *unexpected energy movement* has been reduced to a minimum" (Hall, 1977). ANSI Z241.1-1975 further defined *ZMS* as the mechanical state of a machine in which every power source was locked off that could produce a machine member movement. This included not only electricity but pressurized fluids; compressed air; mechanical potential energy (springs, suspended parts); kinetic energy (drop hammers, die tools); and any workpiece or material supported, retained, or controlled by the machine that could move or cause machine movement (Figure 4-2 gives a full definition and explanation of ZMS). This concept recognized that the problem of dealing with various stored energies was compounded by several factors:

- Relationships between energies or subsystems are often not obvious and can be hidden or camouflaged. Workers may not be aware that some machine components are driven by pressurized fluid while others are run by electricity. Locking off one system may not necessarily neutralize the other system.
- Energy subsystems are often not confined to one location but distributed throughout a system. This makes lockout/tagout procedures more complex and increases the risk of an energy-release incident.
- The energy state of a subsystem can be difficult to assess, making it dangerous to assume that a subsystem has been neutralized when the main system is locked out. For instance, pressurized fluid trapped in a machine line or cylinder could still unexpectedly move parts of the machine.

ZMS procedures followed the familiar concepts found in lockout procedures with a few significant additions, such as locking out multiple-energy sources, blocking movable machine parts, and using multiple locks and tags. These additions made the procedures more exacting and specific to eliminate the hazards posed by unsecured moving parts or multiple-energy systems. The zero mechanical state approach was designed to promote safety consciousness among workers by prompting them to think about the possibility of any unexpected energy release and to take proper precautions. On the other hand, the designers tried to tailor ZMS procedures to real-world conditions and to avoid creating unreasonable safety demands. They did not require, for example, that there be no air or oil under pressure anywhere in a machine, only that there be no air or oil under pressure *that could be released accidentally to cause a machine part to move and to endanger a worker.* Figure 4-3 presents a ZMS policy statement for general machine safety instructions.

ZMS provides for primary and secondary protection of workers from hazardous energy release. *Primary protection* refers to protective means that cannot be deactivated as the result of a malfunction or intentional bypassing or to the prevention of machine movement by directly controlling the power source or using mechanical stops. An example of primary protection is the main disconnect switch used to lock off power to an electric drive motor.

Secondary protection is a device or structure that can be bypassed or might malfunction, thus defeating the planned protection. All ordinary control devices fall into this category because they do not directly prevent the application of power. Instead, they control a device that does prevent power application. Limit switches or operator protective devices are examples

Definition of Zero Mechanical State (ZMS) According to American National Standard Z241.1-1981

Standard Definition	Commentary/Explanation
2.69 Zero Mechanical State (ZMS). That mechanical state of a machine in which:	
1) Every power source that can produce a machine member movement has been locked off;	**E.2.69 Zero Mechanical State (ZMS)**—Over the years, changes in equipment requirements and design have incorporated the use of pressurized air, hydraulics, and electricity, as well as combinations of these media to perform certain functions on the equipment, with the result that the commonly used terminology of "Locked Out" or "Locked Off" does not describe as safe a condition as was originally intended. This committee proposes a new term, **Zero Mechanical State (ZMS),** which represents maximal protection against *unexpected mechanical movement of the machine* when setups are made or maintenance performed on the equipment by maintenance or authorized operator personnel. The ZMS concept does not apply only to foundry equipment, but has significant applications to all types of equipment used in industry. Special attention should be given to the definition and use of the ZMS concept in this standard.
2) Pressurized fluid (air, oil, or other) power lock-offs (shut-off valves), if used, will block pressure from the power source and will reduce pressure on the machine side port of that valve by venting to atmosphere or draining to tank;	This may be accomplished by more than one valve for each power source, provided that each valve can be locked off. This requirement is met by a three-way valve (or equivalent), properly connected. It will also prevent leakage, because of valve malfunction, from reaching the machine.
3) All accumulators and air surge tanks are reduced to atmospheric pressure or are treated as power sources to be locked off, as stated in paragraphs 1) and 2);	See commentary column opposite (5).
4) The mechanical potential energy of all portions of the machine is at its lowest practical value; so that opening of pipe(s), tubing, hose(s), or actuation of any valve(s) will not produce a movement that could cause injury;	Holding a machine member against gravity or a spring force by blocking, by suspension, or by brackets or pins designed specifically for that purpose is permissible to reach **Zero Mechanical State.**
5) Pressurized fluid (air, oil, or other) trapped in the machine lines, cylinders, or other components is not capable of producing a machine motion upon actuation of any valve(s); 6) The kinetic energy of the machine members is at its lowest practical value; 7) Loose or freely movable machine members are secured against accidental movement; 8) A workpiece or material supported, retained, or controlled by the machine shall be considered as part of the machine if the workpiece or material can move or can cause machine movement.	A machine in **Zero Mechanical State** may still have fluid (air, oil, or other) under pressure trapped in its piping. (In instructions for ZMS, proper consideration should be given to test to verify that the desired set of conditions has been reached. For instance, when the motor power disconnect has been locked out [off], this can be verified by pressing the **start** pushbutton to make sure the motor *does not* start. Part of the ZMS procedure should be test actuation of the **start** initiator after lock out.)

Figure 4-2. Definition of zero mechanical state according to American National Standard Z241.1-1981.
Source: Printed with permission from the American Foundrymen's Society, Inc., 1981.

of secondary protection. Workers must understand the limits of these devices and use extreme caution when applying them. Also, these devices should be inspected often because a malfunction may not be noticed until the device is needed.

One key to a successful ZMS program is carefully written instructions. Manufacturers and employers should follow these guidelines:

1. Obtain accurate, up-to-date electrical control schematic drawings and hydraulic and pneumatic flow diagrams for the equipment.
2. Define the position that each movable machine member must be in during maintenance or servicing and any supports or restraints for machine parts that may be required.
3. Describe in detail the method and sequence of locking off each power source and reducing all pressure that may create or cause a machine movement. This step should also take into consideration material in hoppers and workpieces.
4. Test all initiator and valve manual overrides to verify that all power sources have been shut off and all remaining pressure reduced to prevent unexpected machine movement. Special attention should be given to detented and closed center valves.

Eventually, Charles Price, then section administrator of the National Safety Council, urged that the scope of ZMS be broadened to zero *energy* state (ZES) to cover precautions for radiation, thermal, and chemical forms of energy (Price, 1975).

American Petroleum Institute

The American Petroleum Institute (API) has worked to develop safe maintenance and work practices in petrochemical installations. The threat of fire and explosions in this industry prompted early efforts to prevent and reduce energy-release incidents. In 1983, API published the second edition of *Safe Maintenance Practices in Refineries*, which covers safe practices

GENERAL MACHINE SAFETY INSTRUCTIONS

General Instructions

1. Machine installation must be approved by employer before allowing employees to work on equipment.
2. Only trained, authorized, and supervised personnel, designated by the employer, may perform operational adjustment and maintenance functions on equipment.
3. Instructions and drawings provided by the equipment manufacturer must be used when performing adjustments and maintenance functions.
4. All tools and equipment (hoists, slings, chains, blocking, etc.) used in maintenance must be approved by the employer.

Adjustment and Troubleshooting (Equipment *not* in ZMS)

1. Work in this category includes only those tasks which by their nature must be accomplished while the equipment is operating or has *not* been placed in ZMS (Zero Mechanical State). *This condition requires special alertness and training of all employees and must be recognized as extremely hazardous.*
2. During debugging, adjustment, check-out and certain troubleshooting, it may be necessary to remove guards to observe machine functions while the equipment is *not* in ZMS. *This is extremely hazardous and requires special alertness.* Guards must be replaced as soon as the task which required their removal is complete.
3. In restoring equipment to its normal state following a malfunction or jam-up, it is frequently necessary to operate certain powered machine functions by use of manual control or manual valve override. *This must be recognized as an extremely hazardous procedure,* and all persons must know for sure that all persons are clear of the path of possible moving machine members.

Maintenance (Equipment in ZMS)

1. Work in this category includes:
   a. All tasks that require any part of the body of the maintenance employee to be placed in the path of possible moving machine members.
   b. All adjustment and troubleshooting tasks which can be accomplished with the equipment in ZMS (Zero Mechanical State).
   c. When locking off air or hydraulic power sources of the machine being worked on and the adjacent machines, this may cause other non-adjacent machines to become either partially or completely free of power sources. This does not, however, place that machine in ZMS.

   Repair or maintenance activity on any piece of equipment must be started only after the complete ZMS Instructions for the machine to be worked on have been satisfied.
2. Specific ZMS Instructions for the machine unit being worked on must be followed closely before placing any part of the body in the path of possible moving machine members.
3. Specific ZMS Instructions for the machine being worked on may also include instructions for adjacent machines to lock off power sources and reduce the mechanical potential energy of the adjacent components.

Figure 4-3. General machine safety instructions.
Source: Printed with permission from the American Foundrymen's Society Inc.

for maintenance work, specific tasks and equipment, and hazardous materials.

Safety measures begin with work authorization procedures to ensure that proper safeguards are taken. These procedures bridge the communication gaps among departments and employees involved in maintenance work. The authorization permits can be written or oral and define worker responsibilities, time period for the work, emergency procedures, and exceptions to the procedures. Written permits are issued after conditions in the proposed work area have been checked to ensure it is safe to proceed with the job. These permits cover such activities as fire or hot work, entry to a confined space, cleaning, and excavation. Oral permits cover less risky and more routine tasks such as instrument calibration and repair, minor adjustments of equipment, and housekeeping. Finally, tags are used to keep workers associated with the maintenance tasks aware of the status of the equipment being repaired. Tags may say the following: SAFE FOR ENTRY, DO NOT OPEN, READY FOR WORK, DO NOT OPERATE, or OUT OF SERVICE FOR REPAIRS (American Petroleum Institute, 1983).

The API publication emphasizes that because of the hazards associated with refinery equipment, safety must be built into every job. This includes a thorough evaluation of each job to be performed to reveal conditions that might be hazardous. Mechanical and physical hazards peculiar to refinery maintenance include the following:

Preshutdown period. Before beginning work, maintenance personnel should be alert to the risks of doing work in the face of such hazards as working on hot lines and equipment, rotating and reciprocating equipment; difficulties of operating or gaining access to controls; clothing becoming wet with hydrocarbon materials; and slipping hazards.

Primary shutdown period (unit is being opened). During this stage of the shutdown of equipment and processes, workers must be alert to such conditions as lines under pressure; presence of toxic or flammable gases, liquid hydrocarbons, corrosives; and hot lines and equipment.

Secondary shutdown period (unit is down and isolated). This stage completes the shutdown to permit maintenance work to be done safely on equipment and in confined spaces. Workers must take steps to ensure that there is no possibility of accidental energy release that could start the machinery, ignite a fire, or release toxic materials. These procedures include locking out and tagging electrical equipment before work begins and conducting tests to verify that all power sources are secured.

The publication lists specific procedures for the safe performance of electrical work and for blanking or blinding equipment. As a first step, workers must always obtain proper authorization to do the work. For electrical work, the rules describe safety procedures for working on de-energized and energized lines, unprotected elevated locations, and high-voltage lines. Appropriate locks and tags must be placed on the disconnecting device of the circuit on which work is to be performed to prevent the circuit from being energized while work is in progress. When blanking or blinding a line, workers must be sure that lines and equipment are depressurized and opened slowly to ensure there is no buildup of gas or liquid and that block valves are holding. If a line is under pressure, it should be opened only under the direction of the operating supervisor.

These measures, together with other lockout/tagout regulations, have been in force in the petroleum industry for a number of years. Continued revision and updating ensures that workers have the latest safe work procedures and practices to protect them in this hazardous industry.

In January 1990, API published RP 750, *Management of Process Hazards*, which further expands upon their earlier work.

Edison Electric Institute

The Edison Electric Institute (EEI), founded January 12, 1933, is an association of investor-owned electric utilities. Since the 1930s, one of the primary missions of EEI has been to address industry safety issues and accident prevention measures. Its transmission and distribution committee assumes primary responsibility for all safety issues. The association's various technical committees, made up of member-company volunteers, have worked to evaluate new procedures, concepts, and equipment; to review and publish technical papers and reports; and to hold sessions where specialists could exchange information and seek advice from their colleagues on difficult industry problems.

EEI has worked to help its member companies produce a reliable network of electrical service and provide more efficient production of electric power; faster recovery from interruptions of power due to storms, natural disasters, and other emergencies; and research into the production, transmission, and use of electrical energy. Through its committees, EEI has brought electrical safety work issues to the attention of its member companies.

Although EEI has not specifically prepared lockout/tagout standards as an association, the group has participated in the development of the ANSI Z244.1 standard and was heavily involved in the writing of OSHA's final rule 29 *CFR* 1910.269, which is discussed later in this chapter. EEI has not always agreed with federal and industry standards but has acknowledged the need for such procedures and has tried to have some influence on their development. As a prominent organization in the electrical utility industry, EEI is concerned that any lockout/tagout standards and regulations reflect the real-world conditions of electrical work in the field as well as in plants and offices (Crickmer, 1993; *Edison Electric Institute*).

CONSENSUS AND SAFETY ORGANIZATIONS' EFFORTS

Some of the safety organizations that have addressed energy hazard and control issues for more than half a century have been the National Safety Council, National Fire Protection Association, and American National Standards Institute. Since the late 1940s, these organizations have sought to develop energy hazard controls and worker safety requirements to prevent injuries and fatalities from energy-release incidents.

These organizations have emphasized in their publications and meetings that safety must be an integral part of the management process and not simply an add-on that is left to the discretion of workers and line supervisors.

National Safety Council

The National Safety Council began publishing data sheets on lockout methods in the 1950s and has refined and updated these methods ever since. These publications include

- Data Sheet D-Gen. 41, *Methods of Locking Out Electric Switches*, 1950. This data sheet emphasized lockout procedures but did not stress the use of tags.
- NSC B199 1970, *Safety Requirements for a Lockout Program*. In this publication, the National Safety Council attempted to provide guidance for companies and industries that wished to establish a lockout method and for government agencies to aid in the formulation of rules and regulations. NSC defined lockout concepts, described the devices and procedures, and listed common errors made by companies and workers in attempting to secure energy sources.
- Data Sheet 237, Revision B, *Methods of Locking Out Electrical Switches*, 1971. In this update, the National Safety Council began recognizing the greater role that tags could play in securing energy sources.
- Data Sheet 1-237-78, *Methods of Locking Out Electric Switches,* 1978. An update of the previous publication, this data sheet recognized that locks would also be required to secure hydraulic, gas, air, or other energy sources and that blocks may also be needed to prevent gravitational movement.

In the 1970s, work done by members of the National Safety Council, such as Charles Price's zero energy state, helped to advance the concepts of hazardous energy control and to produce more rigorous requirements for safe procedures. In 1973, the NSC was appointed secretariat for the development of the ANSI Z244.1 lockout/tagout standard.

American National Standards Institute

In March 1973, the ANSI Z244.1 Committee held its first organizational meeting in New York to develop a standard on lockout/tagout. The National Safety Council functioned as the secretariat and provided a draft document detailing lockout/tagout procedures that was used as a point of reference for the committee's deliberations.

Frank Rapp of the UAW chaired the committee, which was composed of about 50 members and alternates. The basic draft of the standard was finished, balloting activity was completed, and processing of public review comments was done during 1975. From 1975 to 1982, various administrative delays and procedural complications precluded the standard from being officially released. In March 1982, the publication *American National Standard for Personnel Protection—Lockout/Tagout of Energy Sources—Minimum Safety Requirements* Z244.1 was approved and published.

In the foreword of the standard, the following paragraph is noteworthy:

> Neither the standards committee, nor the sponsor, feel that this standard is perfect or in its ultimate form. It is recognized that new developments are to be expected, and that revisions of the standard will be necessary as the art progresses and further experience is gained. It is felt, however, that uniform requirements are very much needed and that the standard in its present form provides for the minimum performance requirements necessary in developing and implementing a lockout/tagout procedure for the protection of employees. (American National Standards Institute, 1982)

This benchmark document was the result of years of effort involving numerous trade association committees, labor organizations, and safety groups. In its final form, ANSI Z224.1 covered the following topics:

- The scope, purpose, and application of lockout/tagout methods and systems and definitions of all terms used in the document.
- Lockout/tagout policy and procedures, including definition of responsibilities and communication and training of lockout/tagout requirements.
- Minimum requirements for lockout/tagout procedures for the preplanning stage, which covers a survey of energy sources and related exposures, identification of energy-isolating devices, selection and procurement of protective materials and hardware, assignment of duties and responsibilities, and preparation of de-energizing and start-up sequences.
- Implementation of lockout/tagout procedures, which includes preparation for, application of, and release from lockout/tagout.
- Special lockout/tagout considerations, covering interruption of lockout/tagout (testing energized equipment); equipment design and performance limitations; exposure of outside personnel; group lockout/tagout; coordination of shift/schedule changes; authorization for lockout/tagout application and removal; work on energized equipment; and production operations.

During the development phase of OSHA's standard 29 *CFR* 1910.147, Control of Hazardous Energy, the ANSI Z244.1 document was heavily referenced as the principal resource. Almost all of the concepts and principles in the consensus document now exist in the 29 *CFR* 1910.147 standard.

In 1992, the ANSI Z244.1 committee voted to reaffirm the standard, and ANSI approved the action and granted the new status.

National Fire Protection Association

In 1976, the National Fire Protection Association (NFPA) established the Committee on Electrical Safety Requirements for Employee Workplaces, which was to report to the association through the Electrical Correlating Committee of the National Electric Code (NEC) Committee. The Electrical Safety Requirements Committee (NFPA 70E) was formed to help OSHA prepare electrical safety standards that could be promulgated through Section 6(b) of OSHAct.

The new committee was formed after OSHA had attempted to use the latest edition of NFPA 70, *National Electrical Code (NEC)*, as a guide, and several problems arose. First, updating a new edition of the NEC would have required a great deal of effort and expense by OSHA and others and could have resulted in standards that conflicted with other national and local standards. Furthermore, the NEC was intended for those who design, install, and inspect electrical installations and not for employers and employees who had to address safety issues in the workplace. Many of the provisions were far too technical for the average employer and employee to understand. Finally, some of the NEC provisions were not directly related to employee safety, while other requirements for electrical safety-related work practices and maintenance were missing from the NEC.

As a result, a new standard was needed that would fulfill OSHA's needs and still be consistent with the NEC. The new document would be developed by a competent group, representing all interests, and would use suitable portions of the NEC and other documents dealing with electrical safety. With the encouragement of OSHA, the NFPA's Standards Council authorized the formation of the Committee on Electrical Safety Requirements for Employee Workplaces to develop a document that could be used as a basis for evaluating electrical safety in the workplace.

The committee envisioned creating a standard for electrical installations that would be compatible with OSHA requirements for workplace safety in locations covered by the NEC. The new standard would consist of four parts:

- Installation Safety Requirements
- Safety-Related Work Practices
- Safety-Related Maintenance Requirements
- Safety Requirements for Special Equipment.

The parts could be published separately, although it was understood that they all represented equally important aspects of electrical safety. The new standard was named NFPA 70E, *Electrical Safety Requirements for Employee Workplaces*. The first edition, published in 1979, consisted only of Part I. The second edition, published two years later, contained Parts

A. General Requirements. Each employer shall document and implement the general requirements specified in the following Paragraph A(1) through Subparagraph A(4)(e). These general requirements supplement the lockout-tagout procedure specified in Paragraph B(2) through Subparagraph B(7)(b).

(1) Implementation Responsibility. The employer shall be responsible for the implementation of the lockout-tagout procedures. Employees shall be trained to understand the safety significance of their responsibilities in implementing the procedures.

The lockout-tagout procedures shall be complied with when work is performed on or near deenergized circuit parts or equipment in any situation where there is a danger of injury due to unexpected energization of the circuit parts or unexpected start-up of the equipment.

(2) Training. The employer shall provide training to assure that all employees who are assigned to work on or near deenergized circuits or equipment understand the purpose of the lockout-tagout procedure, and understand the requirements of the procedure that apply to their specific work assignments.

(3) Coordination with Other Procedures. The electrical lockout-tagout procedure shall be implemented in coordination with appropriate procedures for safely isolating other energy systems such as hydraulic, pneumatic, thermal, process gases and fluids, chemical, and mechanical, in order to provide for the isolation of all energy sources that could endanger employees.

(4) Lockout-Tagout Device Requirements. The locks, tags, and other hardware that are required to be used to comply with the provisions of Paragraph B(3) shall be provided by the employer, and shall be the only devices used to lock out or tag out for personnel protection. Locks and tags that are used for such personnel protection shall be:

(a) standardized according to one or more of the following: color, shape, size, type, or format;

(b) distinctive in appearance, easily recognizable, clearly visible;

(c) designed to convey all information required for the application;

(d) designed to deter accidental or unauthorized removal; and

(e) designed to withstand the environment to which they are exposed for the duration of their application.

Figure 4-4. NFPA 70E, Part II, Safety Related Work Practices; Chapter 4, Lockout-Tagout. (*continued*)
Source: Reprinted with permission from NFPA 70E, *Electrical Safety Requirements for Employee Workplaces*, copyright © 1988, National Fire Protection Association, Quincy, MA 02269. This reprinted material is not the complete and official position of the National Fire Protection Association, on the referenced subject which is represented only by the standard in its entirety.

B. Procedural Requirements. The lockout-tagout procedure shall be documented by the employer and shall contain requirements to safeguard employees while they are working on or near deenergized circuit parts or equipment in any situation where there is danger of injury due to unexpected energization of the circuit parts or unexpected start-up of the equipment. Where applicable, the following Paragraph B(1) through Subparagraph B(7)(b) shall be included in the procedure to provide safeguards for employees in the specific workplace to which the lockout-tagout procedure applies.

(1) Administrative Provisions. The scope, purpose, and areas of application of the lockout-tagout procedure shall be defined; and administrative requirements such as responsibilities for implementing, compliance, and training shall be specified.

(2) Procedures for Deenergizing Circuits and Equipment. The procedures specified in the following Subparagraphs B(2)(a) through B(2)(d) shall be included in the lockout-tagout procedure to assure that circuits and equipment which are to be deenergized for employee safety will be isolated from all electric energy sources and any stored electric or mechanical energy that might endanger employees will be released or restrained.

(a) Preplanning. Procedures shall require preplanning to determine where and how electric energy sources can be disconnected to safely deenergize circuits and equipment that are to be worked on.

(b) Equipment Shutdown. Equipment shutdown procedures shall be included so that the electric equipment involved is safely shut down before circuits are deenergized in accordance with the following Subparagraph B(2)(c).

(c) Disconnecting Electric Circuits and Equipment. Procedures shall require that: (1) the circuits and equipment to be deenergized are disconnected from all energy sources, and that (2) disconnecting devices are operated only by persons authorized by the employer or under the supervision of an authorized person. Control circuit devices such as push buttons, selector switches, or electrical interlocks that deenergize electric power circuits indirectly through contactors, controllers, or electrically operated disconnecting devices shall not be used as the sole disconnecting means for deenergizing circuits or equipment that are to be worked on.

(d) Stored Energy Release. The procedure shall include requirements for releasing stored electric or mechanical energy which might endanger personnel. All capacitors shall be discharged and high capacitance elements shall also be short-circuited and grounded before the associated equipment is touched or worked on. Springs shall be released or physically restrained shall be applied when necessary to immobilize mechanical equipment.

(3) Procedures for Applying Locks or Tags or Both. The lockout-tagout procedure shall include requirements for applying locks or tags or both to provide that circuits and equipment which have been disconnected from all sources of electric energy will not be reenergized without prior notification to all employees who are assigned to work on or near the deenergized circuits or equipment.

Both locks and tags shall be required to be placed on each disconnecting means that is used to deenergize the circuits and equipment to be worked on. The locks and tags shall comply with the provisions of Paragraph A(4) through Subparagraph A(4)(e) and shall be attached by authorized employees in a manner that impedes operation of the disconnecting means.

The tags supplement the locks in preventing accidental or unauthorized operation of the disconnecting means, and they shall contain a statement prohibiting unauthorized operation of the disconnecting means or removal of the tag. When appropriate, information such as work permit number, work to be performed, identification of circuits or equipment or both to be worked on, name of employee authorizing lockout, name of employee who placed the tag, name of employee authorized to remove the tag, time and date shall be included.

Exception No. 1: Applying Locks Only. Locks only shall be permitted to be placed on the disconnecting means to effect lockout where only one circuit or equipment is deenergized, the lockout period does not extend beyond the work shift, the locks are of a distinctive design that clearly discourages unauthorized removal, and procedures are implemented by trained employees to provide a level of safety equivalent to using both locks and tags.

Exception No. 2: Applying Tags Only. Where locks cannot be applied or are not feasible, or where it can be demonstrated that tagout procedures will provide equivalent safety, tags only shall be permitted to be used provided the tags are of a distinctive, standardized design that clearly prohibits unauthorized energizing of the circuits or removal of the tag; and the use of tags is accompanied by an additional safety requirement such as the removal of an isolating circuit element, blocking of a controlling switch, or opening a disconnecting device; and procedures which define and specify employee responsibilities regarding the use of tags as tagout devices are implemented by employees who have been trained to comply with the procedures.

Exception No. 3: Work Permitted with no Locks or Tags. Where circuits or equipment are deenergized for minor maintenance, servicing, adjusting, cleaning, inspection, operating corrections, and the like, the work shall be permitted to be performed without the placement of locks or tags on the disconnecting device, provided the disconnecting device is adjacent to the circuit parts and equipment on which the work is performed, the disconnecting device is clearly visible to all employees involved in the work, and the work does not extend beyond the work shift.

Figure 4-4. *(continued)*

I and II, while the third edition in 1988 combined the original Parts I and II with a new Part III. The next edition will complete the standard by adding Part IV.

Employers should understand that NFPA 70E applies to an electrical installation or system only *as it is part of an employee's workplace*. This limitation distinguishes NFPA 70E from the earlier NEC standard, which applied to safety issues associated with the design, installation, and modification or construction of electrical systems. As a result, the NFPA 70E standard, while compatible with NEC provisions, cannot be used in place of them.

To determine which requirements of the NEC should be included in Part I, the committee developed

(4) Procedures for Verifying Deenergized Condition. The lockout-tagout procedure shall require the following actions to be accomplished after locks or tags or both have been applied to determine if all the electric energy sources have been disconnected and the equipment is inoperative:

(a) Verify that Equipment Cannot Be Restarted. Operate the equipment operating controls such as push buttons, selector switches, and electrical interlocks, or otherwise verify that the equipment cannot be restarted.

(b) Verify that Circuit and Equipment Are Deenergized. Test the circuits and equipment by use of appropriate test equipment to verify that the circuits and equipment are deenergized. On circuits operating at more than 600 volts, nominal, the test equipment used shall be checked for proper operation immediately before and immediately after this test.

(5) Procedures for Shift Changes. Provisions shall be included in the lockout-tagout procedure to assure coordination during shift changes or changes in employee assignments. Specific procedures shall be included to cover transfer of responsibilities and transfer of lockout-tagout devices. Procedures shall also be included to cover the safe removal of lockout-tagout devices by a designated individual, other than the employee who applied the devices, in case the employee is absent from the workplace due to sickness or other reasons.

(6) Procedures for Restoring Electric Service. Procedures for restoring electric circuits and equipment to service shall be included in the lockout-tagout procedure, and shall cover the following requirements:

(a) Verify Circuits and Equipment Are in Condition to Energize. Before electric circuits and equipment are reenergized, appropriate tests and visual inspections shall be conducted to verify that all tools, mechanical restraints and electrical jumpers, shorts and grounds have been removed, so that the circuits and equipment are in a condition to be safely energized.

(b) Verify Employees Are Clear of Circuits and Equipment. Before circuits and equipment are reenergized all affected employees shall be notified to stay clear of the circuits and equipment. There shall also be a visual verification that all employees are in the clear.

(c) Lockout-Tagout Device Removal. Each lockout-tagout device shall be removed from each electric disconnecting means by the authorized employee who applied the device or under the direct supervision of that employee, except as provided in Paragraph B(5) for employees who are absent from the workplace.

(d) Release for Energizing. Where appropriate, the employees responsible for operating the machines or process shall be notified when circuits and equipment are ready to be energized, and such employees shall provide assistance as necessary to safely energize the circuits and equipment.

(7) Procedures for Testing or Temporary Operation. Procedures shall be included where there is a need for testing or temporary operation of circuits and equipment which have been deenergized and locked out or tagged out. The following requirements shall be included in such procedures:

(a) Verify that the circuits and equipment are in a condition to be safely energized according to Subparagraph B(6)(a).

(b) Verify that employees are clear of the circuits and equipment according to Subparagraph B(6)(b).

(c) Verify lockout-tagout device removal according to Subparagraph B(6)(c).

Figure 4-4. (*concluded*)

the following guidelines, according to which applicable provisions should

- Give protection to employees from electrical hazards
- Be taken from the NEC in a way that preserves their intent regarding employee safety
- Be selected in a way that reduces the need for frequent revision yet avoids technical obsolescence
- Provide for compliance by means of inspections
- Be free of unnecessary detail
- Be written clearly so employers and employees could easily understand them
- Avoid adding requirements not found in the NEC and avoid changing the intent of the NEC provisions if the wording is changed in the new document.

Thus, Part I of NFPA 70E serves a specific need of employers, employees, and OSHA and is not intended to be a substitute for the original NEC standard. Figure 4-4 presents the complete text of NFPA 70E, Part I, Chapter 4, which outlines the scope, application, procedures, and exceptions of an effective lockout/tagout system.

GOVERNMENTAL ACTION: STATE

Before the passage of the OSHAct in 1970, the industrialized states, because of their local conditions, economies, and labor lobbies, had various hazardous-energy-safety requirements in place. However, in most cases the regulations had little consequential influence, and enforcement resources were limited. The states had nothing as comprehensive as the current OSHA regulations on hazardous energy control.

After 1990, the states either adopted federal OSHA standards 29 *CFR* 1910.147 and 29 *CFR* 1910.331–332 verbatim or developed their own standards, which were deemed to meet or exceed the federal requirements. Fifteen states and the Virgin Islands have approved state plans, with all but Hawaii

Table 4-B. OSHA State Plan Status

State	Plan status	Standards
Alabama	In process	Federal
Alaska	Approved 9/84	Federal
Arizona	Approved 6/85	Adopted federal
Arkansas	In process	Federal
California	Certified 8/77	State developed
Connecticut	Approved 10/78	Public employees only
Delaware	In process	Federal
Florida	In process	Federal
Hawaii	Approved 4/84	State developed
Idaho	In process	Federal
Indiana	Approved 9/86	Adopted federal
Iowa	Approved 7/85	Adopted federal
Kentucky	Approved 6/85	Adopted federal
Maryland	Approved 7/85	Adopted federal
Massachussetts	In process	Primarily federal
Michigan	Certified 1/81	State developed
Minnesota	Approved 7/85	Adopted federal
Missouri	In process	Federal
Nevada	Certified 8/81	Federal
New Mexico	Certified 12/84	Federal
New York	Approved 6/84	Public employees only
North Carolina	Certified 9/76	Federal
Oklahoma	In process	Federal
Oregon	Certified 9/76	State developed
Puerto Rico	Certified 9/82	Federal
Rhode Island	In process	Federal
South Carolina	Approved 12/87	Adopted federal
Tennessee	Approved 7/85	Adopted federal
Texas	In process	Federal
Utah	Approved 7/85	Adopted federal
Vermont	Certified 3/77	Adopted federal
Virginia	Approved 11/88	Adopted federal
Washington	Certified 11/88	State developed
West Virginia	In process	Federal
Wyoming	Approved 6/85	Federal

Source: Developed by E. Grund. © National Safety Council.

adopting or using various federal standards verbatim. Nine states have had their job safety plans certified by the Department of Labor (an interim status prior to approval). Four states, California, Michigan, Oregon, and Washington, use state standards for lockout/tagout that are the equivalent or somewhat more stringent than federal regulation. Michigan, with its "locks only" provision, appears to have the most unique of all state lockout/tagout rules. (Table 4-B lists OSHA state plan status.)

California

In a 1982 study, the California Occupational Safety and Health Board found that an average of nearly 30 workers were killed and close to 600 workers were disabled annually because of incidents on the job involving electrical shock alone. Many other states are believed to be experiencing similar losses associated with hazardous-energy-release incidents (California Occupational Safety and Health Board, 1982). The state's Division of Industrial Safety conducted studies of electrical injuries and found that unsafe practices contributed to 7 out of every 10 cases. The most frequently cited human errors were

- Use of defective or unsafe tools or equipment
- Failure to de-energize equipment during repairs or inspection
- Taking up an unsafe position near energized equipment
- Unsafe practices by co-workers and others (California Division of Industrial Safety, 1973).

California passed its own version of an Occupational Health and Safety Act in the 1970s. Originally, California general safety regulations under its Title 8 were somewhat less comprehensive than the corresponding federal OSHA standards. The state lockout or blockout provisions were oriented to electrical and mechanical hazards and did not address all of the issues covered in 29 *CFR* 1910.147. Over the past few years, the state has taken steps to adopt detailed standards that meet or exceed current federal energy control criteria. For example, Title 8 of the California Code of Regulations: General Industrial Safety Orders, Group 2 Safe Practices and Personal Protection, Article 7 Miscellaneous Safe Practice, section 3314, Cleaning, Repairing, Servicing, and Adjusting Prime Movers, Machinery, and Equipment did not prescribe employee training or procedure audits as defined in the federal standard.

In December 1991, however, Title 8 was amended to incorporate provisions of the federal OSHA standard 29 *CFR* 1910.147 in order to make the California code at least as effective as the federal regulations. The amendments adopted the same language concerning requirements for tagout device strength, energy-control procedures, documentation of energy-control procedures, and periodic inspections.

Other amendments required employers to provide a statement of the developed energy-control procedures; to document in writing that safe work practices are followed; to conduct periodic inspections by an authorized employee other than those using the procedure; to certify that inspection(s) have been performed; and to inform all outside service personnel (contractors) of the on-site lockout/tagout procedures before they begin work. The standard is provided in Figure 4-5.

California Code of Regulations

§3314. Cleaning, Repairing, Servicing and Adjusting Prime Movers, Machinery and Equipment.

(a) Machinery or equipment capable of movement shall be stopped and the power source de-energized or disengaged, and, if necessary, the moveable parts shall be mechanically blocked or locked to prevent inadvertent movement during cleaning, servicing or adjusting operations unless the machinery or equipment must be capable of movement during this period in order to perform the specific task. If so, the employer shall minimize the hazard of movement by providing and requiring the use of extension tools (e.g., extended swabs, brushes, scrapers) or other methods or means to protect employees from injury due to such movement. Employees shall be made familiar with the safe use and maintenance of such tools by thorough training.

(b) Every prime mover or power driven machine equipped with lockable controls or readily adaptable to lockable controls shall be locked out or positively sealed in the OFF position during repair work and setting-up operations. Machines or prime movers not equipped with lockable controls or readily adaptable to lockable controls shall be considered in compliance with this order when positive means are taken, such as de-energizing or disconnecting the equipment from its source of power, or other action which will prevent the prime mover or machine from inadvertent movement. In all cases, accident prevention signs and/or tags shall be placed on the controls of the machines and prime movers during repair work.

NOTE: For the purpose of this order, "locked out" means the use of devices, positive methods and procedures, which effectively prevent unexpected or inadvertent movement of the machine or materials.

(c) The employer shall provide a sufficient number of accident prevention signs or tags and padlocks, seals or other similarly effective means which may be required by any reasonably foreseeable repair emergency. Signs, tags, padlocks, or seals shall have means by which they can be readily secured to the controls. Tagout device attachment means shall be of a non-reusable type, attachable by hand, self-locking, and non-releasable with a minimum unlocking strength of no less than 50 pounds.

(d) During repair prime movers, machines, or equipment shall be effectively blocked or otherwise secured to prevent inadvertent movement if such movement can cause injury to employees.

(e) On repetitive process machines, such as numerical control machines, which require power or current continuance to maintain indexing and where repair, adjustment, testing, or setting up operations cannot be accomplished with the prime mover or energy source disconnected, such operations may be performed under the following conditions:

(1) The operating station where the machine may be activated must at all times be under the control of a qualified operator or craftsman.

(2) All participants must be in clear view of the operator or in positive communication with each other.

(3) All participants must be beyond the reach of machine elements which may move rapidly and present a hazard to them.

(4) Where machine configuration or size requires that the operator leave his control station to install tools, and where machine elements which may move rapidly, if activated, exist such elements must be separately locked out by positive means.

(5) During repair procedures where mechanical components are being adjusted or replaced, the machine shall be de-energized or disconnected from its power source.

NOTE: "Participant" shall mean any other person(s) engaged in the repair, adjustment, testing, or setting up operation in addition to the qualified operator or craftsman having control of the machine operating station.

(f) An energy control procedure shall be developed and utilized by the employer when employees are engaged in the cleaning, repairing, servicing or adjusting of prime movers, machinery and equipment. The procedure shall clearly and specifically outline the scope, purpose, authorization, rules and techniques to be utilized for the control of hazardous energy, and the means to enforce compliance, including but not limited to, the following:

(1) A statement of the intended use of the procedure;

(2) The procedural steps for shutting down, isolating, blocking and securing machines or equipment to control hazardous energy;

(3)The procedural steps for the placement, removal and transfer of lockout devices or tagout devices and the responsibility for them; and,

(4) The requirements for testing a machine or equipment, to determine and verify the effective-

Figure 4-5. California Standard 3314 (*continued*).
Source: State of California, General Industry Safety Orders.

ness of lockout devices, tagout devices and other energy control devices.

(g) The employer's hazardous energy control procedures shall be documented in writing.

(h) The employer shall conduct a periodic inspection of the energy control procedure at least annually to ensure that the procedure and the requirements of this section are being followed.

(1) The periodic inspection shall be performed by an authorized employee other than the one(s) utilizing the energy control procedures being inspected.

(2) Where lockout is used for energy control, the periodic inspection shall include a review between the inspector and each authorized employee, of that employee's responsibilities under the energy control procedure being inspected.

(3) The employer shall certify that the periodic inspections have been performed. The certification shall identify the machine or equipment on which the energy control procedure was being utilized, the date of the inspection, the employees included in the inspection, and the person performing the inspection.

(i) Whenever outside servicing personnel are to be engaged in activities covered by this section, the on-site employer's lockout or tagout procedures shall be followed.

NOTE: Authority cited: Section 142.3, Labor Code. Reference: Section 142.3, Labor Code.

HISTORY

1. Amendment filed 10-25-74; effective thirtieth day thereafter (Register 74, No. 43).
2. Repealer and new subsections (a), (b) and (c) and amendment of subsection (d) filed 5-12-77; effective thirtieth day thereafter (Register 77, No. 20).
3. Amendment of subsections (c) and adoption of subsections (f)-(i) filed 12-23-91; operative 1-22-92 (Register 92, No. 12).

Figure 4-5. *(concluded)*

Oregon

Oregon followed suit by adopting 29 *CFR* 1910.147 language for lockout/tagout procedures and adding its own rules to make the guidelines more specific. For example, in section 437-02-154 of the Oregon code, the state added the following state rule: "In addition and not in lieu of the definition contained in 1910.147(b) for 'lockout device,' each person's lock shall have either a key or combination which is unique to that device." A very curious turn of events took place regarding the adoption of 29 *CFR* 1910.147 section (f)(3), group lockout or tagout. This provision was not adopted by the Oregon Occupational Safety and Health Division, thereby prohibiting group lockout in Oregon. This will undoubtedly create some unusual interpretations by the state and unique applications by employers when complex energy isolation tasks with many isolating devices are undertaken (Oregon Occupational Safety and Health Division, 1991).

Washington

The Washington state standard WAC 296-24-110, passed in November 1990, updated existing requirements and is also nearly identical to the federal OSHA regulations. It covers the servicing and maintenance of machines and equipment and establishes minimum performance requirements for the control of hazardous energy. It does not cover construction, agriculture, and maritime employment; installations under the exclusive control of electric utilities; exposure to electrical hazards from work on, near, or with conductors or equipment in electric use installations; or oil and gas well drilling and servicing. Like the federal standards, this state plan requires employers to establish a written program to train employees in the lockout/tagout provisions. In addition, the standard is interpreted for enforcement in WISHA Regional Directive 91-9A.

Washington also addresses hazardous energy control (lockout/tagout) in a number of vertical standards. The most notable—safety standards of sawmills and woodworking operations, Chapter 196-78 WAC—contains several interesting provisions in sections 71501, General Provisions, and 71503, Lockout-Tagout. Under the former, direction is given to stop all power-driven machinery before repair, adjustment, or attempts to remove material or refuse. The lockout-tagout section required padlocks for energy isolation to be in use as of July 1, 1982. The standards are reproduced in Figure 4-6.

Michigan

Because Michigan has long been the site of heavy industrial and automotive manufacturing, the state has adopted strong measures protecting workers on the job. The state's Occupational Safety Standards Commission, Department of Labor, developed energy control–related safety standards that were adopted in November 1969 and amended in October 1971. However, in terms of safeguarding machinery and protecting workers from unwanted energy transfers, the state has tended to focus on lockout procedures rather than tagout techniques in its standards. The United Auto Workers note that Michigan is the only state with a locks-only requirement in regard to preventing inci-

Chapter 296-78 WAC
Safety Standards for Sawmills and Woodworking Operation

WAC 296-78-71501, General provisions.

(1) All machinery or other equipment located or used on the premises of the operation or in the processes incidental thereto, shall be provided and maintained with approved standard safeguards, irrespective of ownership.

(2) Machines shall be so located that each operator will have sufficient space in which to handle material with the least possible interference from or to other workers or machines.

(3) Machines shall be so placed that it will not be necessary for the operator to stand where passing traffic creates a hazard.

(4) Aisles of sufficient width to permit the passing of vehicles or employees without crowding shall be provided in all work areas and stock or storage rooms.

(5) All metal decking around machinery shall be equipped to effectively prevent slipping.

(6) All machinery or equipment started by a control so located as to create impaired vision of any part of such machinery or equipment shall be provided with an audible warning device, where such machinery or equipment is exposed to contact at points not visible to the operator. Such devices shall be sounded before starting up unless positive mechanical or electrical interlocking controls are provided which will prevent starting until all such posts are cleared.

(7) A mechanical or electrical power control device shall be provided at each machine which will make it possible for the operator to stop the machine feed without leaving his position at the point of operation.

(8) All machines operated by means of treadles, levers, or other similar devices, shall be provided with positive and approved nonrepeat devices except where such machine is being used as an automatic repeating device.

(9) Operating levers and treadles on all machines or machinery shall be so located and protected that they cannot be shifted or tripped accidentally.

(10) All power driven machinery shall be stopped and brought to a complete standstill before any repairs or adjustments are made or pieces of material or refuse removed, except where motion is necessary to make adjustments. [Statutory Authority: RCW 49.17.040, 49.17.050 and 49.17.240. 81-18-029 (Order 81-21), 296-78-71501, filed 8/27/81.]

WAC 296-78-71503, Lock out—Tag out.

(1) To avoid accidental activation of machinery, electrical devices or other equipment which could create a hazardous condition while performing maintenance, repair, cleanup or construction work, the main disconnect(s) (line circuit breakers) shall first be locked out and tagged in accordance with the following provisions:

(2) Effective date. Effective July 1, 1982, only padlocks or other equivalent protective devices shall be used for locking out the main (disconnect(s) (line circuit breakers) of machinery, electrical devices or other equipment that is shut down while maintenance, repair, cleanup, construction work or other type of work is done to the equipment. Tags shall be used to supplement the padlocks or other equivalent protective devices, and shall be used only for informational purposes.

(3) Padlocks, tags or equivalent protective devices to be supplied. The employer shall supply and the employee(s) shall use as many padlocks or other equivalent protective devices as are necessary to effectively lock out all affected equipment.

(4) Lock out plan. An effective lock out plan shall be formulated in writing and all concerned employees so informed. The plan shall contain specific procedures for locking out equipment, information to be contained on supplemental tags and specific procedures for unlocking equipment after requires, cleanup, etc., have been completed.

(5) Informational tags. Tags used for providing supplemental information with lock out padlocks or other equivalent protective devices shall contain the name of the person authorizing placement, reason for placing, date, signature of person placing tag and such other relative information as deemed necessary by the person placing the tag.

Figure 4-6. Washington Standards 71501 and 71503. (*continued*)
Source: State of Washington, Safety Standards.

(6) Lock out by pushbutton only. Locking out a machine or item of equipment by use of a pushbutton or other local control device only will not be acceptable as meeting the intent of these rules.

(7) Coordination of locking out devices. When repair, adjustment, cleanup, maintenance or construction work is necessary and the lock out procedures must be followed by any person not familiar with all power sources or material entry sources to any area involved, that person shall consult with the operator, supervisor, or some person that is capable of informing him of proper lock out procedures and supplemental tagging information.

(8) Lock out before removing guards. Equipment shall be stopped and locked out before employees remove guards or reach into any potentially hazardous area. The only exception to this rule will be when equipment must be in motion in order to make proper adjustments.

(9) Removal of lock outs. Each person actively engaged in the repair, maintenance, cleanup, etc., shall lock out the affected equipment and place the informational tag. Upon completion of the work and reinstallation of the guards, that person shall personally remove his lock and tag, except when it is positively determined that an employee has left the premises without removing his lock and tag, other persons may remove the locks and tags in accordance with a procedure formulated by each firm and approved by the division of industrial safety and health.

(10) Valves to be locked and tagged out. Each valve used to control the flow of hazardous materials into, or used to activate the equipment being worked on, shall be locked and tagged out.

(11) Piping systems deactivated. Prior to working on piping systems containing pressurized or hazardous materials, the valve(s) controlling the flow to the affected area shall be locked and tagged out. The piping in the area to be worked on shall be drained and purged, if needed. If the piping contains hazardous materials, the piping shall be isolated from the work area by the insertion of blank flanges in the piping system.

(12) Pipelines without valves. If pipelines or ducts are constructed without valves or closures that can be locked out, the lines or ducts shall be broken at a flange and a blank flange inserted to stop accidental flow of any hazardous material.

(13) Testing after lock out. After locking out and tagging equipment, a test shall be conducted to ascertain that the equipment has been made inoperative or the flow of hazardous material has been positively stopped. Precautions shall be taken to ascertain that persons will be subjected to hazard while conducting the test if power source or flow of material is not shut off.

(14) Temporary or alternate power to be avoided. Whenever possible, temporary or alternate sources of power to the equipment being worked on shall be avoided. If the use of such power is necessary, all affected employees shall be informed and the source of temporary or alternate power shall be identified. [Statutory Authority: RCW 49.17.040, 49.17.050 and 49.17.240. 81-18-029 (Order 81-21), 296-78-71503, filed 8/27/91.]

Figure 4-6. (*concluded*)

dents caused by unwanted transfers of energy. Various changes have occurred over the years regarding hazardous energy control, including the following:

- 1969–71: Guards for Power Transmission, Part 7 of state safety standards R 408.855 of the Complied Laws, requires employers to de-energize machinery during maintenance or repair work and apply lockout procedures.
- 1971–73: Part 7 of the safety standards is restated in revised standards with no change in lockout procedures and no inclusion of tagout methods.
- 1973–74: State standards are revised to require locks only on machinery, with minor exceptions for cord/plug equipment of 110 V or less.
- In 1993, the state adopted standards similar to federal OSHA standards but stipulated that state regulations have precedence when they are more stringent than provisions found in the federal standards. The locks-only provision for energy isolation is still in force.

GOVERNMENTAL ACTION: FEDERAL

Before 1970, no general, all-inclusive consensus standard existed for lockout/tagout or other means of disabling machines or equipment to protect workers who were doing maintenance or service work. Instead, as discussed previously, various employers, labor unions, states, trade associations, and safety organizations had developed their own standards regarding lockout/tagout methods and systems. Some of the more complete standards were established in industrial states such as

Michigan and in more progressive states, particularly California and Washington. However, enforcement was spotty, and compliance inspections were rarely performed. Even though consensus standards developed by employers and associations were created in an attempt to make lockout/tagout methods more uniform, compliance was strictly voluntary.

When OSHAct required the development or adoption of occupational safety and health standards, OSHA did not create a specific section for lockout/tagout procedures in its General Industry standards, 29 *CFR* part 1910, which became effective on August 27, 1971. Instead, OSHA adopted various lockout-related provisions of national consensus standards that had been developed for specific industries or types of equipment. These provisions were scattered in bits and pieces throughout several sections in 29 *CFR* 1910:

- 1910.178 Powered industrial trucks
- 1910.179 Overhead and gantry cranes
- 1910.181 Derricks
- 1910.213 Woodworking machinery
- 1910.217 Mechanical power presses
- 1910.218 Forging machines
- 1910.522 Welding, cutting, and brazing
- 1910.261 Pulp, paper, and paperboard mills
- 1910.262 Textiles
- 1910.263 Bakery equipment
- 1910.265 Sawmills
- 1910.272 Grain handling
- 1910.399 Electrical.

As a result, no uniform coverage was provided by the regulations for locking out and tagging out machines and equipment. Several other weaknesses were evident in the consensus standards that OSHA adopted. Inconsistencies existed among regulations for different equipment and industries and among those for different types of equipment in the same industry. For example, in some instances the regulations require certain equipment to have the capability of being "locked out" but do not require workers to use the lockout control. None of the existing standards addressed the need for written procedures outlining how to apply and release energy-control devices and measures. Neither did they address such issues as considering energy-control measures in the selection of hardware, the need for safety communication, periodic inspections, assignment of duties, documenting procedures, and training employees in energy-control methods. This lack of a general standard and the gaps in existing regulations contributed to industry's ineffectiveness at reducing the number of injuries and fatalities workers experienced related to hazardous energy.

In addition, for enforcement OSHA had to rely primarily on use of the General Duty clause (section 5[a][1] of the act) to ensure that employers provided safe working conditions for their employees to guard them from the release of hazardous energy. Like similar voluntary compliance efforts, this approach met with limited success. If an employer challenged a citation, OSHA bore the burden of proving that a hazard was a "recognized" hazard and that it was causing or could cause serious injury or death.

A review of the lockout/tagout provisions that OSHA adopted from consensus standards seemed to indicate that the consensus groups had one of two primary concerns in mind when originally developing their standards. The concerns involved the need either (1) to provide equipment with the physical means or capability to isolate energy sources during maintenance and repair activities or (2) to choose which control measures (locks or tags) were to be provided and used on specific machines, equipment, or processes covered by the standard. In the first case, the control is specified, but its use is not specifically required; in the second case, the use of locks, tags, or other energy controls are advised but not always in a consistent manner.

To remedy the deficiencies in enforcement and to fill a significant gap in lockout/tagout coverage, OSHA developed and continues to develop comprehensive standards for energy control in general industry. So far, these include 29 *CFR* 1910.147, Control of Hazardous Energy Sources, Lockout/Tagout; 29 *CFR* 1910.331–.335, Electrical Safety-Related Work Practices; and 29 *CFR* 1910.269, Electrical Power Transmission, Generation, and Distribution. OSHA's intention is not to replace existing lockout/tagout requirements in other sections but to supplement and support them by requiring that employers establish a well-defined procedure and training program for workers.

NIOSH Guidelines

Congress established the National Institute of Occupational Safety and Health (NIOSH) to carry on research, educational activities, and other functions related to occupational health and safety issues. NIOSH is under the jurisdiction of the U.S. Department of Health and Human Services. Section 22(d)(2) of OSHAct authorized the director of NIOSH to make recommendations to OSHA concerning improved safety and health standards in the workplace.

In 1983, NIOSH produced a document entitled *Guidelines for Controlling Hazardous Energy During Maintenance and Servicing*. The guidelines were developed by making site visits to industrial facilities, collecting and analyzing information related to maintenance techniques and processes, evaluating accident case studies and incident reports, searching available literature, and reviewing other pertinent data. The emphasis was on the recognition of energy hazards and safe work procedures. Boeing Aerospace created the document, and representatives from safety and training organizations, industry, and labor reviewed it. The agency made an attempt to have all viewpoints represented in the document.

The guidelines were created to provide a "logical system for performing maintenance and servicing activities safely, and recognizes that such activities can be performed safely with energy present, with energy removed, and while re-energizing" (NIOSH, 1983). NIOSH guidelines were among the first to recognize that not all operations could be reduced to a zero energy state. The agency provided a logic tree that presented a step-by-step diagram for controlling hazardous energy that could be used in the formulation of specific maintenance and servicing procedures (see Appendix 3 for the logic tree diagram).

According to the NIOSH guidelines, the basic decision to be made before maintenance work begins is whether the task can be accomplished safely with or without energy present. Factors in this decision include the following concepts:

- Energy is always present.
- Energy is not always dangerous.
- Danger occurs only when the amount of energy released exceeds human tolerances.

Before specific energy-control measures and devices are developed, all energy sources should be identified, analyzed separately, and then analyzed in combination with other energy sources (mechanical, thermal, chemical, and so on) that may be present.

NIOSH concluded the document by stating that while it found no evidence that one type of energy control was clearly superior over another, everyone involved recognized the need to use some means to prevent hazardous-energy-release incidents. Major hazard causes included maintenance activities initiated without de-energizing equipment or controlling the energy hazard and inadequate energy blocking or isolation; residual energy was not dissipated, and energy was accidentally activated. The worker population at risk could not be defined precisely, because employers could make the controls available to all workers, to those engaged in specific activities such as construction or maintenance, or to maintenance personnel only. NIOSH found that at the time (1983) most federal and state hazardous-energy-control regulations were inadequate and inconsistent. State and federal safety standards did not allow the performance of maintenance tasks with power on, even under routine conditions or when equipment could not be completely de-energized.

NIOSH identified areas for additional research in the area of hazardous energy control:

- Studies should be conducted to develop universal procedures for recommended guidelines, to identify problems created by the guidelines, to determine the effectiveness of different hazard control methods in preventing injuries, and to identify the costs of implementing guidelines.
- Guidelines should be developed for other maintenance hazards such as unsafe access routes to and from maintenance areas, toxic or caustic materials that pose a danger to maintenance personnel, and selection and use of ancillary equipment. These and other areas are not covered under guidelines for controlling hazardous energy release.
- Evaluation of the potential hazards in the application of new technology in the workplace should be undertaken, including robotic machines, laser equipment, ultrasonic applications, and explosive forming. The United States must use these new forms of technology if it is to compete on equal footing in the world market. Research must be done to ensure that their application does not endanger workers or that current hazard-control methods are updated to safeguard those who must work on or near new machines, equipment, and processes.

Examples of Alternate Methods of Isolating or Blocking Energy and Securing the Point(s) of Control, from the NIOSH document, is reprinted in full in Appendix 2.

Lockout/Tagout Standard (29 *CFR* 1910.147)

Regulation 29 *CFR* 1910.147, The Control of Hazardous Energy (Lockout/Tagout), was promulgated on September 1, 1989, by the U.S. Department of Labor, Occupational Safety and Health Administration. The standard focuses primarily on procedures for effective control when dealing with potentially hazardous energy sources. It specifies employer requirements to ensure worker safety when servicing and performing maintenance "of machines or equipment in which the unexpected energization, startup, or release of stored energy could cause injury to employees" (29 *CFR* 1910.147[a][1][i]).

This regulation outlines the minimum requirements to establish energy-control programs that "ensure that before any employee performs any servicing or maintenance on a machine or equipment where the unexpected energizing, startup or release of stored energy could occur and cause injury, the machine or equipment shall be isolated from the energy source, and rendered inoperative" (29 *CFR* 1910.147[c][1]). Energy-control programs can be accomplished through methods such as lockout/tagout procedures, blocking and guarding machine parts, periodic inspections, training and communication, lockout/tagout device specifications and application, and so on. OSHA estimated that the standard would prevent about 122 fatalities, 28,400 lost workday injuries, and 31,900 non–lost workday injuries each year.

Industry sectors impacted by hazardous energy risks. The industries that were impacted most by hazardous energy risks included agriculture, forestry, and fishing; mining; construction; manufacturing; transportation and public utilities; wholesale and retail trades; finance, insurance, and real estate; and service sectors. Although employees in all industries are exposed to some hazardous-energy-release risks, research showed that the majority of accidents and injuries occur in manufacturing. In addition to routine maintenance and servicing tasks, workers are in danger when they try to unjam or adjust machines or equipment and put parts of their body in the machinery while it is still running. OSHA cataloged the main causes of accidents:

- Servicing machines while the equipment is still operating
- Failing to ensure that power was off or adequately controlled
- Inadvertently activating machinery or equipment while it was being serviced or worked on
- Employer failing to require workers to use a procedure or control method
- Employer failing to train workers in procedures to control hazardous energy.

Major issues. During the development of the standard, four major issues were raised that OSHA had to address in the final rule. These had to do with the methods of energy control to be used and how detailed the standard should become regarding performance requirements.

1. *Should OSHA require the use of locks, locks and tags, or tags alone to control potentially hazardous energy?* This was the most hotly debated issue in the proceeding. Most favored locks over other methods, although some strongly supported locks and tags as better safeguards for workers. In the Final Rule, 29 *CFR* 1910.147, OSHA determined that lockout was a surer means of ensuring that machinery and equipment remains de-energized and should be the preferred method. However, OSHA also recognized that tagout would need to be used where lockout devices cannot be applied.
2. *Should OSHA require employee participation in the development of lockout procedures and the training programs required by this standard?* Labor unions and others believed that OSHA should require employees and employee representatives to take part in developing and implementing lockout programs. OSHA believed that a specific provision mandating employee involvement was not necessary except in cases where employees would be exposed to toxic substances and harmful physical agents. Under this standard, OSHA places responsibility on the employer for developing procedures and taking steps to create an effective hazardous-energy-control system.
3. *Should OSHA change the scope and application statements of this standard in this Final Rule to cover construction, maritime, agriculture, electric utility, and oil and gas well–drilling industries?* OSHA proposed exempting these industries from the 29 *CFR* 1910.147 standard because of their unique circumstances and work practices. Trying to accommodate these industries would greatly complicate the development of a generic energy control standard for general industry. Nevertheless, OSHA is working on projects to cover the special safety needs of workers in construction, agriculture, maritime, electric utility, and oil and gas well–drilling jobs.
4. *Should OSHA state the requirements of this final standard in performance language?* Some objected to the use of performance language rather than specification terms because they felt it allowed employers too much leeway in complying with the standard. Most of the commentors and OSHA, however, stressed the need for flexibility; the performance language was retained in the final standard.

Industry sectors affected by Final Rule. The Final Rule affects most employment covered by OSHA under part 1910 with the exception of activities specifically excluded from coverage and employment for which OSHA is developing separate coverage (for example, gas and oil field services industry). According to OSHA estimates, the Final Rule affects activities in nearly 1.7 million establishments employing about 39 million workers. These establishments have been divided into high-impact, low-impact, and zero- or negligible-impact groups.

- The high-impact group consists of all manufacturing industries whose workers face significant risks from hazardous-energy-release incidents.
- The low-impact group includes transportation; utilities; wholesale trade; retail food stores; and service industries such as personal services, business services, automotive repair, miscellaneous repair, and amusement services. The risks of injuries or fatalities due to hazardous energy release are much less in these industries.
- The negligible-impact group consists of industries that have little potential for lockout- or tagout-related incidents. These include retail trade, finance, insurance, real estate, service, and public administration firms.

OSHA's analysis focused on the regulatory effects of the Final Rule on high- and low-impact firms.

Population at risk. In estimating the number of workers at risk from hazardous energy releases, OSHA classified "at risk" jobs as those being held by workers who would actually perform lockout or tagout operations. In 1989, OSHA estimated that 2 million workers in high-impact industries and 1 million workers in low-impact industries are at risk of injury from unexpected start-up of machines or equipment or release of stored energy. The risk appears highest for those working as craft workers, machine operators, and laborers, particularly if they work on packaging and wrapping equipment, printing presses, and conveyors. These and other employees at risk may have to alter their work patterns to comply with the Final Rule.

Cost of compliance. In 1989, OSHA estimated that full compliance with the standard would cost 631,000 firms $214.3 million during the first year of implementation and $135.4 million in subsequent years. The costs can be summarized by category:

- For locks, tags, and other hardware, first-year costs are estimated at $18.5 million and recurring annual costs $8.9 million thereafter.
- For voluntary equipment modification to facilitate lockout or tagout, first-year costs may be $27 million with no subsequent annual costs. Work practice modifications are estimated at $102.7 million, and the same amount is estimated for recurring annual costs.
- For planning and implementation of lockout or tagout procedures, OSHA calculated first-year costs at $35.2 million and annual recurring costs at $21 million.
- For employee training, first-year costs are estimated at $31 million, with annual recurring costs of $3.6 million.

For firms not currently using adequate lockout or tagout procedures, first-year costs of compliance for manufacturing firms would range from $120 per firm for those with 20 employees or less to over $28,000 for those with more than 250 employees. Low-impact firms would incur first-year costs of about $169 per establishment.

Definitions of terms. Definitions in the OSHA lockout/tagout standard include the following:

- *Capable of being locked out.* An energy-isolating device is capable of being locked out either if it is designed with a hasp or other attachment or integral part to which, or through which, a lock can be affixed or if it has a locking mechanism built into it.
- *Energy-isolating device.* A mechanical device that physically prevents the transmission or release of energy, such as a manually operated circuit breaker or a disconnect switch. The term does not include a push button, selector switch, or other control circuit devices.
- *Energy source.* Any source of electrical, mechanical, hydraulic, pneumatic, chemical, thermal, or other energy.
- *Lockout.* The placement of a lockout device on an energy-isolating device, in accordance with an established procedure, to ensure that the energy-isolating device and the equipment being controlled cannot be operated until the lockout device is removed.
- *Lockout device.* A device that uses a positive means such as a lock, either of key or combination type, to hold an energy-isolating device in the safe position and prevent the energizing of a machine or equipment. The standard also gives specific descriptions of the types of lockout devices that should be used.
- *Tagout.* The placement of a tagout device on an energy-isolating device, in accordance with an established procedure, to indicate that the energy-isolating device and the equipment being controlled may not be operated until the tagout device is removed.
- *Tagout device.* A prominent warning device, such as a tag and a means of attachment, that can be securely fastened to an energy-isolating device, in accordance with an established procedure, to indicate that the energy-isolating device and the equipment being controlled may not be operated until the tagout device is removed.

Similar to the section on lockout devices, the standard gives specific guidelines for the types of tagout devices that should be used and how they should be attached to machines or equipment.

- *Lockout/tagout.* (i) If an energy isolating device is not capable of being locked out, the employer's energy control program under paragraph (c)(1) of this section shall use a tagout system.

Key elements of the standard. Key elements of the lockout/tagout standard are summarized as follows:

- Scope, application, and purpose paragraphs describe what industries the standard covers, the situations covered and not covered by the regulations, and the purpose for which the standard was designed.
- General regulations describe
  - Energy-control programs that employers must establish
  - Lockout/tagout requirements and when to use locks and tags to control energy sources
  - Full employee protection in a lockout/tagout system
  - Energy-control procedures that must be developed to create an effective lockout/tagout program
  - Protective materials and hardware that can be used or must be used to secure movable parts and ensure that hazardous energy is controlled
  - Periodic inspections to ensure that lockout/tagout devices are in good working order and that employees are following procedures correctly
  - Training and communication of lockout/tagout procedures to all employees

- Energy isolation to be performed only by authorized personnel
- Notification of employees of the application and removal of lockout or tagout devices.

- The standard lists the key steps in the application of effective energy-control devices. Only authorized personnel should apply and remove locks, tags, or other devices to prevent unwanted energy releases.
- The standard also describes procedures for releasing machinery or equipment from lockout or tagout and for testing or positioning of machines, equipment, or components to ensure safe start-up. These additional provisions also cover outside personnel, group lockout or tagout, and coordinating energy control among various shifts or personnel changes.

Court challenge and OSHA's response. When 29 *CFR* 1910.147 was first published, both labor and industry representatives filed petitions for review of the standard in the U.S. Court of Appeals in the District of Columbia Circuit. The court rejected many of the petitions but required OSHA to explain its reasoning on several issues raised by the United Auto Workers (UAW) and National Association of Manufacturers (NAM). Both organizations had found the standard lacking: UAW on the basis that it was too broad to be reasonably enforced and that it did not protect workers sufficiently, and NAM on the grounds that it was not cost-effective. In particular, four objections were raised:

- The criteria used by OSHA in setting safety standards under section 3(8) were unclear.
- The standard preferred lockout over tagout procedures.
- The rule was applied too broadly to all general industry workplaces in which hazardous servicing and maintenance operations take place, ignoring differences among industries. OSHA should have conducted industry-by-industry risk analyses and applied the standard only to those industries for which there were specific findings of significant risk to workers.
- NAM contended that the modest safety gains from the standard did not justify the expense of compliance.

The OSHA agency replied to these objections by using a supplemental statement of reasons for the remand issues raised by the U.S. Court of Appeals, Washington DC.

1. **Criteria for standards.** OSHA reexamined its criteria used to set safety standards and, more particularly, the criteria for the lockout/tagout standard. According to OSHA guidelines, a standard must (a) substantially reduce a significant risk of material harm, (b) be technologically feasible either with existing equipment or with new equipment developed to comply with the standard, (c) be economically feasible, (d) be at least as protective as existing standards, (e) be supported by the evidence in the rule-making record and be consistent with prior agency action, and (f) if different, must effectuate the act's objectives better than the existing national consensus standard. These criteria limit the agency's safety rule–making ability and establish that the OSHAct does not delegate excessive rule-making authority to the agency.
2. **Lockout preferred over tagout.** The standard's preference for lockout over tagout is warranted by the fact that lock-based programs are less vulnerable to human error and can be expected to save more lives and prevent more injuries than tag-based programs.
3. **Application across a general industry spectrum.** Even though available injury data show a wide range of accident rates for different industrial sectors, OSHA contends that "lockout/tagout hazards are so pervasive and arise during such a wide variety of servicing and maintenance activities that any attempt to define the standard's scope by employer sector within general industry would result in the standard excluding some hazardous servicing and maintenance activities from coverage" (FR, 1993).

Simply because there are low reported accident rates or concentrations of equipment in some employment sectors does not mean that workers in those sectors are not exposed to serious risks. For example, railroads, public transit, communications, public utilities, and amusement and recreation services all have servicing and maintenance activities that require lockout/tagout protection. OSHA maintained that erring on the side of caution rather than excluding certain sectors was in the workers' best interests and was consistent with the general duty provision in OSHAct that required all employers to provide safe working conditions for employees. The broad application of the standard across industries emphasized the standard's coverage in terms of performance rather than strictly according to SIC employment codes. The standard does not apply to servicing and maintenance that present minimal and readily controlled risk, such as work on electrical equipment that can be de-energized by simply unplugging it (such as a vacuum cleaner or fan). As a result, each covered employer's burden is determined by the frequency and complexity of servicing actually undertaken.

For similar reasons, OSHA decided not to limit the standard to particular equipment: ". . . the standard applies only when the unexpected energization or release or stored energy could cause injury to employees. Machines and equipment that present no hazard are excluded from coverage." In this way, OSHA responded to some of NAM's concerns without reducing the breadth of coverage provided by 29 *CFR* 1910.147.

4. **Economic cost/benefit relationship.** OSHA evaluated two approaches to regulatory decision making: formal cost-benefit analysis and risk-risk analysis. The agency already performs an extensive analysis that includes estimations of compliance costs and of deaths and injuries prevented as well as supporting rule-making comments, existing consensus standards, and policy changes. This approach, OSHA has stated, addresses any concerns that the standard is too broad or that it imposes large costs for relatively small safety gains.

   In particular, OSHA found that the relationship between the benefits of a lockout/tagout program and the costs it imposes on industries or individual companies is reasonable. According to OSHA research, the standard should save about 122 lives and 28,400 lost workday injuries a year at a cost of $214 million the first year and $135 million the second. The lockout/tagout standard's Regulatory Impact Analysis relied on OSHA's cost-effectiveness studies to demonstrate the economic tradeoffs of the final rule. The agency believed that its approach ensured that the standard reduces significant risk at the least cost to employers.

Electrical Safety-Related Work Practices Standard (29 *CFR* 1910.331–.335)

In its original version, the OSHA standards were inadequate because of incomplete industry coverage and incomplete and inconsistent criteria within the standards. In 1990, 29 *CFR* 1910.331–.335, Electrical Safety-Related Work Practices, was amended to correct many of these weaknesses, and all amended work rules took effect in August 1991. The U.S. Department of Labor, in July 1991, released inspection procedures and interpretive guidelines to ensure uniform enforcement of standards for these regulations.

This new rule was designed to protect more than 3 million workers from electric shock, burns, and other electrical accidents. OSHA predicted that the rule would prevent 78 deaths and 3,400 injuries each year. The new regulations covered workers in manufacturing, transportation, retail and wholesale trades, and finance and service sectors who work with electrical equipment during the course of their jobs and face the risk of injury through electrical shock.

OSHA found that more than half of the workplace electrocutions were due to unsafe work practices. Although most workers know high voltage is dangerous, few may realize that low-voltage alternating current can also cause fatalities by affecting the heart. Working with or near electrical equipment and energy sources requires careful training in safety procedures. To remedy the lack of safe work practices, OSHA's new rule required employers to follow prescribed work practices that include

- De-energizing electrical equipment as the primary way to protect workers
- Locking out or tagging out electrical sources to prevent equipment from being accidentally turned on or from discharging stored energy
- Limiting work on energized equipment to qualified persons
- Keeping vehicles operating near energized overhead lines beyond a minimum safe distance
- Prohibiting the use of portable metal ladders around exposed energized parts
- Requiring special protective measures for portable electrical equipment such as prohibiting the alteration of plugs and outlets.

The rule provides guidelines for electric power and lighting circuits, handling test equipment, safe use and maintenance of personal protective gear, and training requirements that must be tailored according to the worker's risk of being electrocuted or shocked.

OSHA estimated that first-year costs of implementing the rule to total $74.6 million, of which $71.5 million is accounted for by training requirements. Annual recurring training costs are estimated at $17.9 million. The lockout/tagout provisions will make up $3 million of the first-year totals and will cost nearly $2.5 million for each following year. The new rule amended provisions in OSHA's general industry standards, making the regulations more uniform and less redundant by removing those that referred to the 1971 National Electrical Code.

The key elements of the new rule are as follows:

- 1910.331, Safety-Related Work Practices: The provisions of 1910.331–335 cover electrical safety-related work practices for both qualified workers (who have training in avoiding electrical hazards) and unqualified persons (those with little or no such training) who work on, near, or with premises wiring, wiring for connection to supply, other wiring, and optical fiber cable.
- 1910.332, Training: This provision contains training requirements that apply to employees who face a risk of electrical shock that is not reduced to a safe level by the electrical installation requirements of 1910.303–.308. The occupational categories covered by this rule include blue-collar supervisors, electrical and electronic engineers and technicians, electricians, industrial machine operators, mechanics, painters, and welders.
- 1910.333, Selection and Use of Work Practices: This section of the rule requires safe work practices to prevent electric shock or other injury to be used whenever work is performed near or on equipment or circuits that are or may be energized. These work practices must be tailored to the nature and extent of the hazards. They cover working on or near exposed de-energized or energized parts, including the application of lockout/tagout devices, by qualified and unqualified persons.
- 1910.334, Use of Equipment: This rule covers safe work practices for portable electric equipment, elec-

tric power and lighting circuits, test instruments and equipment, and occasional use of flammable or ignitible materials.

- 1910.335, Safeguards for Personnel Protection: This section covers use of protective equipment and alerting techniques to protect employees from hazards that could cause injury due to electric shock, burns, or failure of electric equipment parts.

Electric Power Generation, Transmission, and Distribution Standard (29 *CFR* 1910.269)

Although describing lockout/tagout provisions for general industry, 29 *CFR* 1910.147 had exempted workers in the electric utilities industry, among others. Specifically, the standard made the following exclusions:

> (B) Installations under the exclusive control of electric utilities for the purpose of power generation, transmission, and distribution, including related equipment for communication or metering. (C) exposure to electrical hazards from work on, near, or with conductors or equipment in electric utilization installations which is covered by subpart S of this part.

To more fully protect employees of electric utilities, in January 1993 OSHA published a Final Rule, 29 *CFR* 1910.269, Electric Power Generation, Transmission, and Distribution. The standard affects about 3,200 companies and organizations that make up the electric utility industry along with contractor line crews and line-clearance tree trimmers. OSHA's standard comes nearly 10 years after such provisions were first put forward by labor and industry organizations. OSHA based the new rule on a draft standard that had been jointly submitted by the Edison Electric Institute and the IBEW, recent ANSI consensus standards, and current OSHA requirements for telecommunications and construction workers. Most of the final rule provisions were effective May 31, 1994.

When implemented, 29 *CFR* 1910.269 will benefit an estimated 666,000 employees in utility-related industries. According to OSHA estimates, the proposal would prevent from 24 to 28 deaths a year. The standard may cost industry $20.7 million, with utilities likely to pay $16.3 million a year; contractor line crews, $1.6 million; and line-clearance tree-trimming contractors, $2.8 million.

Key elements of the Final Rule. The standard contains a wide range of safety-related topics affecting training for qualified and unqualified workers, medical services, job planning, hazardous energy control (lockout/tagout), work in enclosed spaces, and trenching and excavation operations. Other provisions include

- Sections on personal protective equipment, ladders, platforms, stepbolts, manhole steps, portable power tools, live-line tools, and material handling.
- Provisions that deal with exposed parts of energized lines or equipment, procedures for de-energizing lines, protective grounding methods, high-voltage and high-power testing lines and equipment, safety requirements for mechanical equipment, and work practices involving overhead lines or equipment.
- Sections on communication facilities (for example, microwave signaling) as they relate to electric power generation, transmission, and distribution, listing safe work practices for walking and working surfaces and special requirements for hazards unique to electric generation, transmission, and distribution.

In March 1994, Edison Electric Institute, Tampa Electric Co, Consolidated Edison Co. of New York Inc. and Florida Power Co. filed a petition for review with the U.S. Court of Appeals for the Eleventh Circuit (*Edison Electric Institute v. OSHA, CA 11, No. 94-2389*). The suit follows an unsuccessful request by EEI to OSHA to delay the effective date of most of the standard's provisions until January 1995.

In addition, 29 *CFR* 1910.137, Electrical Protective Equipment, which discusses insulating blankets, matting covers, line hose, gloves, and rubber sleeves, was revised and reissued as a final standard.

The specific provisions in section (d) hazardous energy control (lockout/tagout) procedures are summarized in Figure 4-7.

Mine Safety and Health Administration

Mine safety regulation has historically been administered by the Bureau of Mines within the U.S. Department of the Interior. However, because this agency was also charged with promoting the mining industry, critics claimed there was an inherent conflict of interest within the department. The establishment of the Mine Enforcement Safety Administration (MESA) in 1973 failed to resolve this conflict. As a result, in 1977, the Mine Safety and Health Act (MSHA) was passed, which in effect transferred primary responsibility for mine health and safety administration to the U.S. Department of Labor.

MSHA makes a distinction between health research and safety research. NIOSH is responsible for the development of standards and for miner health research, in cooperation with MSHA. Miner safety research is the responsibility of the U.S. Department of the Interior, and mine inspector training falls under the authority of the U.S. Department of Labor.

Safety and health standards are found in 30 *CFR*, Mineral Resources, subchapters N and O. More specifically, the standards regarding hazardous energy are placed in various subparts relating to electrical, machinery/equipment, and miscellaneous safety issues. Although MSHA does not prescribe any broad lockout/tagout procedures comparable to 29 *CFR* 1910.147, it does contain a number of specific

(d) Hazardous energy control (lockout/tagout) procedures—(1) Application. "The provisions of this paragraph apply to the use of lockout/tagout procedures for the control of energy sources in installations for the purpose of electric power generation, including related equipment for communication or metering. Locking and tagging procedures for the deenergization of electric energy sources which are used exclusively for purposes of transmission and distribution are addressed by paragraph (m) of this section."

NOTE 1: Installations in electric power generation facilities that are not an integral part of, or inextricably commingled with, power generation processes or equipment are covered under § 1910.145 and Subpart S of Part 1910.

NOTE 2: Lockout and tagging procedures that comply with paragraphs (c) through (f) of § 1910.145 of Part 1910 will also be deemed to comply with paragraph of this section if the procedures address the hazards covered by paragraph (d) of this section.

(2) *General.*

(i) The employer shall establish a program consisting of energy control procedures, employee training, and periodic inspections to ensure that, before any employee performs any servicing or maintenance on a machine or equipment where the unexpected energizing, start up, or release of stored energy could occur and cause injury, the machine or equipment is isolated from the energy source and rendered inoperative.

(ii) describes the requirements for an employer's energy control program.

(iii) requires procedures to be developed, documented and used.

(iv) requires the procedure to clearly and specifically outline the scope, purpose, responsibility, authorization, rules, and techniques for energy control and the measures to enforce compliance.

(v) requires the employer to conduct an inspection of the procedure at least annually.

(vi) describes the requirements for employee training.

(vii) describes employee training requirements on tag limitations.

(viii) describes circumstances where employee re-training is required.

(ix) defines requirements for training records.

(3) *Protective materials and hardware.* This section describes the types of lockout/tagout devices and their design performance criteria.

(4) *Energy isolation.* Lockout and tagout device application and removal may only be performed by the authorized employees who are performing the servicing or maintenance.

(5) *Notification.* Affected employees shall be notified by the employer or authorized employee of the application and removal of lockout or tagout devices. Notification shall be given before the controls are applied and after they are removed from the machine or equipment.

NOTE: See also paragraph (d)(7) of this section, which requires that the second notification take place before the machine or equipment is reenergized.

(6) *Lockout/tagout application.* This section describes the elements, actions, and sequence to be included in established energy control procedures.

(7) *Release from lockout/tagout.* Describes actions by authorized employees to ensure that the work area is clear of non-essential equipment, components are operationally intact, employees safely positioned, and affected employees notified before release.

(8) *Additional requirements.*

(i) describes the sequence of actions of the lockout/tagout devices must be temporarily removed while the machine/equipment is re-energized for testing or positioning.

(ii) describes group lockout/tagout requirements.

(iii) describes shift/personnel change requirements.

(iv) describes contractor/employer communication requirements.

(v) describes requirements for situations where energy isolating devices are installed centrally under the control of a system operator.

Figure 4-7. 29 *CFR* 1910.269, lockout/tagout section digest (*continued*).
Source: Occupational Safety and Health Administration, 29 *CFR* 1910.269, *Electric Power Generation, Transmission, and Distribution.*

requirements to de-energize machinery, equipment, and processes and makes reference to lockout of energy sources. See Figures 4-8 and 4-9 for standards extracted from 30 *CFR*, subchapters N and O, illustrating the surface and underground mining approach to the control of hazardous energy.

SUMMARY

- Lockout/tagout codes of safe practice must deal with activities involving machines, equipment, or processes that occur under energized conditions with guards temporarily removed or where some part of

Part 56—Safety and Health Standards: Surface Metal and Nonmetal Mines

Subpart K—Electricity
Section 56.12016 Work on electrically-powered equipment

Electrically powered equipment shall be deenergized before mechanical work is done on such equipment. Power switches shall be locked out or other measures taken which shall prevent the equipment from being energized without the knowledge of the individuals working on it. Suitable warning notices shall be posted at the power switch and signed by the individuals who are to do the work. Such locks or preventive devices shall be removed only by the persons who installed them or by authorized personnel.

Section 56.12017 Work on power circuits

Power circuits shall be deenergized before work is done on such circuits unless hot-line tools are used. Suitable warning signs shall be posted by the individuals who are to do the work. Switches shall be locked out or other measures taken which shall prevent the power circuits from being energized without the knowledge of the individuals working on them. Such locks, signs, or preventative devices shall be removed only by the person who installed them or by authorized personnel.

Subpart M—Machinery and Equipment
Section 56.14105 Procedures during repairs or maintenance

Repairs or maintenance of machinery or equipment shall be performed only after the power is off, and the machinery or equipment locked against hazardous motion. Machinery or equipment motion or activation is permitted to the extent that adjustments or testing cannot be performed without motion or activation, provided that persons are effectively protected from hazardous motion.

Section 56.14211 Blocking equipment in a raised position

(a) Persons shall not work on top of, under, or work from mobile equipment in a raised position until the equipment has been blocked or mechanically secured to prevent it from rolling or falling accidentally.

(b) Persons shall not work on top of, under, or work from a raised component of mobile equipment until the component has been blocked or mechanically secured to prevent accidental lowering. The equipment must also be blocked or secured to prevent rolling.

(c) A raised component must be secured to prevent accidental lowering when persons are working on or around mobile equipment and are exposed to the hazard of accidental lowering of the component.

(d) Under this section, a raised component of mobile equipment is considered to be blocked or mechanically secured if provided with a functional load-locking device or a device which prevents free and uncontrolled descent.

(e) Blocking or mechanical securing of the raised component is required during repair or maintenance of elevated mobile work platforms.

Part 57—Safety and Health Standards: Underground Metal and Nonmetal Mines

Subpart K—Electricity
Section 57.12084 Branch circuit disconnecting devices

Disconnecting switches that can be opened safely under load shall be provided underground at all branch circuits extending from primary power circuits near shafts, adits, levels and boreholes.

Subpart L—Compressed Air and Boilers
Section 57.13019 Pressure system repairs

Repairs involving the pressure system of compressors, receivers, or compressed-air-powered equipment shall not be attempted until the pressure has been bled off.

Section 57.14206 Securing movable parts

(a) When moving mobile equipment between workplaces, booms, forks, buckets, beds, and similar movable parts of the equipment shall be positioned in the travel mode and, if required for safe travel, mechanically secured.

(b) When mobile equipment is unattended or not in use, dippers, buckets, and scraper blades shall be lowered to the ground. Other movable parts, such as booms, shall be mechanically secured or positioned to prevent movement which would create a hazard to persons.

Figure 4-8. Subchapter N. Metal and Nonmetal Mine Safety and Health.
Source: 30 *CFR*, Subchapter N.

the worker's body is exposed to danger. Today, certain work needs to be done under energized conditions. However, in other cases, limitations or flaws in the machinery can place workers in danger.

- When lockout/tagout procedures are developed, they need to include all energy sources and all instances where workers may be exposed to energized machinery. Maintaining state-of-the-art codes of safe practices requires periodic review of current standards and regulations to determine how they relate to changes in machine design, technology, and/or energy sources.

Part 75—Mandatory Safety Standards—Underground Coal Mines

Subpart F—Electrical Equipment (General)
Section 75.509 Electric power circuit and electric equipment; deenergization

All power circuits and electric equipment shall be deenergized before work is done on such circuits and equipment, except when necessary for trouble shooting or testing.

Section 75.511 Low-, medium-, or high-voltage distribution circuits and equipment; repair

No electrical work shall be performed on low-, medium-, or high-voltage distribution circuits or equipment, except by a qualified person or by a person trained to perform electrical work and to maintain electrical equipment under the direct supervision of a qualified person. Disconnecting devices shall be locked out and suitably tagged by the persons who perform such work, except that in cases where locking out is not possible, such devices shall be opened and suitably tagged by such persons. Locks or tags shall be removed only by the persons who installed them or, if such persons are unavailable, by persons authorized by the operator or his agent.

Subpart H—Grounding
Section 75.705-2 Repairs to energized surface high-voltage lines

An energized high-voltage surface line may be repaired only when

(a) The operator has determined that:

(1) Such repairs cannot be scheduled during a period when the power circuit could be properly deenergized and grounded;

(2) Such repairs will be performed on power circuits with a phase-to-phase nominal voltage no greater than 15,000 volts;

(3) Such repairs on circuits with a phase-to-phase nominal voltage of 5,000 volts or more will be performed only with the use of live line tools;

(4) Weather conditions will not interfere with such repairs or expose those persons assigned to such work to an imminent danger; and

(b) The operator has designated a person qualified under the provisions of Section 75.154 as the person responsible for carrying out such repairs and such person, in order to ensure protection for himself and other qualified persons assigned to perform such repairs from the hazards of such repair, has prepared and filed with the operator:

(1) A general discription of the nature and location of the damage or defect to be repaired;

Figure 4-9. Subchapter O. Coal Mine Safety Standards (*continued*)
Source: 30 *CFR*, Subchapter O.

- There are three ways to address the problem of lockout/tagout standards: mandates, performance criteria, and specifications. Of the three, performance criteria are considered the most effective at establishing uniform guidelines. Mandates are generally too broad and unenforceable, although they set social objectives. Specifications, although detailed, can be too restrictive.
- Part of the problem in developing adequate lockout/tagout codes is the lack of precise and mutually agreed upon definitions of *code*, *regulation*, or *standard*. In addition, standards have inherent weaknesses such as lack of consistency, time and cost to implement, unreasonableness, and so on. Compliance alone cannot guarantee worker safety because too many situations on the job fall outside the codes, procedures, or regulations.
- Employers have attempted to develop lockout/tagout measures to safeguard property and workers during maintenance and repair of machinery. The typical objective in such efforts was to state the problem, establish uniform application of general performance standards, and allow each facility or subsidiary freedom to develop its own specifications.
- Unions also took an active role in developing lockout/tagout procedures. They lobbied for passage of OSHAct, challenged OSHA regulations, and worked to create ANSI guidelines on hazardous energy control.
- Trade associations attempted to develop broader, more uniform safety guidelines and standards regarding lockout/tagout issues with a specific industry perspective. They focused on protecting the operator at the point of operation from contact at places other than the point of operation, while guards are temporarily removed, and when the job involves placing some body part in danger.
- Trade associations also helped to develop the concept of zero mechanical state, which emphasized primary protection to safeguard workers from accidental energy release. Eventually the concept was broadened to include zero energy state. The electrical utilities developed many of their own standards and procedures that applied to special problems in this industry.
- Consensus and safety organizations have sought to develop energy hazard controls and worker safety requirements to prevent injuries and fatalities from energy-release incidents. Among these are the Na-

(2) The general plan to be followed in making such repairs;

(3) A statement that a briefing of all qualified persons assigned to make such repairs was conducted informing them of the general plan, their individual assignments, and the dangers inherent in such assignments;

(4) A list of the proper protective equipment and clothing that will be provided; and

(5) Such other information as the person designated by the operator feels necessary to describe properly the means or methods to be employed in such repairs.

Subpart J—Underground Low- and Medium-Voltage Alternating Current Circuits
Section 75.903 Disconnecting devices

Disconnecting devices shall be installed in conjunction with the circuit breaker to provide visual evidence that the power is disconnected.

Subpart R—Miscellaneous
Section 75.1730 (c),(e) Compressed air; general; compressed air systems

(c) Repairs involving the pressure system of compressors, receivers, or compressed-air-powered equipment shall not be attempted until the pressure has been relieved from that part of the system to be repaired.

(e) Safety chains, suitable locking devices, or automatic cut-off valves shall be used at connections to machines of high-pressure hose lines of three-fourths of an inch inside diameter or larger, and between high-pressure hose lines of three-fourths of an inch inside diameter or larger, where a connection failure would create a hazard. For purposes of this paragraph, high-pressure means pressure of 100 p.s.i. or more.

Part 77—Mandatory Safety Standards, Surface Coal Mines and Surface Work Areas of Underground Coal Mines

Subpart E—Safeguards for Mechnical Equipment
Section 77.404 (c) Machinery and equipment; operation and maintenance

(c) Repairs or maintenance shall not be performed on machinery until the power is off and the machinery is blocked against motion, except where machinery motion is necessary to make adjustments.

Subpart F—Electrical Equipment—General
Section 77.501 Electric distribution circuits and electric equipment; repair

No electrical work shall be performed on electric distribution circuits or equipment, except by a qualified person or by a person trained to perform electrical work and to maintain electrical equipment under the direct supervision of a qualified person. Disconnecting devices shall be locked out and suitably tagged by the persons who perform such work, except that in cases where locking out is not possible, such devices shall be opened and suitably tagged by such persons. Locks or tags shall be removed only by the persons who installed them or, if such persons are unavailable, by persons authorized by the operator or his agent.

Figure 4-9. (*concluded*)
Source: 30 *CFR*, Subchapter O.

tional Safety Council's data sheets, ANSI's Z244.1 guidelines, and NFPA's standards. These documents focused on safety procedures as they applied to conditions in the workplace.

- Government action at the state level has been to follow the federal lead and adopt many of OSHA's standards verbatim. States such as Michigan, Washington, Oregon, and California have modified these standards or developed unique requirements of their own.
- Before 1970, there was no general, all-inclusive consensus standard for lockout/tagout. In OSHAct, passed in 1970, OSHA did not create a specific section for lockout/tagout but adopted various provisions of national consensus standards developed for specific industries.
- The weakness of OSHAct was addressed by the adoption of comprehensive standards for energy control: 29 *CFR* 1910.147, which focused primarily on procedures for control when dealing with potentially hazardous energy sources; 29 *CFR* 1910.331–335, which corrected many weaknesses in previous standards; and 29 *CFR* 1910.269, which extended protection to workers in the electric utility industry and others working with high-voltage lines.
- NIOSH guidelines were developed to provide a logical system for performing maintenance and servicing activities safely, recognizing that such activities can be performed safely with energy present, with energy removed, and while re-energizing.
- The Mine Safety and Health Act of 1977 established lockout/tagout standards in its Mineral Resources Subchapters N and O. Although MSHA does not prescribe any broad lockout/tagout procedures comparable to 29 *CFR* 1910.147, it does contain a number of specific requirements to de-energize machinery, equipment, and processes and refers to locking out energy sources.

REFERENCES

American Foundrymen's Society. *What You Should Know About ZMS (Zero Mechanical State)*. Des Plaines, IL: Am. Foundrymen's Society, 1961.

American Mutual Liability Insurance Co. Boston: American Mutual Liability Insur. Co., 1924; revised 1931, 1933.

American National Standards Institute. *Lockout/Tagout of Energy Sources—Minimum Safety Requirements*, ANSI 244.1. New York: ANSI, 1982.

American National Standards Institute. *Safety Requirements for Sand Preparation, Molding, and Coremaking in the Sand Foundry Industry*, ANSI Z241.1. New York: ANSI, 1975.

American National Standards Institute. *Standards and the Law*. New York: ANSI, 1984.

American National Standards Institute. *Respiratory Protection*, ANSI Z88.2. New York: ANSI, 1992.

American Petroleum Institute. *Management of Process Hazards*, RP 750. Washington DC: 1990.

American Petroleum Institute. *Safe Maintenance Practices in Refineries*, 2d ed., Washington DC: 1983.

American Society for Testing and Materials. *ASTM Glossary*, 2d ed. Philadelphia: ASTM Committee E-8 on Nomenclature and Definitions, 1973.

California Division of Industrial Safety, "Electrical Work Injuries in California, 1972." Sacramento, CA: California Division of Labor & Statistics Research, December 1973.

California. *Notice of Proposed Changes to Title 8, General Industry Safety Orders, Section 3314: Cleaning, Repairing, Servicing, and Adjusting Prime Movers, Machinery and Equipment*. Occupational Safety and Health Standards Board, Department of Industrial Relations, Sacramento, CA: August 1991.

California Occupational Safety and Health Board. *Electrical Safety, Bulletin* S550. Sacramento, CA: CAL-OSHA Communications, August 1982.

California. *Title 8, General Industry Safety Orders*. California Code of Regulations.

Chemical Manufacturers Association (formerly Manufacturing Chemists Association). *Recommended Safe Practices and Procedures: Electrical Switch Lockout Procedure*. Safety Guide SG-8. Washington DC: 1961.

Crickmer B. Edison Electric Institute: The first 60 years. *Electric Perspectives* (May/June 1993): 47–66.

Edison Electric Institute . . . The Association of Investor-Owned Electric Utilities. Brochure. Washington DC: Edison Electric Institute, 1977.

Employers Insurance of Wausau. "Machinery Lockout and the Broader Concept, Zero Mechanical State."

Federal Register. OSHA Supplemental Statement of Reasons on Three Remand Issues of Lockout/Tagout Standard, *FR* 58, 16612, 3/30/93.

Hall B. Zero energy state—Historical background of the concept. *National Safety News* 115, no. 2 (Feb. 1977):68–72.

Merriam-Webster. *Merriam-Webster's Collegiate Dictionary*, 10th ed. Springfield, MA: Merriam-Webster, Inc., 1993.

Michigan. *Safety Standards*. Occupational Safety Standards Commission, Department of Labor, November 15, 1969; amended, October 15, 1971.

Mine Safety & Health Administration.
30 *CFR* Subchapter N, Part 56, Safety and Health Standards: Surface Metal and Non-Metal Mines; Subpart K, Electricity; and Subpart M, Machinery and Equipment.
30 *CFR* Subchapter N, Part 57, Safety and Health Standards: Underground Metal and Non-Metal Mines; Subpart K, Electricity; and Subpart L, Compressed Air and Boilers.
30 *CFR* Subchapter O, Part 75, Mandatory Safety Standards: Underground Coal Mines; Subpart F, Electrical Equipment; Subpart H, Grounding; Subpart, J Low and Medium Voltage A.C. Circuits; and Subpart R. Miscellaneous.
30 *CFR* Subchapter O, Part 77, Mandatory Safety Standards: Surface Coal Mines and Surface Work Areas of Underground Coal Mines; Subpart E, Safeguards for Mechanical Equipment; and Subpart F, Electrical Equipment.

Mitchell RD. Zero mechanical state—Development of the concept. *National Safety News* 115, no. 5 (May 1977):87–92.

MSHA's goal: Zero deaths by 2000. *Minerals Today* (Oct. 1991):16–25.

National Fire Protection Association. *National Electric Code (NEC)*, NFPA 70. New York: National Fire Protection Association 1976.

National Fire Protection Association. NFPA 70E. *Electrical Safety Requirements for Employee Workplaces*, 3d ed. New York: National Fire Protection Association, 1988.

National Institute for Occupational Safety and Health. *Guidelines for Controlling Hazardous Energy During Maintenance and Servicing*. Washington DC: GPO, 1983.

National Safety Council. *Methods of Locking Out Electric Switches,* Data Sheet D-Gen. 41. Chicago: National Safety Council 1950.

National Safety Council. *Methods of Locking Out Electric Switches*. Data Sheet 237, Revision B, Paragraph 29. Chicago: National Safety Council, 1971.

National Safety Council. *Methods of Locking Out Electric Switches*. Data Sheet 1-237-78. Chicago: National Safety Council, 1978.

National Safety Council. *Safety Requirements for a Lockout Program*. NSC Standard B199. Chicago: National Safety Council, 1970.

National Safety Council. *Guidelines for a Lockout Program.* NSC Standard 0411-B1. Chicago: National Safety Council.

Nothstein GZ. *The Law of Occupational Safety and Health.* New York: Macmillan Publishing, 1981.

Occupational Safety and Health Administration. 29 *CFR* 1910.147, The Control of Hazardous Energy (Lockout/Tagout), also in *FR* 54, 36644-36696, September 1, 1989.

Occupational Safety and Health Administration. 29 *CFR* 1910.269, Electrical Power Transmission, Generation, and Distribution; Electrical Protective Equipment; Electrical Safety-Related Work Practices.

Occupational Safety and Health Administration. 29 *CFR* 1910.331-.335, Electrical Safety-Related Work Practices, August, 1991.

Occupational Safety and Health Administration. 29 *CFR* 1910.137, Electrical Protective Equipment.

Occupational Safety and Health Administration. 29 *CFR* 1910.27, Fixed Ladders.

Oregon. *Interim Oregon Occupational Safety and Health Code*, OAR 437, Division 2, General Occupational Safety and Health Rules (29 *CFR* 1910), Subdivision J: General Environmental Controls, Lockout/Tagout (1910.147). Salem, OR: Oregon Occupational Safety and Health Division (OR-OSHA), Department of Insurance and Finance, 1991.

Price C. ZES—Zero energy state: A systems approach to guarding maintenance servicing functions. *National Safety News* 112, no. 6 (Dec. 1975): 56–57.

Rapp FJ, Brooks BE. *Machinery Lockout Procedures.* Warsaw, Indiana: Health & Safety Staff, UAW Society Security Department, 1980.

State of California, Department of Industrial Relations, CAL/OSHA Communications Unit. *Lock-out/ Block-out.* San Francisco, CA: February 1984.

United Steelworkers of America. International Safety and Health Department. *USWA Safety and Health Program.* Pittsburgh, PA: USWA, 1976.

United Kingdom. *Health and Safety at Work Act.* Chapter 37. London, U.K.: 1974.

U.S. Public Law 91-596, Occupational Safety and Health Act, 91st Congress, S. 2193. Washington DC: U.S. GPO, 1970.

U.S. Public Law 95-164, 91 Stat. 1290, *Mine Safety and Health Act*, 95th Congress, Washington DC: 1977.

Washington. WAC 296-24-11001, The control of hazardous energy (lockout/tagout), Chapter 19.17 RCW. 90-20-091 (Order 90-14), Section 296-24-220, filed 10/1/90, effective 11/15/90.

5 International Hazardous Energy Regulations

Although the problem of controlling hazardous energy knows no geographic borders, progress has not been universal. Various parts of the world, particularly in developing countries, have little in the way of governmental regulation, consensus guidance, or professional practices directed at hazardous energy control. However, Western Europe has been very active in addressing this problem during the past 20 years and provides a reservoir of knowledge directed at the prevention of unintended hazardous energy release.

In this chapter, various international initiatives are examined that relate to the control of hazardous energy. Several are significant—despite being in draft form—while others are evolving rapidly due to market forces or national safety priorities.

OVERVIEW

As industrialization and the need for energy to support growth increase, problems associated with the hazardous release of energy have multiplied. Injuries and property damage in developed countries prompted efforts to control these incidents. However, on a global basis, most of the significant progress has been made since the end of World War II. The emergence of multinational companies and the existence of colonial relationships with developing nations has resulted in the transfer of hazardous-energy-control technology and practices to these nations throughout the world. Therefore, it is not unusual to have energy isolation procedures and hardware in use in various industries or establishments where national guidance or regulation is nonexistent. Today, as global communication intensifies, there is an identifiable trend toward the control of hazardous energy becoming more commonplace.

General Practices

Hazardous-energy-control practices and methods appear to be developing on a global basis as features of various standards relating to electrical or machinery safety. The risks to workers in electrical power generation and transmission and the emphasis on electrical equipment development has led to greater progress in protective measures for this energy source than for other sources of energy such as thermal, hydraulic, or pneumatic. Hazardous-energy-release incidents involving machinery are a leading cause of serious harm to people and therefore have received special attention in industrialized societies. Many nations have fragmented references or regulations pertaining to hazardous energy control that have evolved without addressing the issue in any comprehensive way. The standards, codes, guides, and regulations tend to be specification oriented, not performance oriented, and apply only to specific jobs, technology, or industries. However, change is underway, principally in Western Europe, led by groups like the European Committee for Standardization (CEN) and the European Union (EU).

Europe

In Western Europe, most of the progress on hazardous energy control has taken place since 1970, with the United Kingdom, France, Sweden, and Germany providing the impetus. Guidelines and standards were developed topically with a focus on electrical and machinery issues. The British have perhaps had the most profound influence with the British Standards Institution's standard BS 5304: 1988, *Safety of Machinery* and *Safe System of Work,* an edict emanating from the Health and Safety at Work Act of 1974.

At this time, there is not much that is known about the state of hazardous energy control in Eastern Europe and Russia. Circumstances and the political situation seem to suggest that there has been little emphasis on this problem and that the state of the art trails progress made in Western Europe. However, Russia may have something to contribute as access to its industrial base becomes more common. The available technical literature reveals that most of what can be reported from these nations also is found from a variety of other sources.

Asia/Africa

Energy isolation techniques in Asia, Africa, and the Pacific Rim run the entire spectrum from crude to sophisticated with regard to practices and techniques. In Middle Eastern and African states, the techniques employed are believed to be the product of multinational companies engaged in oil production and mineral development and the influence of previous colonial relationships with European nations.

Japan's booming industrial economy and use of advanced manufacturing techniques are combined with a priority for worker protection. The evidence indicates that energy isolation techniques comparable to or even somewhat more advanced than Western standards are employed.

Canada/Australia

Much of what is found regarding hazardous energy control in Canada and Australia is similar to that found in the United Kingdom. The provinces of Ontario, Quebec, and British Columbia, in particular, have been active in defining energy isolation requirements and promoting safe practices. Australia's energy isolation regulatory references exist in various sections of standards pertaining to machinery, chemical, and electrical safety. The Standards Association of Australia produced AS4042.1-1992, *Safeguarding of Machinery—General Principles,* which has a bearing on unexpected start-ups of machinery and energy control.

Central and South America

Practices and techniques used in basic industries such as mineral development, oil production, machinery, and automotive manufacturing have often been imported via multinational company operations. Brazil, Argentina, Venezuela, and Mexico are believed to offer the most advanced application of energy isolation practices for this part of the world.

INTERNATIONAL LABOR ORGANIZATION (ILO)

International efforts to ensure worker health and safety have paralleled efforts in the United States. Between 1945 and 1949, the International Labor Organization (ILO), based in Geneva, Switzerland, developed a *Model Code of Safety Regulations for Industrial Establishments for the Guidance of Government and Industry.* The code was first published in 1949 with other editions printed in 1954 and 1967. Over 500 pages of text were devoted to all aspects of workplace safety and health. The intent and application of the Model Code is described in Chapter I, General Provisions, as follows:

> Regulation 1: Purpose and Scope:
>
> 1. The purpose of this Model Code is to contribute to the elimination of danger to life, and to secure the safety and health of workers in industrial establishments.
> 2. This Model Code constitutes a guide which Governments and industries shall benefit from to such an extent as they may desire when framing measures for the im-

provement of safety conditions in industrial establishments.

3. The provisions of this Model Code apply to construction, equipment, and working conditions in industrial establishments.
4. With a view of securing some uniformity in basic safety measures, it is desirable that measures based on provisions of this Model Code should, as far as possible, be expressed in the wording of this Model Code.

Regulation 118, Electrical Repairs, and Regulation 212, Safety Measures for Maintenance and Repair Work, covered issues associated with the control of hazardous energy. It is interesting to note that in 1945 lockout, blocking for gravitational movement, blanking fluid/gas lines, inspection before restart of machinery, and so on were advocated as standard work practices. This further supports the contention that lockout/tagout techniques quite likely had their formal origin in the 1920s.

Excerpts from the Model Code, regulations 118 and 212, show the nature of the provisions being proposed as international standards to safeguard against energy-release incidents:

Regulation 118—Electrical Repair.

1. Except in cases in which it is absolutely necessary, repair work shall not be undertaken on live electric circuits.
2. So long as definite and satisfactory evidence to the contrary has not been obtained, repairmen should assume that all parts of electric circuits are live.
3. Work on live parts of electric circuits shall only be carried out—
   (a) on direct order from a competent, responsible person; and
   (b) under direct and constant supervision by a competent person thoroughly familiar with the installation to be repaired, the work to be carried out and the risk inherent in this work, and capable of taking immediately all necessary steps to prevent accidents or mishaps during the work.
4. Before allowing work to begin on any electric circuit, machinery or installation, the person in charge shall take all necessary steps to ensure that—
   (a) the circuit, machinery or installation in question is reliably disconnected from any source of power;
   (b) the switches or circuit-breakers controlling the circuit, machinery or installation have been securely locked in the OFF position; and
   (c) such other measures have been taken as may be necessary in each particular case to prevent the current from being switched on again before the work has been completed and the repairmen withdrawn.
5. After repairs, the current shall only be switched on again on the definite order of a competent and authorised person.
6. When repairs have to be carried out on electric circuits, power cables or overhead transmission lines to which current can be supplied from more than one direction, the circuit, cable or line shall be securely disconnected from the power supply on both sides of the place at which the repairs are to be undertaken.

Regulation 212—Safety Measures for Maintenance and Repair Work.
Repair work on machines

11. When repair work has to be undertaken on a machine—
   (a) the machine shall be stopped before the work is begun; and
   (b) adequate measures shall be taken, preferably by locking the starting or controlling device, to ensure that the machine cannot be restarted until the work has been finished and all the repairmen withdrawn: Provided that, when necessary for testing and adjustment, the machine may be restarted by the person responsible for the repair work.
12. Locks used for the purpose referred to in paragraph 11 (b) of this Regulation shall be of a special type, not openable by keys other than those in the hands of the designated persons.
13. If repairs have to be carried out on a machine, any part of which may move without the power having been applied, e.g., under its own weight, such parts shall be securely blocked before the work is begun.
14. When repair work on a machine has been finished, and before the power is again applied for purposes of production—
   (a) all tools, implements, and materials used during the work shall be carefully removed and collected at a safe place outside the machine;
   (b) the machine shall be fully restored to its proper working condition;
   (c) when possible the machine should be turned slowly by hand to ensure that no objects of any kind have been left in such a place or position as to interfere with the safe operation of the machine; and
   (d) the place around the machine should be properly tidied and restored to its normal condition.

15. When repairs have to be carried out near machinery or other dangerous parts that cannot be stopped or switched off, and when repairmen have to pass close to such machinery or parts in such a way or at such places as not to be adequately protected by the ordinary guards, all necessary temporary measures shall be taken for their protection.

Power transmission

16. The repairing and replacing of transmission belts and other parts of mechanical power transmission installations should be carried out only by persons specially trained and selected for such work.
17. (1) when belts or other transmission parts are to be repaired or replaced in rooms in which non-continuous operations are carried on, before beginning the work the repairman shall be sure that—
(a) the machinery to which the belt or other transmission part belongs is shut off from the power supply;
(b) the power control is securely locked in the OFF position; and
(c) such additional precautions have been taken as may be deemed necessary in each particular case.

Boilers, tanks, and vats

23. If tanks or vessels in which repairs are to be carried out are connected to other tanks or vessels, the connecting pipes shall be securely blocked, by either—
(a) closing the valves and securely locking them in the closed position; or
(b) disconnecting the pipe lines and blanking them off by means of blind flanges.
24. If repairs are to be carried out in a tank or vessel in which stirring or mixing apparatus or machinery is installed, before workers are permitted to enter the tank or vessel the stirring or mixing apparatus shall be—
(a) reliably disconnected from its power supply; and
(b) so locked or blocked that no movement can occur that could endanger the workers.

Piping systems

30. Before starting repairs on any piping systems used for transporting corrosive, explosive, flammable or poisonous substances—
(a) all valves shall be closed and locked, the piping shall be drained and sufficient time shall be allowed for any gas to escape; and
(b) if a cutting or welding torch is used, the piping shall be thoroughly washed if necessary with a neutralizing substance, and flushed with steam or boiling water.
31. In opening flanges on piping systems used for transporting hazardous substances—
(a) a shield shall be laid over the flanges so as to protect the operator from possible spurts;
(b) on horizontal pipes the bottom bolts shall be removed first;
(c) the remaining bolts shall be loosened gently until the contents start to drip; and
(d) unless the flanges part readily, they shall be carefully separated by means of a small metal wedge or a special tool designed for the purpose.

As of 1994, the International Labor Organization has not published any current directives, guidelines, or standards specifically relating to the control of hazardous energy (lockout/tagout).

SAFETY ORGANIZATIONS

Internationally, there are many private and quasi-governmental organizations dedicated to improving the safety and health of their citizenry and working populations. Their capacity to act and the depth of their agendas is frequently limited by the availability of financial support. Therefore, limited information exists on certain aspects or topics relating to occupational safety. The Industrial Accident Prevention Association of Ontario, Canada, and the French National Research and Safety Institute represent two such organizations that have actively contributed to the prevention base for the control of hazardous energy.

Industrial Accident Prevention Association (IAPA)

The Industrial Accident Prevention Association of Ontario, Canada, publishes various safety and health materials to assist employees and employers. Their 1990 publication *Lockout* provides guidance concerning hazards, legislation, procedures, training, control, and references.

A general caveat is found in the first section of the document, which conceptually is consistent with other progressive codes. It reads as follows:

> When installing new machinery or equipment, performing maintenance, servicing or repair operations, all connecting energy sources must be cut off. It is important to remember that an energy source may be mechanical, hydraulic (fluids), electrical, pneumatic (air, gas), gravitational, or stored (spring). It is also

important to remember that more than one energy source is often involved, and must therefore be neutralized through a proper lockout system, before proceeding with the maintenance or servicing job.

Seventeen key procedural elements are identified as part of any proper lockout procedure with notification of supervisory and affected parties and the issuance of the required work permit as the triggering events. These elements are not necessarily in ranked order:

1. Know how the equipment functions.
2. Know how to shut down electrically (break circuits).
3. Lock air, gas, steam, or other valves at their source.
4. Drain or bleed any residual pressure.
5. Block or immobilize components subject to gravity.
6. Machines—lock out switches and ensure all components are at rest.
7. Insist on primary protection—not on secondary protection such as control circuitry; achieve zero energy state.
8. Each worker/supervisor should be protected by a personal lock on the energy isolating device(s).
9. Locks should be a key type with only one key per lock permitted.
10. Machine should be locked at all times until work is complete.
11. Identify related systems and ensure their lockout.
12. Use tags to highlight and detail work in progress.
13. Ensure use of appropriate personal protective equipment.
14. After completion of work, each worker should report to the person in charge and sign the work permit—then remove their lock.
15. Only the person in charge should remove the main lock—do not delegate this action.
16. Equipment should be tested using a checklist:
    - Actuate control to determine circuit is de-energized
    - Verify all components are at rest
    - Ensure all personnel are clear of danger zone(s).
17. Restart machinery, process, equipment with a checklist:
    - Guards in place?
    - Restraints removed?
    - Tools accounted for?
    - Valves in proper position?
    - Personnel accounted for?
    - Appropriate parties notified?

In addition, other supporting IAPA publications such as Guideline S01014 (*Work Permits*) and Booklet BO1228 (*Confined Space Entry*) provide further insight regarding the hazardous energy control process.

French National Research and Safety Institute (INRS)

The French National Research and Safety Institute (INRS) is a nonprofit association under the regulating authority and financial control of the French government. Its activities include research, advisory services, training, testing of protective equipment and machines, development of guidelines and standards, production of publications, data base development, and library information resources.

INRS published two recent documents of importance with regard to the control of hazardous energy:

- *Safety Advice for Servicing and Working on Electrical Systems and Facilities (Low Voltage),* ED 539, 1990
- *Closure and Unclosure,* ED 754, 1992 (this covers the equivalent of U.S. lockout/tagout concepts and procedures).

Both documents are in French, and the excerpts provided have been translated from the original.

In *Safety Advice for Servicing and Working on Electrical Systems and Facilities,* section 2, "Servicing Machinery Powered by Low Voltage," includes the following subsections:

- 2.1 The Phases of Servicing
- 2.2 Safety Measures During Phases Requiring the Continued Supply of Electrical and Other Sources of Power to the Machinery
- 2.3 Repairing Machinery. Closures (Figure 5-1)
- 2.4 Putting Machinery Back into Operation

Closure and Unclosure (*Consignations et Déconsignations*), a large booklet by INRS, provides insight to the global need for more effective control of hazardous energy. Figures 5-2 and 5-3 are translations of the introduction that establishes the framework for the document.

The following publications also provide additional guidance regarding closure/unclosure:

- CNMA Recommendations R 276. *Tanks and Reservoirs.* INRS, 1985.
- CNMA Recommendations R 289. *Mechanized Assemblies of Stationary Facilities.* INRS, 1986.
- *General Electrical Safety Instructions.* Publication UTE(C) 18-510 of the Union technique de l'électricité.
- *Principal Safety-related Terms of Traditional Electric Engineering.* INRS, ED 537, 1991.
- *Closure.* CEDECOS Publication, 1988.

Repairing Machinery

This section addresses machinery repair but reveals general concepts of energy isolation as applied in the French environment. It begins as follows:

If damaged parts or electrical appliances need to be replaced in order to repair machinery, such replacement may only be done:

—once **electrical closure** of the system has been performed, if the replacement occurs in an area where **only electrical risk exists**;

—after **electrical closure and closure of the other power sources** if there are **other risks**.

The holder of a BC or BR certification can perform the electrical **closure** of a **system**.

Section 2.3.1 Procedures for Electrical Closure of Low Voltage Systems.

a) General Low Voltage Equipment Closure Procedure

The procedure defined below is only valid for closure of low voltage systems. **To close** a low voltage electrical system means performing the operations designed to achieve all the following:

1) To **separate** this system from all potential voltage sources (there may be several).

The separation must be performed in such a way as to make it fully apparent. For low voltage systems, verification that the separation is fully apparent is performed:

—either by a visual check of the active parts of the separation device if the cutoff is visible;

—or by checking the control mechanism, visually also, provided the mechanical linkage between the contacts and the control mechanism is direct and provided the control mechanism can in no way be subjected to deformation or deterioration, regardless of any forces that might be brought to bear on the control mechanism by manual means. (**Note:** Low voltage is defined as less than 1,000 V AC or 1,500 V DC.)

2) To **Lock** these separation devices to the open position. To lock a device open means to perform the operations required to:

—put or keep it in the open position;

—prevent it from being operated (for example using a padlock);

—post the status of the device (warning sign). **Note:** When it is not possible to immobilize the control device physically (e.g., with a padlock), the minimum required action is to put a warning sign on the device.

3) To **identify** the locked system to be sure that work or repairs are performed on it.

4) To **verify,** in each active lead immediately after the cutoff point or points, that there is no voltage in relation to ground.

The electrical system on which you are working may have electrical connections to circuits of other systems; some parts of the system on which you are working may remain powered up (relays, contactors, distributor electromagnets, relay contacts). **Close these other systems as well.**

b) Streamlined Low Voltage Equipment Closure Procedure. The streamlined system closure procedure involves only:

—power source separation

—identifying the system;

—checking that there is no voltage.

Use the streamlined electrical closure procedure only:

—if you are performing repair servicing;

—if access to the work area is limited only to the operators for the entire duration of the servicing;

—if the s**eparation devices remain at all times visible** from the place where you are working.

Section 2.3.2—Unclosure of a System

The unclosure of a system may only be performed by the person who performed the electrical closure of the system.

Section 2.3.3—Closure of Facilities Using Other Power Sources

If the repair or replacement of a defective device requires you to work in an area where **mechanical risk**s are present, make an inventory of t**he other risks**, such as:

—risks that the mechanism might start moving again due to instability (unbalanced loads being carried are not immobilized);

—risks due to the presence of other power sources such as pressurized fluids (compressed air, steam, gas, liquid); and then, if need be, perform the appropriate closure.

The prevention of risks due to instability or to nonimmobilization of the parts consists of **effectively blocking and immobilizing** those parts that are likely to move.

If a blocking is performed which does not bear on the parts whose slightest movement one seeks to prevent, it is not considered effective.

Preventing risks due to the presence of pressurized fluid is achieved by performing the closure of the pressurized fluid facility.

As in the case of electrical closure, the closure of a pressurized fluid facility includes the following operations:

Figure 5-1. Section 2.3—repairing machinery. closures. (*continued*)

Source: *Safety Advice for Servicing and Working on Electrical Systems and Facilities (Low Voltage),* ED 539, French National Research and Safety Institute (INRS), Paris, France: December 1990.

—separating the power sources by putting devices such as valves or stopcocks in the closed position;
—locking these devices closed by eliminating all potential activation of the controls and posting the status of the devices;
—decompression by purging tanks and vessels;
—verifying that decompression has been achieved (read pressure off the manometer);
—providing a clean atmosphere to the area after decompression (in the case of gases other than compressed air if the amounts involved in decompression and risks attendant to exposure to such gases, so require).

Figure 5-1. *(concluded)*

Introduction

When machines, systems, or facilities are brought down for servicing or maintenance, work accidents may occur that often have serious consequences; such accidents are due to one or more employees coming in contact with:

- bare parts traversed by electrical currents
- hazardous chemicals
- mechanical parts that move unexpectedly
- pressurized fluids.

In most cases, the victim thought he/she was safe, but closure was incomplete.

The purpose of this guide is to assist in establishing closure procedures appropriate to a given situation. Notwithstanding, it should be borne in mind that other methods of ensuring safety exist.

While primarily intended for operators, this guide can also be of use to those involved in the design process.

This guide lays no claim to spelling out all the specific technical solutions that might be implemented to perform closure. It merely states the steps that must be taken in order to perform a correct closure procedure, regardless of the type of risk (electrical, chemical, or mechanical), and gives a few examples of implementation.

NOTE: Some specific risks, such as radiation (ionizing and nonionizing) or biological effects, are not covered in this brochure.

GENERAL STEPS TO BE FOLLOWED WHEN WORKING ON OR REPAIRING MACHINES, DEVICES OR FACILITIES

Before servicing machines, devices, or facilities, or having someone service them, you should:

1. Make sure that the operational modes to be implemented have been defined and that the risks attendant to them have been analyzed.

The detailed analysis must take into account all personnel and equipment safety aspects, including those not directly linked to the servicing in question (presence of other nearby worksites, another part of the workshop still operating, etc.).

2. Take appropriate measures to eliminate such risks or, if there is a justifiable technical reason why this cannot be done, limit any consequences. Such measures include in particular the closure of a device, machine, system or facility.

3. Entrust the servicing only to personnel having the proper skills and real-situation training, and that are cognizant of the safety measures applicable to it.

4. Make available to such personnel the tools required to provide proper servicing and make sure that these tools are properly used.

Those cases where strict application of the closure procedure mentioned above is not possible (adjustments, tests, etc.) are not covered in this guide. The company should seek, however, to define such procedures on a case-by-case basis.

DEFINITIONS

Closure

The set of dispositions that make it possible to make and keep a machine, device or facility safe (if possible by some physical means) in such a way that a change of state (restoring a machine to working order, closing an electrical circuit, opening a valve, etc.) becomes impossible without the active involvement of all personnel working on it. In addition, there are other more specific definitions: electrical closure, machine shutdown closure, etc.

Unclosure

The set of dispositions enabling a machine, device or facility that has been closed to be put back into operation, insuring the safety of both the personnel working on it and the operators.

Service Personnel

Service personnel are those called on to perform predefined work. This may be:

—an individual

—or a skeleton crew whose foreman or task leader is continuously present at the worksite.

Closure Officer

A closure officer is a person with authority, placed by the head of the company in charge of performing the closure and unclosure of a device, machine, or facility, and who is in charge of either personally taking the safety measures called for by it, or making sure they are taken.

Figure 5-2. Introduction to *closure and unclosure.* *(continued)*
Source: *Closure and Unclosure,* ED-754, French National Research and Safety Institute (INRS), Paris, France: 1992.

CLOSURE AND UNCLOSURE PROCEDURES

In order to ensure continuing safety of a situation, the closure of a machine, device, or facility must include four indissociable phases (see general table of typical closure procedures in Figure 5-3).

a) separation
b) locking and posting
c) "purge"
d) verification and identification

The order in which some of these phases are to be performed may vary depending on the case in question, once the risks have been analyzed: in electrical matters, for example, grounding ("purging") must occur after it has been verified that there is no voltage.

The "purge" consists of eliminating any power and residual power, or in evacuating hazardous substances: discharging a condenser, relieving a pressure, emptying a pipe that contains a corrosive liquid, putting a press at bottom dead center, etc.

The term "purge" has been used in this document for purposes of simplification, since it covers an idea common to all types of risk. A detailed explanation of the term will be given for each risk in the appropriate sections.

The separation and the "purge" must be performed as close as possible to the servicing area in order to make the checks easy to perform.

The verification that there is no voltage, pressure, etc. must itself be considered as work being performed under voltage, pressure, etc.

Figure 5-2. (*concluded*)

Typical Closure Procedures

Closure phase	Type of risk: Electrical	Chemical	Mechanical
Separation	Cut electrical power to all power and control circuits, including emergency power supplies, making it fully apparent.†	Cut off entry of all fluids or solids, including auxiliary circuits, making it fully apparent.†	Cut off transmission of all types of power, including emergency sources and power accumulators, making it fully apparent.†
Locking	Lock using a physical device that is hard to neutralize, the status of which can be seen from the outside, and that can be reversed only by a tool customized for each service person.		
Posting	Clear permanent information that the locking has been performed.		
Purge	Ground and short-circuit conductors (operation to be performed after the verification). Discharge condensers.	Empty, clean (scale, etc.). Remove inert or hazardous atmospheres. Ventilate.	Bring down to the lowest power level by: • stopping mechanisms, including power flywheels; • bringing to stable mechanical equilibrium (bottom dead center), or, if this cannot be done, mechanical blocking • bringing to atmospheric pressure.
Verification	No voltage between any conductors (including the neutral), nor between conductors and ground.	No • pressure • flow of fluids Specific checks as required (atmosphere, pH, etc.)	No power: • no voltage • no pressure • no movement • etc.
	If need be, any residual hazardous areas must be marked.		
Identification	The purpose of the identification is to ensure that the work will actually be performed on the facility or on the system that has been closed. To do this, diagrams and marking of the components must be readable, permanent, and updated.		

Figure 5-3. General table of typical closure procedures.

† I.e., either by direct line of sight to the separation device, or by reliably locking the position of the separation device to the external control mechanism reflecting that position.

Source: *Safety Advice for Servicing and Working on Electrical Systems and Facilities (Low Voltage),* ED 539, French National Research and Safety Institute, (INRS), Paris, France; December 1990.

NATIONAL CONSENSUS ORGANIZATIONS

European standards-making bodies have been particularly active in developing minimum requirements for controlling hazardous energy sources. In most instances, the standards are derivatives of machinery- or electrical-based safety criteria and do not address energy isolation comprehensively. However, the principles and concepts are reasonably well defined, and the protective procedures generally are complete. England, France, and Sweden have developed noteworthy guidelines that deserve review. Both European and European Economic Community standards and directives have benefited from the pioneering work of the member countries' technical committees associated with various aspects of hazardous energy control.

British Standards Institution (BSI)

The British Standards Institution (BSI) is the counterpart of the American National Standards Institute (ANSI) and the U.K. member of the International Organization for Standardization. BSI prepares and promulgates consensus standards, offers technical help to exporters, and makes available certain specialist services.

BSI has no standard similar to ANSI's Z244.1, Lockout/Tagout; however, BS 5304:1988, *Safety of Machinery,* is an excellent reference with relevant provisions. Section 5, Machinery Design; Section 6, Selection of Safeguards; and Section 13, Safe Working Practices, offer guidance in various aspects of hazardous energy control. The hardcover book is a particularly useful general reference for machinery safety, with special emphasis on design considerations.

The Machinery and Components Standards Committee (MCE) entrusted the Technical Committee MCE/3 to develop BS 5304:1988 with broad representation from various trade associations, Advanced Manufacturing Technology Institute, Institution of Mechanical Engineers, Trades Union Congress, Institution of Occupational Safety and Health, and Health and Safety Executive. First published as CP 3004:March 1964 and revised as BS 5304:December 1975, this standard identifies the hazards arising from the use of machinery and describes methods for their elimination or reduction, for the safeguarding of machinery, and for the use of safe working practices.

Section 13, of BS 5304, Safe Working Practices, begins with an introduction that is critical to achieving prevention success when dealing with hazardous energy control and machinery. It is directly quoted as follows:

> It is not always possible to eliminate hazards or to design completely adequate safeguards to protect people against every hazard, particularly during such phases of machine life as commissioning, setting, process change-over, programming, adjustment, cleaning and maintenance, where often direct access to the hazardous parts of the machine may be necessary. There are also a number of types of machinery where, at present, it is recognized that complete safeguarding cannot be provided even for operational activities. For some of these types of machinery, safe working practices are specified, e.g., in statutory regulations.
>
> It should be emphasized that safety of machinery depends on a combination of hazard minimization measures, safeguards and safe working practices. These should take account of activities during all phases of the machine's life.
>
> Safe working practices should be taken into account at the design stage, since the provision of jigs, fixtures, fittings, controls and isolation arrangements will frequently be involved.

The standard further comments that situations involving unguarded machinery under power in any phase of operation should be avoided by appropriate design measures whenever technically feasible. The use of completely different types of machines to achieve the same end product may be an alternative.

Section 13.2.3, Practices for Isolation and Dissipation, provides direction with regard to the control of hazardous energy potential. It begins with the premise that although safeguards are provided to preclude access during most phases of machine life, they may be compromised by the need to gain access to hazardous areas, for example, setup, die change, and so on. Interlocked guards may be used to deactivate the machine while short-term activities such as lubrication or adjustment take place. Additional safeguards may be necessary where it is possible to gain access inside the guarding structure with the guards/interlocks reclosed. Presence-sensing devices may be used to supplement the interlocks, but most often safe work practices are necessary.

External isolation and dissipation. Higher-risk nonoperational activities may require more reliable and general means of energy interruption. For example:

- Mechanical power transmission: clutch isolation, removal of chains/belts, or shaft sections.
- Electrical power: de-energizing switches, removal of fuses, removal of plugs from sockets, earthing (grounding).
- Hydraulic or pneumatic power: valve isolation, electrical isolation of pumps, disconnection from pneumatic mains, open venting to atmosphere.

- Services: isolation of steam, water, gas, or fuel supplies.
- Process/material supplies: isolation of lines and line blinding or blanking.
  NOTE: Provision for these capabilities needs to be made at the design stage.

Internal isolation and dissipation. Residual energy or material in the machine, equipment, process may need to be relieved or dissipated as follows:

- Mechanical power: allow rotating parts to run down, use props and catches to support parts which may fall due to gravity.
- Electrical power: discharge capacitors, disconnect stand-by power (batteries), and so on.
- Hydraulic power: discharge accumulators, relieve pressurized piping, and the like.
- Pneumatic power: discharge air pressure except where used for hold up.
- Services: vent, purge, drain residual steam, gasses or fuel.
- Process and material supplies: empty, vent, drain, purge, and clean.
  NOTE: Provision for those capabilities needs to be made at the design stage.

Where controls are remote from the plant or machinery and personnel are not within vision, the effectiveness of energy isolation safety measures should be ensured. Relevant controls and facilities should be lockable or operable only by the use of tools or keys. Each person should apply his or her padlock or key to each relevant control.

Security at interlocking access gates where entry can be gained to a dangerous area should be provided with a device that prevents closing of the gate/door such as a trapped-key bolt, an exchange-key bolt, a captive key with lockable operating handle, a tongue-operated switch, or a gravity- or spring-operated latch.

When tasks must be carried out without full safeguarding and isolation is not possible, restricted machine operation may be necessary, for example, inching or slow speed. Safe practices should be applied, and any internal and external power sources and process/service connections should be isolated where possible. The controls or devices used to restrict the mode of operation should be given the same consideration in terms of reliability and security as controls or devices that provide for complete isolation.

Section 13.3—*Supervising Control* states that:

> Where safety from mechanical hazards is dependent on people carrying out safe working practices, it is essential that an appropriate degree of managerial and/or supervisory control is exercised. Where risk is minimal, verbal instructions may be quite adequate but as risk increases it becomes essential to define procedures in writing in order that they can be supervised more rigorously. Where the risk level is high, e.g., there is a possibility of serious injury or death if the procedure is not followed correctly, the adoption of a permit to work system is regarded as essential. This will normally involve specification of the controls, etc., for isolation and for internal hazard dissipation and supervisory checks that they have been operated and secured and that the plant is free of hazard (or that additional practices, such as use of protective equipment, are followed.)

Another appropriate British Standards reference is BS 2771 *Electrical Equipment of Industrial Machines—Part I 1986.* It applies to electrical equipment of machines (not portable by hand) used in industrial production and operated from a supply up to 1,000 V AC. It states the requirements for safety of personnel and property, uninterrupted productivity, long life of equipment, and the ease and economy of maintenance. It is identical to the European equivalent EN 60204 (CEN/CENELEC) Part I and is related, but not equivalent to, IEC 204.1 and 204.2.

Permit to work systems. These systems require management to identify the hazards to which employees are exposed and to develop a "safe system of work" whereby these hazards are eliminated or at least recognized by the employee(s) so that personal precautions can be taken (the concept of *safe system of work* is discussed later in this chapter). A written or documentary system requires formal action on the part of those doing the work, those responsible for it, and those authorized to sign such permits. Those supervising the work should assume that personnel are identified, are properly trained, and understand the task involved and the necessary precautions to be taken.

Work under potentially hazardous circumstances can be done safely using the "permit to work system." The design of a permit to work (Figures 5-4 and 5-5) will depend on the type and degree of risk, the task complexity, and the nature of the industry.

French Standards Association (AFNOR)

The French Standards Association (AFNOR) is equivalent to the British Standards Institution and serves as the French member of the International Organization for Standardization. AFNOR is also France's official representative on the Joint European Standards Institute (CEN/CENELEC). Much like its British counterpart, it participates in European consensus standards–making activity, prepares and produces standards, and interacts with other world standards-making bodies on appropriate matters.

Machinery and Equipment
Certificate of Appointment

Part I

REQUEST FOR CERTIFICATE OF APPOINTMENT to approach unfenced machinery for the purpose of observation which is found immediately necessary.

Signed: .... (Supervisor)

I.D. No. .... Dept: ....

Dated: ....

A Certificate of Appointment may only relate to one person.

The counterfoil must be handed to the authorised person and retained by him until the permit is returned with Part III completed.

Instructions can then be given to clear Part IV and resume normal operation.

Serial No: ....

Machinery and Equipment – Certificate of Appointment

Part II **Unfenced Machinery/Equipment**

This certificate appoints Name:...................... Badge No: .... Dept:...........

to approach Machine/Equipment:......... BT NO: .... Loc: .... Dept:...........

for observation purposes only of the process or part(s) as detailed herewith:

....

....

....

Signed: .... (Authorised Person) Date: ....

Time: .... am/pm

Note: ONLY PERSONS AUTHORISED BY THE PLANT ENGINEER ARE PERMITTED TO ISSUE THIS CERTIFICATE OF APPOINTMENT

Part III (To be completed by person appointed in Part II)

I hereby declare that *(1) task designated is complete/incomplete.
*(2) all guards replaced or machine/equipment is left in a safe condition.
*(delete whichever is not applicable)

Signed:.................................. (Appointed Person) Date:..................................

Time: .... am/pm

Part IV I hereby declare that this certificate is now cancelled.

Signed: .... (Authorised Person) Date: ....

Time: .... am/pm

THIS CERTIFICATE IS VALID ONLY FOR THE SHIFT IN WHICH IT IS ISSUED OR THE COMPLETION OF THE OBSERVATION (WHICHEVER IS EARLIER).

WHEN COMPLETED MAIL BOTH PARTS TO PLANT SAFETY ENGINEERING DEPARTMENT.

FORM NO:

(Printed on back of certificate)

PRECAUTIONS TO BE TAKEN FOR SAFE ENTRY INTO TRANSFER MACHINES FOR THE PURPOSE OF OBSERVATION

1. A close fitting, single piece overall suit in good repair shall be worn. It shall have no loose ends and no external pockets except a hip pocket. It shall be worn in such a way that it completely covers all loose ends of outer clothing.
2. No guard shall be removed from any part of machinery except when the observation cannot otherwise be carried out and it shall be replaced immediately the observations have been completed.
3. Appointed persons shall make proper use of any appliances provided for the safe carrying out of the observation.
4. Appointed persons shall make proper use of the secure foot-hold and hand-hold where provided as a precaution against slipping.
5. If a ladder is used it shall either be securely fixed, lashed or footed.
6. Another person, who has been instructed as to what to do in case of emergency shall be immediately available within sight or hearing.
7. Where there is a foreseeable risk of eye injury from the machining process, the appropriate eye protection shall be worn.

SPECIAL NOTES:

* Only persons who have been appointed in writing overleaf, shall enter Transfer Machines specified for the purpose of observation.
* Only persons who have obtained the age of 18 shall be appointed.
* An appointed person must not perform any operation other than that specified in Part II of the certificate.
* The appointed person must have been instructed as to the requirements of the "Procedure for the Safe Entry into Transfer Machines" and be sufficiently trained for the work and be acquainted with the dangers from moving machinery.

Figure 5-4. Permit to work form.
Source: *British Standard Code of Practice for Safety of Machinery,* BS 5304: 1988 Extracts from BS 5304:1988 are produced with the permission of the British Standards Institution. Complete copies can be obtained by post from BSI Sales, Linford Wood, Milton Keynes, MK14 6LE.

PROCEDURE FOR THE SAFE ENTRY INTO TRANSFER MACHINES FOR THE PURPOSE OF OBSERVATION

PURPOSE

In certain exceptional conditions, approach to unfenced, that is to say, unguarded running machinery, is unavoidable. To ensure the safety of persons who approach such machinery the following procedure shall be adopted.

SCOPE

Approach to unfenced running machinery may be made only by persons who have been appointed in writing by an authorised person for the purpose of observations which are considered immediately necessary.

APPLICATION

Where it is considered immediately necessary by an "authorised person" for persons to approach unguarded running machinery to observe malfunctions of process or part(s), a "Certificate of Appointment" shall be issued.

PROCEDURE

Where it is established that it is immediately necessary to approach unfenced machinery the Production Supervisor responsible for the machinery in question shall raise a "Certificate of Appointment" and complete Part I requesting approach to unfenced running machinery.

This certificate is to be handed to the authorised Person who will appoint a person to approach the unfenced machinery where he considers approach immediately necessary.

The 'authorised person' shall complete Part II of the certificate and hand the certificate to the appointed person after returning Part I to the relevant supervisor. On completion of the observations the appointed person shall complete Part III before returning the certificate to the authorised person who issued the certificate.

Note: Where the observations are not completed by an appointed person, the authorised person should ensure that the appointment of another person continuing the observation is made on a further certificate. Part IV should only be completed by the authorised person who issued the certificate.

Note: Where more than one certificate is issued for the same task (only one certificate in force at any one time) all certificates shall be kept together and returned to the Supervisor.

The Supervisor on return of the Certificate(s) can resume normal operations and forward all parts of the certificate to the Safety Department.

DEFINITIONS

Supervisor

Where the procedure refers to Supervisor, it shall be the line Foreman or General Foreman of the area in which the machinery that requires observance is located.

Appointed Person

The person named in Part II of the Certificate of Appointment carrying out the observation of parts of the machinery specified having been sufficiently trained. His appointment is only valid for the period of observation of machinery for which the certificate was raised.

Authorised Person

A person who has been trained in and authorised by management to perform specific tasks in connection with the operations of part or whole of a machine or process. He shall be authorised in writing by the Plant Engineering Manager to determine where approach to unfenced machinery is immediately necessary and consequently to appoint and issue "Certificate of Appointment".

TRAINING

Before the delegation of "Authorised Persons" in respect of the operation of this procedure and before the appointment of any person to approach unfenced running machinery:

1. they must be fully and carefully instructed in:
   * the dangers that may arise when approaching unfenced machinery,
   * the precautions to be adopted and the procedures to be followed,
   * the various guarding systems associated with the types of machinery and with the energy systems available on such machinery,
   * the sequence of operation.
2. they must have a thorough knowledge of the process and the operation of the machinery in which observations are expected.
3. be fully aware of the requirements of this procedure.

Note: Persons undergoing training in accordance with the requirements of the above shall only operate under the immediate supervision of a person who:—

a. has a thorough knowledge and experience of the working of the machine

b. has been trained in accordance with the requirements of this procedure

c. has been authorised or appointed in writing.

Figure 5-5. Safe entry procedure.
Source: *British Standard Code of Practice for Safety of Machinery,* BS 5304: 1988 Extracts from BS 5304:1988 are produced with the permission of the British Standards Institution. Complete copies can be obtained by post from BSI Sales, Linford Wood, Milton Keynes, MK14 6LE.

Again, the French have no holistic standard for hazardous energy control such as ANSI's Z244.1, *Lockout/Tagout*, but they do have a variety of standards that are directly or indirectly relevant. Some of the available French standards with energy isolation references are as follows:

- NFE 09-001 (1980): Technical prevention of accidents that may occur due to mechanical and thermal risks generated by machines and systems.
- E 09-002 (1981): Technical prevention of accidents that may occur due to mechanical risks generated by machines and systems (summary of mechanical risks, of the practices and protective devices).
- E 09-051 (1983): Technical prevention of accidents that may occur because of hazards caused by machines and equipment (locking and interlocking devices for use with guards).
- E 09-052 (1986): Technical prevention of accidents that may occur because of mechanical hazards caused by machines and equipment.
- NFC 79-130 (1985): Electrical equipment of industrial machinery (Part I) General Rules.
- NFE 60-250: Safety techniques applied to machines.

Swedish Standards Institution (SIS)

The Swedish Standards Institution is a national consensus standards–making body that represents Sweden on the Joint European Standards Institute (CEN/CENELEC) and the International Organization of Standards (ISO/IEC). It is an EFTA (European Free Trade) country, which is essentially the Scandinavian counterpart of membership in the EU.

No generic hazardous-energy-control standard exists in Sweden as of this writing, but there are a number of standards that provide appropriate guidance. An excellent example is SS 436-04-21, *Electrical Equipment of Industrial Machines: Prevention of Unintentional or Unauthorized Starting of Machines.* The standard does not address pneumatic or hydraulic forms of power but does provide a series of circuit diagrams that illustrate proper installation of equipment for the prevention of starting (EPS). The following information is derived from an unofficial translation of the standard.

The standard's scope addresses unauthorized starting of an electrically driven machine(s) that has been stopped (such as for maintenance or repair). It provides recommendations for designing EPS for facilities where:

- AC voltage is less than 1,000 V
- AC voltage exceeds 1,000 V
- DC voltage is less than 1,500 V
- control circuits (EPS) exist
- industrial machines (remote control) are used
- industrial machines (internal control) are used.

The standard defines *EPS* as an easily accessible hard-operated switching device and associated circuitry that are intended to overrule any starting function and thereby prevent unintended or unauthorized starting of a motor or a machine.

Section 3, General, describes the nature, location, and function of the EPS switch:

> 3.1 The switching device of an EPS shall have facilities for locking and for indicating the position of the contacts, both facilities being in accordance with the requirements of SS 428 06 05 (IEC 947-3).
>
> 3.2. If possible, the switching device of the EPS shall be installed in the vicinity of and within sight from the motor(s)/machine(s), the unintentional or unauthorized starting of which shall be prevented.
>
> 3.3. As a rule, the switching device shall be connected into the main circuit of the electrical equipment (of the motor). However, this standard also deals with EPS connected in control circuits as there are situations when such a solution must be sought. This standard deals with three basic arrangements (Figure 5-6):
>
> - EPS arranged in a main circuit, without interlocking
> - EPS arranged in a main circuit, with interlocking
> - EPS arranged in a control circuit.

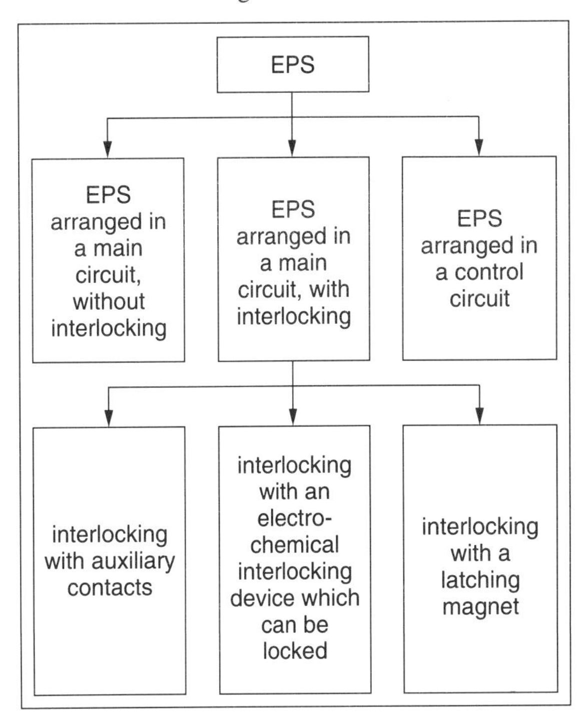

Figure 5-6. Arrangements for EPS.
Source: *Electrical Equipment of Industrial Machines: Prevention of Unintentional or Unauthorized Starting of Machines,* S 436-04-21, Swedish Standards Institution, Stockholm, Sweden: 1989.

3.4. Interlocking is used when the switching device of the EPS has insufficient making or breaking capacity or when it is not intended to break the actual current. Three arrangements for interlocking of the EPS (Figure 5-6) are dealt with in the standard:
- with auxiliary contacts
- with an electromechanical interlocking device which can be locked
- with a latching magnet.

Interlocking of the EPS according to this standard is intended to provide protection in general but the different kinds of interlocking have disadvantages which must be considered when planning and designing.

The use of auxiliary contacts or an electromechanical interlocking device has the disadvantage that there is no return information on whether the switching device closer to the supply is in the OFF or ON position when the EPS is operated. This is, however, not considered to be problem in those cases when one can see from the place where the EPS is installed whether the motor is running or not.

The use of a latching magnet implies that the EPS cannot be operated if there is no control voltage. For this reason, the control supply system of the latching magnets must be so arranged that it is not switched off during normal switching in service.

REGIONAL/GLOBAL CONSENSUS ORGANIZATIONS

Standards harmonization has been the subject of considerable controversy and is the ongoing objective of many individuals, organizations, and countries, for example, worldwide adoption of the metric system. Some of the organizations from around the globe attempting to establish meaningful, uniform standards are the European Committee for Standardization, the International Organization for Standardization, Joint European Standards Institute, World Health Organization, and so on.

Safety and health issues have received greater attention during the past 15 years with regard to regional/global standardization. The emergence of the European Economic Community and its blueprint for the 1992 internal market has added impetus for expansion of broad safety directives, guidelines, and standards.

European Committee for Standardization (CEN)

The European Committee for Standardization (CEN) is a counterpart of the International Organization for Standardization (ISO). It deals with the full field of standardization except for electrical/electronic engineering issues. Membership reflects both EU and EFTA countries as follows:

CEN/CENELEC

EU Countries	*EFTA Countries*
AFNOR (France)	Iceland
AENOR (Spain)	NSF (Norway)
BSI (U.K.)	ON (Austria)
DIN (Germany)	SFS (Finland)
DS (Denmark)	SIS (Sweden)
ELOT (Greece)	SNV (Switzerland)
IBN (Belgium)	
IPO (Portugal)	
Luxemburg	
NNI (Netherlands)	
NSAI (Ireland)	
UNI (Italy)	

A recent (February 1993) CEN initiative that is currently in a draft or provisional form, pr EN 1037, *Safety of Machinery—Isolation and Energy Dissipation—Prevention of Unexpected Start-Up,* is of vital significance with regard to the control of hazardous energy. The draft was prepared by the CEN/CENELEC Joint Working Group (CEN/TC 114), which also prepared EN 418, *Safety of Machinery—Emergency Stop Equipment, Functional Aspects—Principles for Design.* If the draft becomes a European Standard, CEN members are bound to comply with the CEN/CENELEC Internal Regulations that stipulate the conditions for giving it the status of a national standard without any alteration.

The standard's scope involves measures directed at (1) isolating machines from power supplies and dissipating or restraining stored energy, (2) preventing unexpected machine start-up, and (3) allowing safe access in danger zones. It addresses all energy sources such as various power supplies, stored energy, and external influences (such as wind). All parts of the machine workpieces and/or processed material(s) are considered.

The following note seems to acknowledge that there are situations in which energy isolation is neither possible nor practical and that other safeguarding measures must be implemented.

> No guidance is given in the present standard on how to distinguish tasks which are relevant of isolation and energy dissipation from those which are relevant of other means of preventing unintended/unexpected start-up. This distinction can be made only "case after case," on the basis of the risk assessment (including determination of hazards covered) carried out by the machine designer or the C-standard maker. (pr EN 1037)

The introduction to pr EN 1037 states that one of the most important considerations of the safe use of machinery is keeping a machine in a stopped state during human interventions into danger zones. This should be a major goal of the machine designer. Examples of interventions into danger zones are as follows:

- inspections
- corrective actions (clearing blockages, etc.)
- setting, adjustment
- manual loading/unloading
- tool change
- lubrication
- cleaning
- decommissioning
- minor maintenance/repair
- diagnostic work, testing
- work on power circuits
- major maintenance (works requiring significant dismantling).

According to pr EN 1037, machine automation has produced new problems regarding the relationships between operation/motion and stopped state/rest and what was expected from these pathways. Therefore, a machine having its movable component at rest shall be considered as operating if any element can be automatically started.

The pr EN 1037 standard incorporates by reference provisions from other normative publications, several of which are listed here:

- EN 292-1: 1991. Safety of machinery—Basic concepts, general principles for design—Part 1: Basic terminology, methodology.
- EN 292-2: 1991. Safety of machinery—Basic concepts, general principles for design—Part 2: Technical principles and specifications.
- EN 60 204-1: 1992. Electrical equipment of industrial machines—Part 1: General requirements.
- pr EN 983. Safety of machinery—Safety requirements for fluid power systems and components—pneumatics.
- pr EN 982. Safety of machinery—Safety requirements for fluid power systems and components—hydraulics.

According to pr EN 1037, *isolation and energy dissipation* is defined as consisting of four inseparable actions:

1. **Isolating** (disconnecting, separating) **the machine** (or a defined part of the machine) from all power supplies
2. If necessary (for instance, on large machines or in installations), **locking** (or otherwise securing) **all the isolating units** in the isolating position
3. **Dissipating or restraining/containing any stored energy** that may give rise to a hazard

   NOTE: Energy considered in (3) above may be stored in, for example,
   - mechanical parts continuing to move through inertia
   - mechanical parts liable to move by gravity
   - capacitors, accumulators
   - pressurized fluids
   - springs.
4. **Verifying** by a safe working procedure the effect of the measures taken previously.

Of special interest are several statements in the general considerations section directed at machinery designers:

According to an essential safety requirement of EN 292-2, all machines have to be provided with means intended for isolation and energy dissipation, particularly in view of major maintenance work on power circuits and decommissioning. In view of foreseeable necessary interventions for which isolation and energy dissipation are not appropriate, the designer shall provide (per the risk assessment) additional means to prevent unexpected start-up.

According to EN 292-1, the designer should determine as completely as possible the different intervention procedures for the operators, so that appropriate safety measures can be associated with each procedure. This is intended to prevent operators from being induced to use hazardous intervention techniques because of technical difficulties.

Section 5, *Means intended for isolation and energy dissipation*, covers means for power supply isolation, locking, energy dissipation/restraint, and verification/testing procedures.

Section 6, *Other means intended for preventing unexpected start-up*, includes design strategy, protection from accidental generation of start commands, multilevel stop commands, and automatic monitoring of the stopped condition.

Section 7, *Additional measures*, identifies the need for signaling and warning devices for the various machine states and the criteria for immobilization of moving parts.

European Committee for Electrotechnical Standardization (CENELEC)

The European Committee for Electrotechnical Standardization (CENELEC) corresponds to the International Electrotechnical Commission (IEC). It focuses on electrical/electronic engineering standards and includes representatives from both EU and EFTA countries. As mentioned above, CEN and CENELEC collaborated as a joint working group when drafting pr EN 1037. In addition, numerous CENELEC standards exist that deal with electrical hardware criteria, design, specification, and so on that have bearing on electrical energy isolation.

International Organization for Standardization (ISO)

The International Organization for Standardization (ISO) has over 90 members representing the world's principal standards-making organizations. In concert with the International Electrotechnical Commission, (IEC), it produces more than 85% of all international standards in the context of the General Agreement on Tariffs and Trade (GATT) code. The ISO/IEC secretariats are distributed among 31 countries with over 3,000 technical committees and working groups. The ISO and IEC have produced over 10,000 standards, with a similar number in various stages of preparation and approval. At the present time, the ISO has no broad standard dealing with the control of hazardous energy, which explains the lack of reference by the CEN/CENELEC joint working group 114, Safety of Machinery (pr EN 1037).

International Electrotechnical Commission (IEC)

The International Electrotechnical Commission (IEC) is the global counterpart of CENELEC. Its emphasis is on electrical/electronic matters. As mentioned previously, its functions are closely tied to ISO activities, generating approximately one-third of the total ISO/IEC standards and managing about the same number of technical committees/working groups.

Various standards associated with electrical hardware have importance with regard to designs and procedures for energy isolation. Several of particular relevance are as follows:

- IEC 947-3. Safety switches for maximum 1,000 V AC.
- IEC 529. Classification of degrees of protection provided by enclosures.
- IEC 157-1. Circuit breakers.
- IEC 204-1. Electrical equipment of industrial machines—general requirements.
- IEC 265-1. High-voltage switches.
- IEC 408. Air-break switches and disconnects (low voltage).

European Union (EU)

As Europe moves toward the creation of a single market, the member nations of the European Economic Community (EEC), which was renamed the European Union (EU) in 1994, must achieve the free movement of goods across national boundaries. All EU countries have laws regarding product safety, manufacturing, and so on, but different laws can cause technical barriers to trade. The EU governing structure is complex, consisting of a hierarchy of elected or nominated officials and permanent civil servants, mostly centered in Brussels and Luxembourg. The structure is as follows:

- *Council of Europe*. This supreme body consists of the heads of state of the member nations. The council meets twice a year to develop common strategies on foreign policy and internal matters.
- *Council of Ministers*. This is made up of appropriate ministers from member states, the number determined by the topics. It is the major policy-making body of the EU.
- *European Commission*. Consisting of representatives from each member state, this commission proposes new initiatives, advances proposed policies, and implements adopted policies. It is of major importance in the occupational health and safety field.
- *The European Parliament*. Members of this assembly are elected by citizens of each member state. It has no legislative powers, but its main functions are to discuss issues, express opinions, and supervise the commission and council.
- *Economic and Social Committee*. This is a consultative body with 189 delegates from member states representing trade unions, employers, and other interest groups. Its main function is to generate ideas and serve as consultants to other councils, commissions, and committees.

The EU's process of formulating directives and regulations is shown in Figure 5-7.

In May 1985, for example, European Community ministers agreed on what they called the "New Approach to Technical Harmonisation and Standards" as a way to address this long-standing problem for business. These "new approach" directives, or EU laws,

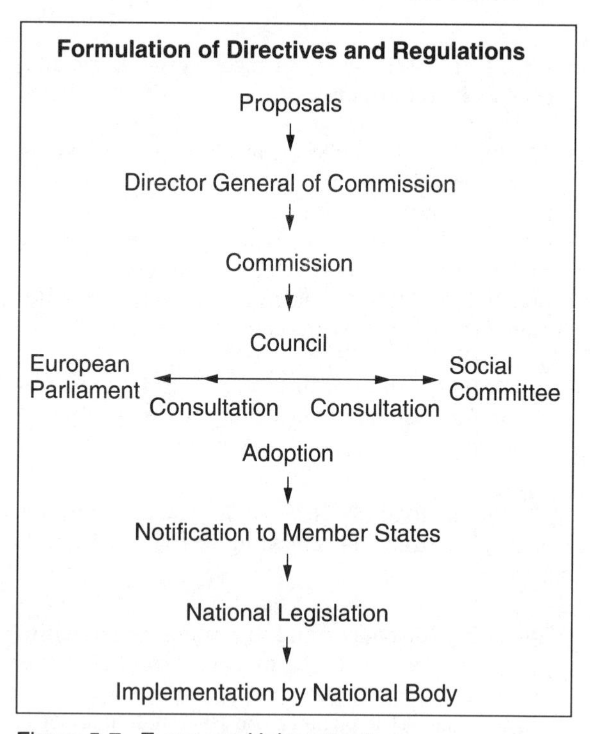

Figure 5-7. European Union structure.
Source: M.S. Hughes, "Europe's Mixture of Diverse Customs Poses Challenge to EEC Safety Goals," *Occupational Health and Safety* (October 1993). Reprinted courtesy of Occupational Health and Safety.

establish "essential requirements" (such as, for safety) that must be met before products can be sold anywhere in the European Union. European standards fill in the details.

In June 1989, the European Community passed Council Directive 89/392/EEC, commonly known as the Machinery Directive, establishing requirements for the safe design and manufacture of machines. Amended in 1991, this directive does not apply to machines for which the risks are mainly of electrical origin; such machinery is covered by Council Directive 73/23/EEC, which is the harmonization of laws of member states relating to electrical equipment designed for use within certain voltage limits (the so-called Low-Voltage Directive).

The Machinery Directive is divided into nine annexes, or parts, as follows:

- Annex A: Machinery excluded from the coverage of the directive
- Annex B: Essential health and safety requirements relating to the design and construction of machinery
- Annex C: Types of machinery subject to special procedures
- Annex D: Technical construction file
- Annex E: EC type examination
- Annex F: EC declaration of conformity
- Annex G: Attestation procedures
- Annex H: Directive on the use of work equipment at the workplace
- Annex J: Standards supporting the Directive.

Annex B: Essential health and safety requirements. Annex B is divided into five parts that cover a wide range of potential dangers to workers and other exposed persons who are within a specified danger zone around machine assembly and manufacture.

Part 1 includes the materials used in construction of the machinery; lighting; controls, stability, fire, noise, vibration, radiation, emission of dust, gases, and so on; maintenance; and instructions. Part 2 has additional requirements for agrifoodstuffs machinery, portable hand-held machinery, and machinery for working wood and analogous materials. Part 3 deals with particular hazards associated with mobility. Part 4 addresses hazards related to lifting, and Part 5 addresses those associated with working underground.

Companies can show they satisfy the essential health and safety requirements by manufacturing in conformance with specified European standards or with the essential health and safety requirements listed in the directive. To show compliance, the manufacturer, or its authorized representative in the EU, must create and keep available the technical construction file described in Annex D or Annex E, depending on the type of machinery constructed. The manufacturer is then issued a certificate of adequacy, and a CE mark is affixed to the machine to indicate it fulfills the requirements of the directive.

Preventing unwanted energy transfers. Section 1.6 of the directive, Maintenance, established safe working procedures for repairing and maintaining manual and automated machinery. The sections dealing with isolating energy sources read as follows:

> **1.6.1 Machinery Maintenance**
> Adjustment, lubrication, and maintenance points must be located outside danger zones. It must be possible to carry out adjustment, maintenance, repair, cleaning, and servicing operations while machinery is at a standstill.
>
> In the case of automated machinery and, where necessary, other machinery, the manufacturer must make provision for a connecting device for mounting diagnostic fault-finding equipment.
>
> Automated machine components which have to be changed frequently, in particular for a change in manufacture or where they are liable to wear or likely to deteriorate following an accident, must be capable of being removed and replaced easily and in safety. Access to the components must enable these tasks to be carried out with the necessary technical means (tools, measuring instruments, etc.) in accordance with an operating method specified by the manufacturer.
>
> **1.6.3 Isolation of Energy Sources**
> All machinery must be fitted with means to isolate it from all energy sources. Such isolators must be clearly identified. They must be capable of being locked if reconnection could endanger exposed persons. In the case of machinery supplied with electricity through a plug capable of being plugged into a circuit, separation of the plug is sufficient.
>
> The isolator must be capable of being locked also where an operator is unable, from any of the points to which he has access, to check that the energy is still cut off.
>
> After the energy is cut off, it must be possible to dissipate normally any energy remaining or stored in the circuits of the machinery, without risk to exposed persons.
>
> As an exception to the above requirements, certain circuits may remain connected to their energy sources in order, for example, to hold parts, protect information, light interiors, etc. In this case, special steps must be taken to ensure operator safety (The Department for Enterprise, 1992).

Council Directive of November 1989. Along with the Machinery Directive, this Council Directive (89/655 EEC) holds employers primarily responsible for the health and safety of workers as regards the provision and use of work equipment, which is defined as "an activity involving work equipment such as starting or stopping the equipment, its use, transport, repair, modification, maintenance and servicing" (The Department for Enterprise, 1992). Work equipment can be anything from automatic car-wash machinery to a hand saw and covers a wide range of power and manual machinery and tools.

Among other provisions, this directive requires employers to ensure that only specifically designated workers carry out repairs, modifications, maintenance, or servicing. Employers must provide information, instruction, and training to their employees regarding work equipment. In addition, there are minimum requirements on specific health and safety hazards with which work equipment must comply such as visible, identifiable, and clearly marked control devices; suitable guarding for all exposed or hazardous parts; work equipment shut down for maintenance or, if this is not possible, protective measures carried out to prevent harm (The Department of Enterprise, 1992).

These directives, which are supported by standards developed by the member countries, are the EU's efforts to establish uniform standards in worker health and safety.

NATIONAL REGULATIONS

Most of the industrialized countries of the world promulgate numerous standards governing workplace health and safety. With regard to the control of hazardous energy, most countries outside of North America have opted to approach the subject of energy isolation by either promulgating machinery/electrical standards or establishing requirements for "safe work systems or permitting." There is no evidence of any treatment of the general subject by other nations that parallels the United States' governmental lockout/tagout approach. However, the United Kingdom's safe-systems-of-work approach and EEC Directives 89/392 (Machinery) and 89/665 (Work Equipment) may result in a European regulation of superior effectiveness.

United Kingdom

In Great Britain, the Health and Safety Executive Agency is the regulatory authority for occupational safety and health. It establishes standards, enforces the statutory requirements, provides consultation to employers, employees, unions, and trade associations and produces various education/training materials for public use.

Although there is no specific lockout/tagout standard in the nation at this time, industries, trade unions and associations, and government use the Health and Safety at Work Act (1974) as a general umbrella for safe practices, along with regulations passed under the Factories Act (1961) and the Offices, Shops and Railway Premises Act 1963 (for example, the Power Press Regulations of 1965 and the Prescribed Dangerous Machines Order of 1964, requiring that new workers receive sufficient training or adequate supervision before working with dangerous machines). The Health and Safety at Work Act, Section 2, charges employers with the duty "to ensure, so far as is reasonably practicable, the health, safety and welfare at work of all employees . . . the provision and maintenance of plant and systems of work that are, so far as is reasonably practicable, safe and without risks to health" [Health and Safety at Work Act, Section 2(1) (2) (a)].

The most extensive of the duties placed on the employer are those of providing a safe system of work, which is a formal procedure that results from a systematic examination of a task in order to identify all the hazards. It defines safe methods to ensure that hazards are eliminated or risks minimized. A safe system of work is needed when hazards cannot be physically eliminated and some element of risk remains. Some of the cases in which a safe system is required could also be applied to lockout/tagout procedures:

- cleaning and maintenance operations
- making changes to work layouts, materials used, or working methods
- employees working away from their regular work stations or working alone
- machine breakdowns or emergencies.

The safe system of work involves five key steps:

1. *Assess the task to be done.* This step involves considering what is used (for example, does potential exist for failures of machinery, electrical hazards, or (automatic or manual controls being operated), who does what task (delegation, training, and human errors), where the task is carried out (localized hazards in the workplace, problems caused by poor lighting or power supplies, and adjacent work operations), and how the task is done (potential failures in work methods and procedures).
2. *Identify the hazards.* Where possible, eliminate the hazards, such as locking and tagging adjacent power supplies, controls, or machinery. Reduce other risks before implementing a safe system of work.
3. *Define safe methods.* A safe system of work can be defined orally, by a simple written procedure, or by a formal permit to work (Figure 5-8). It may include a range of precautions from simple lockout procedures and protective equipment through a full written permit to work. The steps to arrive at a safe system of work are diagrammed in Figure 5-9, which shows the decision-making process that should be followed to help ensure safe working conditions and procedures.

The steps in a safe work system include the preparation and authorization needed at the beginning of the job; clear planning of each work step;

Nacanco Limited

Permit To Work

Mechanical/Electrical MV/LV

PTW No. M ____

1 Issue

To ____________________ Employed By ________________

For the following work to be carried out:

____

____

It is safe to work on the following apparatus provided that all the requirements of the Safety Rules are observed.

____

____

All Other Parts Are Dangerous

] ________________

] ________________

State points of isolation (Steam,] ________________
Water, Air, Gas Valves, etc. shut] ________________
and locked off and isolated from] ________________
their motors.] ________________
] ________________

State special precautions (flames] ________________
or smoking prohibited, and define] ________________
protective clothing required, etc.)] ________________
Nature of tests carried out and] ________________
results.] ________________

Signed ______________ Time ___________ Date ___________
Being an authorised person.

2 Receipt

Note: Following issue, the permit must be signed by the person in charge of the work and retained in his possession until the work is completed.

I accept responsibility for carrying out work on the apparatus detailed on this permit in accordance with the provisions of the Safety Rules (Electrical and Mechanical) and no attempt will be made by me or any man under my control, to carry out work on any other apparatus.

Signed ______________ Time ___________ Date ___________
Being the person in charge of the work.

CC967

Figure 5-8. Nacanco Ltd. permit to work—front. (*continued*)
Source: Courtesy of American National Can Company.

specifying safe work methods for each part of the job; establishing means of access and escape; and planning for the tasks of dismantling, disposal, and/or returning the work area to its original state at the end of the job.

In special cases where a permit-to-work system is needed, management will need to use properly documented procedures. Permits to work should (a) define the work to be done, (b) explain how to make the work area safe, (c) identify remaining hazards

Permit To Work

Mechanical/Electrical MV/LV

3 Clearance

Note: The apparatus mentioned hereon must not be recommissioned until this clearance has been signed and the Permit returned to the person in charge of the work and cancelled.

All men under my charge have been withdrawn and warned that it is no longer safe to work on the apparatus specified on this permit, and all gear and tools are clear, and that all guards have been replaced and loose material removed.

Signed ______________ Time ____________ Date ____________
Being the person in charge of the work.

4 Cancellation

This permit and all copies of it are cancelled.

Signed ______________ Time ____________ Date ____________
Being an authorised person.

BE SURE. BE SAFE.
REPORT ANY DEFECTS

Figure 5-8. Nacanco Ltd. permit to work—back. (*concluded*)
Source: Courtesy of American National Can Company.

and the precautions to be taken, (d) describe inspections to be carried out before normal work can be resumed, and (e) identify the person responsible for controlling the job.

Jobs that regularly require a permit-to-work system include work in confined spaces and on electrical equipment; cutting into pipework containing hazardous substances; and hot work on equipment or facilities containing flammable dusts, liquids, gases, or residues of these substances.

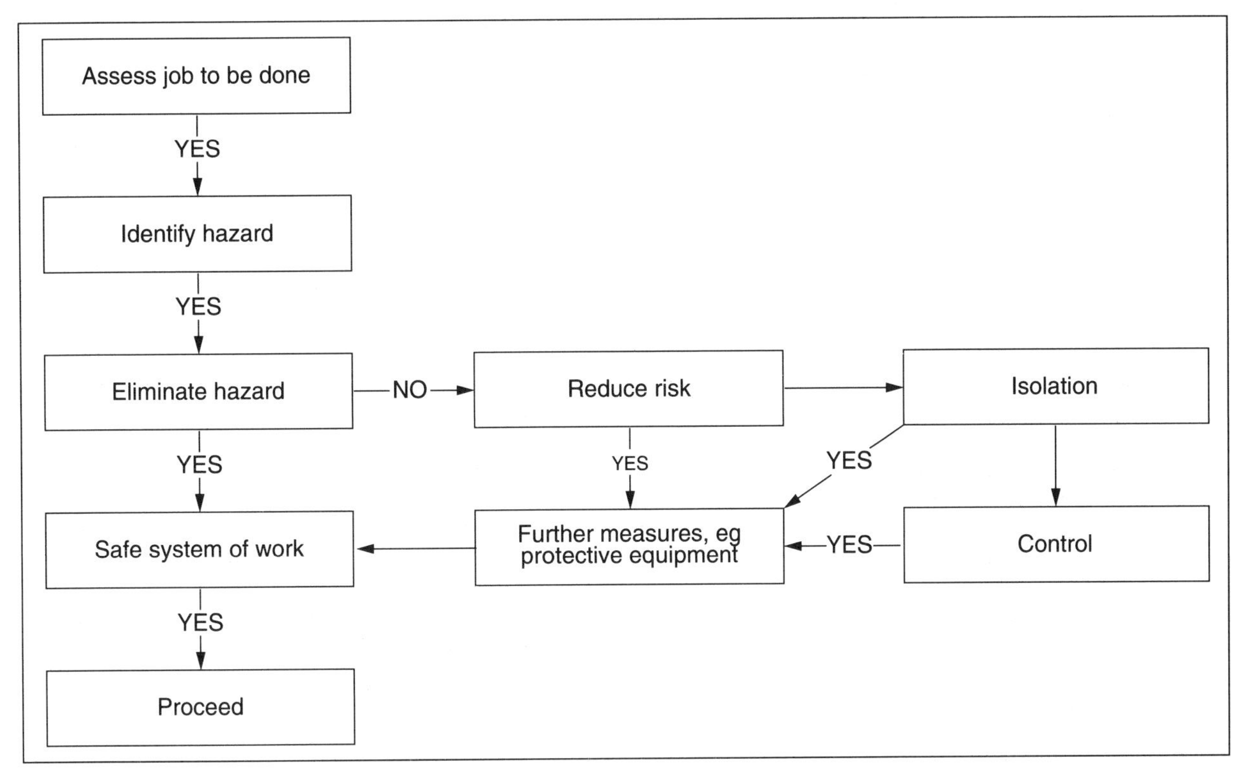

Figure 5-9. Safe-system-of-work logic diagram.
Source: U.K. Health and Safety Executive Agency, *Dangerous Maintenance* (London: HSMO, 1987). British Crown copyright. Reproduced with the permission of the controller of Her Majesty's Stationery Office.

4. *Implement the system.* Once it has been designed, the system must be communicated properly, understood fully by employees, and applied correctly. Supervisors and managers should be trained in safe systems of work and should be fully aware of potential risks and the precautions that may be necessary. Shortcuts, in particular, are to be avoided. If an unexpected problem arises, management should stop the work until a safe solution can be found.
5. *Monitor the system.* Management should periodically check to ensure that employees continue to find the system workable, that procedures established in the safe system of work are being carried out and are effective, and that any changes in circumstances that may require changes to the system are taken into account.

All safe systems of work should be fully documented in terms of hazards, precautions, and detailed safe working conditions. The use of written safe systems of work can be one of the best methods of communicating health and safety information to employees, outside contractors, and self-employed workers. Written safe systems of work should include the following information:

- correct use of facility, tools, and equipment
- details about the working environment—heating, lighting, ventilation, and so on
- in-house safety rules
- formal issue, proper use, and regular maintenance of all necessary protective equipment and clothing
- relevant chemical/product safety literature
- correct and safe working methods, incorporating appropriate safety precautions.

The techniques of job safety analysis and measurement of safety performance can be used to monitor the effectiveness of safe systems of work and other safety issues regulated under the Health and Safety at Work Act.

The U.K. Health and Safety Executive Agency (HSE) also prepares publications in collaboration with industry trade associations that analyze accidents in various industries and how to prevent them. Such booklets as *Deadly Maintenance: A Study of Fatal Accidents at Work*, which documents hazards in general maintenance work, and *Dangerous Maintenance*, which examines maintenance accidents in the chemical industry, provide detailed analyses of case studies and preventive measures.

The HSE produced the Provision and Use of Work Equipment Regulations 1992 (PUWER), which established important health and safety laws for work equipment and its safe use per Directive 89/655/EEC. PUWER amplifies and makes more explicit the general duties of employers, the self-employed, and persons in control to provide safe facilities and equipment. They became effective on January 1, 1993.

Although various regulations contained in PUWER have some bearing on the control of hazardous energy, regulation 19, "Isolation from Sources of Energy," is specific. It states the following:

> (1) Every employer shall ensure that appropriate work equipment is provided with suitable means to isolate it from all its sources of energy.
> (2) Without prejudice to the generality of paragraph (1), the means mentioned in that paragraph shall not be suitable unless they are clearly identifiable and readily accessible.
> (3) Every employer shall take appropriate measures to ensure that reconnection of any energy source to work equipment does not expose any person using the work equipment to any risk to his/her health or safety.

The HSE gives further narrative guidance in its *Work Equipment* publication to help affected parties to understand the intent of the regulations as fully as possible. The section on energy isolation is as follows:

> Isolation means establishing a break in the energy supply in a secure manner, i.e., by ensuring that inadvertent reconnection is not possible. The possibilities and risks of reconnection should be identified as part of the risk assessment, which should then establish how security can be achieved. For some equipment, this can be achieved by simply removing the plug from the electrical supply socket. For other equipment, an isolating switch or valve may have to be locked in the OFF or CLOSED position to avoid unsafe reconnection. The CLOSED position is not always the safe position: for example, drain or vent outlets may need to be secured in the open position. If work on isolated equipment is being done by more than one person, it may be necessary to provide a locking device with multiple locks and keys; each will have their own lock or key, and all locks have to be taken off before the isolating device can be removed.
>
> The main aim of this Regulation is to allow equipment to be made safe under particular circumstances, such as when maintenance is to be carried out, when an unsafe condition develops (failure of a component, overheating, or pressure build-up), or where a temporarily adverse environment would render the equipment unsafe, for example electrical equipment in wet conditions or in a flammable or explosive atmosphere.
>
> There may be some circumstances in which, for particular safety reasons, stopping equipment does not remove all sources of energy, i.e., the power supply is helping to keep the equipment safe. In such cases, isolation could lead to consequent danger, so it will be necessary to take appropriate measures to overcome that risk before attempting to isolate the equipment.
>
> It is appropriate to provide means of isolation where the work equipment is dependent upon external energy sources such as electricity, pressure (hydraulic or pneumatic) or heat. Internal energy which is an inherent part of the materials from which the equipment is made, such as its potential energy, chemical or radiological energy, similarly cannot be isolated from the equipment; nevertheless there should be means of preventing such energy adversely affecting workers, by restraint, barrier or shielding.
>
> Electrical isolation of electrical equipment for work on or near conductors is dealt with by regulation 12 or the Electricity at Work Regulation 1989. Guidance to those Regulations expands on the means of isolating electrical equipment. Note that those Regulations are only concerned with electrical danger (electric shock or burn, arcing and fire or explosion caused by electricity), and do not deal with other hazards (such as mechanical) that may arise from failure to isolate electrical equipment.
>
> Heat energy may be supplied by circulation of preheated fluid such as water or steam. In such cases, isolating valves should be fitted to the supply pipework. Similar provision should be made for energy supplies in the form of liquids or gases under pressure. The performance of such valves may deteriorate over time, and their effectiveness often cannot be judged visually. A planned preventive maintenance programme should therefore be instigated which assures effective means of isolation.
>
> The energy source of some equipment is held in the substances contained within it. Examples are the use of gases or liquids as fuel, electrical accumulators (batteries), and radionuclides. In such cases, isolation may mean removing the energy-containing material, although this may not always be necessary.
>
> Also, it is clearly not appropriate to isolate the terminals of a battery from the chemical cells within it, since that could not be done without destroying the whole unit.
>
> Some equipment makes use of natural sources of energy such as light or flowing water. In such cases suitable means of isolation include screening from light and the

means of diverting water flow respectively. Another natural energy source, wind power, is less easily diverted, so sail mechanisms should be designed and constructed so as to permit minimal energy transfer when necessary.

Regulation 19(3) requires precautions to ensure that people are not put at risk following reconnection of the energy source. So, reconnection of the energy source should not put people at risk by itself initiating movement or other hazard. Measures are also required to ensure that guards and other protection devices are functioning correctly before operation begins.

Canada: Ontario

The province of Ontario passed the Occupational Health and Safety Act and Regulations for Industrial Establishments in 1980. This code, like OSHAct in the United States, covers such areas as definitions of industries, duties of employers, responsibilities of agents such as safety engineers and inspectors, and requirements for safe working conditions and safety training of employees.

Lockout/tagout issues are dealt with under Regulation 692, sections 36 (tumbling mills and dryers), 46 (power supply), 54 (silos, bins, and hoppers), 72–75 (confined spaces), 78–80 (machinery), and 82 (drums, tanks, and pipelines). The following excerpt, sections 79 and 80, concerns the maintenance and repair of machinery:

> **79.** A part of a machine, transmission machinery, devise or thing shall be cleaned, oiled, adjusted, repaired or have maintenance work performed on it only when,
> (a) motion that may endanger a worker has stopped; and
> (b) any part that has been stopped and that may subsequently move and endanger a worker has been blocked to prevent its movement. R.R.O. 1980, Reg. 692, s. 79.
>
> **80.** Where the starting of a machine, transmission machinery, or devise or thing may endanger the safety of a worker,
> (a) control switches or other control mechanisms shall be locked out; and
> (b) all the effective precautions necessary to prevent such starting shall be taken. R.R.O. 1980, Reg. 692, s. 80.

Section 46 offers guidelines on disconnecting and locking out the power supplies of machinery before work is done on them. If the power supply exceeds 750 V and cannot be disconnected, the section describes the protective equipment and safe work procedures to be followed:

> (a) the work shall be carried out by a competent person under the authority of an electrical utility; and
> (b) rubber gloves, mats, shields or other protective equipment, and procedures adequate to ensure the safety of all workers shall be used while work is being performed; and
> (c) a person, other than the worker doing the work, who is trained in the use of artificial respiration, shall be conveniently available while the work is being performed. R.R.O. 1980, Reg. 692, s. 46 (Workplace Health and Safety Agency, 1991).

COMPLIANCE AND PREVENTION

Letter Versus Spirit

Part of the problem in establishing codes of safe practice has been that standards and regulations must cover worker behavior as well as workplace conditions, machinery, and other items. Generally, standards and regulations are more effective when they can be defined specifically regarding either conditions (such as the amount of light, the quality of air, and the amount of noise) or machinery and other items (in other words, machine guards, control panel arrangements, and storage of drums) and when they can be monitored to ensure compliance.

Worker behavior and work practices, on the other hand, are more difficult to standardize because of the variables involved with individual workers. For example, rules of the road for driving are universally accepted and applied with reasonable consistency from country to country. Drivers know they must come to a complete stop at a stop sign. However, there may be hundreds of circumstances in which drivers merely pause at a stop sign before driving through an intersection. They may know the intersection well and feel no need to come to a complete stop. Each driver and each intersection produce a set of circumstances in which conformance to the rules varies greatly.

This leads to the question, Is a standard for behavior really a standard if the majority of people for whom it is developed do not comply? On many major urban highways, for example, most drivers exceed the established speed limit and maintain less than the required distance in car lengths between themselves and the car ahead of them, particularly during commute times. Someone who drives at the speed limit and attempts to leave adequate space between his/her car and a car traveling in front simply creates an opening into which another car will move, eliminating the safety cushion. Technically, there is nothing wrong with the speed limit or the general rule that an appropriate number of car lengths must separate cars on the road to allow adequate space for emergency stopping. But the realities of urban driving make both the speed limit and the safe-distance rule less effective than they were intended to be.

For the same reason, regulations governing worker behavior are much more difficult to establish and effectively monitor and enforce. The performance style of standards development, that is, the British Safe Systems of Work, seem to offer the most promise in focusing more on the end objective than on the means.

Minimalist Approach Versus Best Practice

In the minimalist approach, an employer adopts the minimum requirements for lockout/tagout safety that will qualify as complying with the regulations. For example, if a written policy for locking out machinery is required, the employer may simply list the minimum steps as suggested in the regulations and neglect to tailor the policy to the specific hazards of the particular jobs that workers do. The minimalist approach to hazard control may comply with the letter of the law but violates the spirit in failing to provide the maximum protection for workers.

The best practices method, in contrast, looks at each job individually and evaluates the particular hazards and skills needed to do the job safely. For example, repairing a plastic injection molding machine and repairing a conveyor belt have some hazards in common—both jobs involve locking off electricity and blocking moving parts. But they also have unique hazards that require different safety procedures. For example, the person repairing the molding machine may be exposed to thermal hazards from the mold, whereas the person working on the conveyor belt may have to climb to an elevated position and be exposed to a fall hazard.

This approach also looks at the way people work. It takes into consideration the shortcuts, mistakes, and choices that workers make when doing a job and tries to develop practices that eliminate unsafe work steps and encourage safe work habits. In this way, employers can devise safety measures that recognize how people really work, as opposed to a generic list of work practices based on a hypothetical or "average" worker.

As a result, the best practices approach is most appropriate for ensuring that workers understand how to work safely and prevent hazardous-energy-release incidents. This approach goes beyond mere compliance with the law and addresses the specific hazards to which workers are exposed in their workplace. Most regulations, in fact, contain some type of statement that indicates they are to be used only as guidelines and not as specific work practices. Generic safety procedures are usually not enough to control energy hazards.

Compliance = Accident Free?

The ideal goal of any safety program is to eliminate all accidents in the workplace. This is as true of the evolving lockout/tagout standards as it is of fire prevention or toxic chemical control.

But even 100% compliance with the existing safety requirements in any firm is unlikely to eliminate accidents completely because of a multitude of uncontrolled variables, not to mention the inevitable human mistakes and errors in judgment that occur in any workplace. In fact, as previously mentioned, placing too much confidence in standards compliance can actually make workers less vigilant and, ironically, increase the chances that an incident will occur.

Regulatory compliance should be viewed for what it is—a minimum threshold for safety and not the ultimate solution. National standards and regulations must be written broadly, particularly regarding the dynamics of behavior and how work is actually done. This leaves the employer to address how the specific local circumstances and conditions will be handled so that the required tasks can be performed without incident. This unavoidable fact, therefore, moves all responsible employers beyond compliance. In most cases, regulations and standards serve as a framework for hazard control. It remains the task of the employer to finish the structure.

Commitment to Safety

The most rigorous and detailed regulations, policies, standards, or codes do little to safeguard workers unless there is a strong commitment to safety in a company. This commitment must come from top management and be embraced by every manager down to the line supervisors and lead persons. If safety does not appear to have high priority among management, workers will see little reason to comply with safety measures that in some cases can be time consuming, inconvenient, or difficult.

A commitment to safety includes the following:

- Identifying the energy hazards in a company's workplace
- Identifying unsafe practices and high-risk tasks
- Knowing which lockout/tagout and other hazardous energy control regulations apply
- Developing written policies and best practices for safety
- Training, motivating, and rewarding employees not only to work safely but also to make safety a high priority in all phases of their jobs. Employees must understand the need for safety procedures and be willing to practice them each time they must repair, adjust, or service machinery or work with energized equipment.
- Developing methods to monitor compliance and to measure results.

The standards and regulations regarding lockout/tagout procedures have been painstakingly developed over the past few decades as good tools for employers and employees to use in the workplace. Applying those tools requires dedication, hard work, and a genuine desire on the part of both management and workers to prevent energy-release incidents.

SUMMARY

- Western Europe has been active since the 1940s in national and international efforts to prevent unintended hazardous energy releases. With the emergence of multinational corporations in developed and developing countries, the use of new forms of energy and sophisticated technology has increased. These firms may have control measures in place that do not exist on the national level.
- In Western Europe, the United Kingdom, France, Sweden, and Germany are the leaders in the control of hazardous energy release. The BSI has developed some of the earliest and most influential standards. Little is known about efforts in Eastern Europe and Russia. Developments in Asia, Africa, and the Pacific Rim range from crude to sophisticated. Canada and Australia follow BSI standards, whereas Central and South America have fairly advanced techniques imported through multinational corporations working in oil, mining, and manufacturing industries.
- The International Labor Organization developed a *Model Code of Safety Regulations for Industrial Establishments for the Guidance of Government and Industry*, first published in 1949. The code covered all aspects of workplace safety and health, with special sections devoted to energy isolation and preventing hazardous energy release.
- Various private and quasi-governmental safety organizations have dedicated their efforts to improving the safety and health of citizens and workers. They are often hindered by lack of funds, but they have published numerous pamphlets, booklets, standards, and guidelines for industry. These organizations include the Industrial Accident Prevention Association and the French National Research and Safety Institute.
- National consensus organizations, particularly in Europe, have developed minimum requirements for controlling hazardous energy sources. In most instances, the standards have been derived from machinery- and electrical-based safety criteria and have not addressed energy isolation comprehensively. BSI, the French Standards Association (AFNOR), and the Swedish Standards Institute (SIS) have all developed a number of standards that provide appropriate guidance for most types of hazardous energy control. With the exception of BSI, however, they have no holistic standards devoted to hazardous energy isolation or control.
- On the regional and global level, several consensus organizations have attempted to harmonize standards, but the task has proven difficult. Among these groups are the European Committee for Standardization, the European Committee for Electrotechnical Standardization, the International Organization for Standardization, the International Electrotechnical Commission, and the European Union. The creation of uniform standards is regarded as a major step toward greater economic trade and cooperation among nations.
- Most of the industrialized nations of the world promulgate numerous standards governing workplace health and safety. Most countries outside North America have opted to approach the subject of energy isolation either through machinery/electrical standards or through requirements for safe work systems or permitting. Chief among these nations is the United Kingdom, which works through its Health and Safety Executive Agency.
- The key steps of Britain's safe work system are as follows: assess the task to be done, identify the hazards, define safe methods, implement the system, and monitor the system. The use of written safe systems of work can be one of the best methods of communicating health and safety information to employees, outside contractors, and self-employed workers.
- The province of Ontario in Canada passed the Occupational Health and Safety Act and Regulations for Industrial Establishments in 1980. This code, like OSHAct, covers such topics as definitions of industries, duties of employers, responsibilities of agents, and requirements for safe working conditions and safety training of employees.
- Work in hazardous energy control since the end of World War II has made considerable strides, but much remains to be done. Standards and regulations are fairly effective regarding work conditions or machinery and other tangible items but are less effective in standardizing worker behavior and work practices.
- Adding to the difficulty, many employers adopt the minimum requirements of lockout/tagout safety and fail to provide maximum protection for workers. A best practices approach, in contrast, looks at each job individually and evaluates the particular hazards and skills needed for its safe accomplishment. This approach also takes into consideration the way people actually work.
- Regulatory compliance must be viewed for what it is—a minimum threshold for safety, not an ultimate solution. National standards are written broadly. It is up to the employer to address how specific local circumstances and conditions will be handled. Above all, controlling hazardous energy requires a total commitment to safety from the top management levels to the line workers.

REFERENCES

British Government, Secretary of State. *Health and Safety at Work, etc. Act of 1974*, Chapter 37. London: HMSO, 1974.

British Standards Institution. *British Standard Code of Practice for Safety of Machinery,* BS 5304: 1988. London: Butler and Tanner, Ltd., 1988.

The Department for Enterprise. *The Single Market—Machinery.* London, England: Department of Trade and Industry, 1992.

Eicher LD. *European Community—1992: Regional Standardization, Testing, and Certification: What Effect on Global Competitiveness?* Paper presented at ANSI Annual Public Conference, March 1989.

European Committee for Standardization (CEN). *Safety of Machinery—Isolation and Energy Dissipation—Prevention of Unexpected Start-up*, pr EN 1037 (draft). Brussels, Belgium: CEN, 1993.

French National Research and Safety Institute. *Closure and Unclosure*, ED 754. Paris, France: French National Research and Safety Institute, 1992.

French National Research and Safety Institute (INRS). *Safety Advice for Servicing and Working on Electrical Systems and Facilities (Low Voltage)*, ED 539. Paris, France: INRS, 1990.

Hughes MS. Europe's mixture of diverse customs poses challenge to EEC safety goals. *Occupational Health and Safety* 62, no. 10 (October 1993): 68–70.

Industrial Accident Prevention Association. *Confined Space Entry,* Booklet BO 1228. Toronto, Ontario: Industrial Accident Prevention Association, 1984.

Industrial Accident Prevention Association. *Lockout.* Toronto, Ontario: IAPA, 1990.

Industrial Accident Prevention Association. *Work Permits*, LPAA726. Toronto, Ontario: Industrial Accident Prevention Association, 1993.

International Labor Organization. *Model Code of Safety Regulations for Industrial Establishments for the Guidance of Government and Industry.* Geneva, Switzerland: International Labour Organization, 1954.

Ontario Ministry of Labour. *Lock-out Procedure for Machinery*, Engineering Data Sheet No. 9.02. Toronto, Ontario: Ontario Ministry of Labour, 1980.

Standards Association of Australia. *Safeguarding of Machinery,* AS 4042.1-1992. North Sydney, N.S.W: Standards Association of Australia, 1992.

Swedish Standards Institute. *Electrical Equipment of Industrial Machines—Prevention of Unintentional or Unauthorized Starting of Machines*, SS 436-04-21. Stockholm, Sweden: Swedish Standards Institute, 1989.

U.K. Health and Safety Executive Agency. *Dangerous Maintenance.* London: HSMO, 1987.

U.K. Health and Safety Executive Agency. *Deadly Maintenance: A Study of Fatal Accidents at Work.* London: HMSO, 1988.

U.K. Health and Safety Executive Agency. *Safe Systems of Work*, IND(G)76L. London: HMSO, 1992.

U.K. Health and Safety Executive Agency. *Work Equipment: Provision and Use of Work Equipment Regulations 1992*, Guidance or Regulations L22. London: HMSO, 1992.

Workplace Health and Safety Agency, Ministry of Labour. *Occupational Health and Safety Act and Regulations for Industrial Establishments.* Toronto, Ontario: Ministry of Labour, 1991.

6

The Process Approach

As the momentum builds for initiating corrective action, one tends to believe that establishing standards and regulations will deal a remedial blow to the perceived problem of serious injuries arising from energy-release incidents. Certainly, one can argue that at least the fundamental preparatory work has been done . . . or has it? The relative persistence of the problem seems to indicate a real need to establish benchmarks for what is conceded to be the best (safest) available practice. Guidelines and standards only serve as minimum thresholds for performance and cover broad generalities; as the focus is shifted to the exceptions or extremes, their lack of specifity makes them of little use.

If the standards governing the control of hazardous energy are crafted using a *specification* approach, significant detail is provided, thereby reducing the opportunity for judgment and flexibility. Conversely, if a *performance* approach is used, broad parameters are defined, allowing greater deviation to occur because of interpretation and varying methods of application. However, regardless of the approach, the user has much more to do than adhere solely to the assembled detail.

PROCESS OR PROGRAM

U.S. business is accustomed to addressing issues from a short-term rather than long-term perspective. Integrating ongoing comprehensive activity into the management structure and culture is the exception rather than the rule. Managers are accustomed to engaging high-priority problems feverishly for short time spans until the affected *project* has been completed. What is needed for lockout/tagout success is a commitment to a sustained effort highlighted by continuous improvement.

Merriam-Webster's Collegiate Dictionary, 10th edition, defines *program* as "a plan or system under which action may be taken toward a goal." Its definition of *process*, however, is more deliberate: "a series of actions or operations conducing to an end; *esp.*: a continuous operation or treatment esp. in manufacture." The difference between these two definitions is that the latter implies focused activity without a finishing point.

The process cycle for the control of hazardous energy is depicted in Figure 6-1. It consists of five major steps: (1) design, (2) implement, (3) monitor, (4) evaluate, and (5) refine. In step 1, all of the necessary elements of the hazardous-energy-control system are identified and connected to form a cohesive structure. In step 2, the system is implemented or placed into action. Communication, training, application, and so on are initiated with varying results. Step 3 provides for monitoring or auditing the outcome against the requirements or organizational expectations. In step 4, results are evaluated with some form of measurement

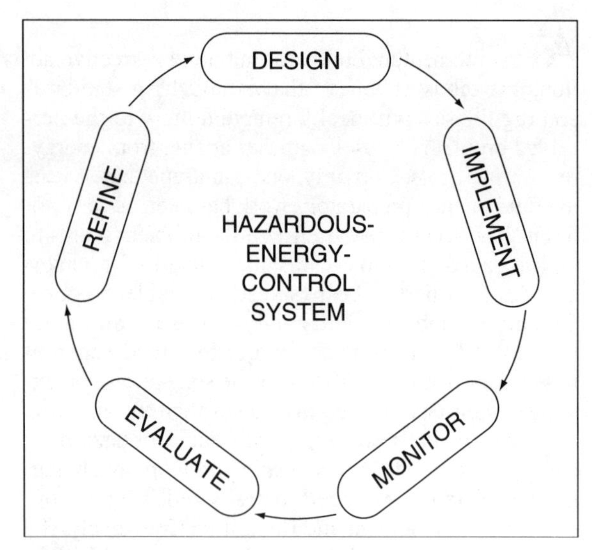

Figure 6-1. The process cycle.
Source: Developed by E. Grund. © National Safety Council.

and assessment of progress or performance. In step 5, the conclusions gained from the previous step are used to shape or refine the system and create a revised, second-generation design. The process cycle is repeated as the hazardous-energy-control system is continuously improved. The concept of an ongoing cycle where adjustment, growth, renewal, and innovation occur repetitively is critical to the prevention of hazardous-energy-release incidents.

The process approach should be adopted so that management fully understands that its first step toward hazardous energy control is the beginning of a long journey. Managers have to take a long-haul, holistic view in which substance is more valued than form.

By using a process approach, each organization can more effectively emphasize prevention while downplaying mere compliance. This does not mean that compliance with regulatory requirements is not an objective or part of the agenda. It means that a process-driven method can generate compliance as a by-product of a more universal strategy for prevention of hazardous-energy-release incidents. To put it another way, managers can be very efficient (compliant) doing things right or very effective (preventive) doing the right things. It is possible to achieve a high degree of compliance with existing standards while not effectively controlling all factors that may contribute to the unintended release of hazardous energy.

SITUATION APPRAISAL

If a hazardous-energy-control system is to be developed and put in place or an existing program is to be rebuilt, an orderly approach is necessary. First, management needs to assess the current situation with respect to hazardous energy control. The assessment might proceed in a number of ways but should address the following aspects as a minimum:

A. Potential for hazardous energy release
   1. Operational demands (machine/equipment/process)
      a. Adjustments
      b. Cleaning (materials, scrap, debris)
      c. Clearing jams
      d. Gauging/calibrating
      e. Cycling
      f. Charging/emptying
      g. Positioning, etc.
   2. Servicing demands (machine/equipment/process)
      a. Cleaning (readying for maintenance, appearance, etc.)
      b. Lubricating/oiling
      c. Fueling
      d. Painting, etc.
   3. Maintenance demands (machine/equipment/process)
      a. Troubleshooting
      b. Repair (breakdown)
      c. Modification
      d. Installing
      e. Testing, etc.

B. Influencing factors
   1. Nature of business/process
   2. History/past practices
   3. Safety culture (status)
   4. Other (see Chapter 3, Causation Analysis)

C. Status of hazardous-energy-control effort
   1. Stage of development
   2. Degree of effectiveness
   3. Employee perception/attitude

D. Compliance picture
   1. Regulatory status
      a. Level of standard conformance
      b. Citation history (if any)
      c. Abatement commitments
   2. Enforcement background (managerial)
      a. Existing policy/rules
      b. Actual practice/documentation

E. Energy-release incident experience
   1. Site/company injury incidents
   2. Noninjury incidents (damage, near-miss, etc.)
   3. Industry data (related experience)

F. Machine/equipment/process hazard analysis
   1. Hazard identification
   2. Hazard evaluation
   3. Safeguard selection.

When examining the potential for hazardous energy release, pay particular attention to the operating, servicing, and maintenance demands. The frequency of activity, scope, complexity, and risk dimensions of these demands will represent a major factor in establishing the priority or criticality for the prevention effort. Knowledgeable personnel, records, and so on can provide the background information needed to more adequately define what specifically needs to be done.

Evaluate the influence factors qualitatively (subjectively) to provide a picture of the general operational and safety environment, using a cross section of selected personnel. For example, if the history of the facility's energy-control activity is less than complimentary, different introduction or reinforcement methods may be necessary. Frequent needs for energy isolation due to machine/equipment malfunction or breakdown will influence the development of essential control tactics. If lack of management credibility is identified as a contributing factor, an action plan will need to be devised to address this barrier to success. The reasons for lack of credibility should be identified, and appropriate steps should be taken to improve the poor performance.

Review the organization's current position with regard to the existing energy-control system, which may be at any stage between plans on paper and a fully developed, mature approach. Past performance and the perception of the work force will provide important insight as to whether fine tuning or major reconstruction is in order. The state of the energy-control system will also impact the treatment of various issues such as resources, schedule, involvement, and the like.

Next, examine the degree of conformance to standards such as existence of procedures, completion of training, documentation records, etc. Obligations resulting from past regulatory citations need to be identified and understood to ensure system completeness and to avoid any likelihood of repeat violations. It is also beneficial to identify the organization's history with regard to enforcement of any existing requirements/rules for the control of hazardous energy. If safe practices and rules are routinely ignored by line supervision and hourly personnel, a different strategy will be necessary when approaching the enforcement issue in the future.

Gathering incident experience from within the organization and from other relevant sources should be the next step. The data will provide a quantitative and qualitative sketch of what is occurring and possibly some ideas as to where failures and deficiencies exist. The assembled information can also be used to change opinions and attitudes regarding the gravity of the situation. It can also be digested and packaged for presentation to employees as an awareness raising activity. In addition, OSHA compliance personnel have been instructed to pay special attention to injuries related to maintenance and servicing operations. No one should know more about the organization's experience with energy-release incidents than the responsible management at the facility.

Finally, when completing the situation appraisal, it is useful to conduct a machine/equipment/process hazard analysis. In OSHA's Instruction Standard 1-7.3, dated September 11, 1990, the compliance safety and health officer is directed to ask the employer for any hazard analysis or other basis on which the local energy-control plan was developed. This is *not* an OSHA standard requirement but will aid in determining the adequacy of the plan. The information previously assembled in items 1 and 5 of the situation appraisal will be particularly helpful in this effort. W. G. Johnson defines *hazard* as:

> The potential in an activity (or condition or situation) for sequence(s) of errors, oversights, changes, and stresses to result in an unwanted transfer of energy with resultant damage to persons, objects, or processes (Johnson, 1980).

A hazard analysis, according to D. B. Brown, consists of three parts: (1) hazard identification, (2) logical procedures for formulating countermeasures, and (3) selection of the best countermeasures to implement. With regard to hazard identification, gathering of historical/experiential data is followed by determining locations/situations with high accident potential, identifying severe hazard consequences, and eliminating from substantial consideration those relatively unimportant hazards (Brown, 1976).

OSHA's view of hazard analysis is expressed more definitively in its rule for process safety management. It requires that an employer perform an initial hazard analysis appropriate to the complexity of the process and then identify, evaluate, and control the hazards involved. The employer is obligated to use one or more of the following techniques to determine the hazards and evaluate them: a hazard and operability study (HAZOP), a failure mode and effect analysis (FMEA), a fault tree analysis, a what-if checklist, or another equivalent method.

The hazard analysis shall reveal (1) the hazards of the process, (2) the identification of prior incidents with catastrophic consequence potential, (3) engineering/administrative controls applicable to the hazards and the consequences of their failure, (4) human factors, and (5) a qualitative evaluation of the possible effects of failure of the employee controls (29 *CFR* 1910.119, 1992).

Manuele, in "On Hazard Analysis and Risk Assessment," states that "to be more precise, professional safety practice requires that hazards be analyzed, that risks be assessed, and that a ranking system be applied when giving advice on multiple hazards." He identifies a five-step approach to this end: (1) hazard identification, (2) exposure description (people, property, or environmental harm or damage), (3) assessment of the severity of consequences, (4) determination of probability of hazard occurrence (subjectively), and (5) assessment of probability and severity (risk). He also mentions a number of other methods that may be used, such as preliminary hazard analysis, gross hazard analysis, hazard criticality ranking, catastrophe analysis, energy transfer analysis, human factors review, hazard totem pole, and double failure analysis (Manuele, 1993).

A variety of hazard analysis and risk assessment techniques/methods exist in addition to those previously mentioned, ranging from simple subjective approaches to complex quantitative exercises. P. L. Clemens, in his article "A Compendium of Hazard Identification and Evaluation Techniques for System Safety Application," provides further resources (Clemens, 1982). For purposes of the preliminary appraisal of the hazardous-energy-control situation, one of the less rigorous approaches (such as logic-tree analysis, brainstorming, what-if analysis, potential causal factor checklist, all outlined in Chapter 3, Causation Analysis) is recommended. See Appendix 3, which contains a logic-tree diagram for controlling hazardous energy. The hazard analysis will be the most demanding aspect of the situation appraisal and could be accomplished as part of the early planning activity, if necessary.

The appraisal should provide management with enough basic information related to hazardous energy control to guide vital decision making and will also serve as the fundamental input when constructing the action plan.

ASSESSMENT AND ACTION

For the purposes of this chapter, we are assuming that the problem to be solved is the absence of a hazardous-energy-control system. However, in most cases the problem being addressed by management/employees will be some subset of this larger issue—for example, company energy-control procedures are inadequate/ineffective; line supervisors and hourly employees' energy isolation conformance is unacceptable; the contractor/company energy-control interaction is deteriorating, etc.

After gathering facts, making observations, and obtaining the views of a cross section of personnel, the key managers must decide upon a course of action. It is assumed that the option of no action is not relevant with regard to hazardous energy control because of existing legal requirements and the potential for severe consequences.

Decision making requires choosing between viable alternative courses of action. Usually this process involves four discrete steps: (1) defining the central problem, (2) developing alternatives, (3) analyzing the alternatives, and (4) making the final decision. Identification of the problem may seem somewhat obvious, but inadequate or inappropriate definition may funnel you away from developing the right set of alternatives for the situation. The element of choice is always crucial to effective decision making, and developing alternatives usually taxes the creative ability of the organization. Whether the organization has the ability to develop new ways of doing things or different approaches to old problems depends on (1) individual creativity, (2) the creative potential of the organization, and (3) the use of creativity-stimulating techniques (Dessler, 1985).

Because the development and implementation of a hazardous-energy-control system is obligatory, much of the analysis of alternatives will be reserved for the planning stage, when numerous options will have to be evaluated while designing the system. The facility manager or top executive, after being briefed on the general situation, will need to give the order to the support staff to put an energy-control system in place. At this point, it is critical not to leap into a free-style mode in which a few individuals are saddled with the entire job but rather to approach it as a discrete task. The organizational leader should direct someone on his/her staff to establish a team that will prepare a plan for designing and implementing a hazardous-energy-control system that addresses all of the relevant issues and contains sufficient detail for future review and decision making.

The assignment should be given to someone on the top manager's staff who has the talent, background, and authority to make the planning process an in-depth creative activity. The top manager should define when the plan should be completed for review and what the expectation is for its quality.

PLANNING BASICS

Those responsible for developing the preliminary plan for a hazardous-energy-control system should have a sound understanding of what the plan's content and process will involve. Every plan (and all of its supporting subplans) exists to facilitate the accomplishment of the defined purpose and objectives. Planning and control have been called the "Siamese twins of management." Unplanned action cannot be controlled because control inherently involves keeping activities on course. Plans, therefore, provide the critical standards for control.

A plan's effectiveness is determined by the degree to which it contributes to achievement of its purpose and objectives as offset by any costs and unforeseen consequences required to place it in action.

Within the process of planning, four basic principles should be acknowledged:

The limiting factor. When evaluating alternatives, the more that can be done to identify and resolve factors that inhibit goal attainment, the more likely the selection of the most favorable alternative.

Commitment. Today's plan encompasses a period of time in the future necessary to visualize, as accurately as possible, the fulfillment of commitments.

Flexibility. Plan flexibility and its costs need to be balanced against the possibility of unexpected events and attendant risks.

Navigational change. The greater the plan commitment toward the future, the more important it is to check your course and make adjustments.

An overlooked aspect of planning is premising: establishing and agreeing to use consistent premises critical to the plan under consideration. *Premises* are defined as the environment in which plans are expected to operate. By examining premises, we should be able to foresee how the plan will actually work in practice and expose internal or external factors that may significantly influence or impact current planning/decision making (Koontz et al., 1984).

Merriam-Webster (1993) defines *plan* as "a method for achieving an end; a detailed formulation of a program of action." Therefore, plans can encompass any course of future action and be divided into more specific types such as missions, purposes, goals, objectives, policies, procedures, programs, projects, and rules (Figure 6-2).

Mission and *purpose* often are used interchangeably in organizations. A mission could be viewed as the grand purpose or reason for existence, which can be very broad or narrow. For instance, DuPont's mission, "Better things for better living through chemistry," is very brief but to the point. A mission statement for hazardous energy control might read:

> Create a facility environment in which unintended energy-release incidents do not occur.

Supporting or amplifying the mission statement or responding to what is to be done, we can add a purpose as follows:

> Establish a control system and utilize procedures to prevent the unintended release or transmission of equipment/process energy.

Goals and *objectives* are also frequently interposed in general practice. In its truest sense, however, a *goal* is a broader end toward which an organization is working, supported by a number of *objectives*, or specific results to be achieved by some definite time. A goal for a facility with regard to hazardous energy control might be

> Reduce energy-release incidents by 50% and increase energy-control safe behavior conformance by 25% by December 1995.

Objectives that reinforce or enable the accomplishment of the above stated goal could be as follows:

1. Establish specific energy-control procedures for each piece of major equipment by September 1994.
2. Identify critical safe behaviors for energy isolation and reinforce them monthly with all authorized personnel before June 1994.

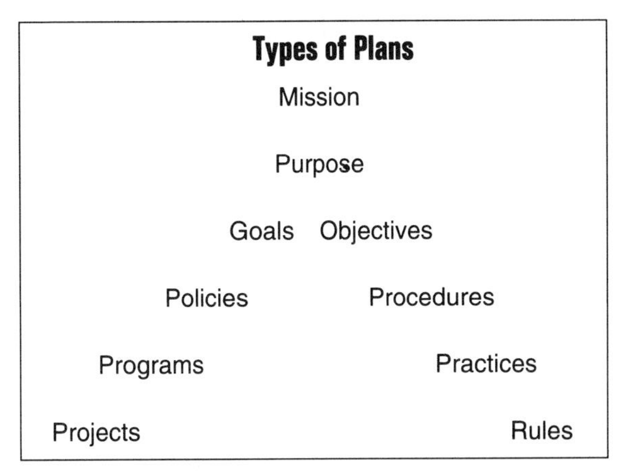

Figure 6-2. Plans can be a specific type for a future action.
Source: Developed by E. Grund. © National Safety Council.

3. Locate and correct all situations where energy-isolating devices do not exist and are needed or exist and are not lockable by December 1994.

Good objectives should be (1) specific and verifiable, (2) attainable, and (3) whenever possible, they should be stated in quantitative terms. However, even when qualitative descriptions are used, objectives can define the nature of the results to be achieved and by what time. Optimally, objectives should be fairly difficult but not unattainable. If they are too easy, they will result in marginal motivational impact and possibly underaccomplishment. If they are too difficult, frustration, superficial effort, and inefficient use of resources can occur. Objectives are supported by subobjectives and so on, thus producing a means-ends chain that depicts the ideal integration of mutually supportive objectives within the organization.

Policies are standing plans that take the form of general statements or understandings used to guide or direct thinking in decision making or to govern actions usually related to repetitive conditions. Policies need to be revised on a periodic basis to ensure that they relate or respond to the conditions and goals that caused them to be developed.

Whereas policies represent guides to thinking or navigating, *procedures* are guides to action or execution. They sequentially detail steps or actions to be taken when handling frequently occurring events with little room for options. They permit execution at lower levels of the organization and provide a means for implementing policy.

Programs, often called *action plans*, are usually single-use plans for achieving specific objectives. They develop the principal steps necessary for goal accomplishment, define the necessary sequence and timing for implementation of the steps, and identify those responsible for performing the activity. Programs vary greatly in scope and size (Gray & Smeltzer, 1989).

Effective programming is comprised of six basic elements:

1. Dividing the activity required to accomplish a goal into basic parts.
2. Determining the relationship among the parts and understanding their sequence.
3. Deciding who will be responsible for each part.
4. Deciding how each part will be accomplished and what resources are required.
5. Determining the time needed to complete each part.
6. Establishing a schedule for each part.

Various techniques such as the GANTT chart, the U.S. Navy's program evaluation and review technique (PERT), and DuPont's critical path method (CPM) have been used extensively to aid in programming (Newman & Logan, 1981). GANTT and PERT charts are illustrated in Figures 6-3 and 6-4, respectively. The GANTT chart represents an incomplete list of actions, with various time lines in weeks, that will be used to control and monitor progress toward the development of a hazardous-energy-control system. The PERT chart shows activity related to training authorized employees in hazardous energy control. The sequential nature, timing, and interrelationships are depicted graphically for managing the developmental process.

Dessler describes 10 principles of effective planning that are useful when preparing to construct a hazardous-energy-control system. Several are more relevant to general business plan development, but all provide insight regarding how to get started right:

1. *Develop accurate forecasts.* What lies ahead that will help or hinder?
2. *Gain acceptance for the plan.* Participation, participation, participation.
3. *The plan must be sound.* Use devil's advocate testing.
4. *Develop an effective planning organization.* Line management responsibility with staff support.
5. *Be objective.* Use realism without dependency on low-probability outcomes.
6. *Market share.* This depends on the ability to satisfy customers' (in this case, employees') needs.
7. *When to abandon.* Know what defines the time to abort and reassess.
8. *Monitoring.* Review regularly, assess, and modify if necessary (flexibility).
9. *Periodic revision.* Plans rarely function without adjustment.
10. *Fit the plan to the situation.* Planning with goals when the environment is predictable and using directional planning otherwise (Dessler, 1985).

In addition to these principles, the planner should be aware of the most common reasons for planning failures. By addressing these issues proactively, more effective lockout/tagout planning can result:

1. *Lack of commitment to planning.* Comfortable with crisis.
2. *Confusion of planning studies with plans.* Plans require decision.
3. *Failure to develop and implement sound strategies.* Direction.
4. *Lack of meaningful objectives or goals.* Clear, attainable, actionable.
5. *Tendency to downplay planning premises.* Uniform and understood.

Actions	August	September	October	November	December
Form Task Group					
Develop Policy					
Develop Resources List					
Define Responsibilities					
Procure Lockout Hardware					
Develop Equipment Procedures					
Weeks ▶	1 2 3 4	5 6 7 8	9 10 11 12	13 14 15 16	17 18 19 20

Figure 6-3. GANTT chart—a hazardous-energy-control system.
Source: Developed by E. Grund. © National Safety Council.

6. *Failure to see the scope of plans.* Differentiate between policies, procedures, practices, etc.
7. *Failure to see planning as a rational process.* Practical exercise in rationality.
8. *Excessive reliance on experience.* The past may not be an accurate predictor of the future.
9. *Neglect of the limiting factor principle.* Select the critical factor in the problem mix and focus attention.
10. *Lack of top management support.* Believe and encourage.
11. *Lack of clear delegation.* Who is doing what and with what authority?
12. *Lack of control and information.* How is the plan working?
13. *Resistance to change.* Comfort zone mentality (Koontz H, O'Donnell C, Weihrich H, *Management.* New York: McGraw-Hill, 1984. Reproduced with permission of McGraw-Hill).

Spending the necessary time on planning and preparation will increase the odds that your hazardous-energy-control system will be effective. The Japanese are quite successful in business because of their emphasis on planning. A typical Japanese organization might spend 70% of its time planning and 30% of its time executing, whereas the typical U.S. organization might allocate its time in an opposite manner. As a result, U.S. organizations often find themselves being surprised by unexpected events, results, and input; repeating activities and actions to make corrections because of oversight or error; or achieving marginal results because of inadequate content, process, or involvement.

PLANNING THE SYSTEM: HAZARDOUS ENERGY CONTROL

Approaching hazardous energy control from a systems perspective introduces the concept of a unified or holistic treatment. This planning strategy is deemed to be far more productive in the long term when compared to the crisis-generated or evolutionary development of lockout/tagout programs. *Merriam-Webster's Collegiate Dictionary*, 10th edition, defines *system* as "a regularly interacting or interdependent group of items forming a unified whole." Lockout/tagout

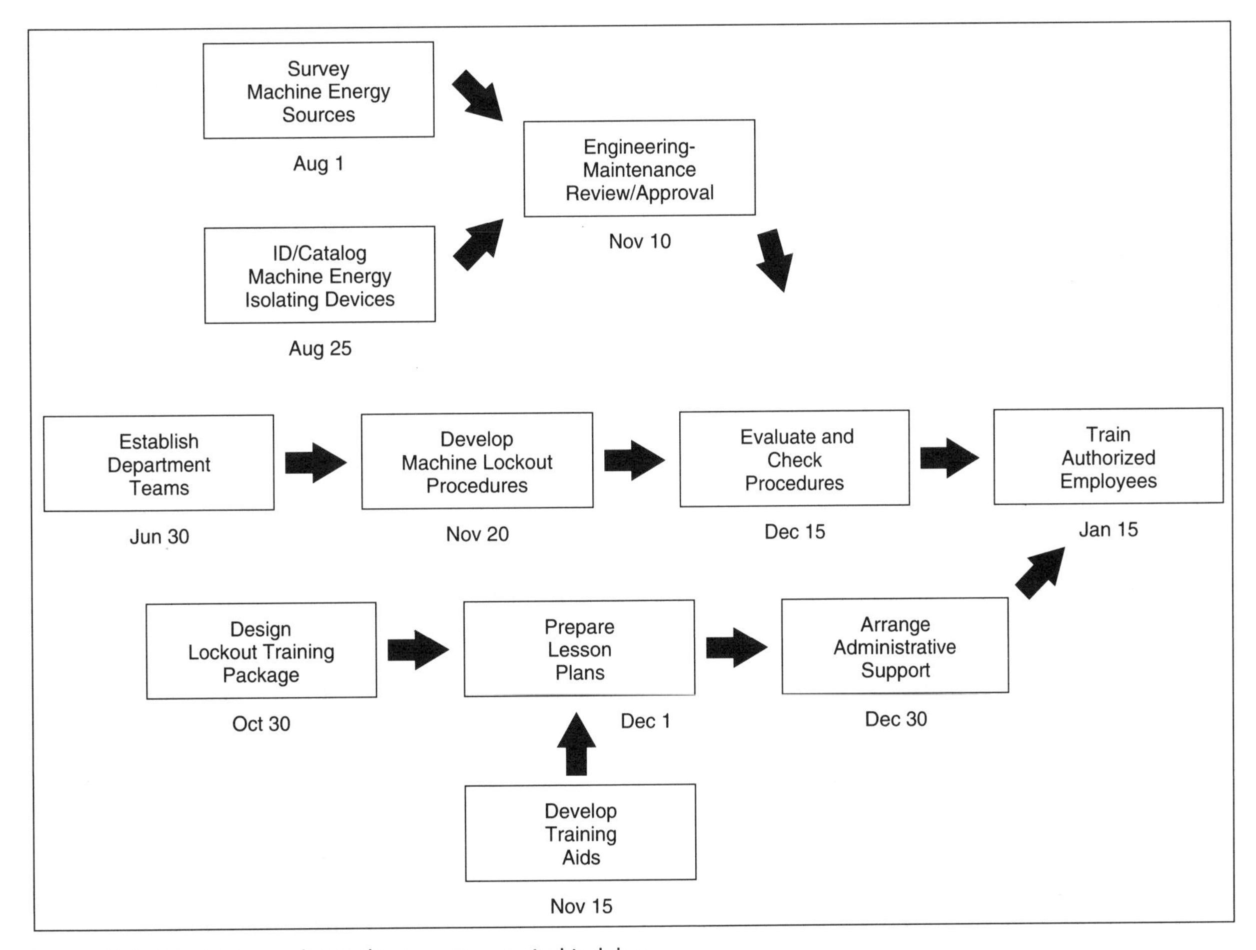

Figure 6-4. PERT chart—hazardous energy control training.
Source: Developed by E. Grund. © National Safety Council.

programs are often established without thorough consideration of *all* of the parts and the relationships between them.

Grose, in *Managing Risk—Systematic Loss Protection for Executives*, defines *system* as follows:

> A system is a composite, at any level of complexity, of operational and support equipment, personnel, facilities, and software (intelligence), which are used together as an entity and capable of performing or supporting an operational role that results in changing known inputs into desired outputs. (Grose, 1987)

He states that "the most fundamental description of a system is an entity for transforming inputs into outputs via resource expenditure." He describes this concept graphically as shown in Figure 6-5.

The system is well defined or has known boundaries. Transformation or conversion of the inputs occurs inside the system, resulting in desired outputs. Labor, materials, capital, and time are expended to accomplish the change. The measurements of the transformation are cost, performance, and schedule. Usually, feedback from the outputs returns to the inputs, which over time adjusts or refines them (Grose, 1987).

From a hazardous-energy-control perspective, Grose's idea of a system could be explained as follows: The known inputs might be the details associated with humans, machines/equipment/materials, environment, and their interaction; the transformation might involve management activity such as planning, training, etc.; and the outputs could then be well-trained workers following effective energy-control procedures.

The hazardous-energy-control system shown in Figure 6-6 recalls a portion of the incident model presented in Chapter 3, Causation Analysis. The graphic is virtually the same with one exception—the interstitial segment identified as *design*. This is added to highlight the critical, and often neglected, role played by engineering design and its influence on the three major system elements: humans, machines, and the environment. Much of the high-risk behavior that precipitates energy-release incidents could be obviated by front-end engineering initiatives.

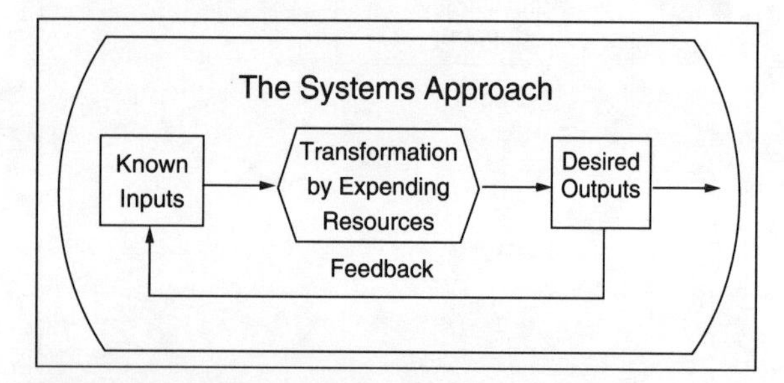

Figure 6-5. The systems approach.
Source: Printed by permission of Dr. Vernon L. Grose.

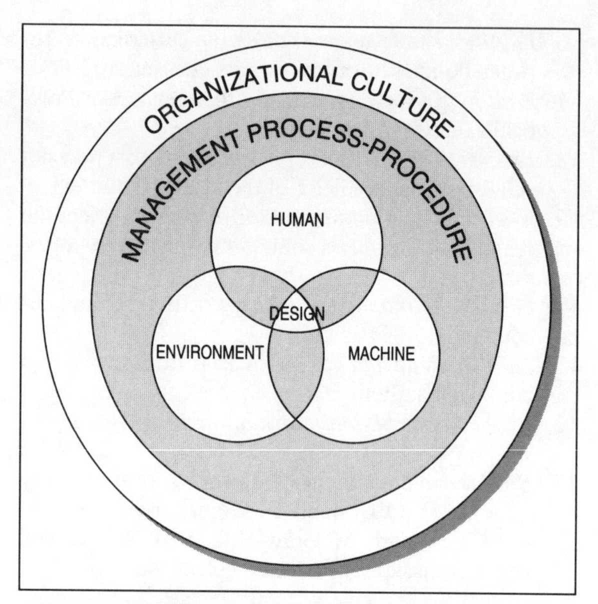

Figure 6-6. Hazardous-energy-control system.
Source: Developed by E. Grund. © National Safety Council.

For planning purposes, we shall view the system from an internal viewpoint. A broader, more encompassing look might include external factors such as legal risks (professional/product liability, third-party claims, etc.), the company's economic health, community issues, international union priorities, insurance coverage, local/state government regulation and politics, and so on. Although these external factors may be relevant, the internal elements are more crucial to the success of our prevention efforts.

The six key elements of the hazardous-energy-control system as presented in the graphic are as follows: (1) organizational culture, (2) management process/procedure, (3) human, (4) machine (equipment/materials), (5) the environment, and (6) design. The organizational culture frames and establishes the set or boundaries within which the system functions. It is made up of beliefs, values, traditions, customs, hierarchies, and the like. These qualities of the culture will influence the way things are done and ultimately the system's success. The remaining system elements are expanded as follows:

Management Process/Procedure.

Plans: Mission, purpose, goals, objectives, policies, procedures, programs, practices, projects, rules.

Actions: Directing, controlling, monitoring, auditing, training, planning, motivating, procuring, maintaining, improving, etc.

Human. Communicating, participating, educating, investigating, leading, coaching, skills, stresses, performance, behavior, attitudes, jobs, tasks, etc.

Environment. Heat, cold, vibration, lighting, noise, dust, weather, toxins, congestion, etc.

Machine (Materials/Equipment/Process. Vehicles, utilities, chemicals, isolating devices, lockout/tagout appliances, piping, energy, machinery, etc.

Design. Ergonomics, controls (human factors), layout, safety features (interlocks, redundancy, etc.), energy isolation capability, machine reliability, isolating device attributes, access, guarding, safety factors, automated systems, self-lubrication, etc.

Core Action Team

After being directed by the facility manager to prepare a plan for designing and implementing a hazardous-energy-control system, the responsible party needs to select and recruit a team to accomplish the task. A suggested approach would be to organize a Core Action Team, or CAT, to collaborate on the plan development. Individuals that assisted with the situation appraisal should be considered for further duty, but more importantly, disciplinary balance and organizational verticality need to be addressed when selecting participants. Management and worker participation should be pursued to ensure early awareness of the activity and to avoid exclusion of individuals close to the heart of the issues. A recommended CAT composition for a manufacturing plant is as follows:

- engineer
- maintenance supervisor
- purchasing agent
- machinery operator
- electrician
- mechanic
- line supervisor
- safety representative.

Other members could be added, such as an industrial engineer, union steward, auditor, and so on, dependent upon the size and complexity of the facility. It should be noted, however, that 6 to 12 team members are usually enough. The focus should be to assemble a small but knowledgeable group with sound background and experience in facility operation. This will facilitate the planning process and provide a diversity of input, which is critical in the plan's formative stages. If discussions touch upon an area where no technical or operational competence exists on the team, additional individuals may be enlisted for brief periods to acquire their contributions.

The team leader should arrange the first meeting and request that the facility manager open the session and discuss his/her expectation for the group. After the big picture is presented and importance emphasized, the manager can leave the team to its deliberations. The team leader can then define the specifics of the assignment, general activity time frame, and expected project completion date. The situation appraisal can be passed out to team members or distributed prior to the meeting as a briefing tool. If the hazard analysis segment has not been done, the team members should address this very early in the process so that they have a complete understanding of what the plan will have to address.

In addition to the materials developed during the situation appraisal, the person in the CAT with the best general understanding of hazardous energy control (lockout/tagout) or the best contacts for finding help should be identified. He or she should be directed to procure general topical information and status/content of lockout/tagout activity from similar businesses for further education/briefing of the team. This effort will help bring all team members to a common level of understanding before engaging the detail. If possible, this is a good time for the team leader to appoint (with concurrence) a two- or three-person task force to conduct a hazardous-energy-control benchmarking study. CAT members may not have enough knowledge to identify which organizations might represent the best of the best with regard to lockout/tagout. This could be done within the specific business community, product line or service, or more broadly. The benchmarking effort might also provide additional input for the team to evaluate prior to reaching any conclusions about the local preliminary plan. The CAT will be, in reality, conducting a planning study or developing a preliminary plan for management that will become the hazardous-energy-control system when the decision is made to approve and implement it.

Mission, Purpose, Goals, and Objectives

The team leader should assign several members of the group to draft statements for the mission, purpose, goals, and objectives related to the facility's hazardous-energy-control system. Again, those selected will concur and ideally have talent for creating the key ideas and expressing them concisely in written form. The section on planning basics will provide further insight as to what needs to be accomplished. These statements or pronouncements may not seem to be the heart of the matter but still serve two important purposes: (1) the process of producing them will help sharpen the team's focus and the system's thrust and (2) the output will define parameters, actions to be accomplished, and expectation or results. The statements are responsive to the basic questions, what, why, how, when, to what degree, and so on.

System Elements

Reflecting on the six key elements depicted in Figure 6-6, the team can identify what needs to be considered when preparing this part of the preliminary plan. The organizational culture can be addressed as a preamble or introduction to the plan because it exists, for better or worse, and will normally be beyond the scope of the planning task.

The following outline of elements and subelements can be used as a blueprint for constructing the hazardous-energy-control system:

I. Management process/procedure
  A. Plans (guidance)
    1. Mission
    2. Goals
    3. Policy
      a. Purpose
      b. Objectives
      c. General procedure
    4. Specific Procedures—machinery, equipment, and process
      a. Power off—regular conditions
      b. Power on—troubleshooting, cycling, hot work, etc.
      c. Special—alternative methods equivalent to lockout/tagout
  B. Actions (results)
    1. Planning—content, tactics, timing, etc.
    2. Inspecting—energy source/isolating device surveys, etc.
    3. Training—awareness, general, specific, authorized, etc.
    4. Procuring—lockout/tagout appliances, signs, etc.
    5. Motivating—promotion, incentives, etc.
    6. Auditing—lockout/tagout application, etc.
    7. Documenting—training/enforcement records, etc.
    8. Improving—feedback, recommendation system, annual review, etc.

II. Human
  A. Communication—announcements, status reports, meetings, etc.
  B. Leadership—defined responsibilities, accountability, visibility, etc.
  C. Participation—involvement strategy, task forces, CAT, etc.
  D. Performance—compliance, meeting requirements, application, etc.
  E. Investigation—energy-release incidents, complaints, recommendations, etc.
  F. Enforcement—counseling, warning, reprimanding, etc.
  G. Praising—positive reinforcement, acknowledgment, etc.
  H. Coaching—peer guidance, on-job-training, lead personnel, etc.

III. Environment (stressors)
  A. Anticipating—early detection of negative factors
  B. Surveying—measuring, inventorying, etc.
  C. Assessing—evaluating, prioritizing, etc.
  D. Optimizing—stressor reduction, improvements, etc.

IV. Machine (materials, equipment, process)
  A. Guarding—adequacy, condition, presence, etc.
  B. Isolation devices—lockable, effectiveness, availability, etc.
  C. Isolatable—all energy sources
  D. Lockout/tagout appliances—normal, special, custom, etc.
  E. Isolation device identification—energy type, magnitude, location, etc.
  F. Piping—color code, material identification, etc.
  G. Special tools—die blocks, tongs, safety chains, etc.

V. Design (engineering)
  A. Project review—design stage
  B. Isolation specifications—devices, location, etc.
  C. Bid package criteria
  D. Internal standards
  E. Design improvements—incident recommendations, suggestions, etc.
  F. Preoperational field review
  G. Contractor competency—lockout/tagout
  H. Modification projects—upgrade, enhancements, etc.
  I. Purchasing collaboration—hardware, etc.

Additional detail regarding the content and features of the hazardous-energy-control system will be provided in Chapter 7, System Elements. It should now be apparent that a significant number of issues need attention and that a continuous systematic approach is therefore necessary.

Resource Requirements

The plan must specify what resources in terms of money, personnel, and time will be necessary to put the energy-control system in place and maintain its existence. In this regard there is a tremendous temptation to guesstimate because reality often is too alarming for fiscal tranquility. A comprehensive energy-control system for a moderate-risk, 500-employee manufacturing facility is likely to cost in the low six-figure area. Of course, we are talking about a system that is predicated upon prevention and not just compliance.

OSHA, in the preamble to its lockout/tagout standard (1910.147 Final Rule), presented a regulatory impact analysis. The Eastern Research Group (ERG), under contract to OSHA, prepared a study entitled "Industry Profile Study of a Standard for Control of Hazardous Energy Sources Including Lockout/Tagout Procedures." The study provided the data for estimating the economic effects of the standard. Based on the study's projection, the U. S. secretary of labor determined that the standard was a "major action" having an annual effect on the economy of $100 million or more. The cost for 631,000 establishments would be $214.3 million for the first year and $135.4 million in subsequent years to comply with the standard.

In examining the Final Rule's effect on high-impact industries (in other words, basically all manufacturing), ERG concluded that 340,000 establishments would spend nearly $143 million for compliance the first year. The average manufacturing facility would spend $420 the first year for coming into compliance with the lockout/tagout standard. The calculation is as follows:

340,000 (establishments) × $420 (average) = $142,800,000

Using these data and their categorical breakdown, the following cost detail per establishment was developed:

1. Locks/tags/hardware	$ 38.00	(9%)
2. Equipment modification to facilitate LO/TO	54.00	(13%)
3. Work practice changes	202.00	(48%)
4. Planning/implementing	67.00	(16%)
5. Employee training	59.00	(14%)
Total	$420.00	(100%)

OSHA, using ERG data, determined that a large establishment (over 250 employees) not currently using *adequate* lockout or tagout procedures would spend $28,172 the first year. These forecasts probably do not reflect the true cost for any manufacturing establishment of coming into full compliance, regardless of what state their hazardous-energy-control system is in currently. In Table 6-A, a budget has been prepared for a moderate-risk, 500-employee manufacturing facility to plan, develop, and implement a comprehensive haz-

Table 6-A. XYZ Manufacturing Company—Hazardous-Energy-Control System: First-Year Budget (*continued*)

Element item	Detail	Salaried employees (#)	Hourly employees (#)	Time (hrs.)	Unit cost ($)	Total cost ($)
I. Management process						
A. General planning	Situation appraisal	4	0	64	20/hr (64)	1,280
	Core action team	5	3	460	20/hr–300	6,000
					15/hr–160	2,400
B. Policy development	Facility—general	5	3	48	20/hr–32	640
					15/hr–16	240
C. Specific procedures						
1. Power off	80 procedures	4	12	230	20/hr–50	1,000
					15/hr–180	2,700
2. Power on	30 procedures	2	4	120	20/hr–30	600
					15/hr–90	1,350
3. Special	20 procedures	2	3	50	20/hr–15	300
					20/hr–35	525
D. Energy source/isolating device surveys	100 machines	2	4	300	20/hr–50	1,600
					15/hr–220	3,300
	20 utilities	1	2	85	20/hr–15	300
					15/hr–70	1,050
	20 other	1	2	70	20/hr–20	400
					15/hr–50	750
E. Employee training	Authorized—2 hrs	15	85	200	20/hr–30	600
					15/hr–170	2,550
	Affected—1 hr	30	300	330	20/hr–30	600
					15/hr–170	4,500
	Other—0.75 hrs	50	20	52.5	20/hr–37.5	750
					15/hr–15	225
	Video purchase				450–3	1,350
F. Purchasing support	Signs, hardware, forms, devices, etc.	2	1	65	20/hr–50	1,000
					15/hr–15	225
G. Motivation	Area meetings,	30	385	207.5	20/hr–15	300
					15/hr–192.5	2,888
	Hand out material,	80	400		1.50	720
	Award incentives	—	—		20 @ 15.50	310

Table 6-A. *(concluded)*

H. Auditing	Application lockout (10% sample —1hr);	150	1,250	1,400	20/hr–150 15/hr–1250 R–1250	3,000 18,750
	General compliance (2 reviews—8 hrs)	2	2	64	20/hr–32 15/hr–32	640 480
I. Record keeping	Training, enforcement, auditing, procedures	5	0	80	20/hr–80	1,600
II. Human element						
A. Communication	Announcements, status reports, special meetings	—	—	—	—	2,500
B. Involvement	Special projects	10	50	220	20/hr–40 15/hr–180	800 2,700
C. Investigation	Incidents, suggestions, complaints	5	12	100	20/hr–20 15/hr–80	400 1,200
D. Enforcement	Rules, counseling, union discussions	10	25	125	20/hr–35 15/hr–90	700 1,350
III. Environment element						
A. Surveys	Lighting, congestion, access, etc.	2	4	105	20/hr–65 15/hr–40	1,300 600
B. Improvements	Area projects, Labor	5	15	450	20/hr–80 15/hr–370	1,600 5,500
	Materials	—	—	—	6,800	6,800
IV. Machine element						
A. Lockout/tagout appliances	Locks, tags, adapters	—	—	—	3,900–tot.	3,900
B. Special equipment	Blocks, chains, valve covers	—	—	—	1925–tot.	1,925
C. Isolation device upgrade	Breakers/valves				20/hr–35	700
	Labor	3	9	160	15/hr–125	1,875
	Materials	—	—	—	4600–tot.	4,600
D. Isolation device identification	Plant devices—15%	1	3	190	20/hr–25 15/hr–165	500 2,475
E. Guards	Interlocking/general improvements	3	8	175		
	Labor	—	—	—	20/hr–35 15/hr–140	700 2,100
	Materials				2,600	2,600
V. Design (engineering)						
A. Contractor	Approval, review, administration	4	0	30	20/hr	600
B. Design	Drawings	3	0	40	20/hr	800
C. Engineering	Review, specifications, field inspection	4	0	80	20/hr	1,600
VI. General administration	Miscellaneous	—	—	—	—	6,000
TOTAL						120,798

Source: Developed by E Grund. © National Safety Council.

ardous-energy-control system. The budgeting premise states that the facility is involved with fabricated metal products (SIC Code 34), is unionized, has 20% of the work force involved with maintenance/service support, is less than 20 years old, has 100 principal pieces of machinery, occupies 350,000 sq ft (32,515 sq m) of buildings, operates continuously with an hourly loaded labor cost of $15/hour and a salary-loaded labor cost of $20/hour, and has a lockout/tagout program that is marginally developed and limited to maintenance employees.

As can be seen from XYZ Manufacturing Company's first-year budget, much more needs to be addressed than the fine elements OSHA considered in its cost/benefit analysis. XYZ's $120,798 is well over the $28,172 that OSHA estimated for a facility of this description. One could easily add several more line items and expand the individual element costs in the budget. Regardless of the final derived dollar amount, the financial part of the plan must be prepared to ensure that funds are available for implementation. It is equally important, after determining the start-up costs, to establish costs of the maintaining the system. Several different elements might be removed or added in the second or third year, for example.

In addition to determining the budget for the hazardous-energy-control system, the CAT should also conduct a personnel resource/talent inventory to identify all personnel with special skills, appropriate background and experience, and personal attributes that would be valuable to the system's development. The CAT members can then develop potential recruit lists for various needed tasks/projects. It is in the best interests of the overall effort to persuade the selected personnel to participate or volunteer instead of drafting them. Remember to achieve an appropriate mixture of hourly/salary and operations/maintenance/staff personnel when developing task assignments and the like. Using external sources of expertise such as headquarters specialists, consultants, or local industrial company personnel who are willing to share their experience or lockout/tagout techniques may also prove valuable.

TASK RESPONSIBILITIES/ASSIGNMENTS

When reviewing the elements of the hazardous-energy-control system or their budget allocations (Table 6-A), numerous actionable items that are ideally suited as assignments for individuals or teams will present themselves. Having accomplished the personnel/resource talent inventory, the CAT should be in a position to assemble the necessary labor to address the various activities. Some of the elements are suited for assignment to the CAT itself, whereas others need input and development by a wide range of other facility personnel.

The following list of topical action items and type/area employee designation is provided as potential CAT assignments for development of key elements of the system:

1. Situation appraisal: CAT task group (selected members with adequate lockout/tagout knowledge).
2. General policy development: CAT
3. Specific machine/equipment procedures: Select group from each area/department; operations/maintenance representation.
4. Energy source/isolating device survey: Select individuals (two to five) to survey entire facility with electrical/mechanical/engineering background.
5. Employee training plan: Human resource specialist and operations/maintenance representatives.
6. Employee training: Combination of area supervision/hourly personnel; human resource train-the-trainer process.
7. Purchasing activity: Purchasing agent, maintenance supervisor, engineer, and maintenance craftsman.
8. Awareness/motivation plan: Cross section of facility personnel with creative talent.
9. Auditing (compliance): Industrial engineer, accountant, quality assurance, operations, maintenance group.
10. Record keeping: Human resource, accounting, quality assurance, production control group.
11. Communication strategy: CAT or special task group.
12. Investigation practice: Cross section of facility personnel; human resource; safety; union representative.
13. Enforcement protocol: Human resource specialist; union official; staff personnel; operations/maintenance representatives.
14. Environmental surveys: Safety; industrial hygiene; engineering; maintenance representative.
15. Special lockout/tagout appliances and equipment: Purchasing agent; maintenance supervision; operations/maintenance personnel/safety.
16. Isolation device identification and upgrade: Engineering; purchasing agent; maintenance supervision.
17. Machine guards: Engineering; maintenance supervision; operations/maintenance personnel; safety.
18. Contractor procedures: Engineering; purchasing; safety; maintenance.

This list of topical action items is not necessarily complete or ideal. It can be used as a reference and stimulus to create a structure for accomplishment of all of the key components of the system.

Kepner and Tregoe, in *Project Management*, introduce several techniques that are of value for managing the task/responsibilities portion of the energy-control plan, two of which are the following:

Work breakdown structure. This identifies and displays the *deliverables* (thing produced or outcome) to be provided and the tasks to be *accomplished* (something achieved). Deliverables and accomplishments can be broken down into as many subcomponents as necessary for completion. The breakdown assists those responsible for the project to understand and monitor all of the involved parts and their relationship to the total. This can be done as a simple listing (Figure 6-7) or in chart form.

Responsibility assignment matrix (RAM). This identifies who will be responsible or where responsibility lies for completing each element of the work breakdown structure and, when multiple contributors are involved, who has the primary responsibility (Table 6-B).

Project: Develop LO/TO Procedures (Department A)

1. Completed list of all machinery and equipment
   - 1.1 Obtained historical roster of machinery
   - 1.2 Conducted physical survey to verify and update
   - 1.3 Prepared current status report
2. Identified all energy sources for current machinery/equipment
   - 2.1 Obtained current machinery/equipment list
   - 2.2 Determined energy types/sources for each unit
   - 2.3 Reviewed available blueprints to verify
   - 2.4 Prepared machine/energy type report
3. Identified isolating devices for each energy source
   - 3.1 Located isolating devices for each machine/energy type
   - 3.2 Verified isolating device effect by actuating
   - 3.3 Marked each device (energy, magnitude, type)
4. Completed procedures for each machine/equipment
   - 4.1 Identified isolation sequence for each unit
   - 4.2 Defined isolation action for each step in procedure
   - 4.3 Defined verification of isolation action required
   - 4.4 Determined lockout technique required for each energy-isolating device
   - 4.5 Established requirements for any additional safeguards
   - 4.6 Listed requirements for residual energy release and restraining gravitational forces

Figure 6-7. Work breakdown structure.
Source: Developed by E. Grund. © National Safety Council.

The completion of the responsibility assignment matrix identifies the work to be done and the departments or individuals who will be responsible or provide the required resources. In all cases, agreement has to be reached regarding the commitment to provide the necessary resources or to assume the direct responsibility for the required work (Kepner & Tregoe, 1989).

Schedule

Gray and Smeltzer, in describing six basic elements of effective programming, identified "establishing a schedule for each part" as one of the elements (Gray & Smeltzer, 1989). There is no practical way to ensure that any plan with more than a few parts will move forward to completion without a schedule. A schedule becomes even more critical when various parts of the overall system are sequentially dependent upon or interconnected to each other; in other words, training lesson plans must be completed and ready before the actual employee training occurs.

Schedules need to define the task, work, or activity to be accomplished with various references to timing such as start date, completion date, time frame, milestones (key dates), etc. Techniques such as GANTT charting, PERT, CPM, and others may be used to develop the planning blueprint.

The CAT leader should be responsible for approving the final schedule content in order to complete the development and implementation of the hazardous-energy-control system by the designated time. Various parts of the total plan may be ahead of or behind schedule for legitimate reasons, but the CAT leader will need to make adjustments, provide additional resources, and so on to keep the project moving toward completion. Task or group leaders should report periodically on their activity status and notify the CAT leader of any consequential disruption or obstacle that will impact their part of the schedule.

Execution/Coordination

The facility manager, after being presented the fully prepared plan and thoroughly briefed on its contents, will have to make a decision regarding its implementation. Additional information might be necessary, but if the team has done its job properly, probably not. Most likely, the debate might be localized to timing, funding issues, personnel utilization, method, etc. Once the remaining issues have been resolved, a kick-off date for the process will be determined and all other subsequent milestones identified. Resources, both labor and financial, need to be available for execution of the various tasks and activities. During the early stages of plan implementation, the CAT leader should verbally and in writing brief the facility manager on the status of all scheduled activity.

The core action team not only plays a vital role in preparing the hazardous-energy-control plan but,

Table 6-B. Responsibility Assignment Matrix (RAM)

PROJECT: DEVELOP LO/TO PROCEDURES (DEPARTMENT A)	Area Engineer A. HOLMES	Prod. Supv. C. JONES	Maint. Supv. H. DAVIS	Lead Mech. F. BILLINGS	Sr. Elect. M. OPPELT	Prod. Oper. W. TRENTON	Safety Coord. H. CWAN
1. Completed list of all machinery and equipment							
1.1 Obtained historical roster of machinery	Central file req.	Verify	Verify				
1.2 Conducted physical survey to verify and update	Assist supvs.	Survey	Survey	Survey-Mech.	Survey-Elect.		
1.3 Prepared current status report		Prep. report				Review report	Review report
2. Identified all energy sources for current machinery/equipment							
2.1 Obtained current machinery/equipment list	Supv. report		Supv. report				
2.2 Determined energy types/sources for each unit				Mach. Inspect.	Mach. Inspect.	Mach. Inspect.	
2.3 Reviewed available blue prints to verify	Central Engr.		Provide input				
2.4 Prepared machine/energy type report	Review	Prep. report	Review				Review/Approve
3. Identified isolating devices for each energy source							
3.1 Located isolating devices for each machine/energy type				Survey	Survey	Survey	
3.2 Verified isolating device effect by actuating	Review/Approve			Tested	Tested	Tested	
3.3 Marked each device (energy, magnitude, type)	Input		Authorize	ID-Mech.	ID-Elect.	Review	
4. Completed procedures for each machine/equipment							
4.1 Identified isolation sequence for each unit				Input	Input	Input	
4.2 Defined isolation action for each step in procedure				Prepare	Prepare	Prepare	Input
4.3 Defined verification of isolation action required	Review/Input	Approve	Approve				
4.4 Determined lockout technique required for each energy isolating device		Approve	Approve	Establish	Establish	Establish	Review
4.5 Established requirements for any additional safeguards		Define	Define	Input	Input	Input	
4.6 Listed requirements for residual energy release and restraining gravitational forces.	Provide input	List	List	Input	Input	Input	

Source: Developed by E. Grund. © National Safety Council.

perhaps more importantly, also guides and coordinates the numerous elemental tasks, projects, and activities that will eventually be launched into motion. The team will need to develop a control method that provides all of the necessary feedback to keep the project moving according to schedule. In addition, there should be mechanisms established to deal with delays, special problems, new input, fresh ideas, etc.

The various task/element coordinators should have a clean understanding of the total plan and the position and sequence their task/element will assume. Reporting requirements need to be defined in order to keep the CAT leader current on progress of all activity.

It is recommended that an overview graphic charting method (GANTT, etc.) be utilized to keep the work force apprised of the key element activity and progress toward completion. This will allow the organization to slowly reach the desired operating speed necessary for accommodation of all of the new or revised energy-control system features.

Implication/Commitment

A prevalent problem with hazardous-energy-control systems or even programs is the reluctance or failure to use the energy and talent of the total organization in developing and executing the approach. Management is too often prone to use a few selected individuals to craft the entire system, conduct a modest review, and announce its introduction without considering the implications or prognosis for success. This occurs for a variety of reasons such as deadline pressures, a view that involvement is too time-consuming and costly, fear of loss of authority or decision-making autonomy, union-management politics, etc.

Without an open, collective, interactive employee process for developing the system, the outcome will be relatively mediocre and deteriorate proportionally to the amount of policing management is able to sustain. Yes, a collective open approach will require more time and expense in the beginning, but the payoff in performance results will more than offset that investment. The control of hazardous energy is too complex and dynamic for a few individuals to conceive, plan, develop, and execute adequately.

Numerous opportunities exist for employees to participate in the process of establishing an effective hazardous-energy-control system by working on such projects as isolating device surveys, procedure development, application (lockout/tagout) audits, employee training, and so on. Sayles and Strauss contend that the manager who uses participative techniques is interested primarily in results and permits the subordinates to work out the details for themselves. The manager sets goals, tells subordinates what he/she wants to accomplish, fixes the limits within which they can work and, in cases where the subordinates are adequately trained, lets them decide how to achieve these goals (Sayles & Strauss, 1966).

The terms *participation* and *involvement* are often used interchangeably. The dictionary, in fact, defines *involve* as "to engage as a participant" (Merriam-Webster, 1993). However, for the purposes of our subject matter, the following definitions should be considered to delineate the relationship between the various degrees of implication and commitment:

- *Participation*: to take part in some activity, event, project, or organization without any consequential depth of engagement or contribution of self.
- *Involvement*: to participate or be absorbed in some activity, event, project, or organization with intense interaction, dedication to purpose, and commitment of self.
- *Empowerment*: that state of involvement in which the committed participants are granted the permission and authority to decide and act in the interest of the group and organization.

As we ascend toward the apex of the employee power pyramid (Figure 6-8), employee involvement, commitment, and freedom to act increase in proportion to a decrease in the traditional tenets of management control. At the apex, employee ownership is high, and strong beliefs usually exist about the value of the product or decision. The question for your organization should not be whether to have or not have participation when developing the energy-control system; the question should be how evolved participation should be. Not every organization is ready or able to function in a state of full employee participation and empowerment. However, a participative approach should be considered a minimum threshold for developing your hazardous-energy-control system.

A number of prerequisites exist for using the participative process successfully:

- The participants must have adequate intelligence and knowledge regarding the issue.
- The participants must be able to communicate with one another.
- The participants should have no fear of any adverse affect as a result of participating.

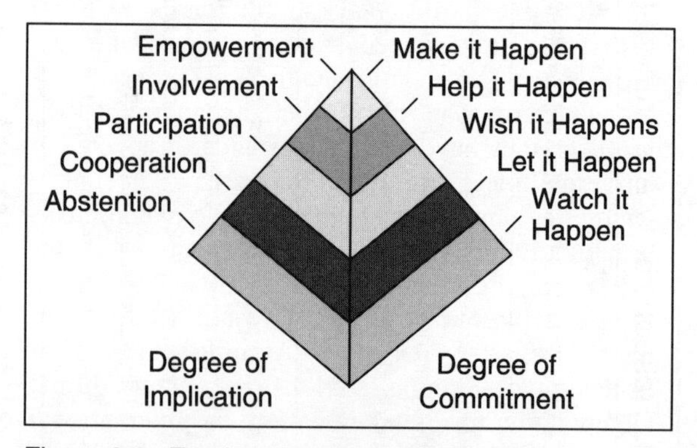

Figure 6-8. Employee power pyramid.
Source: Developed by E. Grund. © National Safety Council.

- The subject, project, etc. must have relevancy to the participant.
- Sufficient time must be provided in which to participate.
- Participation involving decision making must be confined to the area of responsibility of the group.
- The cost of the participation should not exceed the value of the outcome (Davis, 1977).

Whether your team or organization is in the participative, involved, or empowered mode, commitment is essential. Commitment can be described as a pledge or determination to do something; it connotes trust, intensity, and future obligation. A critical mass of mindset and conviction must be generated within the group to change intentions into commitment. Top management's commitment to having a first-class energy-control system is critical. If this is lacking or perceived as superficial, then certainly not much will be accomplished.

SUMMARY

- The process approach is an organized effort involving preparation, planning, systems logic, monitoring, feedback, and continuous improvement.
- The process approach offers a more deliberate, continuous way of developing and maintaining a prevention-oriented approach than one of mere compliance. The process cycle involves five major steps: (1) design, (2) implement, (3) monitor, (4) evaluate, and (5) refine.
- Before putting a hazardous-energy-control system together, an assessment or appraisal of the situation needs to be made. The assessment should address the potential for hazardous energy release, influencing factors, the status of current efforts, the status of compliance, energy-release-incident experience, and machine/equipment/process hazard analysis. According to OSHA, hazard analysis shall reveal (1) hazards of the process, (2) identification of prior incidents, (3) engineering or administrative controls for the hazards, (4) human factors, and (5) a qualitative evaluation of effects of failure of employee controls.
- After evaluation of the information developed in the situation appraisal, the facility manager or top executive will need to give the order to put an energy-control system in place and delegate the authority to a key individual to develop a plan.
- The purpose of any plan is to facilitate the accomplishment of the defined purpose and objectives. Plans come in various forms, such as mission statements, purposes, goals, objectives, policies, procedures, programs, projects, and rules. Plans provide the critical standards for control to ensure that you are going where you want to go.
- The process of planning involves four basic principles that should be acknowledged: (1) the limiting factor, (2) commitment, (3) flexibility, and (4) navigational change. Also consider the premises for planning or the environment in which the plan is expected to operate.
- Effective programming is comprised of six basic elements:

1. Dividing the activity required to accomplish a goal into basic parts.
2. Determining the relationship among the parts and understanding their sequence.
3. Deciding who will be responsible for each part.
4. Deciding how each part will be accomplished and what resources are required.
5. Determining the time needed to complete each part.
6. Establishing a schedule for each part.

- Various techniques such as the GANTT chart, the program evaluation and review technique (PERT), and the critical path method (CPM) are effective in planning and programming control. Dessler (1985) describes 10 principles of effective planning, and Koontz et al. (1984) provide a list of reasons why people fail when planning.
- Utilizing a planning strategy that views hazardous energy control on a systems basis is the desired approach. Six key elements were identified that represent the hazardous-energy-control system: (1) organizational culture, (2) management process-procedure, (3) human, (4) machine (equipment/materials), (5) the environment, and (6) design.
- A core action team (CAT) should be established to develop the facility plan for hazardous energy control. The team should be made up of a cross section of personnel with the right talent for accomplishment of the overall task.
- Resources need to be defined in terms of money, personnel, and time. Budget preparation will ensure a realistic treatment of the costs associated with developing, initiating, and maintaining the system. A personnel/resource talent inventory should be conducted to identify individuals with special skills, appropriate background and experience, and personal attributes that would be valuable to the energy-control system's development.
- A list of topical action items and the types of discipline or function best suited for the responsibility is proposed. The work breakdown structure and the responsibility assignment matrix are tools for managing the system development task in an organized way while staying on schedule. The CAT leader defines the work and timing and keeps the project on schedule.
- The facility manager, after reviewing the preliminary plan, will be required to make a decision

regarding implementation. The CAT team plays an important part in guiding and coordinating the numerous facets of the plan. Task/element coordinators should be well acquainted with the total project and how their particular piece of the plan interconnects.

- The employee power pyramid depicts the degrees of employee implication and commitment and the value of allowing employees to heavily contribute to the final content and shape of the energy control system.

REFERENCES

Brown DB. *Systems Analysis and Design for Safety*. Englewood Cliffs, NJ: Prentice-Hall, Inc., 1976.

Clemens PL. A compendium of hazard identification and evaluation techniques for system safety applications. *Hazard Prevention* (March–April 1982):11–18.

Davis K. *Human Behavior at Work*, 5th ed. New York: McGraw-Hill, 1977.

Dessler G. *Management Fundamentals*. Reston, VA: Reston Publishing Company, Inc., 1985.

Federal Register, Vol. 54, no. 169, September 1, 1989, 29 *CFR* 1910.147, Final Rule, pages 36683-36685. Washington DC: GPO, 1989.

Gray ER, Smeltzer LR. *Management: The Competitive Edge*. New York: Macmillan, 1989.

Grose VL. *Managing Risk—Systematic Loss Prevention for Executives*. Englewood Cliffs, NJ: Prentice-Hall, 1987.

Johnson WG. *MORT Safety Assurance Systems*. New York: Marcel Dekker, Inc., 1980.

Kepner C, Tregoe B. *Project Management*. Princeton, NJ: Kepner-Tregoe, Inc., 1989.

Koontz H, O'Donnell C, Weihrich H. *Management*. New York: McGraw-Hill, 1984.

Manuele FA. *On the Practice of Safety*. New York: Van Nostrand Reinhold, 1993.

Merriam-Webster's Collegiate Dictionary, 10th ed. Springfield, MA: Merriam-Webster, Inc., 1993.

Newman WH., Logan JP. *Strategy, Policy and Central Management*, 8th ed. Cincinnati, OH: South-Western Publishing Company, 1981.

Sayles L, Strauss G. *Human Behavior in Organizations*. Englewood Cliffs, NJ: Prentice Hall, 1966.

U. S. Code of Federal Regulations "Process Safety Management of Highly Hazardous Chemicals." 29 *CFR* 1910.119, Subpart H. Washington DC: GPO, February 1992.

7 System Elements

The hazardous-energy-control system in Figure 6-6 consisted of six elements: (1) Organizational Culture, (2) Management Process-Procedure, (3) Human, (4) the Environment, (5) Machine, and (6) Design. In this chapter, the Lockout/Tagout System Elements are addressed individually and further divided into subelements. The system is graphically presented as a wheel (Figure 7-1) that represents a continuum of discrete activities that interact synergistically. Other activities may be beneficial to the system's success as well, but those presented are viewed as essential. Various subelements may be more critical than others, but that does not diminish the necessity for having each piece fully developed and functioning in order to establish an effective prevention structure.

Energy-release incidents are viewed as flowing from a complex set of causes and effects connected in various sequences. They are multifactorial incidents with elaborate networks of contributory human action or inaction. Simple acts of disobedience or noncompliance are rarely verified as the actual root cause(s) of these incidents. Errors, acts of omission/commission, oversight, error-provocative design, and so on are all part of the dynamic that must be managed effectively. For those entrusted with the responsibility of developing and maintaining a hazardous-energy-control system, the various subelements are discussed in detail.

ORGANIZATIONAL CULTURE

It is not the intent of this text to detail what must be done to establish the ideal organizational culture. However, sufficient evidence exists to suggest that safety initiatives, programs, and processes will not flourish in a climate or culture that is indifferent to them.

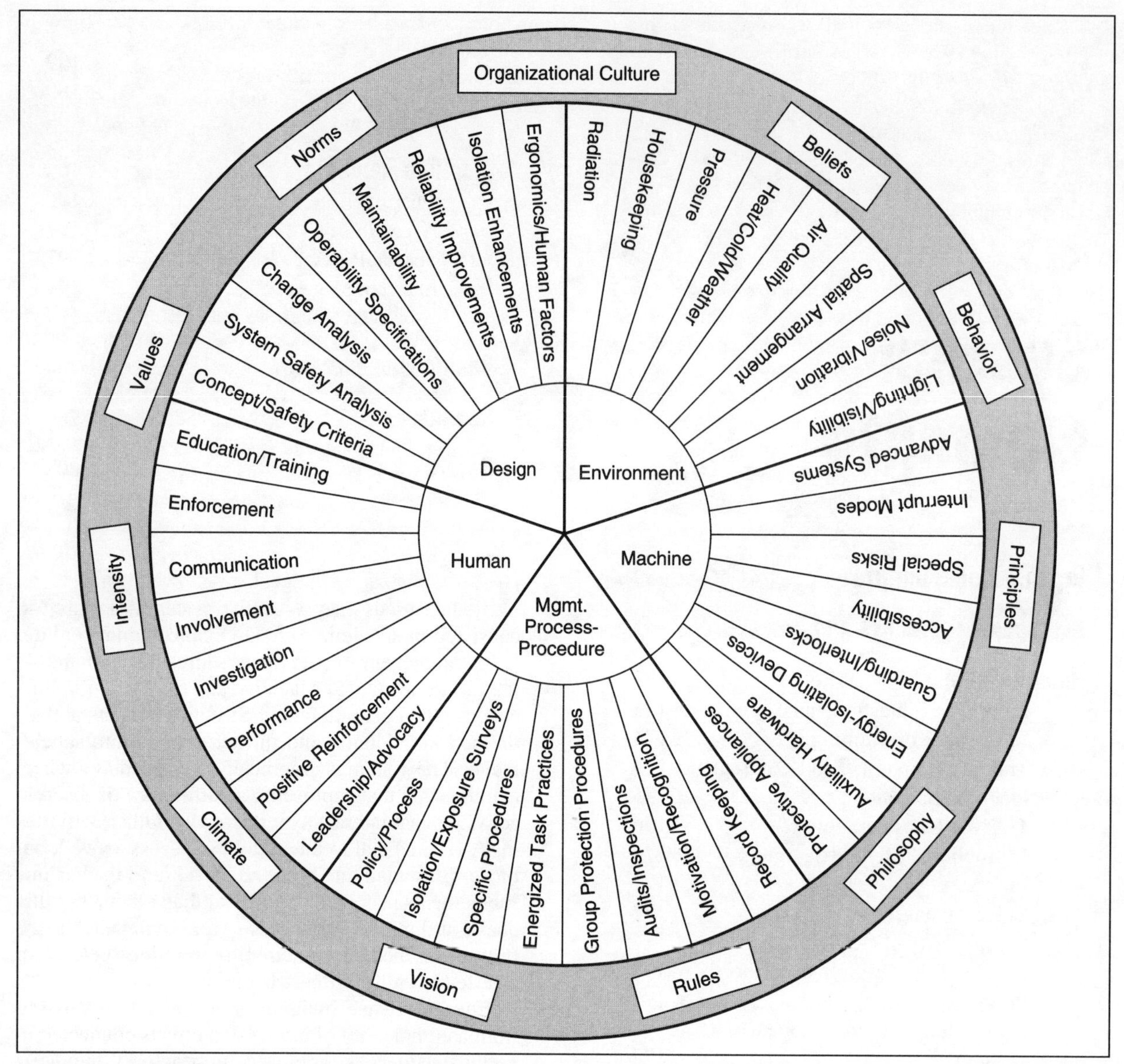

Figure 7-1. Hazardous-energy-control system (lockout/tagout).
Source: Developed by E. Grund. © National Safety Council.

Organizational culture is typically defined as the members' shared beliefs, expectations, norms, attitudes, values, goals, and ideologies. It includes the following components:

- *Observed behavioral regularities* when people interact (rituals, ceremonies, and language)
- *Norms* shared by working groups
- *Dominant values* held by the organization, such as the belief that accidents are preventable
- *Philosophy* that guides organizational policy
- *Rules* by which things get done or one gets along
- *Feeling or climate* that exists in an organization as a product of the physical and psychological environment (Schein, 1985).

The organizational culture may exist on various levels from superficial symbolism to core values. It has aspects that are subjective (values, assumptions, and expectations) and objective (heroes, observed behaviors, and rituals).

Organizational culture is said to develop in response to two major sets of problems that beset every organization: (1) problems of external adaptation and survival and (2) problems of internal integration. The former have to do with how the organization will find its place and cope with the constantly changing environment and the latter relate to how to establish and maintain effective working relationships among group members.

External problems that influence the formation of organizational culture are the organization's mission, strategy, goals, means and methods, structure, measurements, and corrective action for individual or group control. Internal problems are associated with common language and concepts, group boundaries, criteria for inclusion/exclusion, power and status, rules for relationships, rewards, and punishment.

There are two additional influences on the origins of organizational culture that shouldn't be overlooked. The founder of an organization will often imprint it with his/her personal beliefs, values, ethics, and ideas. As time passes, the founder's influence may be intermixed with other significant members' views as the organization matures. This can be seen in the culture that has evolved in DuPont due to the safety beliefs and values of Eleuthère I. DuPont.

Secondly, the dominant culture, customs, and societal norms of the country or region will shape the organizational culture. Industrially underdeveloped or developing countries will present a totally different environment than a leading industrial country, which will influence the evolving safety culture of an organization operating within that geography. The prevailing values of a region or country may impose constraints that dramatically impact the way business is done. Very large organizations may have various subcultures operating within the total company framework.

To change the organizational culture requires using the same methods that serve to maintain it. For example, the culture might be altered by changing (1) the priorities of managers, (2) crisis management methods, (3) reward and punishment criteria, (4) rites and ceremonies, or (5) role models. For example, a culture that tends to reward risk-taking and punish risk-reduction behavior (directly or indirectly) might be deliberately altered by modifying the reward climate or system (Hellriegel et al., 1992).

It may prove difficult to install an effective hazardous-energy-control process or system with a culture that does not provide the necessary supporting environment. However, things are seldom perfect or just right for launching or sustaining new initiatives in the real world. The existing culture must be understood so that the new system is not designed in direct conflict with it. If compromises must be made, the consequences should be understood. For instance, a tough but fair enforcement policy may be impossible in a permissive culture where discipline of any kind has been previously frowned upon or nonexistent.

If key management is interested in changing or improving the culture or if the culture in question is in transition, the following ideas might be beneficial:

- Know the current culture and identify what changes would be beneficial.
- Support those who are willing to act as agents of change and who have ideas or visions.
- Identify any effective subculture that could be the pattern for learning.
- Don't attack culture head on. Find ways to help employees do a better job.
- Use the new vision to guide, not work miracles.
- Cultural change is an extended process (more than five years).
- Action, action, action—live it, don't say it! (Dumaine, 1990)

MANAGEMENT PROCESS AND PROCEDURE

Management process and procedure cover all the classic functions of management that relate to planning, leading, organizing, monitoring, controlling, and improving. Various fundamental activities normally encountered in general business are also necessary for a successful hazardous-energy-control system. Chapter 6, The Process Approach, discussed the planning function in detail, but a more specific treatment of various elements is provided in the following sections of this chapter to guide those involved in the development of their own system.

Policy/program development. The OSHA lockout/tagout rule requires that the employer establish an energy-control program including (1) documented energy-control procedures, (2) an employee training program, and (3) periodic inspections of the procedures. The standard requires employers to establish a program to ensure that machines and equipment are isolated and inoperative before any employee performs servicing or maintenance during which unexpected energization, start-up, or release of stored energy could occur and cause injury.

The purpose of the energy-control program is to ensure that when the possibility of unexpected machine or equipment start-up exists or when the unexpected release of stored energy could occur and cause injury, the equipment is isolated from its energy source(s) and rendered inoperative prior to servicing or maintenance.

OSHA allows employers the flexibility to develop a program and procedures that meet the needs of their particular workplaces and the particular types of machines and equipment being maintained or serviced. No prescribed format for constructing the program has been legally specified, but it must be in written form and available for review. Furthermore, an organization's written lockout/tagout program may have to stand the test of external scrutiny beyond OSHA, such as legal discovery procedures and contractor review.

Various levels of policy/program may be developed. Corporate, company, division, business, facility, and department represent some of the common spans of coverage for organizations. Generally, as the span of influence for the policy/program diminishes, the detail becomes greater. Conversely, as the span increases, broader focus is placed on concepts, principles, key elements, goals, and direction for suborganizations to construct local programs embodying the critical content. A general policy for lockout/tagout suitable for use at a corporate/company level is provided for reference in Appendix 4. The focus of this document would have to be narrowed as its scope is reduced toward the smallest unit of concern, for example, department, group, and so on.

An organizational lockout/tagout policy should identify its purpose and objectives. For example, the following might be appropriate:

- **Purpose:** To establish a lockout/tagout system to prevent the unexpected release or transmission of machine, equipment, or process energy.
- **Objectives:**
  - Prevent unintentional operation or energization of machine, equipment, or processes in order to protect personnel.
  - Establish methods for achieving zero energy state.
  - Use equivalent alternative measures when circumstances warrant.
  - Comply with applicable regulatory standards.

It is also important to include the scope of the policy to identify the range and impact of the document. The following language could be used as a scope for a large multinational company:

- **Scope:**
  - This policy applies to activities such as, but not limited to, erecting, installing, constructing, repairing, adjusting, inspecting, cleaning, operating, or maintaining the equipment/process.
  - This policy applies to energy sources such as, but not limited to, electrical, mechanical, hydraulic, pneumatic, chemical, radiation, thermal, compressed air, energy stored in springs, and potential energy from suspended parts (gravity).
  - All U.S. facilities will comply with the spirit and substance of this policy. If state requirements exceed those found herein, facilities so regulated will comply with the local provisions that are more demanding.
  - International facilities will comply with the substance of this policy or the prevailing national requirements, whichever is more stringent.

Installation/design requirement. After January 1, 1990, whenever replacement or major repair, renovation, or modification of a machine or equipment is performed and whenever new machines or equipment are installed, energy-isolating devices shall be designed to accept a lockout device (29 *CFR* 1910.147[c][2][iii]).

In addition, the following topical sections are recommended for inclusion in the typical policy whether the scope is companywide or at the department level.

Definitions. Identify and list all those terms and their definitions that will have bearing on your energy-control system. Prevailing regulatory or official definitions should be included verbatim even if you disagree with their structure or content. Official interpretations should also be provided when available. Often the original definitions need further explanation or elaboration to communicate the original intent.

For example, the definition of *lockout device* in 29 *CFR* 1910.147 was later amended by adding the sentence, "Included are blank flanges and bolted slip blinds." This addition provides new insight as to the full understanding of what constitutes a lockout device from the federal government's viewpoint.

Internal company definitions may also be included to prevent confusion about local terminology and its meaning. For instance, *block-out* or *lock-in* may be terms specific to your business or culture but alien to the average outsider.

Principles/standards. Establish those key requirements that are unique or distinctive and that form the basis or cornerstone for your energy-control system. In some ways, these requirements reveal the philosophy that provides the system framework. For example, the following principles/standards might be included in your policy:

- Supervisory (all management) and hourly personnel will be required to work under the protection of personal locks/tags.
- The use of lockout devices is the required form of energy isolation.
- A lock(s) may only be removed by the person who applied it; if that person is not available to remove the lock(s), the special procedure for lock removal (A-1) will be followed to restore the operation while ensuring employee safety.
- All facility energy-isolating devices will be identified with the following: (1) energy type, (2) energy magnitude, and (3) device identification number.
- Energy-release incidents will be investigated during the shift in which they occurred by the designated team consisting of hourly craftsmen, safety coordinator, engineering representative, and department head or designee.

Lockout/tagout methods. When instituting an energy-control system or revising an existing one, it will be necessary to decide whether to put in place a method that uses lockout appliances, tagout appliances, or some combination. This decision is quite controversial and hotly debated. There are proponents of all three options who contend their system works best when all other requirements have been satisfied.

OSHA, in the preamble to the lockout/tagout standard 29 *CFR* 1910.147, identified the following major issue: "Should OSHA require the use of locks, locks and tags, or tags alone to control potentially hazardous energy?" A tremendous amount of testimony and comment was directed at this issue. OSHA stated:

> It was unfortunate that attention was focused more on a single aspect of the standard,

> though it is certainly an important one, than on the standard as a whole. The proposed standard was intended to specify that the employer provide a comprehensive set of procedures for addressing the hazards of unexpected reenergization of equipment, and the use of locks and/or tags was intended to be only a single element of the total program.

However, in the Final Rule, OSHA determined that *lockout* is a surer means of assuring de-energization of equipment than tagout and that it should be the *preferred method* used by employees. The agency also recognizes that tagout will nonetheless need to be used instead of lockout where the energy-control device cannot accept a locking device. Where an energy-control device has been designed to be lockable, the standard requires that lockout be used unless tagout can be shown to provide "full employee protection," that is, protection equivalent to that provided by lockout.

Full employee protection. This includes complying with all tagout-related provisions plus implementing additional safety measures that can provide the level of safety equivalent to that obtained by using lockout. This might include removing and isolating a circuit element, blocking a controlling switch, opening an extra disconnecting device, or removing a valve handle/wheel to reduce the potential for any inadvertent energization.

Protective appliances. Determine which types of gear will be used in your lockout/tagout approach. It is suggested that appliances (locks, tags, adapters, fixtures, value covers, chain, cable ties, etc.) that are used to secure energy-isolating devices not be confused with the actual mechanism that interrupts or prevents the transfer of energy.

Auxiliary hardware. Determine what additional protective hardware is required to support the energy isolation effort. These items will be distinguished from appliances in that they do not physically come in contact with the energy-isolating devices themselves. Examples of these items of hardware include keyboxes, control boards, barricades, graphic warning signs, alerting lights, and control panel placards.

Isolation survey. Conduct a survey of principal machines, equipment, and processes to determine if energy-isolating devices are available for each type of energy involved. In addition, cvaluate their condition, their proper identification, their location, and so on. This is a major task and may take a considerable amount of time and effort to accomplish. It could be completed in connection with development of specific isolation procedures. However, there is no effective way to avoid completing this critical activity. Lack of availability of energy isolation devices is often cited as a problem in incident investigations.

Exposure survey. Conduct a survey to determine what tasks are currently done under energized or partially energized conditions, for example, removing jams, checking alignment, cleaning rolls, and tool adjustments. Have the appropriate group review all such tasks to evaluate if it is proper to continue utilizing these methods. Those tasks that receive group approval will require the development of alternative safeguarding measures to be discussed in greater detail later in this chapter.

Responsibilities. Define who will do what for various needs of the lockout/tagout system. At the facility, or department, level, actual names of various individuals or job titles can be used to clarify the responsibilities. The following examples are intended to illustrate the approach:

- *Policy:* facility manager
- *Isolation/exposure surveys:* maintenance superintendent
- *Enforcement rules:* industrial relations manager
- *Specific procedures:* department heads
- *Audit:* union and management committee
- *Appliance and hardware procurement:* purchasing agent.

Specific procedures. Prepare detailed procedures for each machine, piece of equipment, and process that can be used to accomplish effective energy isolation. Procedures can take various forms but should reveal how isolation is achieved for each type of energy involved. Organize the procedures in some manner (such as by department or area) and prepare an inventory for reference and control.

Preparation, application, release practices. Develop the particular work practices or methods that will identify how lockout/tagout will function for the preparation, application, and release phases of machine, equipment, or process energy isolation. Greater detail on this subject is found in Chapter 8, Preventing Energy Transfer.

Group (multiple personnel) protection. Establish procedures for lockout/tagout preparation, application, and release when group, crew, or multiple personnel are involved with machine, equipment, or process energy isolation. The procedures will generally parallel the regular practice for lockout/tagout but will have special provisions for personnel control and accountability, appliance and hardware management, personnel coordination and communication, sequence of events, key control, and shift change. In 29 *CFR* 1910.147, OSHA states that group lockout/tagout shall afford the employees a level of protection equivalent to that provided by the implementation of a personal lockout or tagout device. This will be addressed in greater detail in Chapter 10, Special Situations and Applications.

Alternative measures. The exposure survey previously mentioned will provide information that can be used to prepare the inventory of alternative measures for energized or partially energized tasks. These alternative measures must provide effective employee protection and must only be used when there is consensus among workers and management that the task(s) cannot be reasonably performed under zero energy conditions. Each task conducted under energized or partially energized conditions will require a procedure that identifies the nature of the task, the potential hazards, and the specific safeguards employed to protect the worker.

NOTE: This element is often neglected in hazardous-energy-control systems because of the presumption that lockout/tagout does not apply or that it is impractical. However, management should take the initiative to work with employees to create techniques and methods that are mutually viewed as safe alternatives.

Contractors. Define the actions that will take place when noncompany personnel are used for on-site servicing and maintenance tasks. Such activities might be (1) review of contractor safety capability at the pre-bid stage; (2) lockout/tagout competency; (3) coordination, if jobs involve intermixing of personnel; (4) establishment of clearance procedures; (5) agreement on responsibility for energy isolation, and (6) deficiency correction. See Chapter 10, Special Situations and Applications, for additional detail.

Employee training. Design a hazardous-energy-control training program for employees that includes provision for new or transferred employee orientation, awareness-level indoctrination for employees indirectly involved with lockout/tagout, comprehensive instruction for those affected or closely connected to energy isolation activities, advanced training for authorized employees who actually conduct energy isolation, and specialized instruction for engineering/purchasing personnel and auditors. A training plan should be prepared that defines all parameters.

Audits/inspections. A monitoring or surveillance protocol should be developed to identify what types of audits are to be conducted. The audits can cover administration, application, or performance. They should be formatted rather than random, free-style occurrences. If properly structured, the audits will reduce dependency on the auditor's talent. A team is suggested for audit design in order to examine what is perceived as the more important aspects of the hazardous-energy-control system.

Energy isolation and exposure surveys. Whether beginning an energy-control system or upgrading one, conducting surveys of isolation capability and energized tasks exposure should be on the list of things to do. Management needs to determine in a systematic way if machinery, equipment, or processes can be effectively isolated from the energy sources that feed or power them. Admittedly, these surveys require a considerable investment and commitment of resources, but the information gathered will prove invaluable. Most managers do not like surprises; without the information revealed by these inquiries, too many may occur!

Often, energy-release-incident investigations will identify that an injured employee failed to lock out or tag out a piece of equipment while performing a particular task. However, the real deficiency is likely to have been associated with inadequate or poorly located energy-isolating devices. Many times a tacit understanding exists that the task is to be performed under power without provision for effective alternative safeguarding measures. After the injury incident, those responsible may naively believe that this particular situation should have been accomplished with all energy isolated. Much of this confusion and ineffectiveness could be avoided with comprehensive surveys of energy isolation capability and exposure to energized tasks.

Energy isolation capability survey. The first order of business when conducting such a survey is to develop lists of major machinery, equipment, or processes by department, area, or zone. Most businesses have such inventories or lists for accounting purposes (depreciation schedules, insurance coverage) or maintenance activities (preventive maintenance, parts ordering). Once the lists are prepared in some organized manner, they will be of great use in completing the energy isolation capability survey sheets, an example of which is shown in Figure 7-2. The following procedure is suggested for accomplishing the survey:

- *Organize teams.* Identify knowledgeable individuals, both hourly and salaried, to complete surveys in their areas of responsibility. Maximize participation so that the outcome will be truly representative of the efforts of a sizable cross section of company personnel.
- *Prepare survey sheets.* Use separate sheets for each piece of capital equipment, or group similar machines or units that may be common in a given operational area. Continuous processes may require sets of survey sheets with breakdowns to create a pyramiding structure that includes all components of the process.
- *Conduct survey.* Establish target dates for completion of various segments of the project. Use a coordinator to keep the project on schedule and chart progress. Team leaders are responsible for getting results.
- *Complete survey sheets.* Complete a survey sheet for each piece of equipment that contains all involved energy sources; type, location, and identification of isolating devices; access assessment; residual energy relief features; and comments regarding deficiencies, changes, and improvements.
- *Arrange team leader meeting(s).* Arrange periodic meetings for team leaders to brief the group on problems, ideas, and progress; the coordinator should prepare a summary of key issues, opportunities, and status.

XYZ
MANUFACTURING

Energy Isolation Capability Survey

PLANT: JACKSON   DEPARTMENT: FORGING   PROCESS: HYD. PRESS

Machine/Equipment	Energy Type (Code)	Isolating Device Type and I.D.	Accessibility (Code)	Remarks
Loewy Hyd. Press 250 Ton	E	440 V Breaker (AC) (Col. E 35) #62	W-50	Relabel box and clear materials
	H	Gate valve (3") H4b —Press catwalk	E-6 (W-50)	Relocate valve to floor level; install bleed
	G	Ram/Die safety block(s)	W-50	Paint yellow
	P	Lox Valve (1") 110 P.S.I.—P503	W-50	Install identity tag
	T	Side shields—die area open	W-50	Manipulator use; cool down/thermal blankets

Energy Type Code:
Electrical E; Thermal T; Gravitational G
Hydraulic H; Pneumatic P; Other O
Mechanical M; Chemical C

Accessibility Code:
Within 50' (W-50); Elevated = >6' (E-6)
Over 50' (O-50); Below Grade (BG)

Figure 7-2. Energy isolation capability survey.
Source: Developed by E. Grund. © National Safety Council.

- *Tabulate data.* Organize, record, and prepare the final survey sheets; extract the items or issues for follow-up.
- *Develop deficiency lists.* Develop lists for each area, zone, or department that identify deficiencies in access, identification, location, marking, function, and adequacy of energy-isolating devices.
- *Develop an action list.* Develop a corrective action schedule with priorities based on assessment by team leaders, staff specialists, and coordinator; cost estimates should be included.
- *Prepare a status report.* Prepare progress reports for management indicating performance against the plan; also provide an executive summary of key findings of the survey and overall energy isolation capability by area, department, etc.

Energized tasks exposure survey. The energized tasks exposure survey complements the isolation capability survey. Together they provide invaluable information for completion and refinement of the specific procedures required for machine, equipment, and process energy isolation. However, the energized tasks exposure survey represents a significant opportunity to address a common weakness found in many facility lockout/tagout programs: the failure to examine work practices that take place where energy is present, employees are potentially exposed, and total energy isolation is perceived to be unnecessary or impractical.

Lack or insufficient evaluation of these energized tasks is all too prevalent. This may be occurring because these tasks are often short, transitory occurrences, variable in nature, and viewed as operational distractions; as well, the required energy isolation may take more time than completion of the task. This situation may be deemed acceptable from the perspective of those managing the work but nevertheless represents a continuing energy-control flaw. In order to call attention to this serious deficiency, Figure 7-3 depicts the "limbo zone"—that area of managerial omission or neglect that lies between the de-energized or white state and the energized or black state. It is gray by its very nature and not easily resolved without a commitment to do so.

For example, an organization may clearly define formally or informally how machines, equipment, and process must be de-energized for various repair and service activities. In addition, there also may be formal recognition of various tasks that will take place under energized circumstances utilizing appropriate techniques and safeguards. However, what is suggested by the limbo zone analogy is that numerous tasks and activities exist in many facilities where the equipment is entirely or partially energized and employees are performing work with significant risk. The work for various reasons has fallen into the gray area (limbo zone) and therefore continues without managerial scrutiny and assessment or the application of formal energy-control practices. Employees

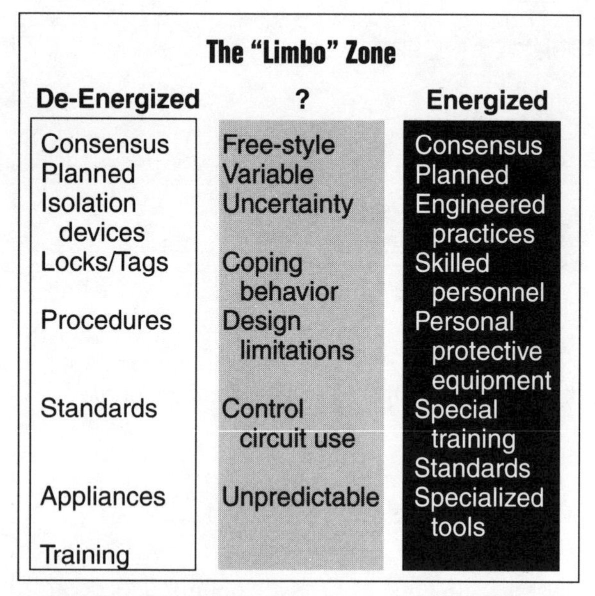

Figure 7-3. The limbo zone—the area of neglect between institutionalized energized and de-energized tasks that have clearly defined safety requirements and characteristics.
Source: Developed by E. Grund. © National Safety Council.

perform these tasks without apparent concern because they don't understand the risk, feel it is convenient, think it will save time, believe it is what management desires, and so on.

For this reason, it is suggested that an energized tasks exposure survey be conducted to identify all of the energized tasks or activities that have escaped formal acknowledgment and assessment. The objective should be to develop either formal de-energized or energized control practices that properly address the involved risks. No tasks, therefore, should remain in the limbo zone representing an uncontrolled hazardous energy situation.

To conduct an energized tasks exposure survey, the following process is recommended:

- *Organize teams.* The same approach or teams may be used as suggested in the energy isolation capability survey; team size and composition will depend on the size and nature of the operation but should always involve hourly employees.
- *Inventory energized tasks.* Use the machine, equipment, and process data from the isolation capability survey as a starting point, and prepare a list of nonsanctioned energized or gray area tasks for each machine, equipment, or process. The inventory can be viewed as Phase I of the energized tasks exposure survey (Figure 7-4). A knowledgeable individual should complete the first two columns of the form; employee interviews and observation will be necessary to identify all such tasks. The draft form of the lists should be reviewed by maintenance and operational personnel to verify completeness.
- *Coordinator assignment.* Assign a coordinator to head up the energized tasks exposure survey project; target dates and schedules should be established to manage the process.
- *Conduct survey.* Teams should interact with maintenance and production workers to identify all energized activities associated with each machine, piece of equipment, or process.
- *Complete survey sheets.* Complete a survey sheet for each major machine and piece of equipment to include energized tasks, energy type, safeguard code, and remarks.
- *Team leader meetings.* Arrange periodic meetings for all team leaders to discuss project progress as well as problems or ideas for improvement; the coordinator should prepare a brief summary of key issues associated with the project.
- *Reevaluate tasks.* Assess energized tasks to determine if work can be performed under isolated conditions; video can be used to evaluate the task with group participation; employees should be involved to determine the best methods for risk reduction.
- *Safe practices (alternative measures).* Develop or schedule time for safe practice completion; these measures should use a combination of techniques, protective equipment, and hand tools to accomplish the task with protection equivalent to that of lockout/tagout.
- *Special projects.* Establish study teams to review difficult tasks that may require engineering improvement or capital expenditure to eliminate the need for the exposure situation.
- *Status report.* The coordinator should prepare periodic status reports for management indicating the progress of the project; the report should highlight any critical issues and those areas where special studies are necessary to finalize the task status.

This process is representative of what might be done to produce the final product. However, the energy isolation capability and energized tasks exposure surveys can be accomplished in other ways. Administratively, the larger the business, the more likely electronic (computer processing) means will be used to generate, store, and process the data from the surveys. Notebook PCs might be used to accumulate the data, which might be later downloaded to more powerful desktop PCs or mainframe systems. The use of electronic systems is advised to improve general administration, provide for continuous updating, and generate various reports and permanent records.

Specific energy isolation procedures. In section (c) general (4) energy control procedure, 29 *CFR* 1910.147 states that

> (i) procedures shall be developed, documented, and used for the control of potentially hazardous energy when employees are

XYZ
MANUFACTURING

Energized Tasks Exposure Survey

PLANT: JACKSON DEPARTMENT: FINISH PROCESS:

Machine/Equipment	Energized Task	Energy Type (Code)	Safeguard (Code)	Remarks
Hofer Cont. Band Saw (6)	Lube saw blades	E	INAD	Procure lube stick holders
	Clear table-flash	E	WP-FIN43	Review
	Blade jam	E	NE	Stop practice; all jams use discon-nect for isolation

Energy Type Code:

Electrical	E	Thermal	T	Gravitational	G
Hydraulic	H	Pneumatic	P	Other	O
Mechanical	M	Chemical	C		

Safeguard Code: **Alternative Measures**

Written practice (WP+#)	Inadequate (INAD)
Not established (NE)	In-process (IP)

Figure 7-4. Energized tasks exposure survey.
Source: Developed by E. Grund. © National Safety Council.

engaged in the activities covered by this section and
(ii) procedures shall clearly and specifically outline the scope, purpose, authorization, rules, and techniques to be used for the control of hazardous energy and the means to enforce compliance, including, but not limited to, the following:
(a) A specific statement of the intended use of the procedure;
(b) Specific procedural steps for shutting down, isolating, blocking, and securing machines or equipment to control hazardous energy;
(c) Specific procedural steps for the placement, removal, and transfer of lockout devices or tagout devices and the responsibility for them; and
(d) Specific requirements for testing a machine or equipment to determine and verify the effectiveness of lockout devices, tagout devices, and other energy-control measures.

If regulatory compliance were the sole objective, there may be some inclination to interpret 29 *CFR* 1910.147(c)(4)(i) and (ii) as a requirement to have a common or generic energy-control procedure. In section (c)(1) energy-control program, the implication is to have an overall approach to energy isolation that includes all of the requisite (training, procedures) elements. However, the exception note in section (c)(4)(i) emphasizes "specific procedures" for particular machines or equipment. The note reads as follows:

Exception: The employer need not document the required procedure for a particular machine or equipment, when all of the following elements exist: (1) The machine or equipment has no potential for stored or residual energy or reaccumulation of stored energy after shut down which could endanger employees; (2) the machine or equipment has a single energy source which can be readily identified and isolated; (3) the isolation and locking out of that energy source will completely deenergize and deactivate the machine or equipment; (4) the machine or equipment is isolated from that energy source and locked out during servicing or maintenance; (5) a single lockout device will achieve a locked-out condition; (6) the lockout device is under the exclusive control of the authorized employee performing the servicing or maintenance; (7) the servicing or maintenance does not create hazards for other employees; and (8) the employer, in utilizing this exception, has had no accidents involving the unexpected activation or reenergization of the machine or equipment during servicing or maintenance.

A cursory review of the exception criteria suggests that very few, if any, machines or equipment will meet all of the stated elements. Simple machines like a drill press, lathe, or circular saw might be able to satisfy the criteria. Therefore, the first sentence of the note implies that specific procedures are required unless the stated criteria can be met.

In any case, the employer should develop specific procedures for energy isolation for *each* piece of machinery, equipment, or process, with prevention as a primary motive. *Even if two machines in close proximity with the same energy type and isolating features can be treated with a common procedure, there is some risk that a small difference in the machines would compromise the lockout/tagout process.*

The energy isolation capability survey previously discussed will provide the essential ingredients for preparing specific energy-control procedures (checklists, safe practices, job safety analyses). Each machine, equipment, or process in the survey should be addressed using a format similar to the checklists in Figures 7-5 and 7-6. The type of information needed to provide employees with what is necessary to do their job safely includes energy type and magnitude; isolating device(s), identification and location; isolating action; special precautions; isolation verification steps; stored energy relief/restraint; and safeguards. Management can also be satisfied that a formal analysis of energy isolation requirements has been conducted for their machines, equipment, or processes.

The specific procedures can be stored as hard copy or in electronic data bases for each area or department of the facility. Establish an index or inventory with date of preparation, review data, and reference identification designation. Assign responsibility for keeping the procedures updated so that as changes occur in the facility, revisions to the procedures will take place. Use the procedures regularly for training new and reassigned employees as well as for briefing employees prior to starting a job. The best procedure will be of little value if employees are not involved in its creation, use, and maintenance.

Safe practices—energized tasks. OSHA acknowledged in its deliberations on the lockout/tagout standard that certain activities, by their very nature, must take place without de-energization. In the electric utility industry, hot or live, line work is done routinely because de-energizing would seriously impact operations and service. However, significant effort has been put forth to establish equivalent safety measures to ensure worker protection. In the application section of 29 *CFR* 1910.147, the following exclusion recognizes the need to work on energized systems:

> (2)(b) Hot tap operations involving transmission and distribution systems for substances such as gas, steam, water or petroleum products when they are performed on pressurized pipelines, provided that the employer demonstrates that (1) continuity of service is essential; (2) shutdown of the system is impractical; and (3) documented procedures are followed, and special equipment is used which will provide proven effective protection for employees.

Furthermore, in the preamble to 29 *CFR* 1910.147, OSHA states, "It is important to note that this standard is intended to work together with the existing machine guarding provision of Subpart O of part 1910, primarily sections 1910.212 (general machine guarding) and 1910.219 (guarding of power transmission apparatus)." OSHA defines at length the phrases "servicing and/or maintenance," "normal production operations," and "repetitive minor adjustments" and explains how they relate to the need for complete energy isolation and lockout/tagout. See "Energized vs. Deenergized Activities" in the preamble to 29 *CFR* 1910.147 for further detail.

LOCKOUT/TAGOUT CHECKLIST FOR ENERGY ISOLATION BM-1-A

Plant: Bogalusa — Machine/Equipment: Contour Shaper

Department: Manufacturing — Energy Involved: Elec ☑ Hydr ☑ Pneu ☑ Thermal ☑ Other______

Energy/Magnitude	Isolation Device/I.D.	Location	Isolation Action
1. Electrical 440 V	Breaker EW8	Column E-4	Open breaker and lock
2. Hydraulic 200 PSI	2½" valve H030	Valve pit-unit base	Close valve, chain and lock, bleed system
3. Air 250 PSI	1" valve AB50	Overhead line near shaping head	Close valve, lock, and relieve residual air
4. Thermal 400 F	1" oil supply valve L33	Base machine—opposite flywheel	Close valve, chain/lock—wait one hour or use insulating pad over heat elements

Special Precautions: (2) Cycle hydraulic controls to relieve system pressure.

Note: (1) Relieve or restrain stored energy (gravity, electrical charges, pressures, etc.); (2) Verify energy isolation (test)

Figure 7-5. Lockout/tagout checklist for energy isolation.
Source: Developed by E. Grund. © National Safety Council.

LOCKOUT : TAGOUT SYSTEM®

CAUTION: Servicing or maintenance is not permitted unless this equipment is isolated from all hazardous energy sources. This is the exclusive responsibility of designated "Authorized Employees" (**see listing at bottom) who must follow the ***complete*** "Lockout/Tagout Procedure" as published by the company. ***This sheet is limited and abbreviated; it must not be considered a substitute for the company's complete Procedure.***

LOCKOUT:TAGOUT CHECK LIST

For Equipment No. 172

Equipment Description: Blivet Sintering Machine with Taurus Automatic Feeder

HAZARDOUS ENERGY		ISOLATING DEVICES			CONTROL DEVICES (Check ✓)			
Type	Magnitude	Type	Location	I.D. No.	Lock	Tag	Both	Add'l Measures (See Below)*
Electric	440V.	Switch	Right-Front Post	S819			✓	
Steam	170psi	Valve	At wall 30 feet west	V47		✓		
Hydraulic	3000psi	Valve	On pump 10 feet east	V16			✓	①

***ADDITIONAL SAFETY MEASURES (Refer To Table Above):**

① After closing and locking valve, drain hydraulic cylinder on machine 172.

****AUTHORIZED EMPLOYEES. Only the following are authorized to undertake the Lockout/Tagout Procedure on this equipment:**

Jerzi Brodofski, Andre Schoenzer
Paul Simpson, Nancy Mahabro
Stanley Cheng, Joe Hershko
Walter Paley
John Wilson

Prepared By: George Murdock **Date:** 8/8/90 **QUESTIONS? PHONE:** 538

 Q-SIGN® V9-11

Figure 7-6. Lockout/tagout checklist.
Source: Courtesy Idesco Corporation, New York, NY.

In 29 *CFR* 1910.147, section (a)(2)(ii), additional insight is provided regarding the application of the standard and the intent concerning the use of alternative measures:

> (ii) Normal production operations are not covered by this standard (See Subpart O of this Part). Servicing and/or maintenance which takes place during normal production operations is covered by this standard only if:
> (a) An employee is required to remove or bypass a guard or other safety device; or
> (b) An employee is required to place any part of his or her body into an area on a machine or piece of equipment where work is actually performed upon the material being processed (point of operation) or where an associated danger zone exists during a machine operating cycle.
> NOTE: Exception to paragraph (a)(2)(ii): Minor tool changes and adjustments, and other minor servicing activities, which take place during normal production operations, are not covered by this standard if they are routine, repetitive, and integral to the use of the equipment for production, provided that the work is performed using alternative measures which provide effective protection (see Subpart O of this Part).

The challenge for employers who believe certain activities should take place under less than zero energy state conditions is substantial. The following must be demonstrated:

1. The activity requires that it be done under energized conditions because
   - zero energy state conditions will have excessive impact on production and customers
   - it is more hazardous to de-energize
   - there is no technological alternative.
2. No changes are possible that would reduce the overall risks encountered under the energized method.
3. Alternative measures have been developed involving the personnel performing the work.
4. Alternative measures developed are capable of delivering a level of safety reasonably equivalent to that provided by machine guarding and/or energy isolation practices.
5. Employees are trained, knowledgeable, skilled, and have the ability to execute the alternative measures successfully.
6. The local or general experience with the alternative measures substantiates their effectiveness.

The employer's obligation, therefore, is to provide, as the British term it, "a safe system of work." The priorities are clear:

1. Secure the machine, equipment, or process to preclude exposure to energy releases.
2. Attempt to convert as many tasks done under partial or total energization to the lockout/tagout state or mode.
3. Design alternative measures that are reasonable, capable of a high degree of safety, and effective.

The energized tasks exposure survey will produce the inventory from which a more thorough assessment can be done. The second and third priorities just mentioned will have to be applied for all energized tasks that are currently known to officially exist. Random or rogue tasks that are just as real may exist informally or without management's official condonement. These must be discovered and dealt with, ideally with the willing cooperation of the hourly work force.

The true test will come when managers and hourly production/maintenance personnel prepare the Safe Practices for Energized Tasks form (Figure 7-7). These practices and the process that produces them will add immeasurably to safety when dealing with hazardous energy potential. Job safety analysis (JSA) can be used effectively for this purpose as well (Figure 7-8). The prepared JSAs should be responsive to satisfying items 3 and 4 mentioned above in the employer's challenge or justification for energized activity.

Auditing. In 29 *CFR* 1910.147(c)(6), Periodic Inspection, the OSHA regulatory position on auditing is revealed. It focuses solely on lockout/tagout application and not on an assessment of the hazardous-energy-control process in general. Basically, the same language is found in ANSI Z244.1, *Lockout/Tagout*, in section 3.2.2, Periodic Inspections, with a recommendation to "include random audit and planned visual observations of compliance with lockout/tagout procedures." OSHA does specify that the periodic inspections occur at least annually with appropriate documentation. The process of monitoring, measuring, and assessing the hazardous-energy-control system is addressed in detail in Chapter 9, Monitoring, Measuring, and Assessing.

Record keeping. OSHA does not have a specific section in 29 *CFR* 1910.147, Lockout/Tagout, that addresses record keeping as such. However, there are numerous inferences and requirements that appear throughout the text of the standard that obligate employers to document much of their energy-control activity. The following discrete areas are subject to potential regulatory review and inquiry:

- *General Energy-Control Program.* The energy-control program should be in written form and specify what precisely is done at the local level; particular attention should be placed on the removal of lockout/tagout devices as a program element.
- *Specific Procedures.* All specific energy isolation procedures for the facility should be in written form covering major pieces of machinery, equipment, and process.

- *Alternative Safe Practices.* Safe practices that define the equivalent measures used to ensure worker safety should be available for all energized tasks.
- *Employee Training.* Records that define the training content and its duration should be available; in addition, the names of the trained employees, their designation (authorized, affected and other), training dates and any retraining documentation that was necessary because of inspection deficiencies must be included.
- *Inspection.* Written records of all lockout/tagout application inspections should be retained; they should include the identity of the machine or equipment where inspection was conducted, the energy-control procedure used, the date of the inspection, the employees included in the inspection, and the person performing the inspection. Inspection data should be retained for a minimum of three years.
- *Contractor Interaction.* Records (minutes, specifications) regarding employer interaction with contractors about lockout/tagout requirements should be available.
- *Group Lockout.* Specific written control procedures for safeguarding employees at shift changes under lockout/tagout conditions should be available.
- *Equipment.* A management directive should be available indicating that after January 1, 1990, whenever major replacement, repair, renovation, modification, or installation of machines or equipment occurs, associated energy-isolating devices shall be designed to accept a lockout device. Maintain any correspondence with equipment manufacturers or suppliers that relates to efforts to meet this requirement.
- *Accident Experience.* All recordable injury cases involving energy-release events and the investigative detail should be maintained in a separate file for possible review by OSHA personnel; in addition, all energy-release incidents of any nature should also be retained.

Safe Practices For Energized Tasks
(Alternate Measures — Lockout/Tagout)

Plant: Durango   Department: Forming   Area: B

Job Task: Extracting Former Jams		Machine/Equip./Process: Shell Former	
Task Sequence	**Energy/Movement Hazards**	**Safeguards**	**Key Points**
1. Actuate emergency stop – control console	Machine forming ram	Machine guarding	Do not use guard interlock for STOP
2. Turn key selector switch to INCH	Machine forming ram	Machine guarding	Remove key and retain
3. Open ram guarding shell	Tooling components	Guard interlocks (2)	Do not attempt to remove jam by hand
4. Extract jam	Tooling components	Extraction tool/tooling block	Insert safety block; use jam pliers; if unsuccessful, use INCH
5. Inch tooling	Tooling dies	INCH control position	No two person jamwork; if unsuccessful – de-energize former
6. De-energize – follow forming lockout procedure DF-4			

Safeguards: Custom tools, blocks and wedges, interlocks, control circuit locks, jog only control, clearances, protective equipment, special gear, cleaning fixtures, insulated equipment, etc.

Prepared by: V. Johnson
Review/Approval: S.J. Pinder

Figure 7-7. Safe practices for energized tasks.
Source: Developed by E. Grund. © National Safety Council.

JSA
Job Safety Analysis

Job	JSA Identification Number	Facility
	Type (General or Specific)	Department

Job Titles of Involved Employees

Required and/or Recommended Personal Protective Equipment

BASIC JOB STEPS	HAZARDS (UNSAFE ACTS/CONDITIONS)	CONTROL MEASURES
Break the job down into its basic steps. Describe and record the job steps in their normal order of occurrence. For example, the job of "using a fire extinguisher" may be broken down as follows: 1. Remove fire extinguisher from hanger. 2. Carry to fire in an upright position. 3. Remove seal and pin. 4. Hold extinguisher in one hand, hose in other hand. 5. Apply chemical to base of fire. 6. Return extinguisher to empty rack.	What are the hazards for each job step listed. You can get your answers by (1) discussing with the employee (2) observing the job steps (3) reviewing past accidents and injuries or (4) a combination of all. Record the hazard by using the letter for the type of accident and by the description of the unsafe act or condition. Caught in (CI) Contact with (CW) Caught on (CO) Fall (F) Struck against (SA) Overexertion (O) Struck by (SB) Exposure (E) Contacted by (CB) Caught between (CBE)	How should the employee do the job to avoid the potential hazard? Review each job step and hazard you have listed. Discuss the hazards with the employee. List each precaution that the employee is to take to avoid the hazard you have shown. Number each separate recommendation with the same number given to the hazard. Answers must be specific and concrete if procedures are to be effective. General precautions — "be careful", "use caution", "or be alert" — should not be used. Answers should state precisely what to do and how to do it safely.
	(*) Critical Hazards	

Prepared By	Other Approvals	Safety Approval	Date Originated	Date Revised

Figure 7-8. Job safety analysis.
Source: Developed by E. Grund. © American National Can Company.

- *Enforcement.* The written employer policy on rules for enforcement of energy isolation infractions should be available; records regarding warnings, suspension, discharge, etc. for lockout/tagout violations should also be readily accessible.

Contractors. Another element of the facility lockout/tagout system that must be considered is the contractor procedure. OSHA requires that the host and outside employers/contractors inform each other of their respective lockout/tagout procedures. The host must ensure that its employees understand and comply with restrictions and prohibitions of the outside employers' energy-control procedures. Numerous other issues exist that should also be addressed from the point where the decision is made to engage external employers. This detail will be discussed in Chapter 10, Special Situations and Applications.

Motivation—recognition. An often-overlooked aspect of any lockout/tagout system is management's responsibility to motivate employees and recognize their efforts. *Motivation* can be defined as the force or internal state that activates, directs, and maintains a person's behavior; any motion that prompts someone into action; or the act of getting someone to desire to do a particular thing.

Motivation is most effective when the objectives:

- are clear, definite, and supported
- are within the employee's capability
- provide for real participation and freedom to act
- are easily understood and communicated
- challenge the individuals
- are relatively short range (Widner, 1973).

Management also needs to recognize what motivates employees and causes them to behave the way they do. Most people act or react in a certain way to satisfy their basic drives or desires. It is the task of management to provide for legitimate satisfaction of these drives or desires so that employees are *willing* to do what is necessary to have an effective energy-control system. This does not mean that management needs to enroll in a psychology course; on the contrary, all that is needed is a conscious planned effort to accentuate the positive and recognize that workers will be led but not driven.

In order for motivation to occur, a nurturing climate must exist, as mentioned briefly in the discussion of organizational culture earlier in this chapter. To be more specific, the following factors will enhance the prospects for improved individual performance:

- *Goal focus:* shared vision or commitment to purpose
- *Openness:* opportunity to participate, be heard, and influence things
- *Delegation:* sharing responsibility for making the right things happen
- *Freedom:* multidirectional informal communication (informational power sharing)
- *Security:* reasonable confidence in the business prospects and stability (we're in this boat together—partnering).

Getting employees to desire to do something involves a certain amount of promotional activity. Many managers are uncomfortable with this type of activity and believe it is too theatrical or circuslike. Although anything can be taken to extremes, it is better to think of promotion as selling, which has a more legitimate place in the art and science of management.

A task group or team can be assigned the responsibility for developing the "sales strategy" for the lockout/tagout system—in short, how do we promote and stimulate the health of the system and get the customer (employees) to buy? The central theme should be high interest, enthusiasm, spirit, and fun—yes, fun, even in the workplace! Heavy involvement from hourly employees is a prerequisite for the development of the strategy. Also do not confuse selling or promoting with incentive gimmicks or giveaways. Incentives at best can be considered a small subset of your sales campaign; a lockout/tagout promotional plan will work with incentives or without them. A sound, well-executed sales strategy can take a good lockout/tagout system and make it excellent.

HUMAN ELEMENTS

A number of items are reserved for discussion under this heading: employee education and training, enforcement, communication, involvement, and investigation. These items could easily have been considered under management process/procedure, but they have unique facets that are also effectively viewed from an individual or interpersonal perspective. They are undeniably critical parts of the total hazardous-energy-control system and therefore need to be integrated thoughtfully.

Employee Education and Training

During the balance of the 1990s and into the new millennium, demands on workers for greater productivity, world-class quality, and error- and accident-free performance will continue to grow. Managers and their work forces are destined to have to deal effectively with:

- technological acceleration
- shorter supplies of younger workers
- decreasing technical skill levels of entry workers
- expanding diversity in the work force
- rapidly changing markets and customer requirements
- major demands for worker training and retraining.

In the expanding technoenvironment of the future, safe performance will be predicated upon a thorough working knowledge of equipment and processes and more sophisticated operational skills. The National Association of Manufacturers reported in a 1989 survey that it was "fairly to very difficult" to find employees to operate complex machinery. Employers will need to approach training on more than a superficial basis, concentrating on the quality of the effort and what the employee must learn and be able to apply.

Importance of Training

OSHA, in its preamble to the lockout/tagout standard 29 *CFR* 1910.147, made a strong case for the criticality of employee training when dealing with hazardous energy. In the Bureau of Labor Statistics' Work Injury Report Study "Injuries Related to Servicing Equipment," 61% of those injured had not received lockout instruction. The UAW reported that in 72 lockout fatality incidents between 1973 and 1988, inadequate employee training had been the cause of 68% of the events. OSHA also reported that industries with effective tagout systems relied heavily on detailed procedures, extensive training programs, refresher training reinforcement, and discipline. OSHA identified training as one of the five key causal factors associated with the occurrence of lockout/tagout incidents and stated that proper training in procedures—understanding how they work and why they were important—was critical for safeguarding employees.

Training Emphasis and Requirements

In January 1989, OSHA issued its Safety and Health Program Management Guidelines, which identified four elements of an effective occupational safety and health program: (1) management commitment and employee involvement, (2) worksite hazard analysis, (3) hazard prevention and control, and (4) safety and health training. Further, in the OSHA reform bills S.575/H.R.1280, introduced to Congress in March 1993, training and education of employees was again given priority treatment and emphasis as follows (Sec. 27 Safety and Health Programs):

> (2) Regulations on training and education. The regulations of the Secretary under paragraph (1) with respect to an employer's safety and health program shall—
>
> (A)(i) provide for training and education of employees at the time of employment, in a manner that is readily understood by such employees, concerning safety and health hazards, control measures, and the employer's safety and health program;
>
> (ii) provide for the dissemination of information to employees at the time of employment, in a manner that is readily understood by such employees, regarding employee rights and applicable laws and regulations; and
>
> (iii) provide for training and education of employees who are selected to be safety and health committee members, at the time of their selection, that is necessary to enable such employees to carry out the activities of the committee under section 28; and
>
> (B) require that refresher training and dissemination of information be provided on at least an annual basis and that additional training and dissemination of information be provided to affected employees and to safety and health committee members when there are changes in conditions or operations that may expose such employees to new or different safety or health hazards or when there are changes in safety and health regulations or standards under this Act that apply to the employer.

In OSHA's Final Rule on the control of hazardous energy in section (C)(7), Training and Communication, the employer's obligations for training are defined as follows:

- Provide training to ensure that the purpose and function of the energy control program are understood by employees and that the knowledge and skills required for the safe application, usage, and removal of energy controls are acquired by employees.
- *Authorized* employees shall be trained in the type, magnitude, and sources of energy; methods/means for energy isolation and control.
- *Affected* employees shall be instructed in the purpose and use of the energy control procedure.
- *Other* employees shall be instructed about the energy control procedure and to not attempt to restart locked or tagged out equipment.
- Employees will be trained in the *limitations* of tags if tagout systems are used.
- Retraining shall be provided for authorized and affected employees when changes occur in their jobs, new hazards are introduced, or procedures are altered.
- Retraining shall be provided when audits or other means reveal there are deficiencies in employee knowledge or use of energy control procedures.

As can be seen by examining the OSHA requirements, an obligation exists to conduct initial training with somewhat varying content depending upon the employee's degree of involvement with hazardous energy control. In addition, retraining based upon changes or detected deficiencies is often required. These minimum criteria are quantitative in nature without clear indication of frequency or duration, in other words, what is the cycle for training and how long should it last? In section (C)(6), Periodic Inspection, there is a requirement under lockout systems to review the energy-control procedure annually with authorized employees in conjunction with the inspection. If tagout systems are in use, the annual review will include both authorized and affected personnel. However, because of the standard's performance style, the employer has latitude to decide the frequency of periodic training, the duration, the training method, the criteria for proficiency, the training plan, the instructor's qualifications, etc. All of the qualitative training decisions are left to the employer, based upon an assessment of what is needed to achieve the wanted end result.

Planning of Training

A facility's hazardous-energy-control training plan will need to be developed as a subset or element of its overall safety/health training program. The following elements should be addressed to create the right structure for learning:

- trainee population/audience
- subject matter/content
- training outlines/lesson plans
- method (lecture, case study, etc.)
- support materials (printed matter, visuals, etc.)
- schedules/training module duration
- style (classroom, on-the-job training, contact, etc.)
- budgets/resources
- testing (before/after)
- retraining criteria
- participant feedback
- training records/data.

NOTE: From a learning perspective, the message is clear—the greater the activity of the participant in the learning process, the greater the retention over time (Figure 7-9).

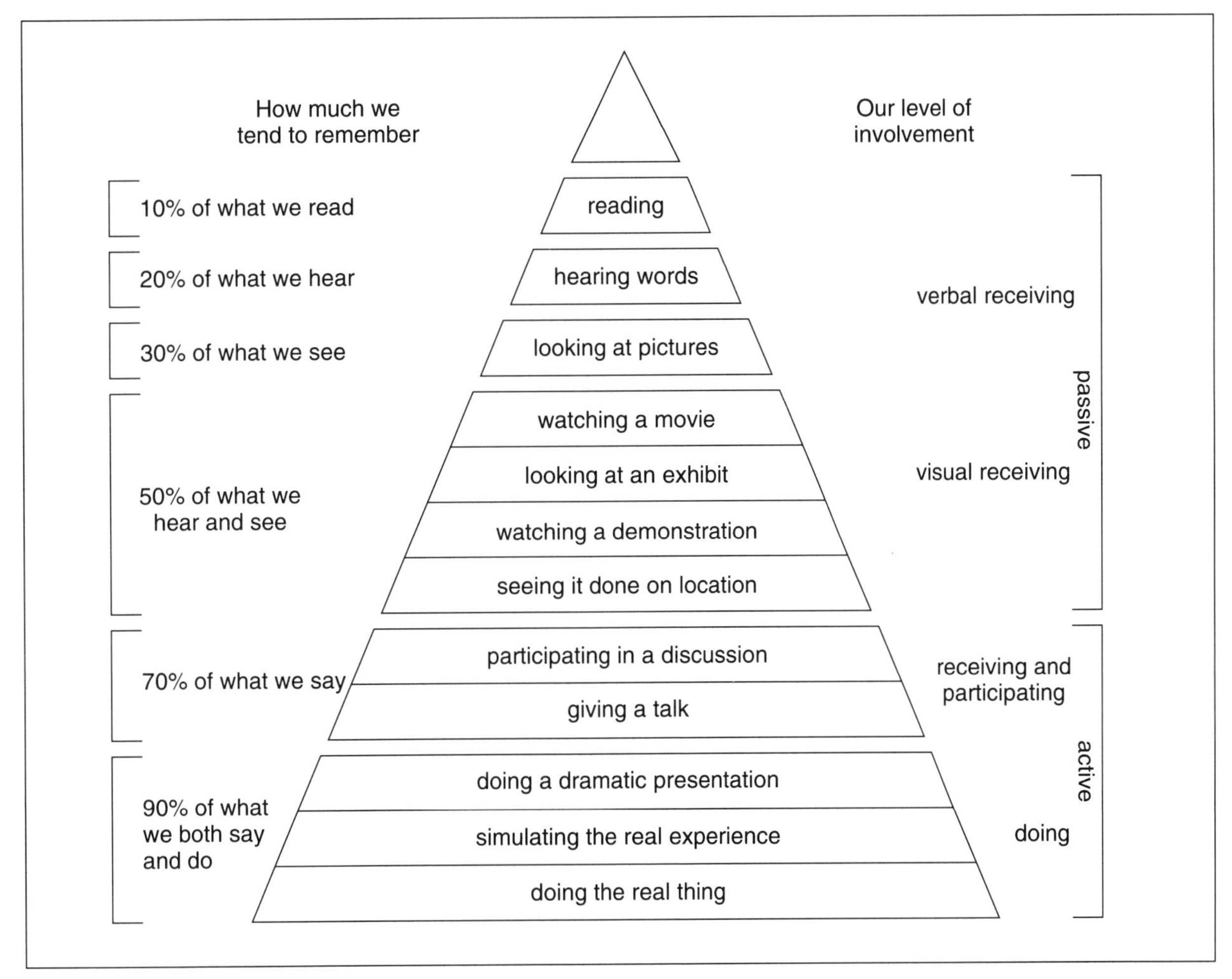

Figure 7-9. Retention of learning.

Training Considerations

When approaching hazardous-energy-control training, consider several distinct opportunities for employees to acquire the required knowledge/skill. First, new employees can be exposed to lockout/tagout principles in their orientation training. Second, initial/basic energy-control training can be provided for various classes of employees—authorized, affected, or other as defined by OSHA. Third, periodic training can be scheduled to reinforce what was provided in the orientation and basic phases and expanded experientially on the job. Fourth, retraining/remedial instruction can be provided to correct flaws identified by testing, observation, audits, inspections, etc.

If the data regarding energy-release-incident causation are accepted and the training of employees to effectively apply control methods/procedures is viewed as critical, then management's approach and expectation for training results should be sensitive to the following realities:

- Employee training based on compliance requirements alone will be far from adequate.
- Application/execution effectiveness depends on the training foundation provided beforehand.
- Training modalities will be by necessity shifted from the passive to active states.
- Authorized employees must be trained to function at the application or synthesis levels of learning.

Authorized employees, whether operations or maintenance personnel, should be educated and trained in hazardous-energy-control principles, techniques, and practices in order for them to effectively deal with the hazards associated with all of the specific tasks for which they are responsible. Ideally, this will result in employees who exhibit fluency and mastery and function at the synthesis level of learning for energy isolation and control. This upper rung of the learning ladder (Figure 7-10), an oversimplified illustration, suggests a logical progression through which employees must pass to achieve full competency.

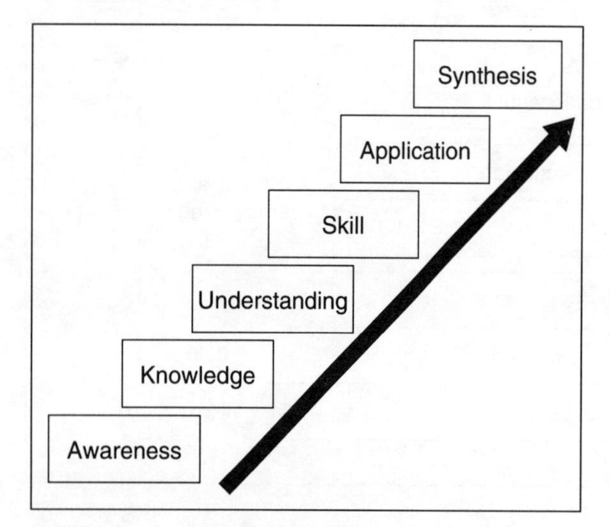

Figure 7-10. The learning ladder.
Source: Developed by E. Grund. © National Safety Council.

At the entry level, *awareness* or familiarization takes place to acquaint the trainee with main points or concepts. The employee can then logically progress to the *knowledge* level, where additional facts, methodology, and job-related information is conveyed and absorption must occur. At the next level, *understanding* takes place, which provides meaning to what was learned. Progression to the *skill* level now provides the opportunity to use the knowledge in executing or performing. Further ascendancy moves the trainee to the *application* level, where what has been learned can be applied under routine, similar, or somewhat different circumstances. A further expansion of this level might involve the employee's ability to effectively analyze situations and apply the appropriate measures. Finally, at the *synthesis* level, the competent employee is able to apply the concepts, principles, and techniques to new, unique, or diverse situations while achieving the desired results. At this point, the employee is creatively addressing the hazard issues and resolving them congruently with energy isolation and control requirements.

Instructional Methods

According to the October 1988 issue of *Training*, the magazine of human resources development, an ever-growing number of instruction methods for employee training are in use in U.S. industry. Organizations used videotape (87.1%), lectures (83.1%), and one-on-one instruction (70.6%) most frequently of all methods. However, since 1988 new developments continue to reshape the way training is delivered. The camcorder explosion has permitted employers to film high-quality video and with modest editing expense personalize the training experience. This technology provides unlimited opportunities for hazardous-energy-control training of employees.

Four other emerging training methods gaining momentum are as follows: computer-assisted instruction (CAI), computer-managed instruction (CMI), computer-based training (CBT), and interactive video training (IVT).

Computer-assisted instruction. CAI is an outgrowth of programmed learning methods; the computer presents information and asks for trainee responses moving or branching in the software program depending upon the answer.

Computer-managed instruction. CMI is a more complex method that assesses the trainee's initial level of competence and provides customized exercises and learning modules; trainee performance is assessed frequently and training content modified continuously to suit the learner.

Computer-based training. CBT is a method that includes both CAI and CMI technology; trainees use their individual computers to access and interact with the material but not with each other; consistent quality of instruction, faster learning, immediate feedback, and flexibility are some of the advantages.

Interactive video training. IVT is essentially CBT with the addition of a videodisc player and color TV monitor; it combines the advantages of CBT with the bonus of sound and high-quality motion pictures to stimulate the student—several lockout/tagout products are currently commercially available in interactive video format with excellent feedback regarding their reception (Fisher et al., 1990).

Enforcement

In its promulgation of the lockout/tagout standard, OSHA introduced, without fanfare, an interesting dimension found in no other standards currently in effect—the enforcement or disciplinary feature! Section (C)(4), Energy Control Procedure (ii), states that "the procedures shall *clearly* and *specifically* outline the scope, purpose, authorization, rules, and techniques to be utilized for the control of hazardous energy, and *the means to enforce compliance* including, but not limited to. . . ." In effect, employers in their energy-control procedures must clearly and specifically define their practice or means to enforce compliance with lockout/tagout requirements. No other mention of this enforcement feature can be found in 29 *CFR* 1910.147.

However, there are two other places where OSHA speaks about enforcement. They are as follows: (1) the preamble to 29 *CFR* 1910.147, and (2) the OSHA Instruction Standard 1-7.3, *The Control of Hazardous Energy (Lockout/Tagout)—Inspection Procedures and Interpretive Guidelines.*

In the lockout/tagout standard preamble, four elements necessary for effective tagout system use are described. OSHA makes this comment to the locks versus tags issue:

> However, it is the fourth common element, *discipline* (author's emphasis) which appears to be the most critical to the success of these programs; the companies with effective tagout programs apply various types of disciplinary action to both supervisors and employees who violate the tagout procedures.

Additionally, the OSHA Instruction Standard 1-7.3 (9-11-90), which guides compliance personnel on how to enforce the 29 *CFR* 1910.147 standard, has this notation:

> H. Inspection Guidelines
> (3)(d) Evaluate the employer's manner of enforcing the program (with reference to (c)(4)(ii)).

Section 5(b) of the U.S. Occupational Safety and Health Act specifies, "Each employee shall comply with occupational safety and health standards and all rules, regulations, and orders issued pursuant to this Act which are applicable to his or her own actions and conduct." However, no federal or state enforcement mechanism or statute exists that places the employee at risk with respect to noncompliance. Regulatory initiatives of any consequence are rarely if ever taken against employees—section 5(b) provides no effective basis for employers to handle the enforcement issue.

Enforcement Issues

With this background, the employer must now decide what is to be done and how. Employers with unions may have to approach this question differently from nonunion establishments. In either case, the employers will need to establish an enforcement policy, communicate it, and apply it fairly and consistently. If no specific policy of enforcement with respect to lockout/tagout exists, an appropriate provision may be found under a facility's general rules of conduct, safety rules and codes, or other such controlling language. In union environments, establishing new work practices or conditions of employment may be subject to collective bargaining. The following questions are relevant to responsibly addressing the enforcement issue:

1. Should there be a separate set of requirements and procedures for enforcing lockout/tagout provisions?
2. Should there be a unique disciplinary scheme for lockout/tagout?
3. Are all violations of lockout/tagout requirements of equal gravity?
4. Does discipline apply if employees are injured in the incident and violations of requirements are substantiated?
5. Who should determine if there were violations subject to discipline?
6. Does the progressive discipline approach apply?
7. Will management receive the same discipline as workers for infractions?
8. Is accelerated discipline appropriate for grave or life-threatening violations?

Is Enforcement Needed?

Many can recall the three Es of safety postulated with fervor during the 1950s and 1960s: (1) education, (2) engineering, and (3) enforcement. Or perhaps the Theory X practitioners who "spoke loudly and carried a big stick" come to mind. Enforcement and discipline were basic management tools used with great frequency during the industrial revolution and through the post–World War II period. However, their use (or the fear of their use) did very little to change safety performance during that period. Why bother with a tactic that is so tarnished from past abuses?

Obviously, things have changed—but not completely. Unfortunately, there are still managers who really like the good old days and what they had to offer, but most now appreciate that instead of enforcement being a primary tool, it should be used sparingly and carefully. All of the proactive initiatives such as positive reinforcement, effective training, involvement, recognition, and so on serve to make enforcement or discipline more of a tool that is available but ideally not used. It should be viewed as a tool of last resort and never as a substitute for progressive management practices. However, if enforcement is to be used at all (and OSHA, in 29 *CFR* 1910.147, says it should) it must be done well and fairly.

Nature of the Disciplinary Process

Discipline can be considered the force that establishes the authority of any manager. The simplest to the most complex organizations have always found it necessary to establish rules or requirements. Some rules are institutionalized, embedded in doctrine, or interwoven in culture. Others function by punishment or its threat, scorn, isolation, or intimidation.

Discipline is a personal matter that involves real people with personalities, emotions, needs, attitudes, and flaws. The loss of face (warning), the loss of pay (time off), or the loss of a job (dismissal) can be totally life altering or a minicrisis, depending on the individual and the circumstances. Generally, the employer will have to carry the heavier burden of proof in any formal administrative or arbitration proceeding. The issues will be both procedural (due process) and substantive (just cause) and will determine whether the applied disciplinary action was legitimate. The employee has a right to a defense, constitutional protections and guarantees, and a full and competent review or hearing. Unions are obligated to provide their members fair representation under disciplinary circumstances.

Frequency	Penalty	
	Minor Violation	Major Violation
1st Offense	Verbal Warning	Suspension or Discharge Based on Gravity of Offense
2nd Offense	Written Warning	
3rd Offense	Suspension (1 Day)	
4th Offense	Suspension (3 Day)	
5th Offense	Discharge	

NOTE: The following Lockout/Tagout violations are major: unauthorized removal of lock/tag; unauthorized defeat of safety devices; failure to follow established energy isolation procedure; unauthorized energization of equipment under lockout/tagout control.

Figure 7-11. Progressive discipline structure.
Source: Developed by E. Grund. © National Safety Council.

Due process involves the existence and consistent use of an administrative procedure that defines the natural rights and obligations of the parties. It establishes safeguards against willfulness and erratic application of the organization's protective framework. Dutiful compliance with due process criteria is no guarantee that the employer will prevail. However, failure to adhere to the requirements almost always produces negative results.

Just cause addresses the employer's ability to sustain the action taken upon review by an impartial third party. It examines whether the employer's disciplinary decision was arbitrary, capricious, unreasonable, or discriminatory to the extent it constituted an abuse of managerial discretion. The following questions form the basis for ascertaining whether abuse has occurred. If one or more are answered in the negative, just and proper cause may not have existed.

1. Did the company provide notice or warning of the possible consequences of the employee's conduct?
2. Was the company's rule reasonable and related to the safe operation of the business?
3. Did the company attempt to determine whether the employee actually violated the rule before applying discipline?
4. Was the company's investigation fair and objective?
5. Was the proof of the infraction substantial?
6. Has the company applied the rule fairly and consistently to all employees?
7. Was the degree of discipline appropriate for the seriousness of the offense and the employee's record?

Progressive Discipline

Management should establish or supplement its existing disciplinary structure with provisions for lockout/tagout enforcement. The emphasis should be on having the mechanism for discipline in place and understood and not on how frequently it is applied. Because of the potentially grave consequences of hazardous-energy-control failures, there should be no vacillation, and "firm but fair" should be the order of the day. A structure for progressive discipline is provided in Figure 7-11 for reference and consideration.

Establishing the Enforcement Element

When the employer establishes a structure for enforcement of safety rules, requirements, or standards, logical sequence and composition must be considered. Often the task is to repair or rehabilitate what is currently in place. Such a situation could be characterized by supervisors conditioned to look the other way and employees performing without concern for the rules and their intent. In cases of fixing the existing enforcement structure, it may be necessary to establish a moratorium or clemency period while the organization and its employees regroup. The employer should ensure that the following major steps are taken when designing the enforcement structure:

1. Determine what is to be enforced.
2. Prepare written general and/or specific rules/requirements.
3. Communicate the requirements (post, hand out, discuss) and negotiate if necessary.
4. Establish the discipline criteria and communicate.
5. Explain to employees why discipline is necessary.
6. Accept feedback and evaluate.
7. Train supervision in "firm but fair" enforcement practices.
8. Install a grace period prior to effective date.
9. Review enforcement process periodically and improve.

A final thought on the subject of employee discipline: If the enforcement of safety/health rules is generally poor on issues such as personal protective equipment use, forktruck driving practices, use of machine guards, etc., there can be no effective specific enforcement of lockout/tagout requirements. Uniform enforcement of *all* safety rules should be the objective, regardless of the topic. However, it should be remembered that the progressive application of positive prevention measures will make the use of discipline a rare event.

Communication

In personal and organizational relationships, communication is often identified as inadequate. *Communication* can be defined as a process by which senders and receivers interact in a given social context or as the art of engaging, informing, and exchanging.

With regard to lockout/tagout, a communications segment of your overall implementation plan is a necessity. The lockout/tagout communications plan should thoroughly identify and explain. The message can be in print or written, visual (posters, charts), audiovisual, (films, video, slides), or verbal and delivered in a variety of ways.

The following items all qualify for consideration when communicating about your hazardous-energy-control system:

- goals/objectives
- training plan
- policy
- specific/special procedures
- employee involvement
- system status/accomplishments
- action schedules
- inspection results (audits)
- safe behavior performance
- engineering energy isolation improvements
- promotional activity
- graphic isolation procedures
- employee recognition
- energy-release incidents
- investigation/survey results
- energy isolation ideas.

Guides to effective communications. The following guidelines from J. Kelly's *Organizational Behavior* are provided for reference when constructing the communication plan:

1. Before communicating, a person should analyze his or her problem in as much detail as possible in order to determine exactly what he or she wishes to communicate.
2. The purpose of the communication should be defined. Specify your intention, then define your aim.
3. The physical and human environment should be considered, that is, timing, location, social setting, and previous experience.
4. Communication is not exclusively verbal. Consult with others if this is thought to be necessary.
5. Objectivity is not always necessarily a criterion of good communication in every circumstance; sometimes two-sided messages are useful.
6. Try to influence the person with whom you are communicating; try to see things from his or her point of view. Remember, different people have different perceptual slants.
7. Assess the effectiveness of the communication if this is possible; this is usually done by encouraging feedback.
8. Choose carefully the type of communication process best suited to your purpose.
9. Many executives seem to believe that it is possible to manage by exclusively vertical forms of communication, but research reveals that a great deal of communication is lateral.

Checklist for communications. For the purpose of analysis, it is useful to look first at

- *The Message*. What is this supposed to be? What language is it to be put in? What information does it contain?
- *Communicators*. Who are they? What are their roles? Where do they stand on the status scale? Is there a status gradient? Does either or both have a vested interest in communicating the message? Are their personalities likely to interfere with the communication process?
- *Media*. What form should be used? What are the mechanics of the information handling? What is the density of the communication system? What is the time pattern?
- *Environment*. What are the circumstances of the communication? Who must know and who must not? Is there a protocol that is appropriate on this occasion? What about situational factors, especially social setting?

- *Effects*. How effective is the system? How capable is it of adaptation? What are its aims—are they being achieved? (Kelly, 1969)

Involvement

The involvement element of the lockout/tagout system is a *major* ingredient of critical consequence. Too many energy-control systems in organizations have been conceived, developed, and implemented by management with the worker as a bystander or disconnected end user. All too often, workers view the established rules and procedures as some managerial initiative to make life difficult. It should be understood that support and commitment are commensurate with the degree that participation/involvement occurs by those affected by the requirements. See Chapter 6, The Process Approach, for additional detail concerning involvement.

OSHA, Lockout/Tagout, and Participation

OSHA, in the preamble to 29 *CFR* 1910.147, identified the following *major* issue: "Should OSHA require employee participation in the development of lockout procedures and the training programs required by this standard?" OSHA, after public comment, responded with the following statement:

> OSHA has determined based on input and education that a specific provision dealing with employee participation in the development of the employer's lockout or tagout procedure is not necessary for the effective implementation of the Final Rule. For standards dealing with exposure to toxic substances and harmful physical agents under section 6(b)(5) of the OSHA Act, section 8(c)(3) of the Act spells out specific requirements for employee involvement in compliance activity. In particular, it requires that employees or their representatives have the opportunity to observe air monitoring and to have access to monitoring records. By contrast, there is no such specific statutory mandate for the present standard. Although OSHA agrees that active employee involvement may enhance understanding and cooperation, the Agency believes that it would be inappropriate to require such involvement in this standard.

OSHA undoubtedly was struggling with the issue of mandating participation and the political ramifications that were associated with this decision. Despite this position there is little doubt that OSHA would support the concept of substantial employee involvement in all aspects of the energy-control process as a voluntary initiative.

National Safety Council Survey

The National Safety Council's Research and Statistical Services Group conducted a survey in 1991 of safety program components' relative importance as viewed by the experts. The Council's Industrial Division members participated by rating the importance of each safety component on a six-point scale from "most important" to "least important."

Of the top 10 components, 8 related to participation, involvement, and activism of top and middle management, supervisors, and employees. The survey results were published in the National Safety Council's newsletter, *Managing the Process*.

Involvement Plan

Opportunities abound for participation, involvement, and empowerment with regard to designing, implementing, and maintaining an effective energy-control system. However, these opportunities can't be realized unless a plan and a climate in which it can occur are present. Each organization varies as to its readiness to accommodate various levels of employee activism. Whatever the degree of readiness, the plan should be designed to fit.

As organizations move toward having empowered employees, greater freedom is necessary for meaningful interaction and decision making. Design your involvement plan with the following goals in mind:

- maximizing vertical and horizontal involvement
- utilizing special skills of employees
- scheduling to avoid activity peaks and valleys
- ensuring plan coordination
- using nontraditional groups (sales, accounting, etc.)
- permitting workers to act as group leaders/chairpersons
- using experts (safety managers, etc.) in support roles
- recognizing employee contributions—often.

Investigation

Accident or, more appropriately, incident investigation has been called an "after-the-fact remedy" for the undetected, the unexpected, or the uncontrolled. Even when anticipation and progressive, proactive prevention initiatives are the rule, the after-the-fact process of investigation should be triggered when failures occur.

Incident investigation. This is a systematic process in which the causative factors of an incident are recognized and understood and action is presented for their elimination or control:

- An effort is made to correlate the basic elements of the incident: what, where, when, who, how, and why.
- A systematic attempt is made by the investigator to evaluate and establish all relevant facts and opinions gathered from all sources.

- The investigation represents the investigator's best judgment of what happened, how it happened, why it happened, and what must be done to prevent such an incident from happening again.

The purpose of incident investigation is the discovery of the causative factors (the hazardous conditions and practices) that brought the incident about so that the necessary action may be taken to prevent a recurrence. If the causative factors of an incident are not discovered and either eliminated or controlled, the stage is set for a repeat performance.

Energy-release incidents may present special challenges to the investigator but always have causes that are potentially identifiable in spite of the all-too-frequent references to "phantom" machinery cycles, "rogue" electrical circuits, computer control "glitches," and "runaway" chemical reactions. The investigation of these incidents takes the same diligence, depth of inquiry, and systematic approach that any other occurrence worthy of examination requires. Remember, the focus is on fact finding, not fault finding.

The fact finding and analysis phase of investigation and the expertise of the investigator is central to the process. Knowing what facts to seek through interviews, video or photographic recording, mapping techniques, failure identification, and physical evidence collection will expedite a high-quality investigation. Analysis of the assembled facts requires the ability to detect sequence gaps, anomalies, changes, contradictions, unusual circumstances, and inconsistencies. The process of fact finding and analysis is depicted in Figure 7-12 where the progression leads to conclusions regarding sequences of events involving the major system elements and subelements; contributing factors and probable causes; determination of what happened or should have happened; the assessment of all input

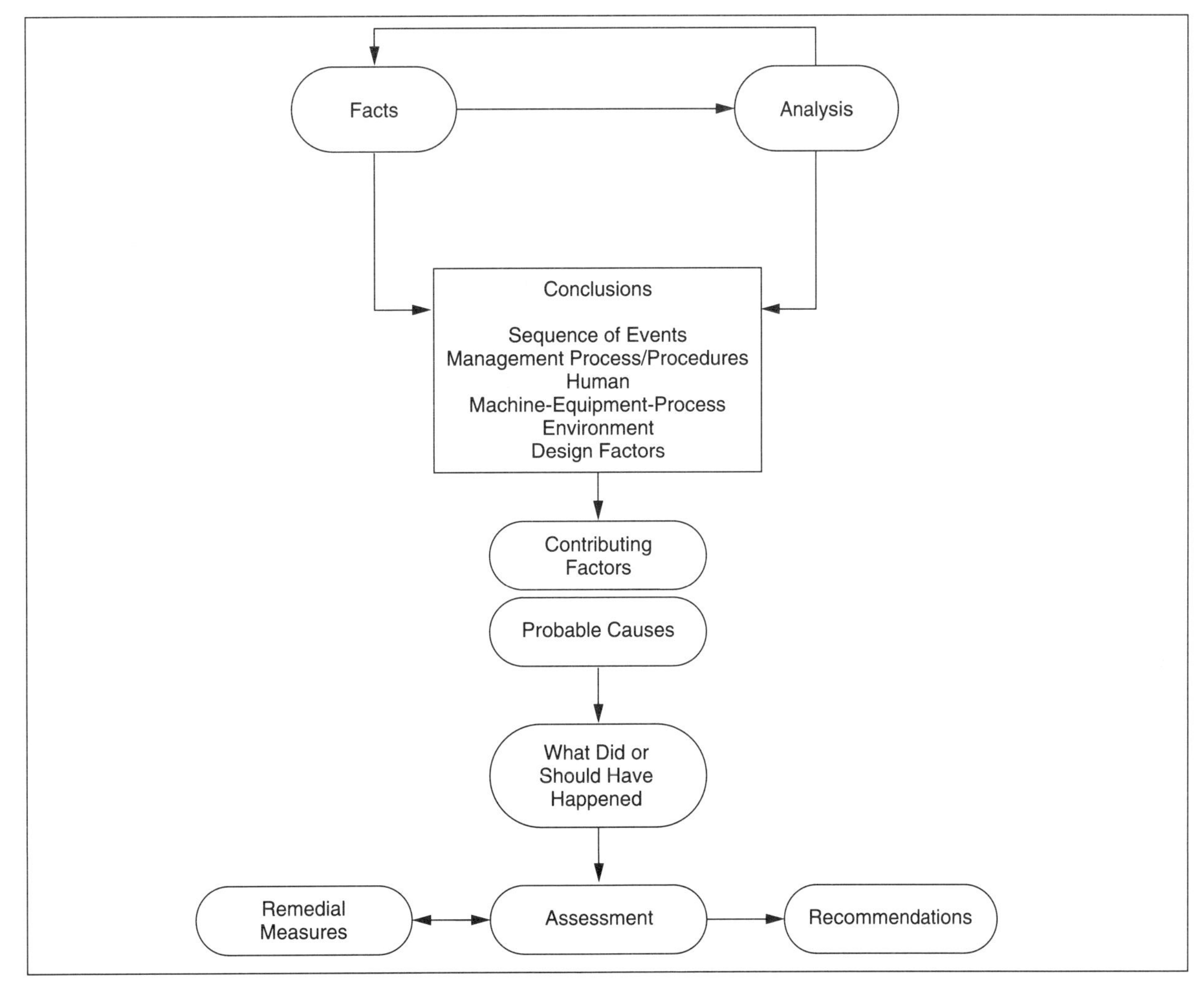

Figure 7-12. Fact-finding and analysis process.
Source: Adapted from W. G. Johnson, *MORT Safety Assurance Systems* (New York: Marcel Dekker, Inc., 1980). Reprinted by courtesy of Marcel Dekker, Inc.

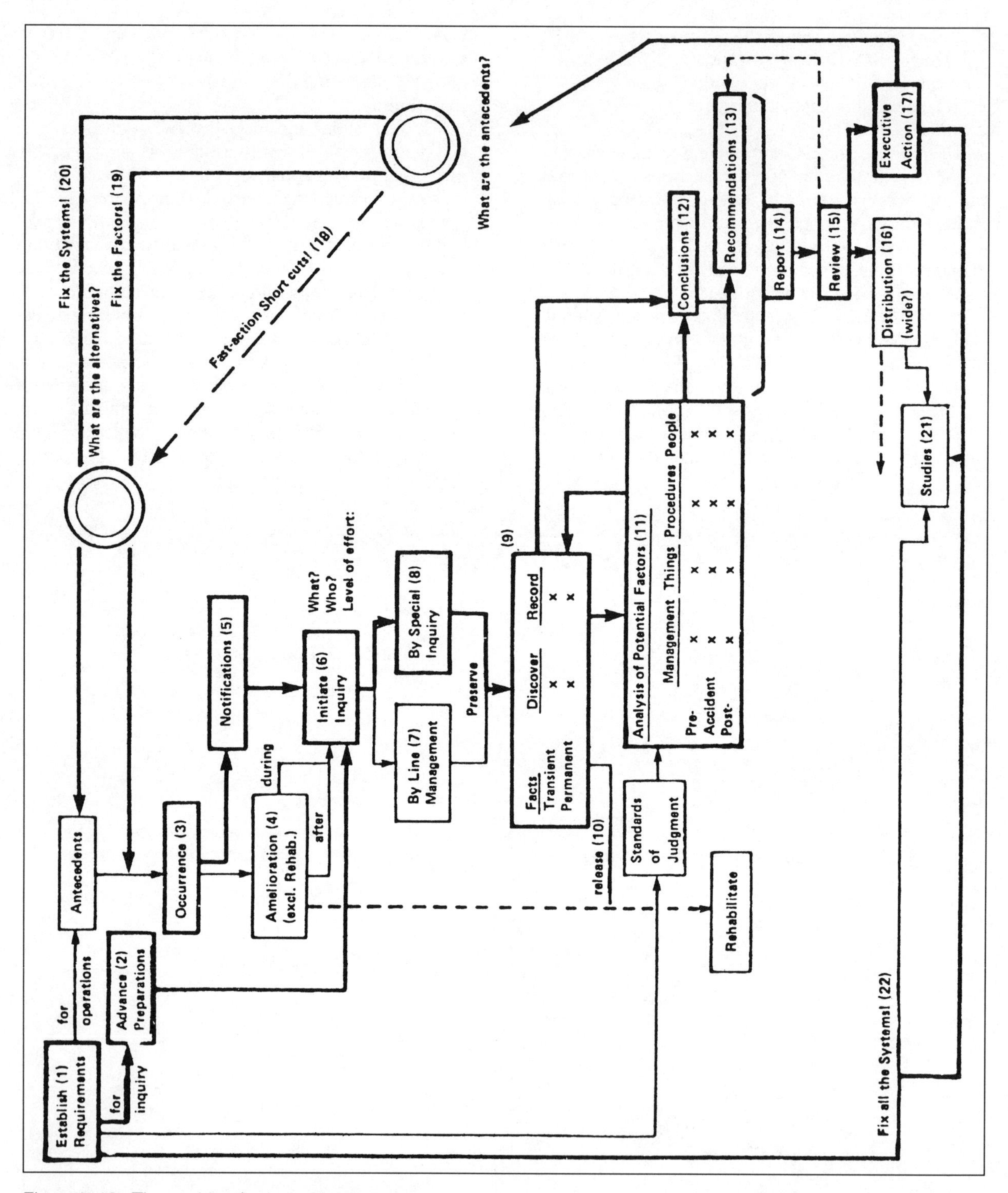

Figure 7-13. The accident/incident inquiry process.
Source: W. G. Johnson, *MORT Safety Assurance Systems* (New York: Marcel Dekker, Inc., 1980). Reprinted by courtesty of Marcel Dekker, Inc.

and possible remedial measures; and development of final recommendations.

In Johnson's *Accident/Incident Investigation Manual*, the accident/incident inquiry process is depicted (Figure 7-13) and described as follows:

The accident inquiry process requires (1) prior establishment of requirements, and (2) advance plans. Immediately after an occurrence, amelioration (4) is the primary activity. Notifications to higher authority are made. The inquiry is initiated (6) as soon as the scene is stabilized. Line management must *secure* the *scene* and initiate collection of transient evidence while awaiting the investigator or board. The fact finding and analysis proceed (9 and 11) until conclusions are warranted. Note that executive action (17) on the report should fix immediate causal factors (18 and 19) and fix the system (20) which allowed the factors to come into being. Executive action should also implement any studies recommended (21).

Various types of energy-release incidents require investigation such as:

- deviations from procedure
- employee complaints
- injury/damage events
- machine/equipment malfunction
- close calls/near misses
- observed infractions
- audit deficiencies.

The potential causal factors listed in Figure 3-13 will prove helpful in identifying system deficiencies and prompting the required corrective action. It is recommended that a special review team be formed to either evaluate or investigate energy-release incidents. The team should be comprised of individuals with appropriate background and skills who can ensure that the proper depth of investigation takes place to identify the root causes of incidents. The team can also act as the clearinghouse for all energy-release-incident history.

Performance

Performance as a component of the energy-control system can be viewed from an individual, group, or organizational perspective. Hazardous-energy-control performance is affected by various external influences such as criticality, level of technology, urgency, complexity, and so on, as mentioned in Chapter 3, Causation Analysis. It is important to know what impacts performance, what can be done to improve it, and how to monitor and measure it.

With regard to individual performance, certain internal factors (psychological and physiological) and external factors (situation, task, etc.) must be acknowledged and addressed (Table 7-A) (Swain, quoted in Hammer, 1976). Managers need to be more familiar with and sensitive to the human factors that influence performance. There should be no cause for surprise when an employee who is working a double shift under process breakdown pressures commits an error that induces an energy-release incident. Anticipation and appreciation of human performance variables is necessary when constructing the energy-control system.

Group and organizational performance is a multiplicative expansion of what was mentioned for individuals with the added complexity of group dynamics, cliques, politics, etc. In *Improving Performance*, Rummler and Brache state:

> Whether the concern is quality, customer focus, productivity, cycle time, or cost, the underlying issue is performance. In our opinion most American managers have been unable to respond effectively to the challenges because they have failed to create an infrastructure for systemic and continuous improvement of performance. We believe that their shortcoming does not lie in the understanding of the problem, in the desire to address the problem, or in the willingness to dedicate resources to the resolution of the problem, rather, the majority of managers simply do not understand the variables that influence organization and individual performance.

They suggest that an organization is a system with three levels of performance: (1) organization, (2) process, and (3) job/performer that must be managed in order to get consistent, high-level output. In order to get desired performance, measurement is necessary. Without measures, managers have no basis for:

- communicating performance expectations
- knowing what is happening
- identifying performance problems
- providing feedback (performance versus standard)
- identifying performance for recognition
- making decisions regarding resources, plans, etc.

Without measures, employees at all levels have no basis for:

- knowing what is expected from them
- monitoring their own performance
- achieving self-satisfaction and knowing what performance is noteworthy
- identifying improvement areas.

Rummler and Brache recommend that this sequence be followed when you want to measure performance or output:

1. Identify the most significant outputs of the organization, process, or job.
2. Identify the critical dimensions of the output.
3. Develop measures for each critical dimension.
4. Develop goals or standards for each measure-performance expectation (Rummler & Brache, 1991).

For example, if organizational performance with regard to energy-release-incident investigation was

Table 7-A. Human Performance Shaping Factors

Performance Shaping Factors		
Extra-individual		**Intra-individual**
Situation characteristics	**Psychological stresses**	**(Organismic) Factors**
Temperature, humidity, air quality	Task speed	Previous training/experience
Noise and vibration	Task load	State of current practice or skill
Degree of general cleanliness	High jeopardy risk	Personality and intelligence variables
Manning parameters	Threats (of failure, loss of job)	Motivation and attitudes
Work hours/work breaks	Monotonous, degrading, or meaningless work	Knowledge of required performance standards
Availability/adequacy of supplies	Long, uneventful vigilance periods	Physical condition
Actions by supervisors	Conflicts of motives about job performances	Influence of family and other outside persons or agencies
Actions by co-workers and peers	Reinforcement absent or negative	Group identifications
Action by union representatives	Sensory deprivation	
Rewards, recognition, benefits	Distractions (noise, glare, movement, flicker, color)	
Organizational structure (e.g., authority, responsibility, communication channels)	Inconsistent	
Task and equipment characteristics	**Physiological stresses**	**Job and task instructions**
Perceptual requirements	Fatigue	Procedures required
Anticipatory requirements	Pain or discomfort	Verbal or written communications
Motor requirements (speed, strength, precision)	Hunger or thirst	Cautions and warnings
Interpretation and decision making	Temperature extremes	Work methods
Complexity (information load)	G-force extremes	Shop practices
Long- and short-term memory	Atmospheric pressure extremes	
Frequency and repetitiveness	Oxygen insufficiency	
Continuity (discrete vs. continuous)	Vibration	
Feedback (knowledge of results)	Movement constriction	
Task criticality	Lack of physical exercise	
Narrowness of task		
Team structure		
Man-machine interface factors: design of prime equipment, job aids, tools, fixtures.		

Source: Swain, AD. *Sandia Human Factors Program for Weapon Development,* SAND76-0326, Sandia Laboratories, Albuquerque, NM. Used with permission.

identified as a significant output of the system, a measurement tool such as that shown in Table 7-B could be developed.

Mager and Pipe, in *Analyzing Performance Problems,* discuss how to find solutions to the problems of human performance. They guide the reader through a very efficient procedure for determining the nature and cause of these problems and whether the performance discrepancies (difference between actual and desired) are skill related or the result of other factors. The concepts of *performance punishing*, in which desired performance produces negative outcomes as

Table 7-B. Performance Measurements

Outputs	Critical Dimensions	Measures	Standards
Incident investigations conducted/ reported	Quality	% acceptable	90% submitted
	Quantity	Number of incidents identified	100% incidents investigated
	Timeliness	Occurrence to report approval time	5 working days

Source: Developed by E. Grund. © National Safety Council.

perceived by the performer, and *non-performance rewarding*, in which avoidance of the desired behavior produces positive outcomes as perceived by the performer, are introduced.

A performance analysis flowchart and quick reference checklist are provided in Figure 7-14 and Table 7-C, respectively, to lead the performance problem analyst in asking the right questions and focusing on appropriate solutions (Mager & Pipe, 1984).

Positive Reinforcement

Management and supervision have become conditioned and accustomed to focusing on problems and reacting to them rather than on successes when dealing with employee performance. In the area of hazardous energy control, greater emphasis needs to be placed on defining what is expected, communicating it, and reinforcing it when it occurs. A crew completing a difficult energy isolation maintenance task is more likely to be hassled about some minor point than praised for a job well done. A planned effort will be necessary to change the old paradigms. Krause, Topf, McIntire, White, and others cited in Chapter 3, Causation Analysis, provide insight as to what should be done.

Leadership/Advocacy

When developing and implementing an energy-control system, one should not overlook or underestimate the importance of leadership. Often programs and systems flounder because of a lack of this crucial ingredient. Leadership can be provided from any point in an organization. However, the higher the point from which leadership flows in the management structure, the greater its likely impact. For instance, the maintenance superintendent could provide effective lockout/tagout leadership but not have significant influence on operating departments. However, if the facility manager provided the leadership, it is likely the total organization would be affected. We should also keep in mind that high rank does not necessarily provide leadership.

The PAR group of Atlanta, Georgia in their Leadership and Teamwork course, state that all of the improvements that are sought by business boil down to two critical skills:

1. The skill to implement change . . . as in *leading* the effort and getting others on board.
2. The skill to work as a team . . . . whether leading or following in a given situation.

Further, *leadership skill* is defined as the ability to obtain whole-hearted followership for a given course of action. *Followership* refers to people willingly thinking, feeling, and behaving in concert toward a common goal. This same definition is used for *teamwork*, thereby implying that teamwork is the inevitable consequence of leadership. Whenever commitments are given or received, real leadership and as a result teamwork (co-ownership of a common goal) is experienced.

The issues of involvement and empowerment are central to any discussion of leadership/followership, not the sheeplike mind-set that is associated with the leadership of dictators or autocrats. Again, the difference between being led or guided and being driven is stressed.

Findlay and Kuhlman, in *Leadership in Safety,* identify seven steps to effective safety leadership that still seem applicable when viewed from a microperspective such as hazardous energy control. They have been slightly modified as follows:

1. reduce emotional response/increase reasoned response
2. convey organizational commitment
3. organize for an effective system
4. allocate resources for performance
5. use staff effectively
6. control by performance measurement
7. balance training and motivation (Findlay/Kuhlman, 1980).

Energy-control systems need advocates and leadership because they are complex, ongoing processes and represent significant potential risk to the organization. After a serious energy-release incident, it is not uncommon to find a lockout/tagout program with substantial deficiency or very unpredictable application. With the right leadership, this outcome is much less likely to occur.

ENVIRONMENTAL ELEMENTS

The environmental element addresses the external or situational characteristics such as lighting/visibility, noise/vibration, spatial arrangement, air quality, heat/cold/weather, pressure, housekeeping, and radiation that affect the worker while performing maintenance and servicing tasks while controlling hazardous

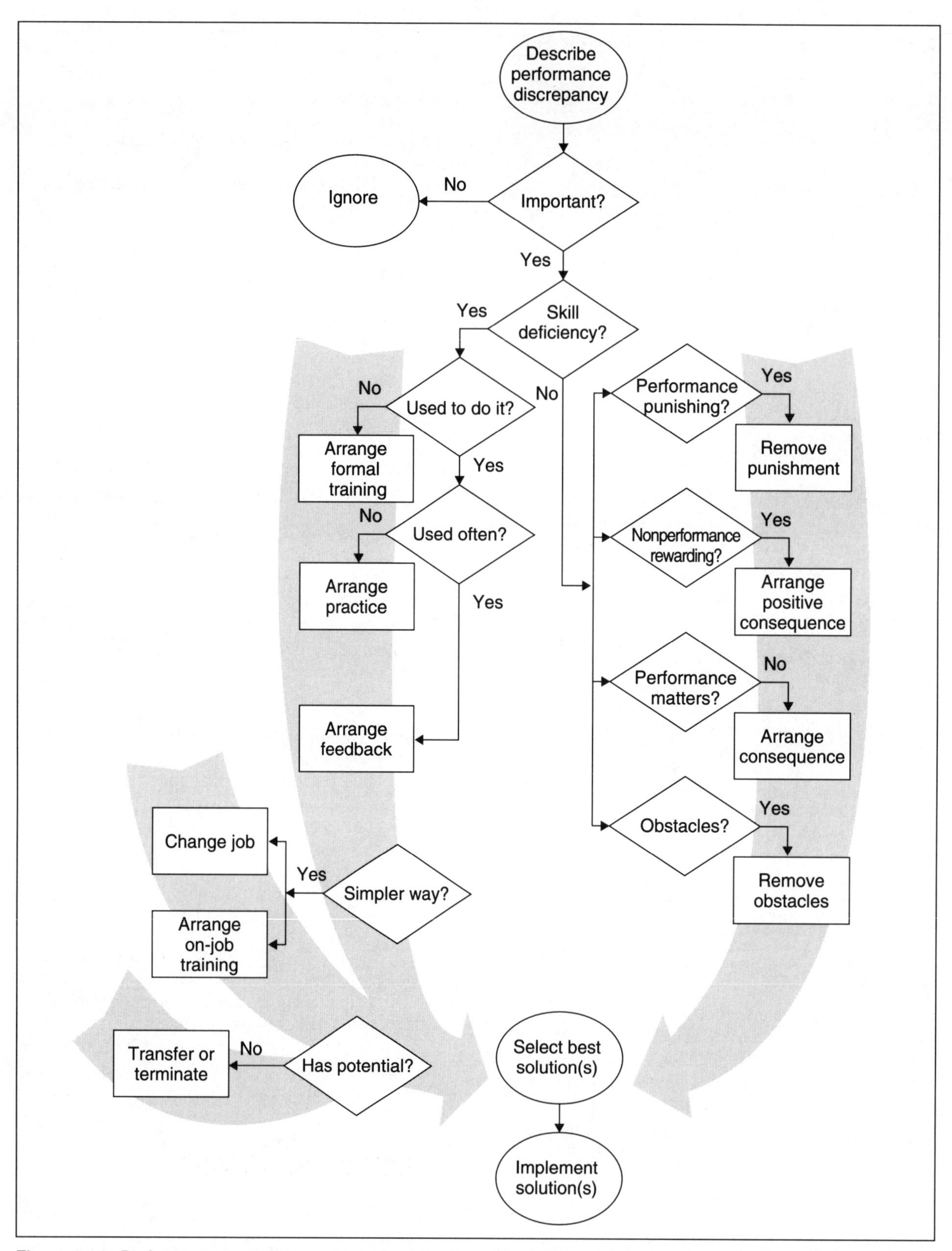

Figure 7-14. Performance analysis flowchart. Note: Caution sequence is not intended to be rigidly applied; consider multiple solutions shown in rectangles.
Source: Mager & Pipe, 1984. Reprinted with permission from Lake Publishing Company.

Table 7-C. Quick Reference Checklist (*continued*)

Key Questions to Answer	Probe Questions
I. They're not doing what they should be doing. I think I've got a training problem.	
1. What is the performance discrepancy?	• Why do I think there is a problem? • What is the difference between what is being done and what is supposed to be done? • What is the event that causes me to say that things aren't right? • Why am I dissatisified?
2. Is it important?	• Why is the discrepancy important? (What is its cost?) • What would happen if I left the discrepancy alone? • Could doing something to resolve the discrepancy have any worthwhile result?
3. Is it a skill deficiency?	• Could the person do it if really required to do it? • Could the person do it if his or her life depended on it? • Are the person's present skills adequate for the desired performance?
II. Yes. It is a skill deficiency. They couldn't do it if their lives depended on it.	
4. Could they do it in the past? 5. Is the skill used often?	• Did the person once know how to perform as desired? • Has the person forgotten how to do what I want done? • How often is the skill or performance used? • Is there regular feedback on performance? • Exactly how does the person find out how well he or she is doing?
6. Is there a simpler solution?	• Can I change the job by providing some kind of job aid? • Can I store the needed information some way (in written instructions, checklists) other than in someone's head? • Can I show rather than train? • Would informal (such as on-the-job) training be sufficient?
7. Do they have what it takes?	• Could the person learn the job (is the individual trainable)? • Does this person have the physical and mental potential to perform as desired? • Is this person overqualified for the job?
III. It is not a skill deficiency. They could do it if they wanted to.	
8. Is desired performance punishing?	• What is the consequence of performing as desired? • Is it punishing to perform as expected? • Does the person perceive desired performance as being geared to penalties? • Would the person's world become a little dimmer if the desired performance were attained?
9. Is nonperformance rewarding?	• What is the result of doing it the present way instead of my way? • What does the person get out of the present performance in the way of reward, prestige, status, jollies? • Does the person get more attention for misbehaving than for behaving? • What event in the world supports (rewards) the present way of doing things? (Am I inadvertently rewarding irrelevant behavior while overlooking the crucial behaviors?) • Is the person "mentally inadequate," doing less so that there is less to worry about? • Is this person physically inadequate, doing less because it is less tiring?
10. Does performing really matter to them?	• Does performing as desired matter to the performer? • Is there a favorable outcome for performing? • Is there an undesirable outcome for not performing? • Can the person take pride in this performance as an individual or as a member of a group? • Is there satisfaction of personal needs from the job?

Source: Mager & Pipe, 1984. Reprinted with permission from Lake Publishing Company.

Table 7-C. (*concluded*)

Key Questions to Answer	Probe Questions
11. Are there obstacles to performing?	• What prevents this person from performing? • Does the person know what is expected? • Does the person know when to do what is expected? • Are there conflicting demands on this person's time? • Does the person lack - the authority? - the time? - the tools? • Are there restrictive policies, or a "right way of doing it," or a "way we've always done it" that ought to be changed? • Can I reduce interference by - improving lighting? - changing colors? - increasing comfort? - modifying the work position? - reducing visual or auditory distractions?
IV. What should I do now?	
12. Which solution is best?	• Have all the potential solutions been identified? • Does each solution address itself to one or more problems identified during the analysis (such as skill deficiency, absence of potential, incorrect rewards, punishing consequences, distracting obstacles)? • What is the cost of each potential solution? • Have the intangible (unmeasurable) costs been assessed? • Which solution is most practical, feasible, and economical? • Which solution will add most value (solve the largest part of the problem for the least effort)? • Which remedy is likely to give us the most result for the least effort? • Which solution are we best equipped to try? • Which remedy interests us most? (Or, on the other side of the coin, which remedy is most visible to those who must be pleased?)

Source: Mager & Pipe, 1984. Reprinted with permission from Lake Publishing Company.

energy. As these subelements decline from optimum or standard conditions, the complications to energy isolation and/or control increase. It is not in management's best interest with regard to system effectiveness to disregard their existence or influence. They will be addressed briefly to focus attention on the problems they may present if not managed.

Lighting/Visibility

Lighting as an issue in energy-release incidents is often overlooked. For instance, marginal lighting conditions may cause an employee to improperly identify energy-isolating devices. Impaired visibility may prevent employees from effectively communicating. Dust, physical structures, and steam may cause employees to take shortcuts when communicating about energy isolation or temporary test maneuvers. Under low–light level conditions, an employee may overlook a tag that is on an energy-isolating device such as an electrical breaker and activate it. Portable lighting should be used where substandard conditions exist. Surveys can identify problems with lighting and visibility associated with lockout/tagout work.

Noise/Vibration

Excessive noise may create conditions in which employees are unable to communicate verbally and are forced to use hand signals and/or visual cues. This presents special problems because such efforts at communication are often misinterpreted. Noise levels in excess of 90 db will make communication quite difficult at distances over 6 ft (1.8 m). There may not be an immediate cure for the noise situation, so special instruction and/or provisions should be made to ensure that communication occurs and is understood. In certain environments, it may be necessary to provide electronic communication equipment to facilitate the necessary contact. Vibration may present special problems for the integrity of energy-isolating devices. Electrical disconnects, valves, and so on may be subjected to stresses that could cause them to malfunction. Normally, this type of problem can be dealt with by conducting vibration analysis surveys and implementing appropriate engineering solutions, such as, isolation mounting, component balancing, decoupling, and damping.

Spatial Arrangement

Layout and arrangement of equipment may prevent employees from accessing energy-isolating devices without great difficulty. Where such situations exist, employees may be tempted to deviate from requirements to avoid the discomfort or inconvenience of using the devices. In addition, where tight quarters present special problems for the worker, there may be an inclination to improvise with practices that add risk. Arrangement of equipment can also present communication problems because of impaired lines of sight.

Air Quality

Air contaminants, whether hazardous or nonhazardous, influence employee performance with regard to hazardous energy control. Employees may be exposed to odor, nuisance, or irritation and therefore may attempt to avoid or hurry the activity that is required. Toxic air contaminants may require employees to wear protective equipment, which adds to the demands associated with the isolation of machines, equipment, and process. Contaminants such as carbon monoxide may impair an employee's judgment with regard to the tasks at hand. The contaminant and/or its concentration may be such that no warning is provided to the employee with regard to what is being encountered. Poor air quality may also influence an employee's decision to avoid entering the area, thereby compromising the procedural requirements that are in place. Decomposing organic materials may produce foul odors. Air quality issues need to be identified and corrected to eliminate them as a deterring factor in the energy isolation process.

Weather

Temperature extremes are another environmental influence that may create problems with the energy isolation. For instance, hot conditions may not only fatigue an employee but may also make him or her more irritable and uncooperative. The heat may cause the employee to take shortcuts that otherwise would not be taken. The physiological effects of the heat will also place the employee under additional stress while high-risk tasks, such as lockout/tagout, are being performed. "Getting in and getting out" may be emphasized, which does not usually bode well for machinery isolation. Complex lockout/tagout procedures may also be more likely to be violated under extremely hot conditions.

Extreme cold creates a similar human reaction in terms of avoidance and psychological response. The cold may make handling metal parts and immobilizing equipment with energy-isolating devices very difficult. Valves and other energy isolation devices may be frozen in place or resistant to total closure.

Outdoor energy isolation tasks can become unbearable when conditions such as heavy rain, snow, and high winds are affecting the work in progress. Conditions at the job site such as flooding, frozen objects, blowing materials, and the like may cause employees to do things that they would not under other, more temperate circumstances. Weather conditions play a critical role in construction, oil and gas production, mining, agriculture, lumbering, marine operations, and so on and need to be dealt with routinely as a special energy isolation issue.

Pressure

Generally, atmospheric pressure variations ranging from 0.10 ATM (standard atmosphere) to 15 ATM have little consequential effect on humans. A standard atmosphere is equivalent to being under approximately 33 ft of seawater (fsw) or 14.7 lb/in^2 gauge pressure (psig). Capsule atmospheres, diving, and tunneling (caisson work) involve variations in pressure from 0.25 to 60 ATA. Normally, these conditions present little direct influence on the application of energy isolation practices. However, physical conditions may be affected by factors such as high temperatures, humidity, and helium diffusivity in pressurized environments. In the Sealab habitat at an ocean depth of 200 ft (61 m), environmental conditions caused electrical short circuiting. Psychological and physiological effects involving headaches, muscular incoordination, concentration loss, mood swings, confusion, myalgia, sinal ailments, and so on have been associated with pressurized environments and could affect workers' ability or motivation to control hazardous energy. (Behnke, 1978)

Of more immediate concern is the control of system or process pressures or vacuums that result from various sources such as thermal or chemical reactions, pneumatic and hydraulic pumps, etc. Pressures may exceed 10,000 psig and present significant danger to workers attempting to neutralize or relieve them. Numerous devices (relief valves, reservoirs, blow-down tanks, safety valves, vacuum breakers, etc.) and established safe practices exist for dissipating residual or excess system pressure. Often system pressures must be relieved in such a way that no release to the external environment occurs. Any external relief of pressure to achieve zero energy state must be done in such a way that personnel are not exposed to the escaping contents. In addition, the pressure may be acting as a restraining force or counterbalance in the system and cause component movement when relieved. Personnel should be aware that under "jam" conditions, system pressures may be extraordinarily high, and forces could cause objects to collapse destructively or be ejected violently.

Housekeeping

Order in the work area can have a direct bearing on the application of energy isolation procedures. Pits that are filled with debris and oil and that house valves may be avoided by employees who need to isolate them. Leaking machinery, excessive scrap, improper storage, and the like may produce situations where employees will avoid or are prevented from following the prescribed procedures. In one case, an employee was unable to gain access to an electrical disconnect because of several pallets of 55-gal. (208.2-l) drums. He performed the work using the limited protection of control circuitry rather than going through the hassle of having the pallets moved. Fortunately, the job was completed without further incident.

In certain instances, a complete cleaning of the area may be required before any maintenance or service work can even begin. For example, when using steam-cleaning equipment around electrical devices, employees would need to be protected by energy isolation or other effective alternatives. Often facilities have certain areas or pieces of equipment that employees would rather avoid than perform maintenance or service functions on them. These onerous conditions have the ability to affect employee attitude and willingness to follow the appropriate energy isolation practices. Good housekeeping sends a strong, continuous signal regarding management's priority on safety.

Radiation

Alpha, beta, and neutron particles and x or gamma rays are the most common types of ionizing radiation. Alpha, beta, and neutron particles are associated with phenomena of the nuclei of radioactive atoms, whereas x rays are emitted spontaneously from radioactive materials. Distance/separation, shielding, and control areas are used as measures to protect employees from exposure. Special isolation procedures must be employed in research establishments, nuclear power generation facilities, nuclear-powered vessels, and other places where energy magnitudes are substantial. Low-level radioactive sources are more commonly found in industrial establishments. Thickness gauges, tracer elements, high-voltage electronic equipment, and radioactive isotopes for nondestructive testing have potential for employee harm if not adequately controlled during equipment maintenance and servicing.

Nonionizing radiation includes the ultraviolet, visible, infrared, microwave, radiowave, radar, laser, and power transmission frequencies of the electromagnetic spectrum. Each type of energy covering some region of the electromagnetic spectrum presents particular safety hazards that must be controlled during inspection, servicing, repair, etc. Energy-control methods ranging from simple electrical deactivation and lockout to shielding to safe discharge of capacitors must be utilized. Various carefully designed safe practices are used when "energized" work is required. Ideally, the radiation or energy output is halted and lockout/tagout type controls are applied by the servicing personnel while they are exposed.

MACHINE ELEMENTS

The machine element is the foundation or focus of the hazardous-energy-control system. The machine, materials, equipment, or process represents that which must be inspected, serviced, maintained, unjammed, or repaired. Under ideal circumstances, it is cooperative and susceptible to safe intervention, but all too often, under less than proper circumstances, it stands ready to resist and inflict pain, damage, and disruption. The challenge is to recognize the machines' potential, understand the hazards and their consequences, and adequately control the identified hazards. If the machine element of energy control is addressed properly, few energy-release incidents are likely to occur. The following subelements are critical to the accomplishment of hazardous energy control.

Protective appliances. Protective appliances are those items that either immobilize/secure an energy-isolating device in place or restrain machine, equipment, or process movement. These appliances should not be confused with lockout/tagout auxiliary hardware or energy-isolating devices. Examples of protective appliances include the following:

Lockout device. A device that utilizes a positive means such as a lock to hold an energy-isolating device in the safe position (gate valve split cover, ball valve lever restraint, plugout, plug hugger, fuse blockout fixture, lockout hasps, adapters, lockout chain/cable, etc.); the devices typically are portable and utilized on a variety of energy-isolating devices where their design is compatible (Figure 7-15 a–c)

Tagout device. A prominent warning device such as a tag and a substantial means of attachment that can be securely fastened to an energy-isolating device that prohibits the operation of the isolating device and the equipment it supplies (laminated plastic, heavy cardstock, plastic and leather thong, metallic, etc.); the devices may be disposable or permanent and affixed with wire, nylon lock straps, polycord, and so on (Figure 7-16 a–b).

Restraint device. A device that has been designed for a particular application that prevents machine or equipment movement (safety block, prop, pins, wedges, safety chain/cable, etc.). These devices are normally portable and positioned or engaged manually; however, they often remain at or near the point of application. Restraint devices generally are used for gravitational forces and/or as a protective safeguard for required activity under less than zero energy conditions (Figure 7-17 a–b).

Auxiliary hardware. Auxiliary hardware are those items and equipment that are used in a complementary manner to facilitate or enhance the lockout/tagout process. They do not come in contact with energy-isolating devices and/or protective appliances under normal circumstances. Examples of auxiliary hardware include the following:

Control circuit protectors. These are fixtures that

Figure 7-15a. Ball valve lockout device and gate valve lockout.
Source: Photo courtesy of W. H. Brady Co.

Figure 7-15b. Single pole circuit breaker lockout or fuse blockout devices.
Source: Photo courtesy of W. H. Brady Co.

Figure 7-15c Lockwrap personalized ID.
Source: Photo courtesy of W. H. Brady Co.

Figure 7-16a. Electrical and lockout tags.
Source: Courtesy of Idesco Corp., New York, NY.

Figure 7-16b. Danger tags.
Source: Courtesy Idesco Corp., New York, NY.

are temporarily or permanently affixed to toggle switches, push buttons, selector switches, control panels, and so on and normally prevent the control from being cycled by their design and use with a lock or fastener. They are not lockout devices and should not be used in lieu of the application of a lockout device on an energy-isolating device. They may be used as a redundant or complementary safeguard or in situations where less than zero-energy-state conditions exist for special tasks where alternative measures are utilized. (Figure 7-18).

Lockout boards and boxes. These are normally used for crew lockout applications where complex

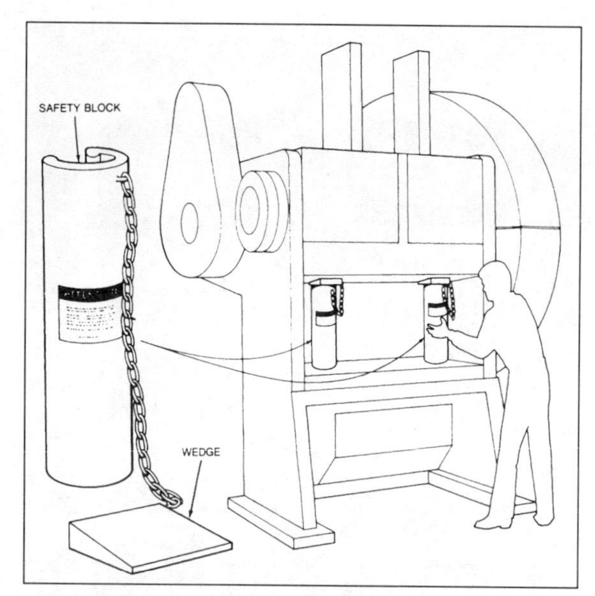

Figure 7-17a. Safety block to prevent potential movement.
Source: *Safeguarding Concepts Illustrated*, 6th ed., National Safety Council, 1993.

Figure 7-17b. Support stand for employee protection.
Source: Reprinted with permission from Institut National de Recherche et de Securite.

equipment is involved and pyramid and supervisory control procedures are employed. Stationary boards usually have machine, equipment, or process components identified with various types of energy-isolating devices. Lockable boxes are attached to the various component sections of the board where supervisory/equipment keys from group lockout devices are placed. Employee lock(s) are then used to secure the component box(es) appropriate to the exposure of the individual; lockout boxes are generally portable and carried to the job site, where keys from all secured energy-isolating devices are placed. Employees then

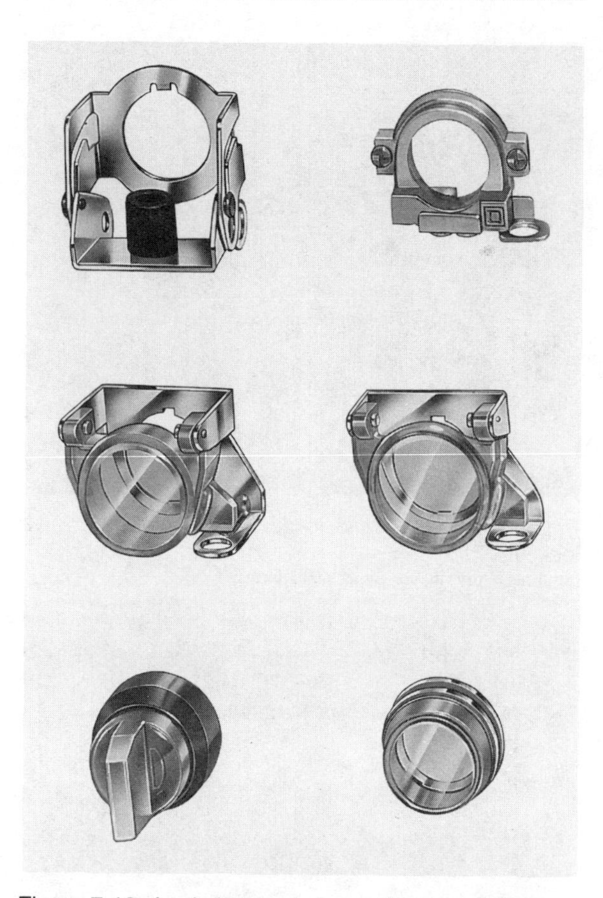

Figure 7-18. Lockable control circuit components.
Source: Courtesy Square D Company.

Figure 7-19. Group lock box.
Source: . Photo courtesy Panduit Company.

place their personal locks on the box hasp to secure the keys used in the application of the general purpose locks. Further detail will be found in Chapter 10, Special Situations and Applications (Figure 7-19).

Signage/graphic displays. These are visual materials intended to warn, apprise, or direct personnel regarding lockout/tagout activity; signs may be portable

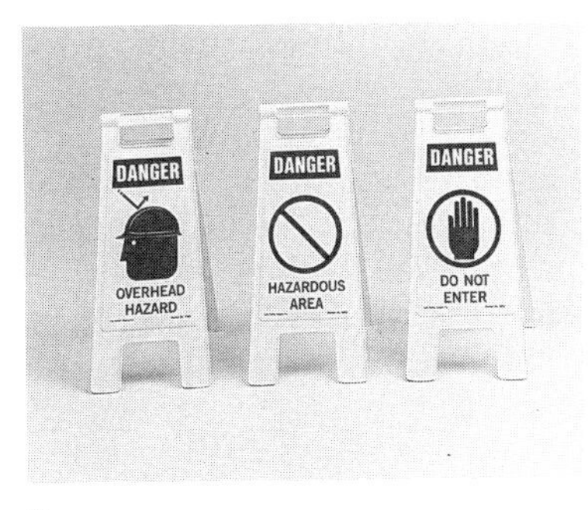

a

b

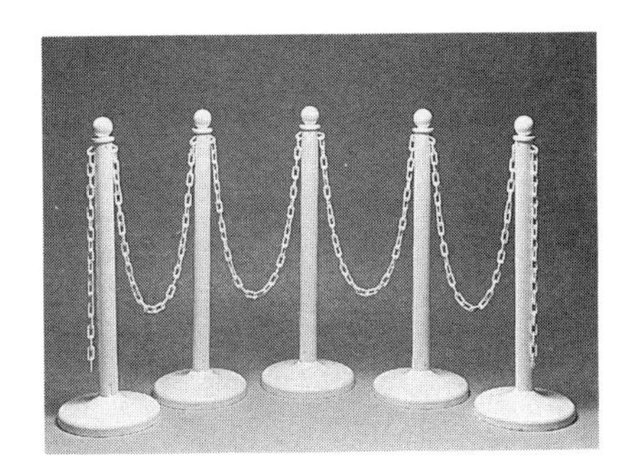

c

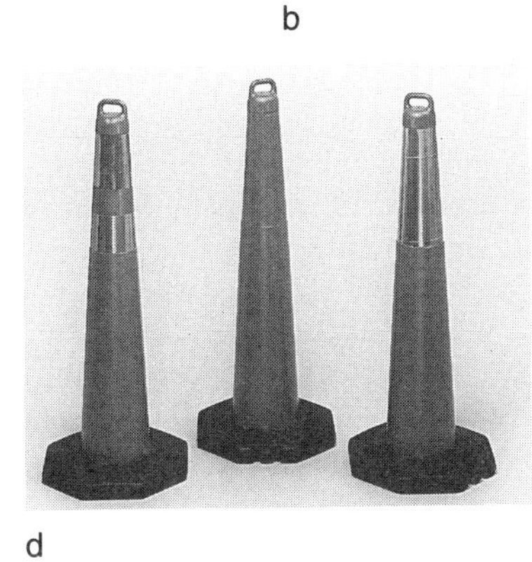

d

e

Figure 7-20a-e. Floor stand sign, guideline stanchions and chain, disposable warning cones, and stacker cones.
Source: Photos courtesy of Lab Safety, Inc., Janesville, WI.

Figure 7-21a. Central storage board for employee lockout/tagout gear.
Source: Photo courtesy of E. Grund.

or permanently installed to limit access, locate energy-isolating devices, communicate lockout/tagout requirements, etc. Graphic displays may be in print form and laminated or electronic; the displays are often used to communicate sequence of energy isolation, identification of energy sources and isolation points, special precautions, and so on.

Barriers. These are items and equipment designed to control access to areas where lockout/tagout work is in process; examples of this type of equipment would be cones, tripod portable warning devices (blinker flashers, etc.), printed high-visibility tapes, plastic chain and post systems, etc. (Figure 7-20 a–e).

Lockout/tagout centers. These are wall-mounted displays of various design with high-visibility graphic treatments for storage of locks, tags, hasps, adapters, etc. (Figure 7-21 a–b).

Figure 7-21b. Safety lockout center.
Source: Photo courtesy of Prinzing Enterprises.

Energy-isolating devices. Mechanical devices that physically prevent the transmission release of energy, these include but are not limited to the following: a manually operated electrical circuit breaker; a disconnect switch; a manually operated switch by which the conductors of a circuit can be disconnected from all ungrounded supply conductors and no pole can be operated independently; a slide gate; a slip blind; a line valve; a block; and any similarly designed devices used to block or isolate energy. Control circuit devices are *not* valid energy-isolating devices.

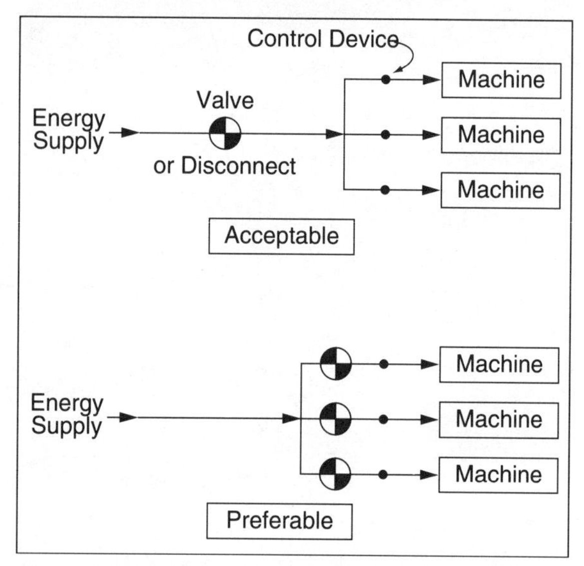

Figure 7-22. Energy-isolating device arrangment.
Source: Developed by E. Grund. © National Safety Council.

Ideally, energy-isolating devices should be provided and installed in such a way that each machine, piece of equipment, and process or system component can be individually isolated for service or repair (Figure 7-22). Energy-isolating devices that control the supply of energy to multiple machine units or pieces of equipment often stimulate the use of control circuitry by personnel attempting to perform work under exposed conditions. The following are four general categories of energy-isolating devices:

Valves. These are various mechanical devices by which the flow by liquid, gas, or loose material in bulk may be started, stopped, or regulated by a movable part that opens, shuts, or partially obstructs one or more parts or passageways; they may be manually or power operated and are used for flow control of chemicals, steam, gas, hydraulic fluids, air, water, etc. Valves are found in common service on lines from 0.25 inch to 5 feet (0.6 cm–1.5 m) in diameter; some valves have design features that permit locking out and/or release of trapped line pressures while many others must be secured/restrained by various protective appliances (Figure 7-23 a–d).

Electrical breaker/disconnect devices. These are units of an electrical system that are intended to distribute or control but not consume energy; they are a means by which the conductors of a circuit can be disconnected from their source of supply. They can close and/or open one or more electric currents.

Breakers and disconnects may be used in AC or DC service and are commonly used with 100–420 V

Figure 7-23a. A chain is used to prevent valve operation.
Source: Photo courtesy of E. Grund.

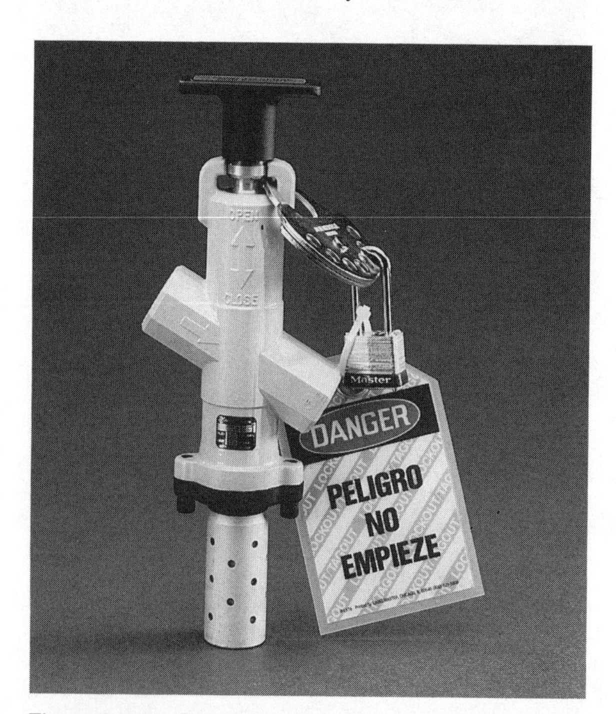

Figure 7-23b. Spool type lockout valve with t-handle operation. Note warning in Spanish.
Source: Photo courtesy of Labelmaster and Norgren.

systems. They are also used in systems exceeding 4,200 V that are associated with power generation and distribution networks. They are generally lockable, with certain exceptions for equipment of older design and breakers grouped in panels (Figures 7-24 a–d).

29 *CFR* 1910.147 (c) general (2) (iii) requires that after October 31, 1989, whenever major replacement, repair, renovation, or modification of machines or equipment is performed and whenever new machines or equipment are installed, energy-isolating devices for such machines or equipment shall be designed to accept a lockout device.

Automatic block/pin type devices. These devices are usually found on stationary machinery/equipment components that automatically engage and prevent mechanical movement due to gravity or cycling commands. They are designed to withstand the forces involved and are most often used when employees are performing tasks with less than zero energy conditions. Examples of these devices are automatic die blocks for presses; pallet elevator insertion safety pins, and conveyor dogs/pins (Figure 7-25 a–b).

Blinds (blanks, spades, slip plates/rings, spectacle plates). These devices are inserted between flanges or placed over the end of lines that have been broken/opened; blinds should be equivalent to the engineering design standard for the piping system and appropriate for the line size and pressure; tags should be attached to the blinds to identify their position; use, and details of the application.

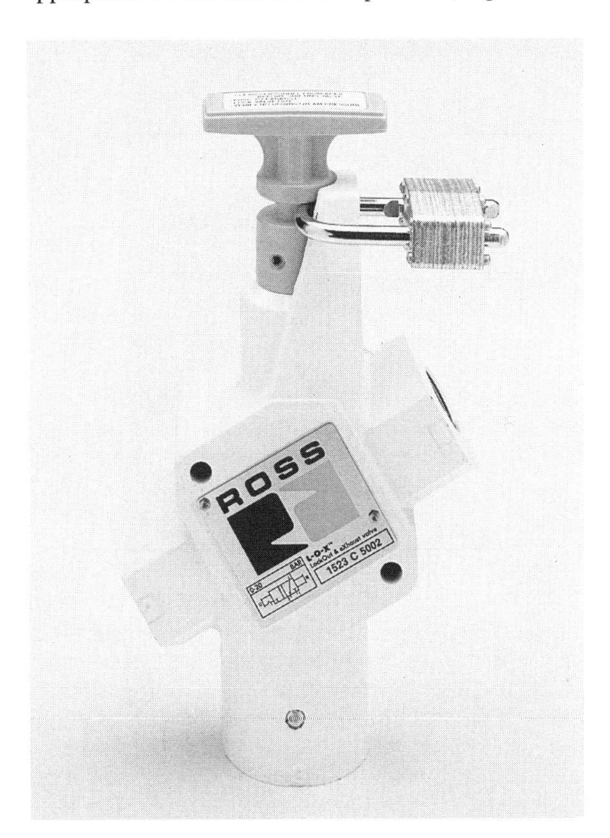

Figure 7-23c. Pneumatic valve with lockout and exhaust feature.
Source: Photo of L-O-X® lockout/exhaust valve courtesy of Ross Operating Valve Company. Used with permission.

Disconnection and misalignment are methods of isolation in which a line is broken (usually at a pair of flanges) and a valve is closed and locked or a blank is inserted on the upstream or pressurized side of the process; the combination is superior to using a valve alone because valves are susceptible to leaking. The point of disconnection or misalignment should be such that personnel are not exposed to potential process leakage (Figure 7-26).

Identification of energy-isolating devices is an important dimension of the lockout/tagout process. If employees are expected to isolate machines and equipment, they should be able to identify the isolating device's function without relying on memory. OSHA's belief that employee training is a substitute for energy-isolating device identification is a serious flaw in 29 *CFR* 1910.147. It is also inconsistent with other requirements found in 29 *CFR* 1910 Subpart S Electrical. For example, section .303 requires the following:

> (.303)(f) **Identification of disconnecting means and circuits.** Each disconnecting means required by this subpart for motors and appliances shall be legibly marked to indicate its purpose, unless located and arranged so the purpose is evident. Each service, feeder, and branch circuit, at its disconnecting means or overcurrent device, shall be legibly marked to indicate its purpose, unless located and arranged so the purpose is evident. These markings shall be of sufficient durability to withstand the environment involved.

The identification of permanent energy-isolating devices should provide the following basic information:

- *Breakers/disconnects.* Energy (AC or DC); magnitude (voltage/amperage); function (machine, equipment, process controlled); identification number (local code or designation for device).

Figure 7-23d. Lockable lever type valve.
Source: Photo courtesy E. Grund.

Figure 7-24a. Breaker panel.
Source: Photo courtesy of E. Grund.

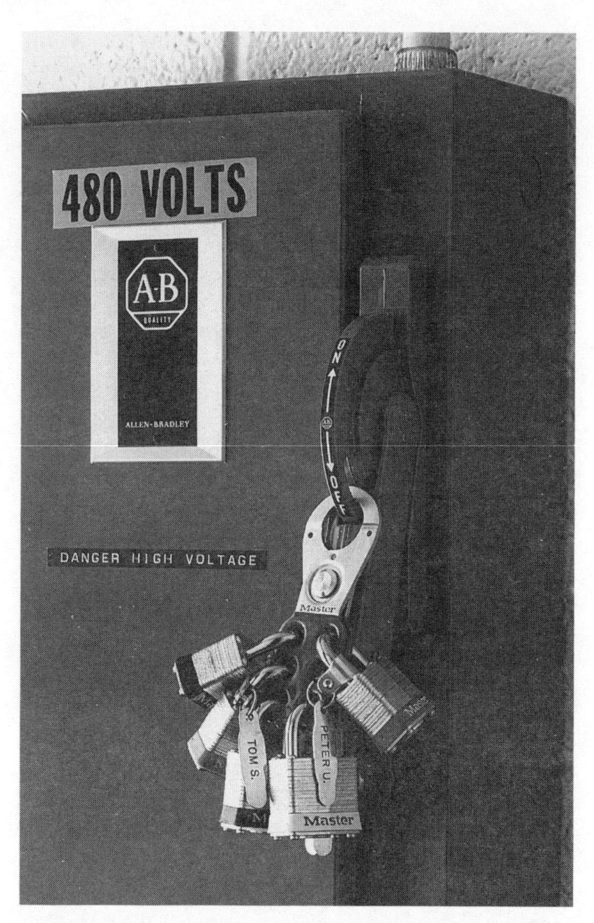

Figure 7-24b. Locked out 480 V disconnect.
Source: Courtesy of Master Lock Co.

- *Valves*. Service (gas, water, steam, air, etc.); magnitude (pressure, temperature, etc.); function (machine, equipment, process controlled); identification number (local code or designation for device).

This information can be placed on permanent (for example, metal or plastic) tags or plates, engraved or stenciled. If tags are used, they should be securely fastened to the device with wire, nylon, etc.

Guarding/Interlocks

In 29 *CFR* 1910.147 Lockout/Tagout section (a)(2) application, the following is found:

> (i) This standard applies to the control of energy during servicing and/or maintenance of machines and equipment.
>
> (ii) Normal production operations are not covered by this standard (See Subpart O of this Part). Servicing and/or maintenance which takes place during normal production operations is covered by this standard only if;
>
> (A) An employee is required to remove or bypass a guard or other safety device; or
>
> (B) An employee is required to place any part of his or her body into an area on a machine or piece of equipment where work is actually performed upon the material being processed (point of operation) or where an associated danger zone exists during a machine operating cycle.
>
> NOTE: Exception to paragraph (a)(2)(ii): Minor tool changes and adjustments, and other minor servicing activities, which take place during normal production operations, are not covered by this standard if they are routine, repetitive, and integral to the use of the equipment for production, provided that the work is performed using alternative measures which provide effective protection (See Subpart O of this Part).

In the preamble to the standard, OSHA further explains that lubricating, cleaning, unjamming, and making minor adjustments and simple tool changes are activities that often take place during normal production operations but may expose employees to equipment energy release. These activities are considered to be "servicing and/or maintenance" for purposes of this standard. When servicing and/or maintenance takes place during "normal production operations," it is important to note that 29 *CFR* 1910.147 is intended to work together with the existing machine-guarding pro-

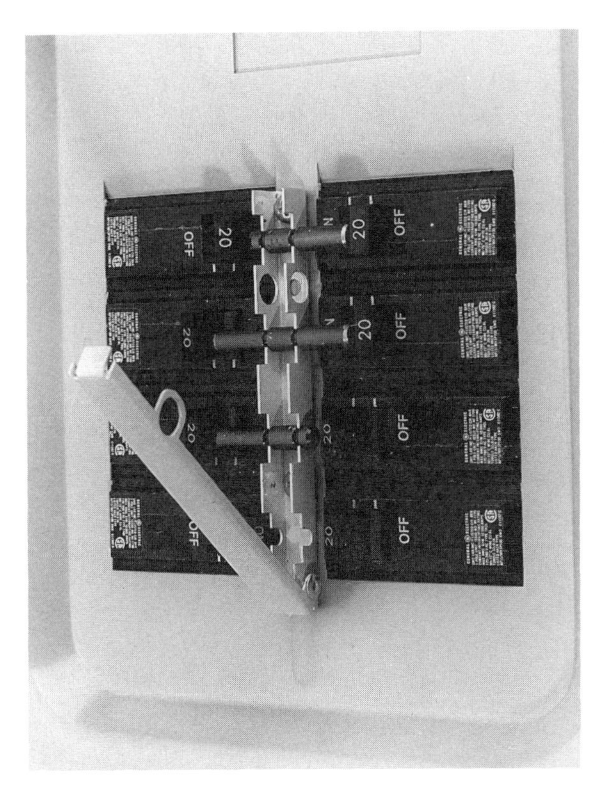

Figure 7-24c. Breaker locking device.
Source: Photo courtesy of Toloc System, Inc.

visions of Subpart O of Part 1910, primarily sections .212 (general machine guarding) and .219 (guarding of power transmission apparatus).

Interlocks (devices that break the electrical power circuit when actuated) are often used in conjunction with machine/ equipment point of operation guarding. They are *not* intended to be energy-isolating devices but function more like control circuit interrupters or mechanisms. They perform a legitimate safety function but are often viewed as an acceptable alternative to using a primary (disconnect) means of protection, which they are not. Interlocks are intended to be access control devices that ensure the guard/barrier is in place before the machine/ equipment is capable of running.

However, interlocks may be used in conjunction with other measures that provide effective protection for production activities that are routine, repetitive, and integral to the use of the equipment, that is, alternative measures to complete de-energization. Additional information will be provided on interlocks in Chapter 10, Special Situations and Applications.

Accessibility

Energy-isolating device accessibility is an often-overlooked aspect of the lockout/tagout process. When employees make value judgments regarding whether or not to de-energize, they unfortunately factor into their decision the access or convenience of use of the isolating device. Isolating devices in crowded and dirty pits,

Figure 7-24d. Control panel with disconnect.
Source: Photo courtesy of E. Grund.

in elevated positions requiring ladder access, and at unreasonable distances are susceptible to employee disregard. A review of energy-isolating device position relative to the worksite should be conducted.

Energy-isolating devices may be located properly yet obstructed by portable materials, equipment, and objects. These unacceptable housekeeping practices contribute to the perception that if using these devices were important, they wouldn't be obstructed—that management isn't really serious about lockout/tagout. Rigidly enforced internal housekeeping standards are needed with clear criteria as to the required clearance distances or paths of access.

In other instances, the question of accessibility is a matter of relationship between the work task, energy-isolating device, machine equipment, or process. For example, if the isolating valve must be reached by portable ladder, if the task requires moving a floor plate and entering a pit, or if the machine and its controls are not proximate to each other for cycling, verification action, and so on, then the arrangement may prove to be error provocative.

Special Risks

Although energy isolation principles remain the same for machines, equipment, and processes that represent special risks, the approach to lockout/tagout may be

Figure 7-25a. Press automatic locking device (disengaged).
Source: Photo courtesy of E. Grund.

Figure 7-25b. Press automatic locking device (engaged).
Source: Photo courtesy of E. Grund.

more complex. Extremely high voltages, pressures, and temperatures often require unique procedures and safeguards. Hazardous materials that are highly reactive, toxic, corrosive, or radioactive will dictate purging, cleaning, line breaking, and personal protective equipment requirements that are beyond those used for normal isolation procedures.

For example, nonsparking tools would be a critical feature for line breaking or blinding in the flammable/explosive benzol process environment; valve leak-

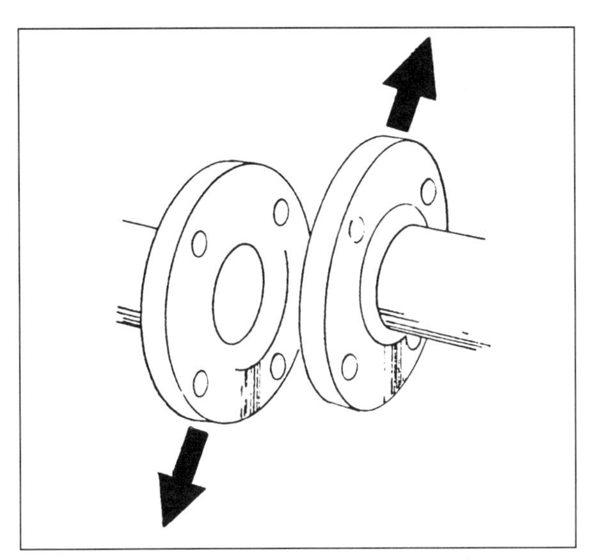

Figure 7-26. Chemical piping system (separation and misalignment).
Source: UAW-GM. *Lockout Energy Control* (Training Workbook for Authorized Employees). UAW-GM National Joint Committee on Health and Safety.

age might be anticipated in a process producing anhydrous hydrogen fluoride due to the severe corrosive service; carbon monoxide around blast furnace stoves might dictate use of a breathing apparatus even during nonconfined maintenance and service tasks.

Elevated work, confined spaces, and excavation/trenching activity often involve energy isolation requirements that are unique or exceed those needed for work with lower inherent risk. Hot tapping (penetrating/welding on pressurized lines), although not covered by 29 *CFR* 1910.147, requires special equipment and procedures to provide effective equivalent protection of workers. Additional information on this topic will be provided in Chapter 10, Special Situations and Applications.

Interrupt Modes

An assessment should be made of machines, equipment, and processes to identify the interrupt or abnormal modes. These deviations contain the seeds of future trouble if not detected and addressed. Much of management's attention is placed on what occurs 85% of the time—on what usually happens. However, very often it is the exception and not the rule that produces undesired results. In fact, energy-release incidents are deeply connected to nonroutine, infrequent, and unplanned tasks.

For example, a machine that jams twice per hour may be tolerable from a manufacturing standpoint. After all, at three minutes per jam, only six minutes per hour is lost to downtime. A high product margin may relegate the jamming problem to a low management priority. However, this means that 16 times per shift, the operator has to cope with clearing the jam and deciding whether to de-energize the machine or to depend on control circuitry for protection. In the absence of any direct order to do otherwise, the operator may take the path of least resistance and reduce the time to two minutes per jam, leaving out the de-energizing step. After all, supervision should be pleased at the reduction in downtime . . . shouldn't they? This hypothetical situation represents the world of interrupt or upset in which employees with good intentions make value judgments every day in an attempt to cope with exceptional situations.

Only when management addresses these abnormal modes and their root causes can the unwanted incidents that they generate be reduced. The following partial listing of interrupt or upset modes can be used to stimulate the needed inquiry:

- equipment start-up
- testing conditions
- breakdown
- jamming
- shutdown
- approval trials
- troubleshooting
- flooding
- material problems
- process modification
- overtime
- energy interruptions
- vacation disruptions
- unusual weather
- changeovers
- new product run
- off-schedule work
- presence of contractors
- rebuilding
- experimentation.

Advanced Systems

Advanced machinery systems, computer-aided manufacturing, and automated programmable assembly are all part of the new wave of production processes appearing in the workplace. Assembly robots, programmable pick-and-place systems, automated handling mechanisms, and hard automation are being combined with programmable logic controllers and computers to produce the goods of tomorrow. Optical scanners, spread-spectrum radio wireless technology, fiber optics, and servomechanisms are representative of the language and technology that will become commonplace in the factory of the future. With these advances come new challenges for controlling energy-release incidents and safeguarding workers.

Commands emanating from computer software may activate advanced machinery systems without warning and override conventional safeguards under less than zero energy conditions. Protective language may be needed in software programs that forbid/prevent certain actions from taking place. Protective or safeguard-dedicated computers may be required to monitor a system and prevent any inputs or outputs that might compromise safety. The Automated Manufacturing Research Facility at the National Bureau of Standards has done work on a "watchdog safety computer" to monitor robot-related operations.

Robots and robot systems are increasing in number with electrical (440 V), hydraulic (2,000 psi; 136 atm), and pneumatic (100 psi; 6.8 atm) forms of energy. Unlike other machines and equipment, the movement and action of these "steel-collar" workers may occur without warning. Whenever possible, maintenance and repair tasks should be accomplished under lockout/tagout protection. However, some required tasks inside the work envelope may require robots to be powered, for example, teaching operations, adjustments/troubleshooting, and possibly side-by-side employee/robot production activities. These tasks need to be performed under special procedures that also include auxiliary safeguards. Additional detail on robot safety will be provided in Chapter 10, Special Situations and Applications.

There will be no further attempt here to address the many possibilities for energy-release incidents that might occur in high-tech/advanced systems. The new and emerging technology will have to be specifically assessed to determine what energy isolation techniques and alternative measures will be necessary to adequately protect workers performing maintenance, service, or production tasks.

DESIGN ELEMENTS

The design of the hazardous-energy-control system represents the *first* and *best* opportunity to prevent energy-release incidents. At the design stage, machines, equipment, and the process are vulnerable to changes that could permanently alter the way workers choose or are required to interact with them. In most safety or hazard control hierarchies, elimination of the hazard/risk is given first preference. The consensus on prevention tactic priority is usually presented as follows:

1. Eliminate the hazard.
2. Apply safeguards.
3. Warn.
4. Train/Educate the individual.
5. Use personal protective equipment.

It is in the first position of the hierarchy that the design opportunity is paramount. This is not to suggest that all danger or risk can or should be eliminated at this stage. When provided with the employee requirements, interrupt modes, incident histories, potential dangers, etc., designers will produce a product that will best serve the safety interests of the user.

According to NASA, the major effort throughout the design phase must be to ensure inherent safety through the selection of appropriate features such as fail-safe devices, redundancy, and increased ultimate safety factor. Known hazards that cannot be eliminated through design selection should then be reduced to the acceptable level through the use of appropriate safety devices as part of the system, subsystem, or equipment (National Aeronautics and Space Administration, 1970). With a greater emphasis on risk reduction through design, management demands for complex procedures (and the need to train employees about them) will be eliminated. In addition, the opportunity for employee exposure to energy release and the frequency or need for application of lockout/tagout will be diminished. The following subelements of design should be considered when developing a hazardous-energy-control system:

- ergonomics/human factors
- isolation enhancements
- reliability improvements
- maintainability
- operability specifications
- change analysis
- system safety analysis
- concept/safety criteria.

Each of these will be discussed in detail.

Ergonomics/human factors. This subelement of design is the key link between the human and the machine, equipment, and process elements of the hazardous-energy-control system. Enhancing human/machine compatibility should eliminate hazards and/or reduce risks associated with tasks involving energy isolation. Karl Kroemer views the goal of ergonomics/human factors as ranging from making work safe and increasing human efficiency to creating human well-being. He further states:

> Human factors [ergonomics] specialists are united by a singular perspective on the system design process: that design begins with an understanding of the user's role in overall system performance and that systems exist to serve their users, whether they are consumers, system operators, production workers, or maintenance crews. This user-oriented design philosophy acknowledges human variability as a design parameter. The resultant designs incorporate features that take advantage of unique human capabilities as well as build in safeguards to avoid or reduce the impact of unpredictable human error. (Kroemer, 1988)

Figure 7-27 shows how ergonomics/human factors interact with related applied disciplines and sciences (Kroemer, 1988).

Kroemer promotes six principles of general workstation design:

1. Plan the ideal, then the practical.

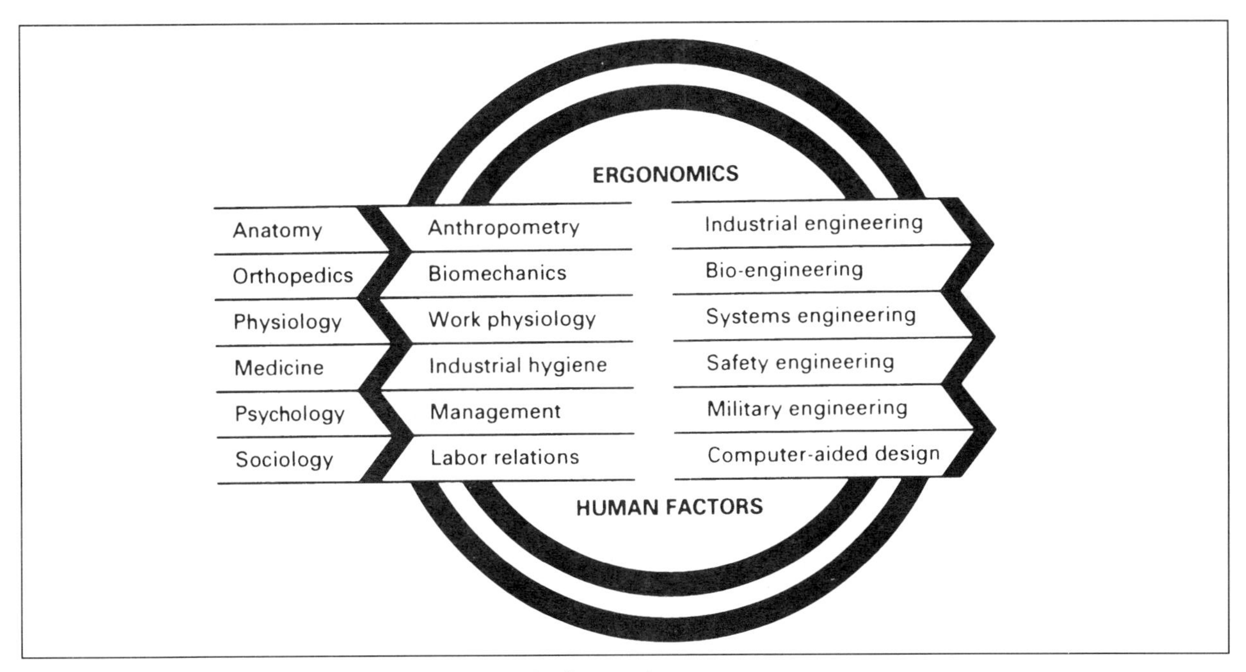

Figure 7-27. Origins and applications of ergonomics/human factors.
Source: Kroemer, Kroemer, Kroemer-Elbert, *Ergonomics: How to Design for Ease and Efficiency*, © 1994, p. 8. Reprinted by permission of Prentice Hall, Englewood Cliffs, NJ.

2. Plan the whole, then the detail.
3. Plan the work process and equipment around the system requirements.
4. Plan the workplace layout around the process and equipment.
5. Plan the final enclosure around the workplace layout.
6. Use mockups to evaluate alternatives and to check the final design.

These general principles can be used as a guide when focusing on machine, equipment, and process energy isolation requirements. Other design aspects of importance are clearances and egress, postures and positions, force requirements, control operation, visual field and displays, visual contact with co-workers, auditory information and feedback, and the like.

Machine, equipment, and process control–related energy-release incidents occur far too frequently. Although controls are not acceptable energy-isolating devices, they play an intimate role in the de-energization process. They are used in the shutdown step, the cycling/testing activity, the positioning task, the zero energy verification step, the start-up sequence, etc. As an example, two mechanics were working on a large press making an adjustment. Near the end of the job, they attempted to cycle the equipment to check their work. One mechanic asked the other to cycle the ram and inadvertently cycled the coil feed; the controls were similar in shape/color and positioned near each other on the panel. The mechanic narrowly escaped having his hand caught in the pinch point between feed rolls. Controls should be designed and located so that potential for visual/tactile confusion is minimized and that susceptibility to inadvertent contact is eliminated. Particular emphasis should be placed on controls that initiate major component movement, transmission of forces, or change of status.

One of the following methods may be used for protection of controls from accidental activation:

1. Locate and orient controls to preclude unintentional contact during normal sequences.
2. Recess, shield, or contain the controls by physical means.
3. Cover or guard the controls.
4. Provide the controls with interlocks or protective movement features.
5. Provide resistance controls that require sustained application.
6. Provide controls with interrupt features or delays when moving through sequences.
7. Install rotary action controls.

Another area deserving human-factors design attention is the location, placement, and force requirements associated with energy-isolating devices. When overlooked, these factors can negatively impact the worker motivation to use isolating devices. Devices should not be located or positioned in such a way that access or posture is awkward or places the employee at direct risk, for example, mounting an electrical disconnect so that the employee has to stand in front of the device when actuating (disconnects can fail violently). Valves by design may require significant manual force to cycle or close completely or become difficult due to the service conditions. This could lead to employee avoidance, partial closure, or unauthorized use of force magnifiers such as pipe levers, large wrenches, slug wrenches, etc. In some cases, the planned or required procedure for valve closure is to use mechanical force, which is acceptable. The point to remember is that the operation of the valve(s) should be well understood and the proper response should be planned and controlled rather than having employees reacting and coping with the circumstances. Powered valves can often be used as substitutes when human capabilities are at their limits. The energy-isolating devices should always be assessed for their human compatibility in conjunction with their location and position. A valve in a pipe rack 30 ft (9.1 m) overhead presents a different demand on employees than one at thigh height or ground level.

Much more could be detailed regarding ergonomics/human-factors issues related to energy isolation that can be found in various texts on the subject. Lockout/tagout systems should include a mechanism to assess the ergonomic aspects of the various machines, equipment, and processes and their related energy isolation tasks.

Isolation enhancements. The design and functioning effectiveness of energy-isolating devices is an area that has significant influence on workers' attitude and motivation toward machine, equipment, and process de-energization. Historically, the installation of breakers, disconnects, valves, and so on was driven by the needs of the *process*, not the *person*. They were often located in such a way as to protect them, not make access to them easy and likely. They often were hard to operate or to secure from unwanted movement. They were neither identified nor installed in such a way to facilitate the relief of system pressure. With such a track record, opportunities abound to enhance the situation with improved design and engineering.

Energy-isolating device enhancement can be approached in several ways that involve internal and external resources. Management may (a) define its requirements for devices and their placement and have an outside engineering firm provide the design and installation, (b) order off-the-shelf devices of superior design and replace existing ones with internal or external work forces, (c) internally design devices and fabricate them or have contractors make them, or (d) use local workers or outside contractors to install the devices. If devices are designed by internal engineering personnel and/or fabricated by inside maintenance personnel, product liability risks are incurred. These risks can be substantial with regard to negligent design or fabrication.

Examples of energy-isolating device enhancement are as follows (Figures 7-28 and 7-29):

- Convert manual power press die block placement to automatic die block insertion.
- Replace simple pneumatic gate valves with lockable, pressure indicating, self-relieving valves.
- Modify gate or globe valves with permanent locking hardware.
- Automate valve operation where appropriate.
- Install devices for each machine/equipment rather than for group services.
- Install pressure-relief features in piping systems, accumulators, etc.
- Relocate or position devices to make their function more apparent and access more natural.

Reliability improvements. Machine, equipment, and process reliability has significant potential for impacting the frequency of exposure of maintenance, service, and production personnel. *Reliability* is defined as the ability of the component or machine to perform its function for a described period of time without failure. Machinery with excessive downtime (which can be as comparatively little as 30 minutes a day) may require workers to intervene to rectify the trouble under zero energy conditions. Management's best strategy for reducing or eliminating energy-release incidents is often to focus on improving machine/equipment/process performance, hereby minimizing worker exposure or opportunity for injury.

Each component of a machine has a reliability factor (predicted, known, unknown) that collectively

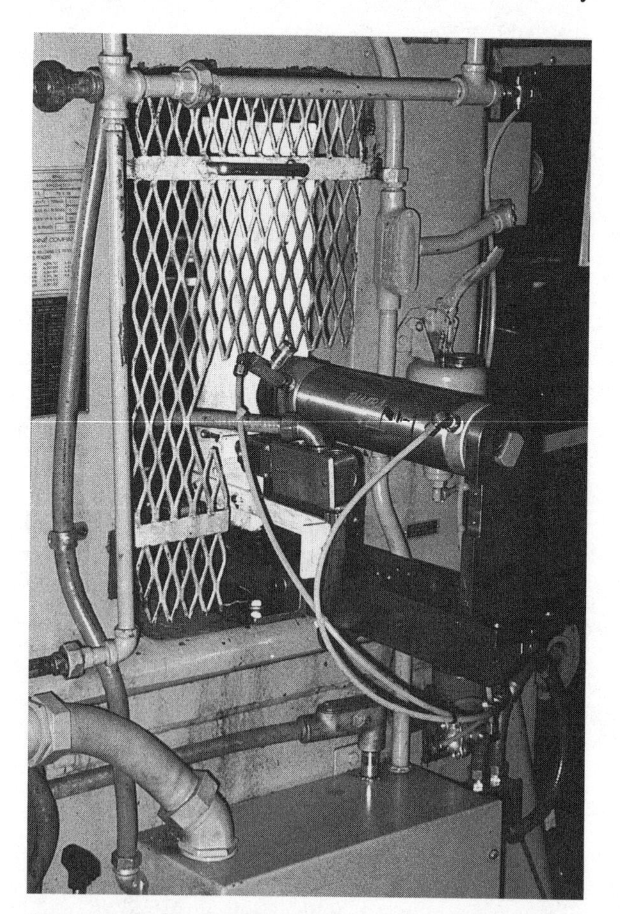

Figure 7-28. Press with automatic die block feature.
Source: Photo courtesy of E. Grund.

determines the overall reliability of the machine. The component failure rates can be expressed in cycles per unit time, hours of continuous or intermittent operation, number of actuations, etc. Improved design can upgrade reliability and contribute not only to better performance but to safety as well. Life cycle analysis can be viewed as an extension of reliability thinking in which the system in question is appraised to determine where unwanted events may occur in time. Factors such as severity of service, preventive maintenance, and overhaul schedule will ultimately affect the life cycle.

Designing for reliability should go beyond the sheer performance of system components individually or collectively, without regard to specific user requirements. Reliability of energy-isolating devices can be improved by upgrading the seal of valves, reducing mechanical linkage failures in disconnects, and so on. Machinery reliability should also be extended to include how raw material is processed or the product is made—for example, frequent jamming of the machine may require workers to intervene and make judgments regarding the energy isolation approach. A machine's ability to stay at a desired speed, to remain at a set temperature, and to produce at tolerances also impacts activities such as adjusting, troubleshooting, checking, etc. The greater the frequency of these types of tasks, the greater the opportunity for error and the occurrence of energy-release incidents.

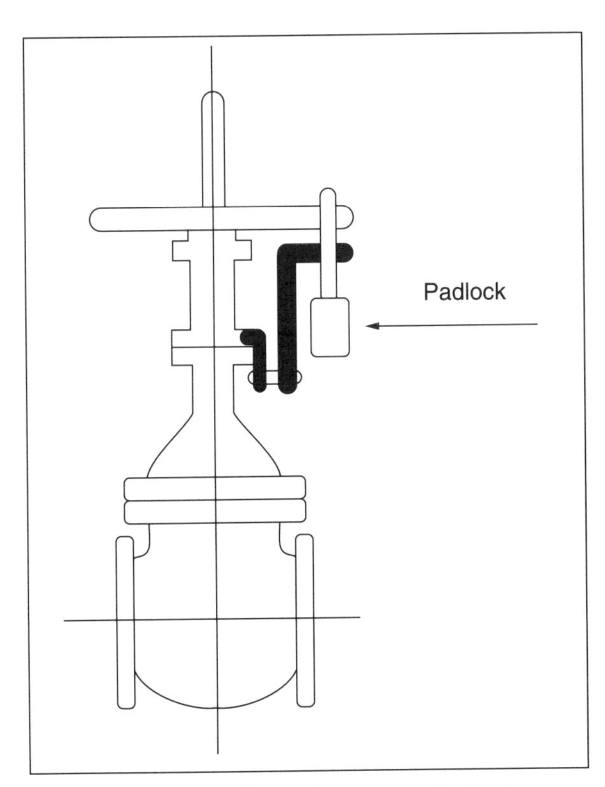

Figure 7-29. Gate valve with permanent adaptor.
Source: Developed by E. Grund. © National Safety Council.

Maintainability. Maintainability relates to the ease, efficiency, cost, and safety associated with keeping machines, equipment, and processes operating according to their design parameters. The average maintenance or service worker may feel that design sensitivity to maintainability and the tasks they are compelled to perform has been neglected. Lubrication, component replacement, tightening, adjusting, and similar tasks put maintenance workers at risk for energy releases. The nature of the task and predetermined procedure for its execution may dictate a power ON, partial power, or power OFF status. Adequate safeguards will be required for employee protection, regardless of the power status.

Designing for maintainability can enhance worker safety in ways that have a direct bearing on potential energy-release occurrences. For instance, piping lubrication fittings outboard of machinery guards will minimize the potential for contact with moving parts and maximize the likelihood of guards remaining in position. Installing dual pressure-relief valves and appropriate piping on a blow-down tank can facilitate changeout and service with less exposure to personnel. Thinking should be extended from a design standpoint to accesses, platforms, and catwalks that permit use of energy-isolating devices without additional risk. Designs that permit sectioning or separation of pieces of equipment should be considered so as to allow repair work to proceed on large installations/processes during their operation while still providing protection for engaged workers.

Professor R. L. Barnett, in "The Doctrine of Manifest Danger," explains how the dangers associated with safeguarding failures are minimized by using four basic tools: (1) reliability design, (2) preventive maintenance, (3) fail-safe design, and (4) manifest danger. Minimum danger is achieved by reducing the number of safeguard functional failures and/or by making designs more forgiving or fault tolerant. The techniques of reliability design and preventive maintenance (Figures 7-30 and 7-31) may be used to effectively reduce the frequency of functional faults (Barnett, 1992).

Operability specifications. Appropriate employees of the design organization and operations management need to work as a team in the preparation of operability specifications. These specifications will define parameters and procedures for installation, support, testing, operating, maintaining, monitoring, emergency response, and so on.

Operating manuals that include guidance on set-up, sequences, precautions, control functions, abnormal conditions, fault detection, cleaning, speeds, pressures, and shutdown need to be prepared by the engineering organization in collaboration with the operational personnel.

Maintenance guidelines/procedures reflecting troubleshooting methods, malfunction indicators, lubrication, service, adjustments, and energy isolation are

RELIABILITY DESIGN - SAFETY SYSTEMS

Objective of Reliability Design-Safety Systems: To control the probability of system failure.

Important Definitions

Failure: The termination of the ability of an item to perform a required function.

Reliability (Definition 1): The ability of an item to perform a required function under stated conditions for a stated period of time.

Reliability (Definition 2): The probability that an item will perform a required function without failure under stated conditions for a stated period of time (non-repaired items).

Covert Faults: A hidden or latent failure of an item that can only be revealed by inhibiting a process demand or by proof testing. These fail-danger faults may inhibit safe action.

Overt Faults: Failures of items that are self-revealing.

Mean Time to Failure (MTTF): Average time to failure of a non-repairable product or the average time to first failure of a repairable product.

Mean Time between Failures (MTBF): Average time between successive failures of a repairable product.

Hazard Rate: For a group of similar items, the hazard rate at any given time is the percentage of first failures per unit time.

Failure Rate: For a group of similar items that are repaired each time they fail, the failure rate is the percentage of all failures per unit time.

Bathtub Curve: A bathtub-shaped curve resulting from the plotting of Hazard or Failure Rate against the time period over which equipment is used:

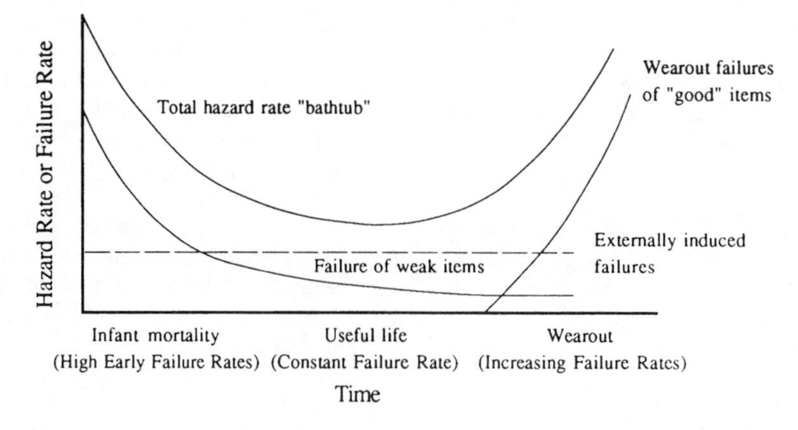

Typical Reliability Tools

Safety Factor: An experience-driven multiplier which ensures that the generalized loads do not exceed the generalized resistance of items.

Quality Control: Ensuring conformance to specifications and tolerances.

Burn-In: Operation of parts under failure-provoking conditions for a time before delivery. The idea is to eliminate items in the range of infant mortality on the "bathtub" curve.

Proof Loading: Subjecting items to a single failure-provoking load condition after manufacture to eliminate any "weak sisters" from the product stream. For example, grinding wheels are operated at 150% of rated speed before shipping.

Redundancy: The existence of more than one means for accomplishing a given function. The various means need not be identical.

Derating: Use of derated parts to assure that the stresses applied are lower than the stresses the parts can normally withstand.

Environmental Control: Protect safety devices from the operating environment. For example, potting electronic components to protect them against climate and shock.

Proven Technology: Use of standard safety devices whose reliability has been established by actual field use.

Figure 7-30. Reliability design—safety systems.
Source: R. L. Barnett, "The Doctrine of Manifest Danger," *Safety Brief—Triodyne Inc.* (September 1992).

PREVENTIVE MAINTENANCE PM

Objective of PM: To retain a system in an operational or available state by preventing failure from occurring.

Important Definitions

Maintainability: The ability of an item, under stated conditions of use, to be retained in, or restored to, a state in which it can perform its required functions, when maintenance is performed under stated conditions and using prescribed procedures and resources.

Mean Time to Repair (MTTR): The total corrective maintenance time divided by the total number of corrective maintenance actions during a given period of time.

Maintenance, preventive: The actions performed in an attempt to retain an item in a specified condition by providing systematic inspection, detection and prevention of incipient failure by repairing or replacing it. This concept must be contrasted with corrective maintenance which restores an item to a specified condition after it has failed.

Reliability Centered Maintenance (RCM): A systematic approach to maintenance planning which takes reliability aspects into consideration.

Typical Preventive Maintenance

1. Scheduled Replacement

a. Definition: The replacement of parts before failure at predetermined times.

b. Note: When the onset of failure of an item cannot be determined, scheduled replacement is the only PM strategy available.

c. Note: Reliability analysis techniques are required to develop replacement schedules (maintenance intervals) that will minimize the number of items which fail.

d. Note: All effective replacement strategies using new parts are applied in the "wear out" or increasing hazard rate range. Unfortunately, there are cases where no new part replacement schedule can be found that will decrease the failure probability.

e. Note: If a failure-free life exists for an item, it may always be replaced within this lifespan before failure occurs. This makes the failure probability zero.

2. Diagnosis

a. Definition: Those maintenance strategies which reveal incipient failure of items such as safety devices.

b. Inspection: Detection of self-revealing deteriorating conditions of items such as safety devices and systems.

c. Nondestructive Testing (NDT): A body of testing techniques and methods which will not compromise the item tested. This includes regular manual testing (e.g., X-ray) and automatic monitoring (e.g., light curtain circuit checking at the end of each cycle). The NDT determines whether the behavior of safety devices falls outside of performance and tolerance limits.

3. Servicing Strategy

a. Definition: Scheduled servicing (not part replacement) to prevent safety device and system failure.

b. Types of Servicing
- Cleaning
- Lubrication
- Calibration
- Adjustment
- Repair (not replacement)

Figure 7-31. Preventive maintenance.
Source: R. L. Barnett, "The Doctrine of Manifest Danger," *Safety Brief—Triodyne Inc.* (September 1992).

also needed. Drawings, sketches, data sheets, and manuals need to be prepared for operational support and reference.

These activities are directed at ensuring that the machine, equipment, or process performance goals and objectives are attained by integrating the human and machine elements in an orderly and controlled manner.

In the case of new or unique equipment, the design/maintenance/operations organizations can work together to determine what is needed to isolate energy during various anticipated tasks. On subsequent generations of existing machines/equipment, the history of operability deficiencies needs to be addressed by the design process to determine if hardware changes, operating manual revisions, job step/sequence improvements, etc. are necessary.

Change analysis. Changes, whether related to form, fit, function, raw materials, product, application, sequence, or any other factors, need to be examined for their impact and their relationships to the design of machines, equipment, and processes. More specifically, are energy-control (lockout/tagout) procedures/techniques being affected by changes that were planned or unplanned or whose impact was not anticipated or understood?

For example: When a planned change in raw material is introduced, are the valves in service more likely to leak? When the speed of a machine is increased by 30%, can product quality checks still be made under energized conditions? When system pressure and temperature are increased, can relief practices permit atmospheric venting? Will mechanics avoid using isolating valves in pits because replacement pumps leak excessively? Will a new computer-controlled system now require energized status for certain tasks? Are the new powered valves more difficult to lock out than the old valves?

Engineering organizations *must receive* design feedback regarding unusual occurrences, warning signs/signals, close calls, and performance obstacles/impediments. If energy isolation is complicated or compromised by design limitations, by designs that did not anticipate current conditions, or by designs that were not intended for current applications, then changes have occurred that need to be addressed.

System safety analysis. Various system safety analysis methods can be employed when evaluating machines, equipment, and processes for hazards and their causes, effects, and countermeasures. These methods such as preliminary hazard analysis (PHA), failure modes and effect analysis (FMEA), fault-tree analysis (FTA), and procedures analysis (PA) are valuable for general system safety assessments or analysis of a specific dimension of the system, for example, energy isolation for maintenance/servicing work. The methods can be quite vigorous and quantitative or can be used with emphasis on logic and the qualitative dimensions or aspects.

These methods are particularly beneficial for appraisals conducted early in the developmental stages (conception and design) of a project. This approach emphasizes that energy isolation requirements for various tasks are examined and machine design is influenced by the identified requirements. Procedural, training, and inspection demands are also identified during the process.

Concept/safety criteria. At the conceptual stage of a project, whether the objective is to create new technology or to build another improved generation of the same machine, equipment, or process, input should be provided regarding what is expected from a safety standpoint. More specifically, design should be affected by what is required to de-energize the equipment safely for all of the anticipated or known tasks. Where zero energy conditions are possible, appropriate isolating devices, safeguards, and system pressure-relief devices need to be incorporated into the design. Where partial or complete power tasks are required, various protective features with equivalent safety should be included.

During the early phases of engineering, operational and maintenance personnel need to provide their views regarding what the design should address to improve overall performance and safety. For example, can a mechanism be built into the machine that would clear jams automatically without worker exposure? Can barrier guards be automated instead of being manually positioned? Will backup equipment modules be incorporated to permit off-line repair/service while the process is diverted?

From a hazardous-energy-control perspective, all data relating to incident experience, repair/service problems, system isolation deficiencies, access limitations, and so on should be collected and synthesized to create the criteria for shaping the machine, equipment, or process design. Attempting to correct lockout/tagout limitations by retro-fit/modification during the operational phase is more expensive and sometimes not reasonable or possible.

SUMMARY

- Hazardous energy control is a system consisting of six major elements: (1) organizational culture; (2) management process/procedure; (3) human; (4) the environment; (5) the machine, equipment, or process; and (6) design.
- The system is graphically depicted as a wheel with organizational culture surrounding the five segments of the circle. This is intended to convey the pervasiveness of the existing culture(s) and its influence on all aspects of hazardous energy control. *Organizational culture* is defined as the members' and employees' shared beliefs, values, expectations, norms, attitudes, goals, and ideologies.
- Culture, although often deeply rooted, can be changed, but the process takes considerable time and effort. Actions that can contribute to cultural change might be: (1) changing managerial priorities, (2) changing methods of managing, (3) changing reward and punishment criteria, (4) changing rites and ceremonies, and (5) changing role models. Perhaps Dumaine (1990) says it best: "Action, action, action—live it, don't say it!"
- The remaining five major elements were divided into 40 subelements that identify the critical ingredients of a fully developed prevention system. Each organization, based on the nature of the business,

operational parameters, culture, prevention program maturity, and degree of sophistication, will need to determine which are the dominant pieces. Organizations that address most or all of the subelements will likely enjoy a high degree of hazardous-energy-control success.

REFERENCES

Barnett RL. The doctrine of manifest danger. *Safety Brief—Triodyne Inc.* 8, no. 1 (Sept. 1992):9.

Behnke AR. *Patty's Industrial Hygiene and Toxicology*, Chapter Nine, vol. 1. New York: John Wiley & Sons, Inc., 1978.

Buskik D. Training today's workforce for safe productivity. *Best's Safety Directory* (1991), pp. 198–201.

Comprehensive Occupational Safety & Health Reform Act (COSHRA) S.575 & H.R. 1280, Kennedy, Metzenbaum, & Ford.

Dumaine B. Creating a new company culture. *Fortune* (January 1990).

Federal Register, vol. 54, no. 169, The Control of Hazardous Energy (Lockout/Tagout) 29 *CFR* 1910.147.

Federal Register, vol. 54, no. 16. Safety and Health Program Management Guidelines. January 26, 1989.

Findlay JV, Kuhlman RL. *Leadership in Safety*. Loganville: Institute Press, 1980.

Fisher CD, Schoenfeldt LF, Shaw JB. *Human Resource Management*. Boston: Houghton Mifflin Company, 1990.

Hammer W. *Occupational Safety Management and Engineering*. Englewood Cliffs, NJ: Prentice Hall, 1976.

Hellriegel D, Slocum JW, Woodman RW. *Organizational Behavior*, 6th ed. St. Paul: West Publishing Company, 1992.

International Union, United Automobile, Aerospace and Agricultural Implement Workers of America. Health and Safety Department. *Occupational Fatalities Among UAW Members: A Fourteen Year Study.* Detroit, MI: UAW, Mar. 1987.

Johnson WG. *MORT Safety Assurance Systems*. New York: Marcel Dekker, Inc., 1980.

Kelly J. *Organizational Behavior*. Homewood, IL: Irwin-Dorsey, 1969.

Kroemer KHE. Ergonomics. In Plug B, *Fundamentals of Industrial Hygiene,* 3rd ed. Chicago: National Safety Council, 1988.

Major RF, Pipe P. *Analyzing Performance Problems*, 2nd ed. Belmont, CA: Lake Publishing Co., 1984.

National Aeronautics and Space Administration. System safety. *NASA Safety Manual* 3(1970).

National Safety Council, Safety and Health Management Services. Safety program evaluation. *Managing the Process* (1992).

Occupational Safety and Health Administration. Instruction Standard 1-7.3, *The Control of Hazardous Energy (Lockout/Tagout)—Inspection Procedures and Interpretive Guidelines,* August 11, 1990.

Occupational Safety and Health Administration. Preamble and Final Rule for *Electrical Safety-Related Work Practices* [55 *FR* 31984, August 6, 1990].

Rummler GA, Brache AP. *Improving Performance—How to Manage the White Space on the Organization Chart*. San Francisco: Jossey-Bass Publishers, 1991.

Schein EH. *Organizational Culture and Leadership*. San Francisco: Jossey-Bass, 1985.

Widner JT. Selected readings in safety. In Lambie HK, *Accident Control Through Motivation.* Macon, GA: Academy Press, 1973.

Preventing Energy Transfer

Management and employee success is measured in terms of its consistency in preventing the unexpected transfer of energy. All of the actions and elements discussed in Chapters 6, The Process Approach, and 7, System Elements, collectively should contribute to the successful control of hazardous energy. However, great preparation does not necessarily translate into great execution!

The major components of any hazardous-energy-control system are as follows without regard to order or priority:

- planning
- system elements
- employee involvement
- training
- execution/application
- designing for safety.

The first four components were addressed in earlier chapters. The focus of this chapter is the execution or application of isolating techniques and practices to prevent unexpected energy transfer. Designing for safety is covered in greatest detail in Chapter 11, Beyond Compliance, where it is offered as the most promising initiative for significant progress in hazardous energy control.

THE CONUNDRUM

Too often serious incidents occur where employers have in place apparently fully developed hazardous-energy-control systems. The weak point in these systems frequently relates to what did or did not happen at the point of execution. Invariably the warning signs were apparent or readily detectable. In a quality process analogy much energy can be spent on exhortation, forming of teams, statistical process control, specifications, requirements, etc., but if no one examines the goods or services (product) or receives feedback from the customer about them, little success can be expected. Similarly, in hazardous energy control, managers and employees will need to examine organizational energy-isolation execution (product) if they want to achieve any degree of success.

The following case histories illustrate the need for increased managerial emphasis on the product (energy-isolation execution).

Failure to Isolate

Two senior maintenance employees were assigned a repetitive task (several times per shift) of making a metal strip feed die check on a high-speed automated power press. The main electrical power supply plug-type disconnects and die blocks (two) were within 10 ft (3 m) of the press. Fixed barrier guarding was removed, and the checks and adjustments were made with one mechanic working on each side of the press. They communicated through the die area and at the side of the press. One mechanic started the press momentarily to check the output, believing his co-worker was clear. The dies caught part of both hands of the other mechanic as they came together. At the last instant, he had decided to remove a small piece of scrap simultaneous with his partner's cycling the equipment. Both individuals had been involved with specific safety procedure training 30 days before the incident, and neither had made any use of the prescribed power-isolation disconnects or die blocks.

Failure to Restrain

A mechanic and his supervisor were evaluating the scope of repairs on a scrap shear. Examination of the conveyor drive disclosed that the bearing cap bolts for the south drive shaft were partially missing or damaged. They decided to tighten nuts on the new stud and get other crew members to replace the balance of the missing or damaged studs. After de-energizing the electrical main and control switches to the shear and its auxiliaries, the mechanic crawled under the conveyor. He and the supervisor proceeded to tighten the nuts on a stud with an air impact wrench. Almost immediately, the south counterweight rotated downward, striking the mechanic in the chest and pinning him under the table. He sustained crushing fatal injuries. No provision had been made to prevent this movement.

Failure to Notify

Two maintenance employees were installing space heater vent pipes in a shop area. They were using the overhead traveling crane runway to position a section of vent pipe when the overhead crane entered their work area while attempting to make a lift for another shop employee at floor level. As the crane passed unexpectedly, it crushed the pipefitter between the crane end trucks and a building column. The crane operator had not been notified, safety bumpers were not in place, and the crane had not been locked out or a watchman stationed to protect the exposed personnel.

In these cases and countless others, experienced personnel were tragically injured or killed because of failure to properly execute or apply lockout/tagout. A "conundrum" is defined as an intricate and difficult problem. The circumstances surrounding the aforementioned incidents are not that intricate or difficult; however, the real conundrum is why managers and workers appear to be unable to develop the right formula or motivation to prevent these unwanted energy transfers. Unfortunately, the solution to this intricate and difficult problem has not been fully appreciated by most organizations. It will take superior development of the actions outlined in Chapters 6, The Process Approach, and 7, System Elements, to accomplish what is so desired. A total transformation in employee and management mind-set will be necessary. To do it right takes a continuous commitment and process.

ENERGY ISOLATION

Numerous terms are in common use to define the act of isolating various forms of energy. Many are the offspring of specific equipment applications, and others are related to a particular industry, for example "cutting out," "tagging out," "block out," "lockout," "holding off," "blanking," "blinding," "double block and bleed," "lockoffs," "lock, tag, and try," "safety clearance," and "hold-up." Although the terms vary, the intent is the same: the isolation of energy to prevent any transfer that might negatively affect personnel or property.

The specific techniques and applications for energy isolation are countless, but all incorporate basic principles such as positive separation of energy from personnel, use of reliable energy-isolating devices, individual protection of personnel, and security of isolating devices. Although the procedural controls for energy isolation appear to be infinite when specific applications are considered, common denominators exist. The scope of these energy control common denominators and their relationship to each other provide the framework for preventing energy transfer.

ACTION CYCLE: ENERGY ISOLATION

Frequently energy-isolation methodology is described in terms of steps or sequential actions that employees follow to safely accomplish the maintenance and servicing of equipment, machinery, or process. Figures 8-1 through 8-5 illustrate the steps or actions recommended in several U.S. and European standards or regulations. The nature of the work usually determines the steps and possibly to some degree their order. In certain industries, such as electrical generation and distribution and chemical processing, sequences and their related networks are often more critical from an order of occurrence standpoint. Com-

CFR Energy Isolation Procedure

1. Preparation for shutdown
2. Shutdown
3. Isolation
4. Lockout/tagout device application
5. Stored energy relief, etc.
6. Verification of isolation
7. Perform work
8. Inspect work area/equipment
9. Clearance and notification
10. Lockout/tagout device removal
11. Re-energize
12. Restart

Figure 8-1. Energy isolation: application of control.
Source: 29 *CFR* 1910.147, Lockout/Tagout, Sections (d) and (e).

ANSI Lockout Procedure

1. Notification
2. Shutdown
3. Isolate energy sources
4. Relieve/restrain stored energy
5. Lockout energy-isolating devices
6. Verification and clearance of personnel
7. Perform work
8. Inspect/check area
9. Remove locks/re-energize
10. Restart

Figure 8-2. Sequence of lockout procedure.
Source: This material is reproduced with permission from American National Standard *Safety Requirements for Lockout/Tagout of Energy Sources,* ANSI Z244.1, copyright 1982 by the American National Standards Institute. Copies of this standard may be purchased from ANSI, 11 W. 2nd St., New York, NY 10036.

plexity, scope of work, numbers of personnel, magnitude of equipment, machinery, process, degree of risk, etc. all influence the action cycle and its progression toward energy isolation.

Consignations et Déconsignations (Closure and Unclosure)

1. Separation
   - Shutdown
   - Isolation
2. Locking
   - Posting (warning)
3. Purging
   - Relieve, ventilate, restrain
4. Verification
   - Testing
   - Identification (diagrams, component, markings, etc.)

Figure 8-3. Closure and unclosure.
Source: Closure and Unclosure, ED 754, French National Research and Safety Institute (INRS), Paris, France, 1992.

Safe System of Work

1. Assess the task
   - What, who, where, how
2. Identify the hazards
3. Define safe methods
   - Preparation/authorization
   - Planning of job sequences
   - Specify work methods
   - Means of access/escape
   - Dismantling, disposal, etc.
   - Permit to work procedures
4. Implementing the system
   - Communicated properly
5. Monitoring the system
   - System workable
   - Procedures are followed and effective
   - Changes are adjusted for

Figure 8-4. Safe system of work.
Source: British Health and Safety at Work etc. Act 1974. British Crown copyright. Reproduced with the permission of the Controller of Her Majesty's Stationery Office.

A format for a nominal action cycle for energy isolation is presented in Figure 8-6. The graphic beginning at the top reflects a given operational status or homeostasis for equipment, machinery, or process. The progression flows clockwise to where the operation is not conforming to requirements. A decision is then made where intervention action initiation occurs that may involve inspection, adjustment, cleaning, repair, and lubrication. The intervenor(s) proceeds through the 12-step action cycle until intervention action completion is achieved. This intervention has

Energy Control and Power Lockout: Major Elements

1. Believe it!
2. Check it!
3. Prep it!
4. Lock it!
5. Release it!
6. Verify it
7. Use it!

NOTE: Each element is expanded upon in detail to provide a comprehensive approach to energy control and power lockout.

Figure 8-5. Major elements of energy control and power lockout.
Source: United Automobile Workers–Ford National Joint Committee in Health and Safety, Detroit: *Energy Control and Power Lockout Mini-Manual,* 1987.

ideally restored the operation to where it is conforming to requirements and therefore the equipment, machinery, or process is now in a normal or stable state. Minor or basic energy-isolation tasks may require less than the 12 discrete steps illustrated; energy-isolation work with moderate complexity and risk could easily follow the clock pattern; high-risk tasks with complex isolation demands may require a few additional steps or substeps to provide the necessary level of worker protection. The action cycle offered does strongly suggest that developers of hazardous-energy-control systems need to consider a comprehensive framework and logical progression for application or execution.

The 12-step process or action cycle is described to acquaint the user with the nature and activity of each step involved in achieving energy isolation. The complexity of the isolation activity will dictate whether actions beyond those detailed will be necessary. The steps reveal the intended general approach to be adjusted by the user to address the process of isolation. The examples are intended to assist understanding but are not intended to cover every situation or requirement.

1. *Notifying*. When the decision is made to intervene with respect to the machinery, equipment, or process (MEP), various notification actions should be triggered. The necessity for intervention may range from the need for simple adjustment to a more complex reactive response to an emergency or breakdown. The nature of the intervention and whether personnel will be potentially exposed to energy transfer will determine if energy isolation is necessary. If isolation of energy is appropriate for the required task(s), a chain of communication (notifying) action should occur. The following actions may be necessary:
   a. If the intervention is planned/scheduled, the individuals responsible for the MEP will be notified by external personnel or receive writ-

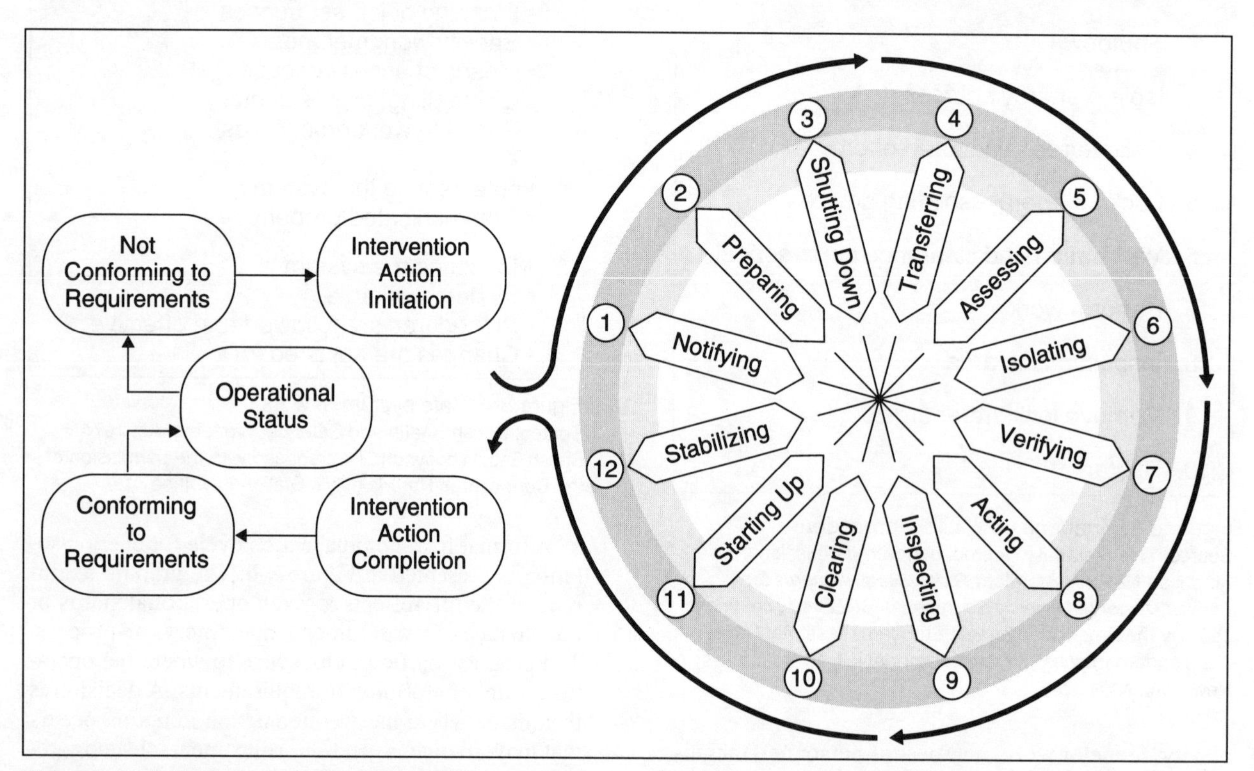

Figure 8-6. Action cycle of energy isolation.
Source: Developed by E. Grund. © National Safety Council.

ten direction. In either case, those responsible for the MEP must agree to the release of the MEP to those who will restore the unit to where it is conforming to requirements.

b. If the intervention is an unexpected/unplanned event, those responsible for the MEP will either initiate their own corrective action or notify appropriate personnel for the type of problem anticipated, e.g., mechanical, electrical, or systems.
c. The individuals responsible for the MEP or the external parties should also notify those upstream and downstream of the operation if there will be any impact due to the isolation of energy or other reasons.
d. Other personnel such as assigned MEP personnel indirectly affected, production control personnel, and maintenance schedulers, may also need to be notified.
e. Personnel directly involved (both production and maintenance) in the repair, adjustment, servicing, etc., will need to be assigned and contacted regarding the nature of the work.
f. Notification activity may involve face-to-face, telephone, electronic media, or page system methods and be documented using sign off, release, or permit forms of written/electronic communication.
g. If outside contractors are involved, their entry and integration at the worksite may be controlled. In many instances, purchasing, engineering, or maintenance groups may be involved in their contact, assignment, and coordination.

The objective of the notification step is to ensure that all affected parties will fully understand the nature and scope of the operational intervention and respond in such a way as to ensure the safety of everyone involved. The issue of timeliness is implied so that information has been received by the right individuals in time for them to respond appropriately with regard to employee safety and MEP protection.

2. *Preparing*. The preparation stage for energy isolation involves organizing for the necessary work before its commencement. Distinct activities are often required by various groups, such as production, maintenance, and engineering. After proper notification, all involved or affected parties should be aware of the intervention action and their respective roles relative to its completion.
   a. *Operating personnel*. Those responsible for the MEP will bring the operation to the right status in preparation for transfer if external personnel are involved. If external personnel are not involved, this activity must still occur to the degree that is necessary to accomplish the intervention. However, the transfer step will be unnecessary. The following actions may be necessary:
      (1) Complete the product run; empty reservoirs, hoppers, feeders, etc.; remove jigs/fixtures; discharge in-process material, etc.
      (2) Cleanup of scrap materials/product without employee exposure; relocate tools, objects, etc. that may impede activity/access.
      (3) Discuss sequences/special requirements with co-workers or supervisors.
   b. *Maintenance personnel*. After notification, external personnel (nonproduction) will begin to prepare for the nature of the required intervention. They will already be aware of the general problem and what type of response might be needed. Precisely what the cause of the problem is and what will be needed to rectify the situation may be unknown. The following actions may occur to get ready for the anticipated work.
      (1) Contact with operational personnel to discuss the technical issues related to the repair or service.
      (2) Assembly of tools, equipment, and materials required for the anticipated work.
      (3) Gathering of energy-isolation protective appliances and auxiliary hardware.
   c. *Other personnel*. Depending on the nature of the work, other personnel (engineers, automated systems specialists, quality control technicians, inspectors, etc.) may have been notified and will need to prepare for their specific role in problem identification and resolution. They may be acting in a distant consultative role or eventually need personal protection if they are exposed while performing their duties.
3. *Shutting down*. Typically the MEP is shut down by operational personnel. If external personnel are authorized to shut down the operation or collaborate with operators, the required actions remain the same. Shutting down actions are:
   a. Notify the involved personnel and others impacted by the MEP shutdown that it is going to occur (when); alarms or signals often are activated as an additional warning, usually with a delay, before actual stoppage is experienced.
   b. In simple MEP situations involving small pieces of machinery/equipment and limited personnel, step a. may be unnecessary.
   c. MEP controls will then be cycled to the OFF or NEUTRAL position after the appropriate operational sequence.
   d. Operators will determine that the MEP has responded to the control commands and that all components of the equipment (e.g., fly wheels) have come to a stop or rest position.

e. Operators may remove process materials/product where personal exposure is not necessary; other controls may be cycled to discharge, transfer, or release system objects/materials that must be removed before the intervention tasks can take place. These controls will then be shut off or neutralized.
f. Operators will then notify the responding personnel or be prepared themselves to proceed to the next action step; when external personnel are involved, a well-planned method should be used to declare the MEP is shut down and that the responding personnel formally acknowledge this fact.

4. *Transferring*. Transfer of responsibility will not be required if the operational personnel will be accomplishing the intervention (inspection, adjustment, servicing). If external personnel will be used, the transfer authorizing their presence and action is necessary. These actions are often part of this process:
   a. Communication. Verbal or written authorization to take charge of the MEP and accomplish the intended work; dialogue regarding the current status of the MEP.
   b. Separation. Movement of personnel into and out of the work area; various arrangements may be possible where operational personnel stand by and assist, leave the immediate area, or are reassigned.
   c. Authority. Some circumstances may require a transfer of supervisory authority where external supervision may direct operational personnel if they are required to assist in the maintenance or servicing.
5. *Assessing*. After transfer, assuming this step was necessary, the incoming personnel should make an appraisal of the status of the MEP situation. This stage provides time for a number of activities that will ensure the smooth progress of the necessary work. The following actions are representative of what typically might be addressed:
   a. The supervisor(s) in charge of the repair/service may assign personnel to various tasks and review the specific isolation procedure for the MEP.
   b. A survey of the worksite can be conducted to determine if any obstructions, gear, or materials will impede the defined work.
   c. A determination can be made regarding the need for operational assistance and to what degree and purpose.
   d. A quick survey can be made to determine if all energy-isolating devices can be located and are accessible and if protective appliances beyond those anticipated will be necessary.
   e. If "crew" lockout/tagout procedures will be necessary, the assessment phase provides opportunity to review the process as assignments are made.
   f. Finally, supervision may determine if everything is in order and place before commencing the isolation phase.
6. *Isolating*. Personnel, after determining the MEP is actually shut down, will begin the process of actuating the energy-isolating devices to reduce the worksite to ideally a zero energy state. There may be a defined sequence for this to occur with various MEP situations (e.g., electrical isolation before pneumatic/hydraulic or main drives before auxiliary components). The following are examples of the work that can be involved:
   a. Properly *identify* the specific energy-isolating devices involved in the de-energization process; matching identification numbers, reading tags and labels, checking color coding, and tracing lines may be involved.
   b. Actuate intermediate/primary electrical disconnects and breakers and apply lockout devices and/or appropriate tags; determine if the energy-isolating device is actually in the open or neutral position whenever possible.
   c. Close pneumatic/hydraulic/steam/chemical valves and apply protective appliances, locks, and tags to secure them and prohibit cycling.
   d. Insert blanks/blinds or disconnect piping at flanges and tag to prohibit removal or reconnection.
   e. Install or connect supports/restraints to prevent gravitational movement and secure with locks/tags.
   f. Install safety blocks, pins, etc. as protection from mechanical movements.
   g. Relieve pressurized systems/stored electrical energy from MEP.
   h. Drain liquids (not under pressure) both hazardous and nonhazardous from systems, ensuring their safe disposition.
   i. Wait for thermal energy to be reduced or provide shielding/personal protective gear.
   j. Disconnect mechanical linkages (chains, shafts, connecting rods, belts) that may transmit power and lock/tag.

Whether proceeding with isolation work under simple or complex (single machine/elaborate systems, one person/large groups) conditions, the concept of *personal protection* should be paramount, i.e., regardless of the system used, each individual worker (including supervision) potentially exposed must be secure from inadvertent energy transfer. Therefore, all energy types and levels present are controlled by various means to allow personnel to perform the necessary work with minimum risk from unexpected energy release.

7. *Verifying*. Verification that energy isolation has been achieved is often overlooked as a critical

step in the "action cycle." The assumption that isolation of energy has occurred because isolating devices have been actuated may place personnel in jeopardy. Testing is necessary to ensure that available energy has been properly controlled and that nothing such as mechanical/electrical defects, improper identification/positioning, or oversight has compromised the isolation objectives. Activities may involve:

   a. Actuating various control devices that direct movement, flow, etc. and monitoring any reaction/response.
   b. Examining that restraints/blocks are extended and/or in contact with surfaces to prevent slippage or ejection under pressure/movement.
   c. Using test equipment (e.g., volt meters) to determine if currents have been grounded/discharged and zero potential exists.
   d. Manually actuating overrides on valves to relieve internal pressures or in some cases opening lines at fittings if not done in previous isolation actions.

Verification action checklists can be developed and used to guide the process of assuring that the steps taken actually achieve energy isolation. Personnel can methodically move sequentially through the list(s), making sure that all planned verifying practices/actions have been taken. This technique becomes more important as the size of the job and general complexity increases.

8. *Acting (maintaining/servicing).* When supervision or personnel with appropriate authority determine that the MEP has been properly isolated, work may commence. Depending on circumstances, the intervening parties may or may not know exactly what will have to be done. Inspection, disassembly, testing, etc. may be necessary. The work may proceed sequentially or simultaneously, with coordination coming from lead personnel or supervision. Maintenance or servicing actions may involve operations and/or support personnel and include the following:
   a. Review of drawings, manuals, and plans to analyze/guide the work.
   b. Examination of product to ascertain what may need repair/adjustment.
   c. Removal or opening of guards, covers, and housings to access the internal components.
   d. Taking the actions to return the MEP to the operational state where it is conforming to requirements.
   e. Reassembling or restoring the MEP to an operating status (including refill of accumulators or reservoirs if drained).
   f. Removal of tools, equipment, and replaced components from all work areas.
   g. Cleaning the work area to return it to a normal operational condition.
9. *Inspecting.* The task of post–job inspection is identified as a stand-alone step of the total action cycle. Often it is treated as a final act of the maintenance/servicing step. However, when treated in this fashion, it can be overlooked in the press to quickly wrap up and return the MEP to operational status. Scrap, welding rod, or bolts/nuts can be left in machinery, causing costly damage and downtime during restart. In other instances, objects have been violently ejected from high-speed machinery or hazardous substances released from loose connections, injuring personnel. Typical inspection activities might include:
   a. Organizing the inspection so that all appropriate sections of the MEP or areas involved in the repair/service/maintenance are covered.
   b. In vertical situations (e.g., repair/service above grade, towers, balconies, machines with elevated access platforms) checking to see nothing remains behind from the intervention work; if miscellaneous items are found unrelated to the repair but can pose a danger to personnel or process, they should be reported formally to operational supervision.
   c. Checking all work areas for any unnecessary items related to the repair/servicing and remove all that are found.
   d. Determining that the MEP is ready for the clearing step to begin.
10. *Clearing.* Clearing describes those actions that relate to returning energy-isolating devices to their normal operating position and transitioning the MEP to operational readiness. Individual personnel remove personal locks/tags from isolating devices or crew lock boxes where keys may be stored for "company/supervisory" locks. There may be a predetermined sequence for returning the isolating devices to their normal position (often the reverse order of the isolation sequence). These actions may be undertaken:
    a. Remove locks/tags and return all isolating devices to the run or operational mode.
    b. Remove blinds, blanks, restraints, and blocks and reconnect any piping or mechanical linkages.
    c. Use specific energy-isolation checklists to systematically note the proper repositioning, removal, or reconnection of devices to reduce the chance of omission.
11. *Starting up.* Operational and/or intervening personnel, after ascertaining readiness for restart, may commence powering up the MEP. This step can involve time for pressures and temperature to rise to operating levels, accumulators/reservoirs to be pressurized, and raw materials to be charged

or flow. It is desirable to determine if the intervention action has corrected the diagnosed problem as soon as possible during the restart process, i.e., power-up to the point that this can be done without completely reaching total operational status. If the intervention or fix was successful, the balance of restart actions may be taken. If not, the process of isolation may need to occur again. However, in many cases total operational status may need to be achieved and product produced or output experienced to determine if a successful resolution occurred. The following actions could be taken:

   a. Actuate/cycle energy-controlling devices and/or place in state of readiness.
   b. Input appropriate computer commands to prestage MEP or establish appropriate operational conditions.
   c. Monitor control panels, indicators, displays to determine if required conditions exist to move to run/operate mode.
   d. Actuate controls to achieve run status and monitor/observe instrumentation or actual performance.

12. *Stabilizing*. During the initial stages of return to production, operational or intervening personnel must ensure that everything is acceptable and conforming to requirements. External work forces begin to prepare for withdrawal and movement to the next assignment. Required paperwork acceptance signatures may be finalized. These actions are often taken:
   a. Operational personnel will determine if the intervention has produced the desired result or if a tolerable fix occurred pending some more desirable future outcome.
   b. Maintenance forces/supervision will determine if the intervention outcome meets their expectations for task completion.
   c. Adjustments not requiring energy isolation may be made.
   d. Monitor the operation for a reasonable period to ensure that the correction appears to be permanent.
   e. Complete documentation, releases, and permit procedural steps.
   f. Remove barricades, warning devices, and special control equipment from work area.
   g. Evacuate area and remove all gear brought to the site; arrange for appropriate assistance for repairable machinery/equipment, scrap that may require mechanized removal from area.

These 12 steps describe what might be viewed as an average conceptual approach to energy isolation. The following breakdown reflects the energy status during the various progressive steps in this cycle:

Energized: Steps 1 through 3
Transition: Steps 4 through 7
De-energized: Steps 8 through 10
Energized: Steps 11 and 12

It is important that personnel interaction and control be properly managed so that no one is exposed by prematurely or delinquently performing assigned corrective work, i.e., working outside of steps 8 through 10. The circumstances involving contractors, permit systems, and group lockout/tagout will be covered in detail in Chapter 10, Special Situations and Applications.

A functional flow diagram for implementation of lockout/tagout requirements is provided in Appendix B of OSHA Instruction Standard 1–7.3 (Figure 8-7). It provides general guidance regarding the nature and sequence of steps in energy-isolation procedures. Each application situation may require something somewhat different than the diagram presented.

The GM-UAW National Joint Health and Safety Committee's training publication *Lockout* provides an excellent flowchart for visualizing the lockout process. Rectangles indicate a step or action and diamonds a point at which a decision must be made. The steps in the flowchart reveal the thought pattern for the general process (Figure 8-8).

PERSONAL PROTECTION

An underlying principle in the application of energy-isolation procedures is the requirement for the protection of *each* exposed individual. This is accomplished in the simplest isolation situation by the exposed worker applying his/her lock and/or tag to a single energy-isolating device (e.g., a milling machine with a wall-type 220-V AC disconnect). A question of principle arises as the complexity of energy isolation increases. The credo "one person–one lock" evolved in industry to support the individual's right and obligation for self-protection from energy transfer. Is it workable under all circumstances? If viewed literally, "one person–one lock" may seem to suggest that each individual will lock out each energy-isolating device involved in the machinery, equipment, and process situation with a number of personal padlocks. This is certainly acceptable and endorsed in simple energy-isolation applications but totally unworkable and impractical under more complex circumstances, e.g., a continuous rolling mill or chemical process with over 100 energy-isolating points.

Managers and workers need to jointly determine when isolation practices transition from the application of personal locks/tags on every involved energy-isolating device to "control or supervisory" locks applied under a different method. However, no compromise in safety or breech of principle or credo need occur if control systems are designed so that employees are effectively protected regardless of the complexity of energy isolation.

As the complexity and numbers of exposed personnel increase, do exposed employees have positive

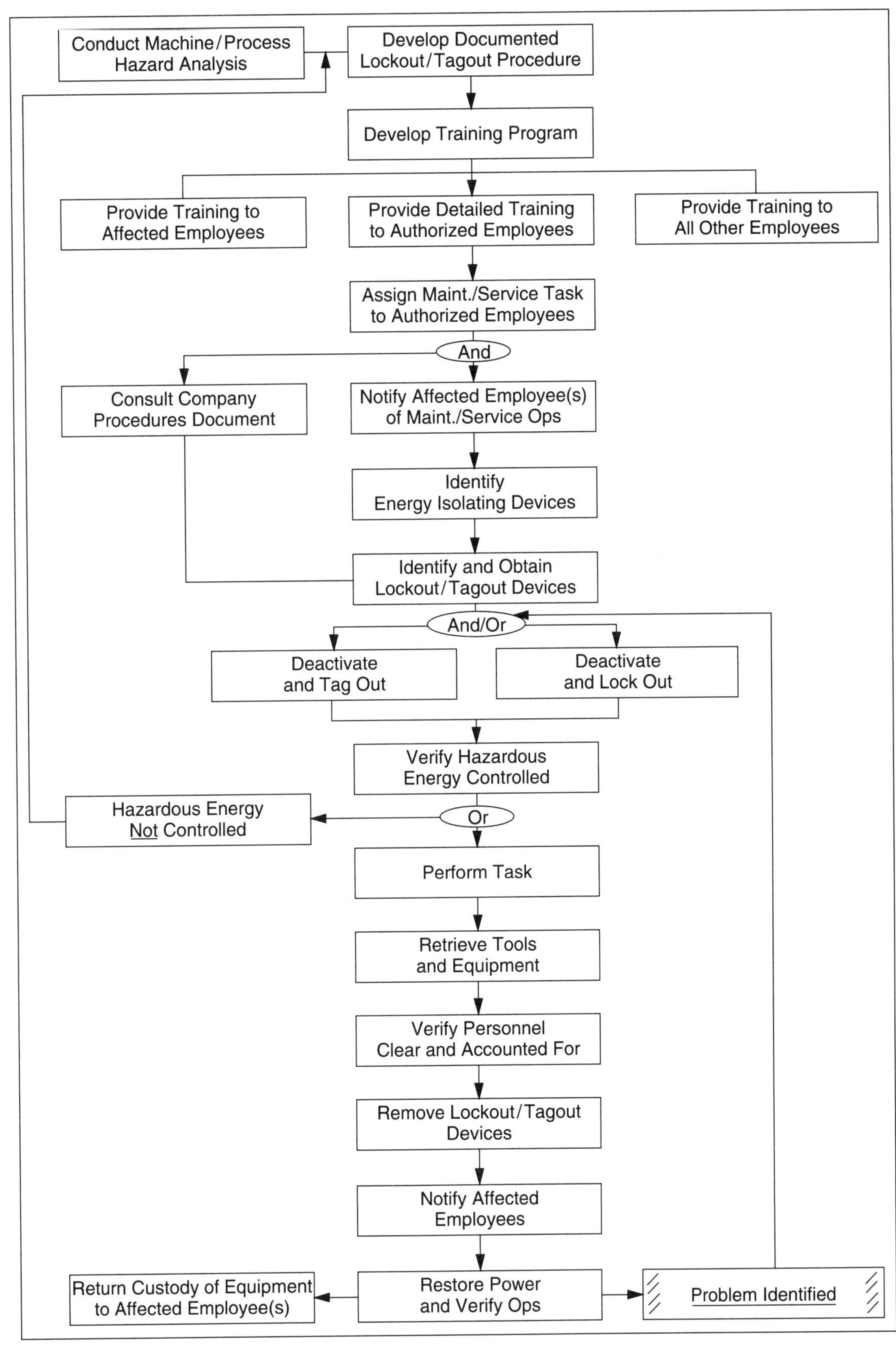

Figure 8-7. OSHA functional flowchart for implementation of lockout/tagout requirements.
Source: OSHA Instruction Standard 1-7.3, Appendix B.

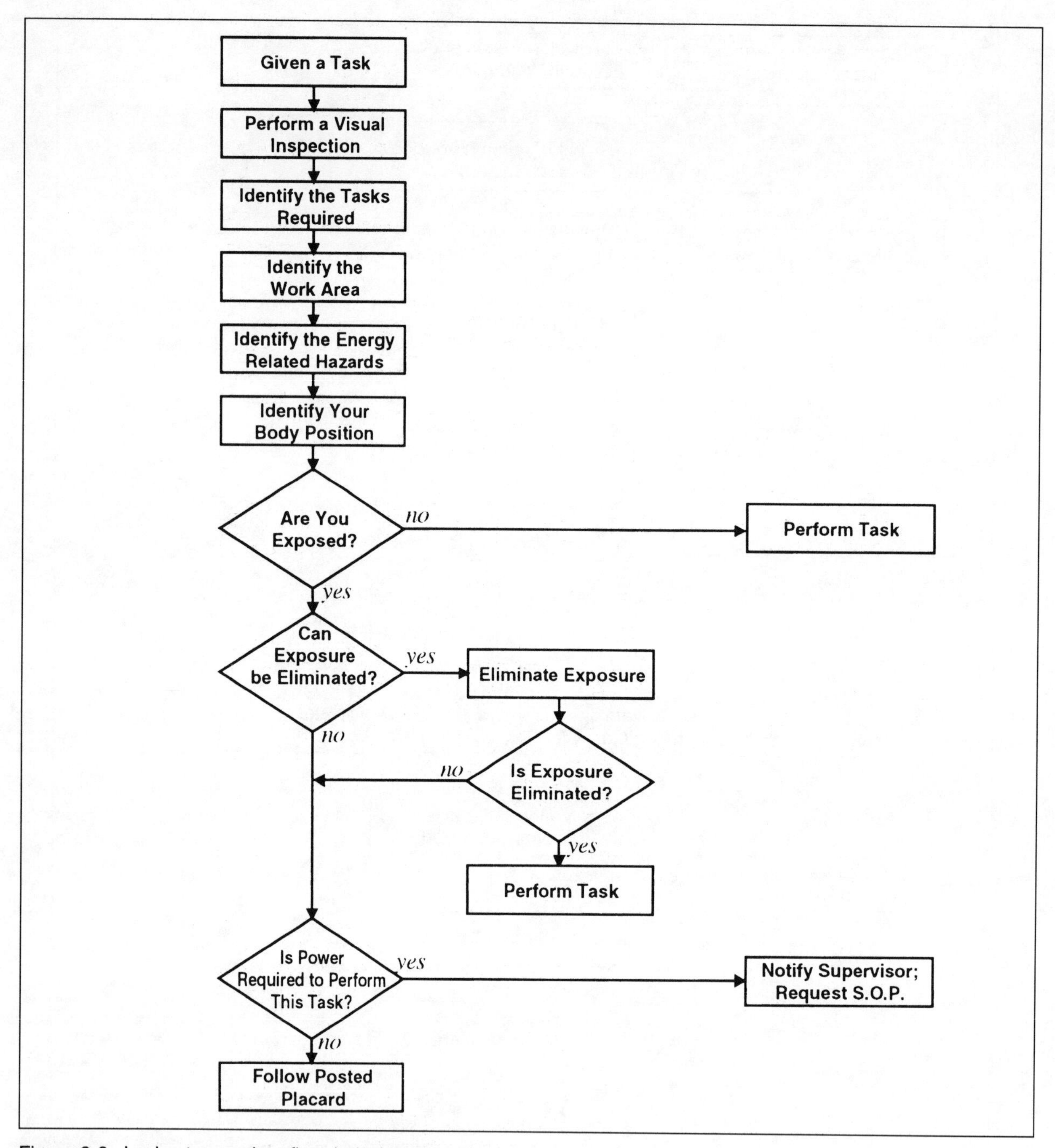

Figure 8-8. Lockout procedure flowchart. (*continued*)
Source: United Automobile Workers–General Motors Corporation. *Lockout Energy Control* (Training Workbook for Authorized Employees). National Joint Committee on Health and Safety, 1994.

control over the keys that operate all locks that were used in the isolation process? Further, an additional supporting axiom that should be a part of any energy-isolating system is that "only the employee who applies their personal lock/tag shall remove it." Ideally this will mean no second set of keys, mastering of keys, loaning of keys, etc. Removal of locks by someone other than the applier under special circumstances is discussed in detail in Chapter 10, Special Situations and Applications. Under systems that function exclusively as "tagout" or where blinds/blanks are used, control methodology is somewhat different and not viewed as secure as the lock/key approach. Each operational setting will likely require an application practice that addresses the unique characteristics of the machinery, equipment, or process and best serves the protective interests of those involved.

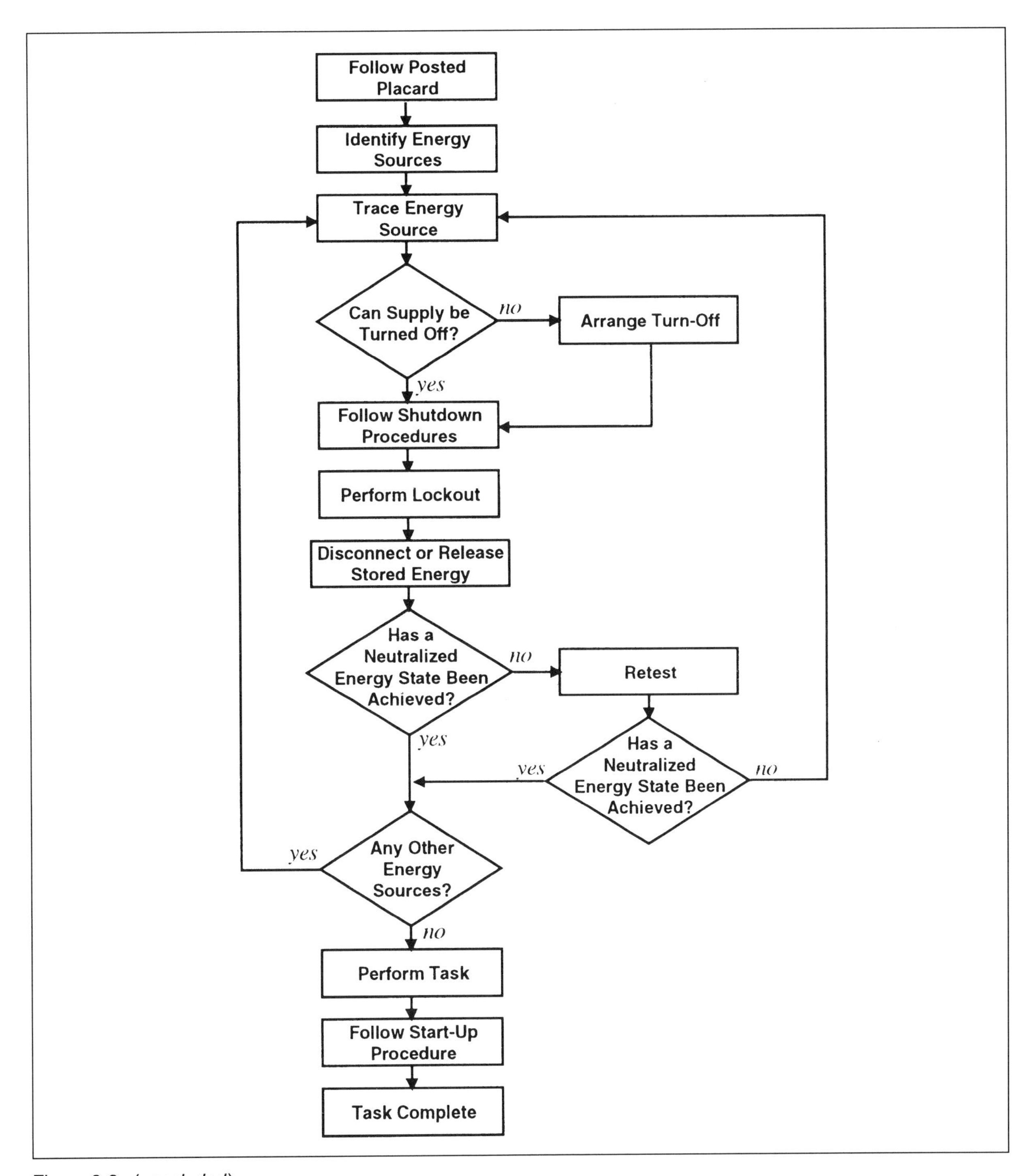

Figure 8-8. (*concluded*)

POSITIONING AND TESTING (MEP)

Frequently there is a need to position, cycle, or test the MEP during maintenance and servicing activities under lockout/tagout conditions. Extreme care is required to re-energize the MEP temporarily while such positioning and testing takes place. In some instances total re-energization is necessary, and in others only partial restoration may be required. In either case, a specified procedure is necessary to ensure that personnel are clear, the sequence is followed, and adequate safeguards are being used. OSHA generally defines the substance of this special circumstance and the sequence of steps to follow in Section (f)(1) of the 29 *CFR* 1910.147 standard:

(1) Clear the machines or equipment of tools and materials.
(2) Remove employees from the machines or equipment area.
(3) Remove the lockout or tagout devices as specified in the standard.
(4) Energize and proceed with testing or positioning.
(5) De-energize all systems, isolate the machine or equipment from the energy source, and reapply lockout or tagout devices as specified.

Two procedural steps/actions are critical during the process of temporary re-energization: (1) accounting for personnel and (2) control of personnel activity. Various methods such as check sheets, physical head counts, and signature rosters are used (1) to determine that all exposed personnel are accounted for; (2) to understand the basis for re-energization; (3) to determine its likely duration; (4) to determine what, if any, duties they may perform, and/or (5) to determine where restrictions may apply. If temporary re-energization is necessary a number of times while work progresses, supervisory vigilance should increase to ensure that oversight or shortcuts do not occur.

SHIFT CHANGE AND WORK INTERRUPTION

The installation, repair, and servicing work often continue for extended periods, i.e., beyond the duration of a typical work shift. In other cases, servicing and maintenance tasks by circumstance occur near the end of one shift and carry over into the next. This "window of transition" by its very nature increases the risks attendant to the isolation and control of hazardous energy. Also, the potential for communication, coordination, and continuity oversights increases as the number of involved personnel grows. Each individual represents a complex variable that is susceptible to failure.

OSHA in 29 *CFR* 1910.147(f)(4) comments on this recognized danger as follows:

> Shift or Personnel Changes. Specific procedures shall be utilized during shift or personnel changes to ensure the continuity of lockout or tagout protection, including provision for the orderly transfer of lockout or tagout devices between off-going and on-coming employees, to minimize exposure to hazards from the unexpected energization, start-up of the machine or equipment, or release of stored energy.

OSHA's reference to "specific procedures and orderly transfer" signals the employer to provide a well-planned approach to the high-risk changing of the lockout/tagout guard. Several of the more generic bridging techniques are presented to illustrate how this orderly transfer might be accomplished:

- *Handshake.* Off-going personnel are not permitted to leave the job site until the on-coming personnel attach their locks to the in-use appliances (e.g., adapters) before off-going personnel remove their locks/tags. Verbal and/or written information may be exchanged relative to the work in progress and the energy-isolation status; provisions exist for situations where workers are either late or absent.
- *Craft control.* Off-going craft personnel (pipefitters, electricians, welders) remove their lockout devices after craft supervision has installed a distinctive control lock to bridge the time period before the on-coming craft personnel install their personal lock/tag devices; in some situations, the craft control lock may be left in place if there is no requirement for that skill on the next subsequent shift.
- *Custodian.* Each machine, equipment, or process has an assigned custodian (process control operator, etc.) who has total responsibility for the area/operation; the custodian applies "special design" padlocks, tags, or control devices for the duration of the job; service/maintenance personnel attach their personal locks/tags on the energy-isolating devices or lockboxes and clear them each time a shift change occurs; the custodian is responsible for seeing that a zero energy state is maintained during the work period and transition times.
- *Log and lock.* A hybrid technique that combines craft, custodian, or supervisory methods with the requirement for personnel to log in and out on a master job control register; the logging process aids accounting for personnel and reinforces the orderliness of the transition period.

Interruption in the continuity of service/repair work occurs frequently under actual business conditions. Under these circumstances the activity may be stopped and the personnel withdrawn without completion of the intended objectives. These interruptions could be caused by the lack of necessary repair parts, absence of necessary skills/personnel, need to fabricate replacement components, delivery delays, time needed for analysis/diagnosis, changes in customer requirements/orders, and emergency breakdowns in other areas of greater criticality.

Supervision will need to have an established practice or method for dealing with these occurrences. The paramount concern will involve in what state can the machine, equipment, or process be left and under what form of control or security? Although the personnel who were working on the job have been withdrawn and are no longer exposed, certain hazards need to be addressed:

- Do exposed parts represent a danger?
- Can someone attempt to start the equipment without knowing its state?
- How will all those affected be warned?
- Have conditions been generated that represent new risks for the area workers unrelated to energy isolation?
- Do any of the isolation measures affect peripheral equipment that needs to be remedied?

In most cases, supervision will need to lock/tag the energy-isolating devices with distinctive out of service hardware that uniquely warns of the conditions. Comprehensive communication (ongoing) must be initiated to ensure that all potentially affected personnel are properly apprised of the conditions and duration of the interruption. Periodic planned inspection should occur to verify that no deterioration or changes have occurred to the "out of service" machines, equipment, or process. The longer the interruption extends, the greater the need for periodic inspection.

HUMAN FACTORS

In Chapters 3, Causation Analysis, and 7, System Elements, the human component of the energy-control system was addressed with regard to incident causation and influence on error-free performance. The importance of a properly trained and skilled worker and his/her personal readiness to execute is worth reiterating. No amount of training or skill will be sufficient if the worker who approaches the energy-isolation task is impaired, stressed, preoccupied, or improperly motivated. Two examples illustrate the point:

1. An employee in a German metals fabricating plant had a child who had become very ill. The repairman knew the heavy press was down at the plant and that they needed his skills for the urgent repair of the machine. Although he was mentally torn, he proceeded to the plant to assist in the rush job. During the final stages of repair, he reached into the die area after having told his co-workers he was clear. Much of his right hand was mangled. Everyone, including the injured worker, believed the act was totally out of character and not readily explainable. At the time, no one knew of the repairman's crisis at home.
2. A maintenance mechanic was attempting to troubleshoot a large vacuum metal plate lifting device supported by a traveling overhead crane. He instructed the crane operator to position the lifting device and its load about 1 ft (0.305 m) off the floor. During his inspection and adjustment, he inadvertently released the vacuum on the lifting device, causing the plate to fall on his forward foot. The crushing injury required amputation of several toes. It later was determined that the employee had arrived late to work and had been reprimanded by his supervisor for repeated late reporting. It was also learned that he had been building a house and working 16 to 18 hours per day for several months trying to finish the shell before the hard winter weather arrived.

These cases and many others demonstrate that it is not enough to have good plans, procedures, hardware, and rules. The worker has to be prepared as well from both a knowledge and personal state. Fatigue, prescription and nonprescription drugs, stress, and poor health may place a worker in great danger with regard to energy-release potential. Psychological factors, such as fear, pride, bravado, ego, overconfidence, lack of concentration, and preoccupation may have a significant influence on error incidence during energy-isolation work.

During the process of troubleshooting, the service/maintenance employee has to be in control and not become frustrated or impatient to proceed before understanding the consequences. Observing while maintaining clearance and the proper safety margin must be ingrained in the thought patterns of repair and operating personnel. The desire for rapid problem diagnosis and resolution needs to be tempered with the mandate for safe pursuit of the work.

Anticipating and sizing up the situation before acting should be a reflexive response before any work has begun. Assessing the potential and actual flow/movement into, over, around, and out of the work space from fixed and mobile sources is a productive tactic that should not be overlooked. The personnel involved in the service/repair work should be encouraged to huddle for a short time to get everyone on the right wavelength before work commences. Contrary to some views, this brief pause actually saves time in most cases and, more importantly, underscores the requirement to proceed safely.

ELECTRICAL ENERGY

Because of its extensive use and the potential for electrocution from relatively small current flows, electricity was the focus of the early energy-isolation movement. The electrician, as a specialist, was a central figure in isolating activities that were intended to safeguard all personnel. The use of switches, breakers, and disconnects to isolate electrical energy evolved with the need to distribute and use this power effectively. The techniques for securing, safeguarding, and warning not to move these isolating devices developed as a response to energy-release incidents where workers were seriously harmed or fatally injured.

Far more energy-release incidents involve low voltages. This is likely a function of magnitude of use, numbers of exposed personnel, and degree and sophistication of training. Low voltage in the United States is normally considered 600 V or less. Additional information regarding high-voltage practices will be found in Chapter 10, Special Situations and Applications.

Safe Work Practices

In 29 *CFR* 1910.331–.335, Subpart S Electrical, *Electrical Safety-Related Work Practices* cover the requirements for both qualified (those who have training in avoiding the electrical hazards of working on or near exposed energized parts) and unqualified persons (those with little or no such training) working on, near, or with the following installations: (1) premises wiring, (2) wiring for connection to supply, (3) other wiring, and (4) optical fiber cable. The standards cover employee training, lockout/tagout procedures, use of nonconductive ladders, prohibition of conductive clothing and jewelry, proper illumination of work areas, good housekeeping, safe use of electrical test equipment, inspection of power tools, safe procedures for energizing/de-energizing circuits, use of safeguards for personal protection, and safe clearance distances.

In 29 *CFR* 1910.333 Selection and use of work practices (a) general, it states:

> Safety-related work practices shall be employed to prevent electric shock or other injuries resulting from either direct or indirect electrical contacts, when work is performed near or on equipment or circuits which are or may be energized. The specific safety-related work practices shall be consistent with the nature and extent of the associated electrical hazards.
>
> **(1) Deenergized parts.** Live parts to which an employee may be exposed shall be deenergized before the employee works on or near them, unless the employer can demonstrate that deenergizing introduces additional or increased hazards or is infeasible due to equipment design or operational limitations. Live parts that operate at less than 50 volts to ground need not be deenergized if there will be no increased exposure to electrical burns or to explosion due to electric arcs.
>
> **NOTE 1:** Examples of increased or additional hazards include interruption of life support equipment, deactivation of emergency alarm systems, shutdown of hazardous location ventilation equipment, or removal of illumination for an area.
>
> **NOTE 2:** Examples of work that may be performed on or near energized circuit parts because of infeasibility due to equipment design or operational limitations include testing of electric circuits that can only be performed with the circuit energized and work on circuits that form an integral part of a continuous industrial process in a chemical plant that would otherwise need to be completely shut down in order to permit work on one circuit or piece of equipment.
>
> **(2) Energized parts.** If the exposed live parts are not deenergized (i.e., for reasons of increased or additional hazards or infeasibility), other safety-related work practices shall be used to protect employees who may be exposed to the electrical hazards involved. Such work practices shall protect employees against contact with energized circuit parts directly with any part of their body or indirectly through some other conductive object. The work practices that are used shall be suitable for the conditions under which the work is to be performed and for the voltage level of the exposed electric conductors or circuit parts.

When working on electrical circuits or equipment, the electrician or repairman has to be prepared personally for the nature of the encountered risks. In NFPA 70E, *Standard for Electrical Safety Requirements for Employee Workplaces* (1988 edition), Chapter 1, Section B.(2), the following is provided:

> **(a) Alertness.** Employees shall be instructed to be alert at all times when they are working near exposed energized parts and in work situations where unexpected electrical hazards may exist.
>
> Employees shall be instructed not to reach blindly into areas which may contain energized parts.
>
> An OSHA fatality case history illustrates the importance of this requirement:
>
> A printing machine operator was having problems restarting the motor on a gilder machine. He opened the right door on the 480-volt gilder relay control panel. With the left door closed and the disconnect handle in the on position, he

tried to reset the relays. When he could not find the tripped relay on the right side, he reached behind the closed left door. He made contact with live parts energized at 480 volts and was electrocuted.

Employees shall not knowingly be permitted to work in areas containing exposed energized parts or other electrical hazards while their alertness is recognizably impaired due to illness, fatigue, or other reasons. . . .

(c) **Conductive apparel.** Conductive articles of jewelry and clothing, such as watch bands, bracelets, rings, key chains, necklaces, metalized aprons, cloth with conductive thread, or metal headgear shall not be worn where they present an electrical contact hazard with exposed energized parts unless such articles are rendered nonconductive by covering, wrapping, or other insulating means. [NOTE: Identical to OSHA 1910.333 (c)(8).]

In addition, 29 *CFR* 1910.335 *Safeguards for Personnel Protection* describes the use of personal protective equipment appropriate for the specific parts of the body to be protected and for the work to be performed. It states that employees shall wear nonconductive head protection and eye/face protective equipment when there are dangers from electrical contact, arcing, flashes, or explosions. It further mentions that alerting techniques, such as safety signs and tags, barricades, and attendants, should be used to warn and protect employees from electrical hazards.

Actuating Current-Interrupting Devices

Employees should be trained and conditioned to actuate electrical breakers, switches, and disconnects with the understanding that these devices can malfunction or fail violently. The following case illustrates this point:

An electrician was servicing a 440-V power supply for a furnace. After locating a blown 100-amp fuse in the electrical switch box, he removed the fuse and replaced it with a new one. He closed the electrical box with his left hand and pushed the lever to the ON position. A short in the circuit caused the fuse to arc and flash. The force of the arc blew the box open, exposing the electrician to the arc, causing burns to both hands and wrists and to the left side of his head, ear, and face. His eyes were not injured by the flash because he looked away when the circuit was energized.

Whenever possible, stand to the side of the device (torso not in line with front of device) when opening or closing the circuit. The handle or lever on disconnect type devices is normally found on the right side as you face the disconnect. Therefore, it is best to operate the lever with your left hand while turning your face away from the device. Some circumstances may not permit this tactic, so variations might be necessary that would reduce any exposure to the frontal outward release of energy (sparks, fire, projectiles). Remember that machinery and equipment should be shut down first so that the disconnect may be actuated without an electrical load on the circuit (exception: some devices are designed for opening circuits under load).

Protective Techniques (Tagout)

When tagout systems are used, one or more additional safety measures must be taken to provide appropriate protection for personnel. The additional measures used must either (1) ensure that the closing of the tagged single switch would not re-energize the circuit on which employees are working or (2) virtually prevent the accidental closing of the disconnecting means. OSHA's Preamble and Final Rule for *Electrical Safety-Related Work Practices* provides the following examples of protective techniques that are commonly used to supplement tags and protect workers:

- removal of a fuse or fuses for a circuit
- retraction of a draw-out circuit breaker from a switchboard (i.e., racking out the breaker)
- placement of a blocking mechanism over the operating handle of a disconnecting means so that the handle is blocked from being placed in the closed position
- opening of a switch (other than the disconnecting means) which also opens the circuit between the source of power and the exposed parts on which work is to be performed
- opening of a switch for a control circuit that operates a disconnect that is itself open and disconnected from the control circuit or is otherwise disabled
- grounding the circuit upon which work is to be performed (*FR*, 8/6/90).

The additional safety measure is necessary because, at least for electrical disconnecting means, tagging alone is significantly less safe than locking out. A disconnecting means could be closed by an employee who has failed to recognize the purpose of the tag. The disconnect could also be closed accidentally.

Fuse Removal

Fuses are not a substitute for an energy-isolating device. They may be used as a supplementary safeguard or where there are no other alternatives. Once removed, they are subject to replacement. However, use of a custom fuse block-out device will make this event very unlikely.

Remember to extract the fuse with an insulated fuse puller. If the fuse is not protected with a switch or disconnect, pull the supply end of the fuse out first. When replacing the fuse, put the supply end in first.

Hardware Issues

Due to the infinite variety of electrical hardware and the broad scope of application, the safety ramifications of each installation and the manner in which it will or may be used must be understood. Defective disconnects and switches, jumpered circuits, inaccurate prints and equipment labels, controls that can be bypassed through relay panels, and remote activation of circuits produce a complex array of variables that must be controlled effectively. The following items are relevant to improving energy isolation effectiveness:

Overhead and gantry cranes. In 29 *CFR* 1910.179 (g) Electric equipment (5) switches, the following requirements are found:

> (i) The power supply to the runway conductors shall be controlled by a switch or circuit breaker located on a fixed structure, accessible from the floor, and arranged to be locked in the open position.
> (ii) On cab-operated cranes a switch or circuit breaker of the enclosed type, with provision for locking in the open position, shall be provided in the leads from the runway conductors. A means of opening this switch or circuit breaker shall be located within easy reach of the operator.
> (iii) On floor-operated cranes, a switch or circuit breaker of the enclosed type, with provision for locking in the open position, shall be provided in the leads from the runway conductors. This disconnect shall be mounted on the bridge or footwalk near the runway collectors. One of the following types of floor-operated disconnects shall be provided:
> (a Nonconductive rope attached to the main disconnect switch.
> (b) An undervoltage trip for the main circuit breaker operated by an emergency stop button in the pendant pushbutton in the pendant pushbutton station.
> (c) A main line contactor operated by a switch or pushbutton in the pendant pushbutton station.
> (iv) The hoisting motion of all electric traveling cranes shall be provided with an over travel limit switch in the hoisting direction.
> (v) All cranes using a lifting magnet shall have a magnet circuit switch of the enclosed type with provision for locking in the open position. Means for discharging the inductive load of the magnet shall be provided.

Cranes and hoists. In 29 *CFR* 1910.306(b)(1) disconnecting means, the following requirements are found:

> (i) A readily accessible disconnecting means shall be provided between the runway contact conductors and the power supply.
> (ii) Another disconnecting means, capable of being locked in the open position, shall be provided in the leads from the runway contact conductors or other power supply on any crane or monorail hoist.
> (a) If this additional disconnecting means is not readily accessible from the crane or monorail hoist operating station, means shall be provided at the operating station to open the power circuit to all motors of the crane or monorail hoist.
> (b The additional disconnect may be omitted if a monorail hoist or hand-propelled crane bridge installation meets all of the following:
> (1) The unit is floor controlled;
> (2) The unit is within view of the power supply disconnecting means; and
> (3) No fixed work platform has been provided for servicing the unit [Figures 8-9 and 8-10].

Elevated disconnects. Often energy-isolating disconnects/breakers are installed in elevated positions. The reasons for this positioning decision vary from protection of the devices to reduced cost of installation. Unfortunately, individuals making these judgments do not understand the direct impact on the workers who need to use the devices for their personal protection. When energy-isolating devices are installed in such a way that employees must use ladders, climb stairs, use auxiliary lifting equipment, etc., their effective use is diminished. In addition, the conditions at these elevated positions such as working space, lighting, and footing are often substandard.

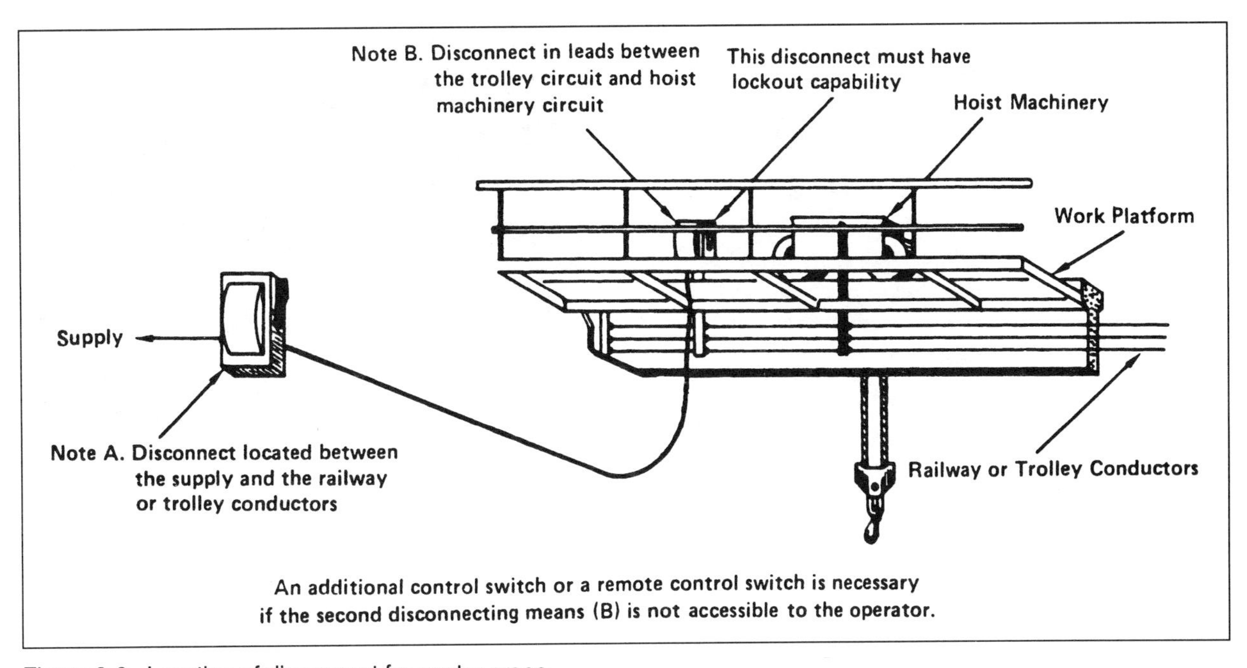

Figure 8-9. Location of disconnect for gantry crane.
Source: U.S. Department of Labor, Occupational Safety and Health Administration. *An Illustrated Guide to Electrical Safety,* OSHA 3073. Washington DC: Occupational Safety and Health Administration, 1983, p. 92.

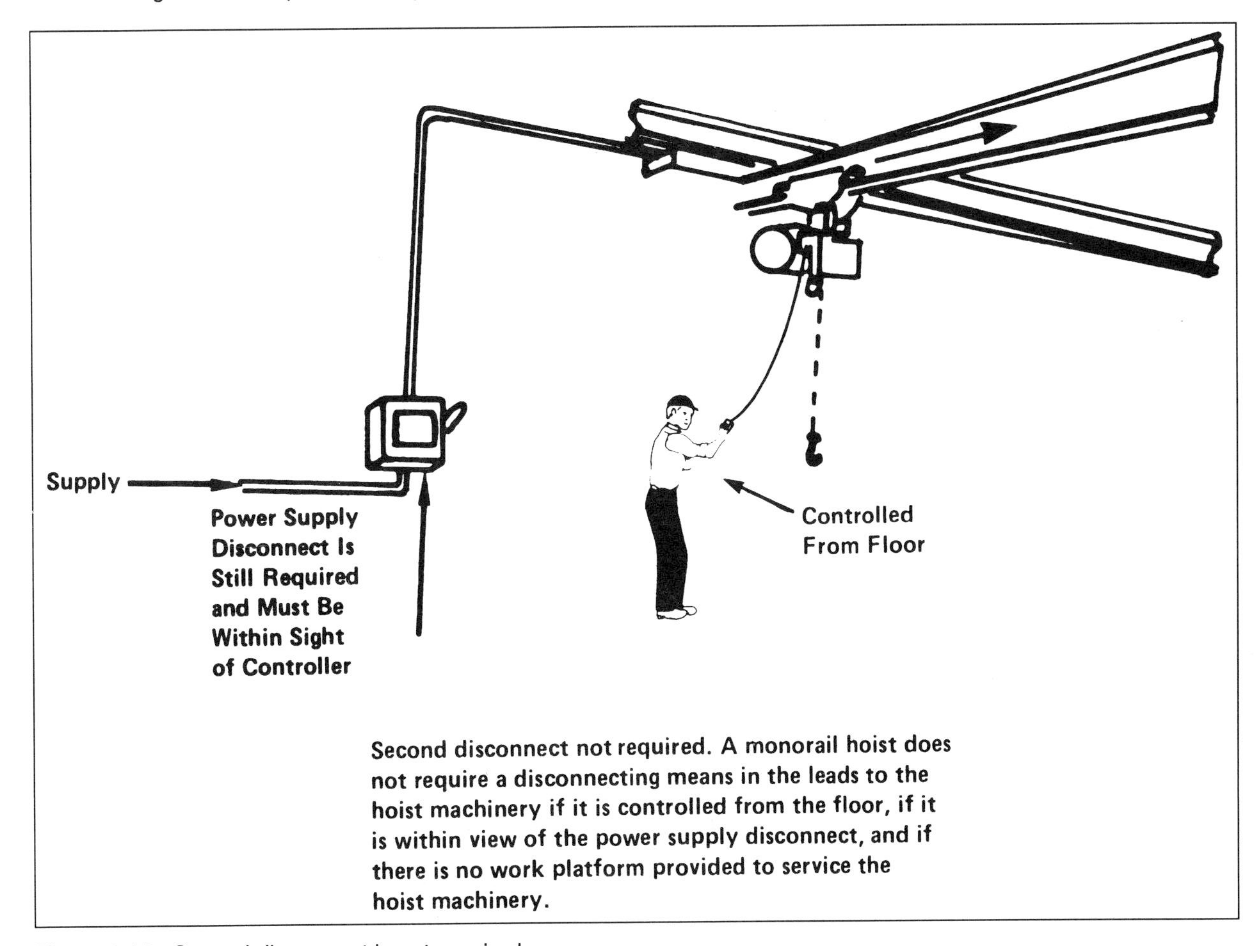

Figure 8-10. Second disconnect is not required.
Source: U.S. Department of Labor, Occupational Safety and Health Administration. *An Illustrated Guide to Electrical Safety,* OSHA 3073. Washington DC: Occupational Safety and Health Administration, 1983, p. 93.

Figure 8-11a. Elevated electrical bus and disconnect.
Source: Photo courtesy of E. Grund.

Figure 8-11b. Elevated electrical bus and disconnect.
Source: Photo courtesy of E. Grund.

For example, consider the installation in Figure 8-11 a–b showing an elevated bus with a bank of disconnects. The operating lever is approximately 12 to 14 ft (3.6 to 4.3 m) from the floor and would normally be actuated using a nonconductive push-pull pole. However, to lock out these disconnects requires the use of a portable ladder. A worker would be inclined to raise a tag on the pole and hang it wherever possible or tag the push-pull pole. Neither of these approaches is very effective. It might be more likely to open the disconnect and go about the task with the assumption that no one would tamper with the device.

Figure 8-12. Elevated motor with disconnect and plug/receptacle fittings for easy removal.
Source: Photo courtesy of E. Grund.

In contrast, the situation shown in Figure 8-12 reveals an elevated drive motor installation enhanced to reduce risk to the worker when troubleshooting or servicing. In this European plant, the 380-V, 1-kw motor was rewired to include a disconnect device and a plug/receptacle fixture. The maintenance mechanic now isolates the motor at the point at which the work is done, breaks the connection at the receptacle, and removes the motor. A replacement motor is installed and the defective motor repaired while on the shop bench. Downtime is reduced, and mechanics spend minimal time working at elevated positions.

These conditions can be identified as employees complete the energy-isolation capability surveys or develop specific MEP lockout/tagout procedures.

Identification of Disconnecting Means

Requirements for identifying disconnecting means and circuits are defined in 29 *CFR* 1910.303(f) as follows:

> Each disconnecting means required by this subpart for motors and appliances shall be legibly marked to indicate its purpose, unless located and arranged so the purpose is evident. Each service, feeder, and branch circuit, at its disconnecting means or overcurrent device, shall be legibly marked to indicate its purpose, unless located and arranged so the purpose is evident.

These markings shall be of sufficient durability to withstand the environment involved. For example, on a

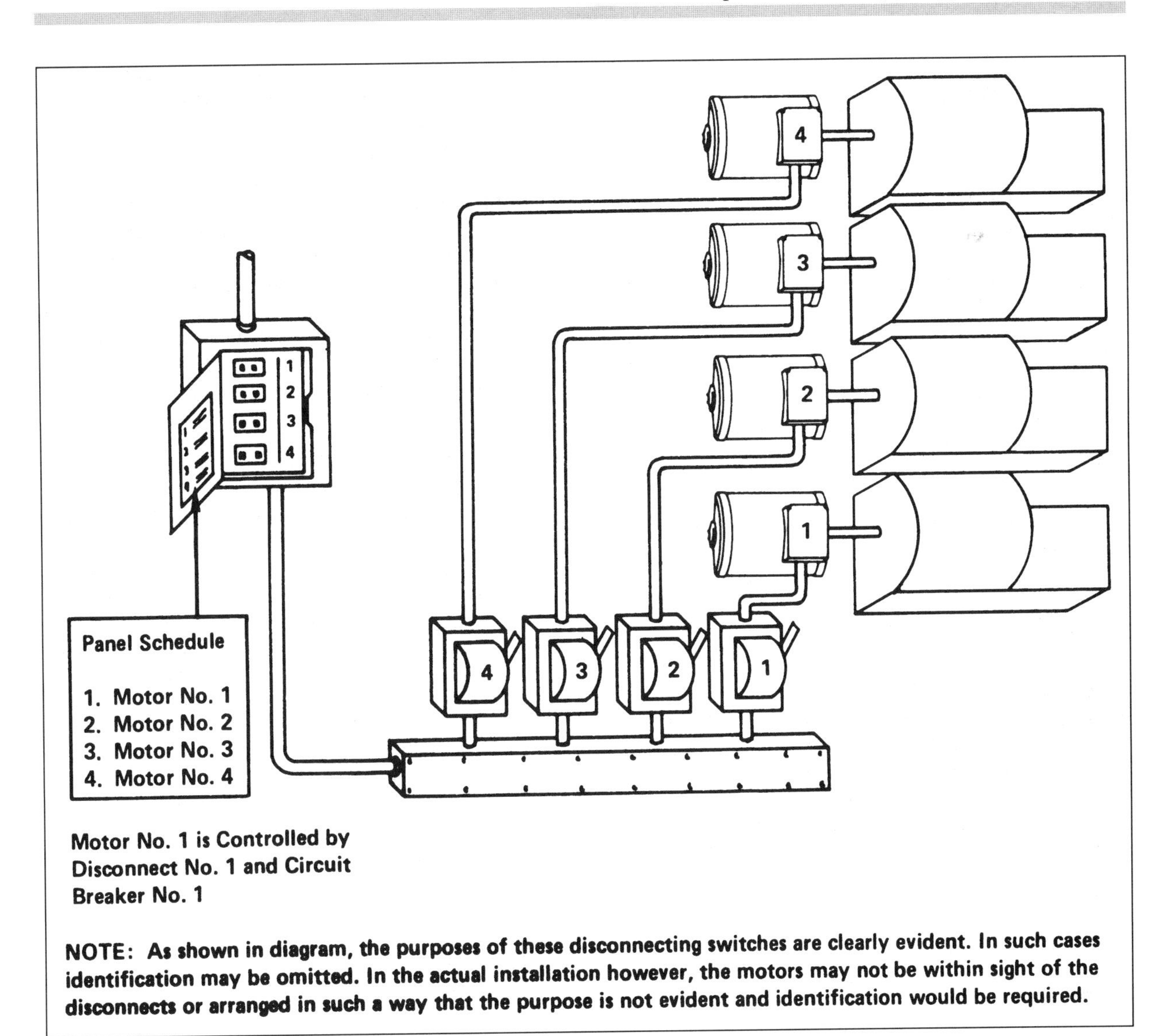

Figure 8-13. Each disconnect and circuit requires identification.
Source: U.S. Department of Labor, Occupational Safety and Health Administration. *An Illustrated Guide to Electrical Safety,* OSHA 3073. Washington DC: OSHA, 1983, p. 5.

panel that controls several motors or on a motor control center, each disconnect must be clearly marked to indicate the motor to which each circuit is connected. In Figure 8-13, the number 2 circuit breaker in the panel box supplies current only to disconnect number 2, which, in turn, controls the current to motor number 2. This current to motor number 2 can be shut off by the number 2 circuit breaker or the number 2 disconnect.

Identification should be specific rather than general. A branch circuit serving receptacles in the main office should be labeled as such, not simply labeled "receptacles."

If the purpose of the circuit is obvious, no identification of the disconnect is required (Figure 8-14).

Remote power switching. Various electrical loads, e.g., ovens, water heaters, fans, strip heaters, HVAC units, and power drops can be controlled/managed remotely. The systems consist of remote-controlled circuit breakers, power supplies, system network interface, breaker interface modules, and wiring. User-supplied control devices may be time clocks, photocells, programmable logic controllers, push buttons/sensors, momentary/maintenance switches, personal computers, etc. These systems need to be evaluated individually and specific procedures generated to ensure that personnel understand their capability and effect and know how to isolate them for personal protection.

Switches. Different types of switches can present problems for personnel who do not understand their specific function. Double-throw switches used in multiphase/DC electrical systems can be confusing. They are designed to reverse the voltage phases when the operating handle is moved to the up or down position.

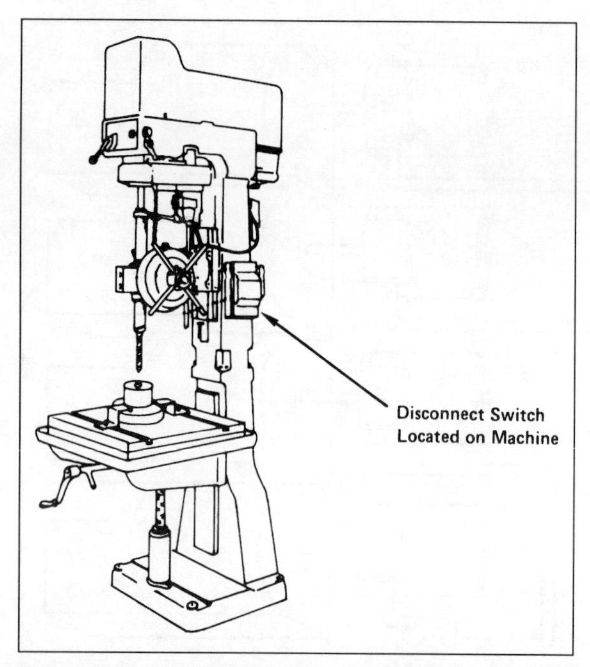

Figure 8-14. Disconnect switch located on machine: no label is required.
Source: U.S. Department of Labor, Occupational Safety and Health Administration. *An Illustrated Guide to Electrical Safety,* OSHA 3073. Washington DC: OSHA, 1983, p. 7.

NEUTRAL, OFF, or OPEN is found in the center position. Therefore, isolation of the device would require it to be secured in the center position.

29 *CFR* 1910.305(c)(i) knife switches states that:

> Single-throw knife switches shall be so connected that the blades are dead when the switch is in the open position. Single-throw knife switches shall be so placed that gravity will not tend to close them. Single-throw knife switches approved for use in the inverted position shall be provided with a locking device that will ensure that the blades remain in the open position when so set.

Single-throw knife switches have one energized (closed or ON) position and one open (dead or OFF) position. The switch must be designed so that when it is in the open position, the blades are not energized (i.e., the blades must be connected to the load side, not the supply side of the circuit). The switches must also be installed so that if the switch falls downward, it will not fall into its energized position. However, some single-throw knife switches are designed to be installed so that they open upward. To be approved for this type of installation, they must have a latch or other locking device (e.g., a spring-loaded device)

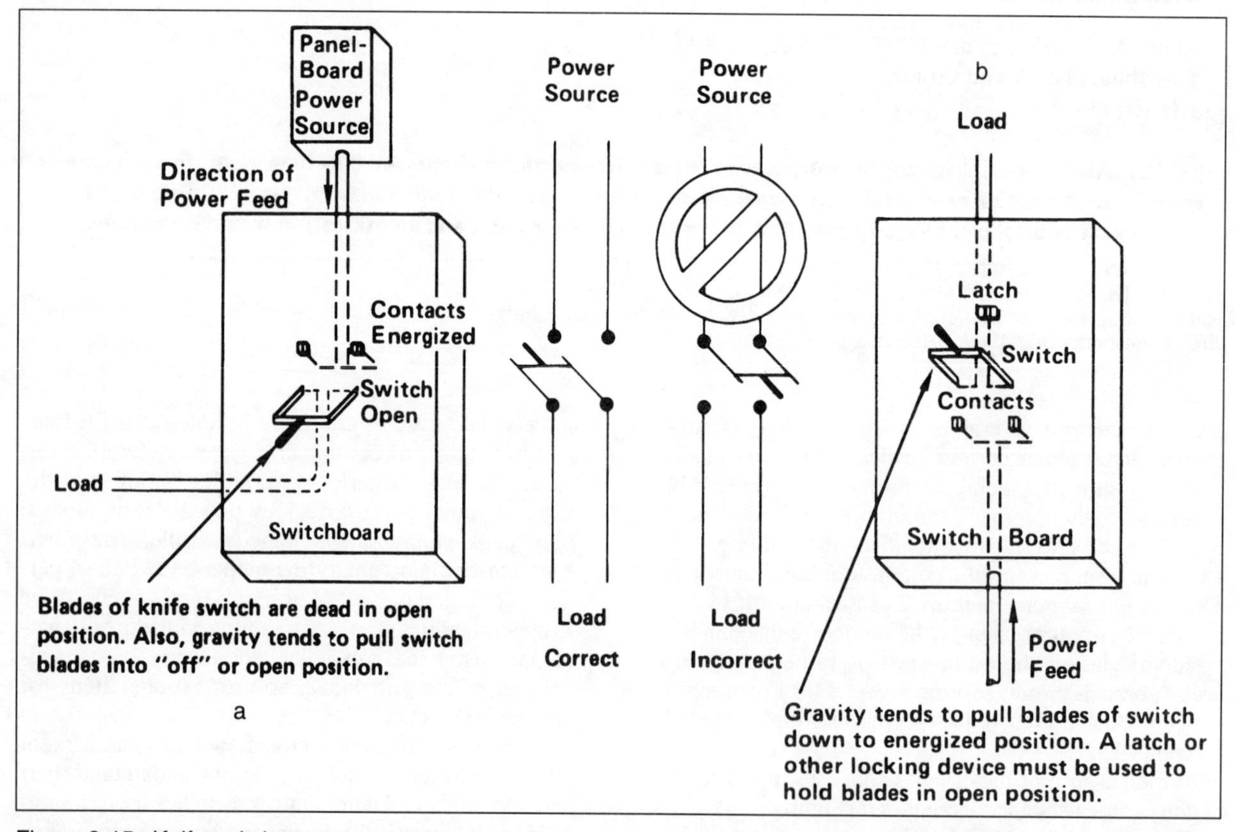

Figure 8-15. Knife switches.
Source: U.S. Department of Labor, Occupational Safety and Health Administration. *An Illustrated Guide to Electrical Safety,* OSHA 3073. Washington DC: OSHA, 1983, p. 69.

used to secure the switch in the open position. Figure 8-15a shows a single-throw knife switch connected so that the blades are dead when the switch is open. In addition, Figure 8-15b shows a latch arrangement that holds the blade in the open position and will prevent gravity from pulling the switch closed.

Double-throw knife switches may be mounted so that the throw will be either vertical or horizontal. However, if the throw is vertical, a locking device shall be provided to ensure that the blades remain in the open position when so set.

Double-throw knife switches are knife switches that have two energized (closed or ON) positions and one open (dead or OFF) position. These switches can be mounted vertically so that they are moved up and down or horizontally so that they are moved back and forth. If switches are mounted vertically, they must have a locking device (e.g., a spring-loaded device) that will hold the switch blades in the open position (Figure 8-16).

Toggle switches can be secured in a fixed position but if they are a three- or four-way switch used for multiple locations, a false sense of security may be induced. All points in this wiring circuit arrangement would have to be immobilized.

Motors. OSHA specifies in 29 *CFR* 1910.305(j)(4) the requirements for disconnecting means for motors, motor circuits, and controllers as follows:

> (i) *Insight from.* If specified that one piece of equipment shall be "insight from" another piece of equipment, one shall be visible and not more than 50 ft (15 m) from the other.
> (ii) *Disconnecting means.*
> (a) A disconnecting means shall be located in sight from the controller location. However, a single disconnecting means may be located adjacent to a group of coordinated controllers mounted adjacent to each other on a multi-motor continuous process machine. The controller disconnecting means for motor branch circuits over 600 volts, nominal, may be out of sight of the controller, if the controller is marked with a warning label giving the location and identification of the disconnecting means which is to be locked in the open position.
>
> Explanatory note: A motor controller is a device, such as a switch or circuit breaker, that controls power to a motor. The controller turns the power off and

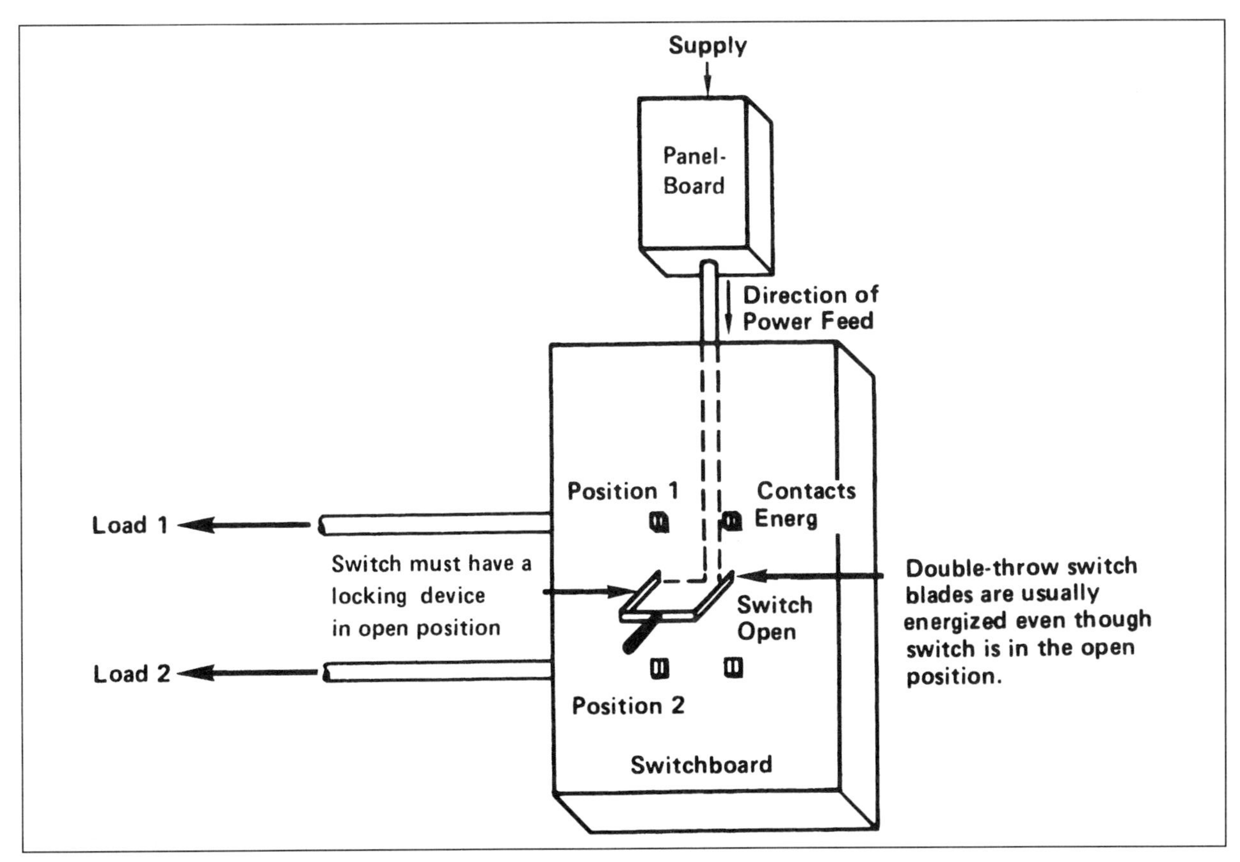

Figure 8-16. Double-throw knife switches with locking device.
Source: U.S. Department of Labor, Occupational Safety and Health Administration. *An Illustrated Guide to Electrical Safety,* OSHA 3073. Washington DC: OSHA, 1983, p. 70.

on and limits the current flow. A disconnecting means for the controller must be within sight from the controller (i.e., visible from the controller and located within 50 ft [15 m] of the controller) [Figure 8-17]. If a group of controllers are located together and are used to control power to more than one motor on a single continuous process machine, a single disconnect switch, located with the controllers, can be used. It should be noted that it is possible for a switch or circuit breaker to serve as both a controller and a disconnect. This depends on where the switch is located and the rating of both the motor and the switch. Detailed specifications on motor disconnecting means are given in Article 430 of the *National Electric Code*.

For larger capacity motors operating at voltages greater than 600 volts, the disconnect can be out of sight of the controller if the controller and the disconnect are labeled. Warning labels on the controller should indicate where the disconnect is located and that it is to be locked out for maintenance. The disconnect must be labeled with identification such as a number to ensure that the correct disconnect is opened or de-energized [Figure 8-18].

(b) The disconnecting means shall disconnect the motor and the controller from all ungrounded supply conductors and shall be so designed that no pole can be operated independently.

(c) If a motor and the driven machinery are not in sight from the controller location, the installation shall comply with one of the following conditions:
  (1) The controller disconnecting means shall be capable of being locked in the open position.
  (2) A manually operable switch that will disconnect the motor from its source of supply shall be placed in sight from the motor location.

Explanatory note: Usually, a motor and the equipment it drives should be within sight of the controller. If they are not within sight of the controller, one of two conditions must be met: (1) The controller disconnect must be designed so that it can be locked in the open or deenergized position to protect persons working on the motor or equipment [Figure 8-19] or (2) a switch that can be manually (not magnetically) operated must be located within 50 ft [15 m] of, and must be visible from, the motor (Figure 8-20].

Troubleshooting/testing. When attempting to diagnose electrical deficiencies, personnel are often required to examine/check energized systems. Each individual authorized to perform such work must be well trained and conditioned to proceed in such a way that risk is minimized at every opportunity. For example, insulated mats, protective gloves, nonconductive tools, test equipment, and partial lockout can be used.

In 29 *CFR* 1910.333(b)(2)(iv) requirements for verifying that the correct circuits have been de-energized are found. The first step is to operate the MEP controls. Of course, operating the equipment controls is not a completely reliable indication that the circuit has been de-energized. It is possible to interrupt a portion of the circuit so that the equipment will not operate even though the rest of the circuit is still alive. Therefore, the standard requires a qualified person to use test equipment to ensure that all parts of the circuit to which employees will be exposed are de-energized. Because it is also possible, under certain conditions, to feed circuits from the "load" side, the test is required to check for any voltage backfeed that might be present.

Paragraph (c) of 29 *CFR* 1910.334 describes the requirements for use, rating, and inspection of electrical test instruments and equipment. Because the use of test instruments can expose employees to live parts of electric circuits, paragraph (c)(1) requires testing work on electric circuits or equipment to be performed by qualified persons.

To prevent injuries to employees resulting from exposed conductors or other defects in the test equipment, paragraph (c)(2) requires the visual inspection of such equipment before use. Of course, employees would not be permitted to use defective or damaged equipment until it has been repaired.

Using test equipment in improper environments or on circuits with voltages or currents higher than the rating of the equipment can cause the equipment's failure. Because employees can be injured as a result of this failure, paragraph (c)(3) requires test equipment to be used within its rating and to be suitable for the environment in which it is to be used.

When testing electrical systems, personnel should determine that limit switches and/or interlocks are closed to prevent a false test response. After cycling any start-type function switches, the reset or stop switches must be activated to prevent unexpected restart on MEP re-energization.

Stored electrical energy. Various pieces of equipment such as induction heaters, transformers, magnets, and welders may possess electrical potential in certain components even after being shut down and isolated. Discharging stored electrical charges contained in capacitors, condensers, etc. will be necessary to reach an energy state where work can proceed safely.

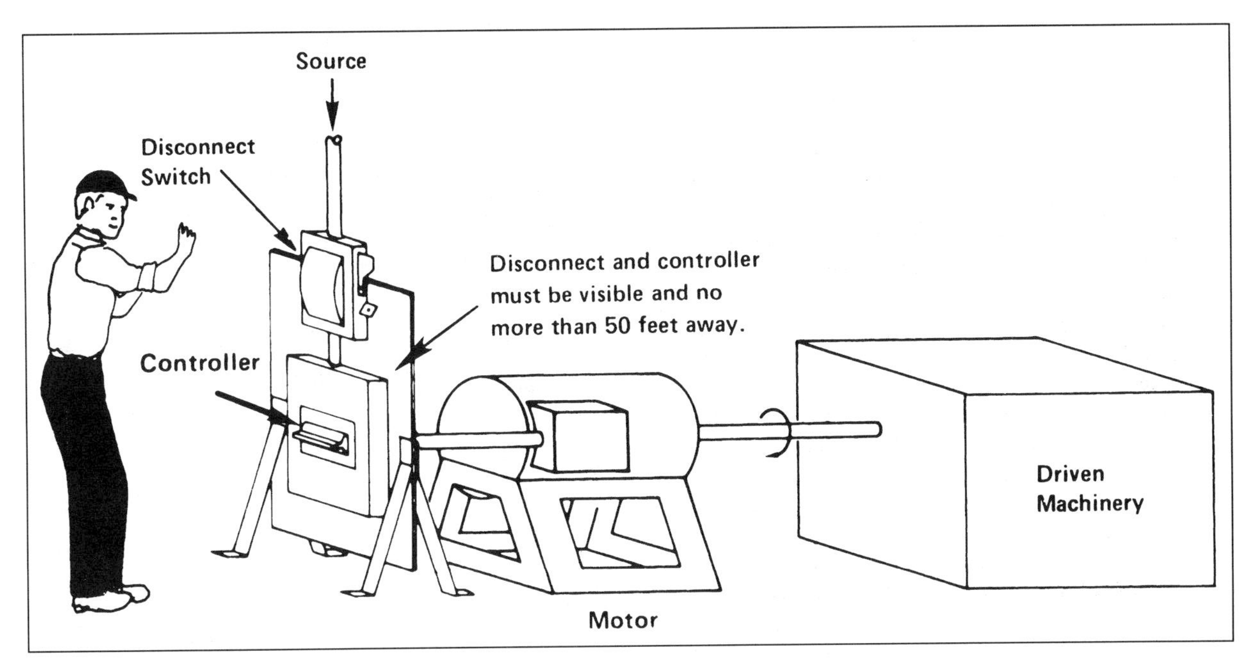

Figure 8-17. Motor disconnecting means.
Source: U.S. Department of Labor, Occupational Safety and Health Administration. *An Illustrated Guide to Electrical Safety,* OSHA 3073. Washington DC: OSHA, 1983, p. 78.

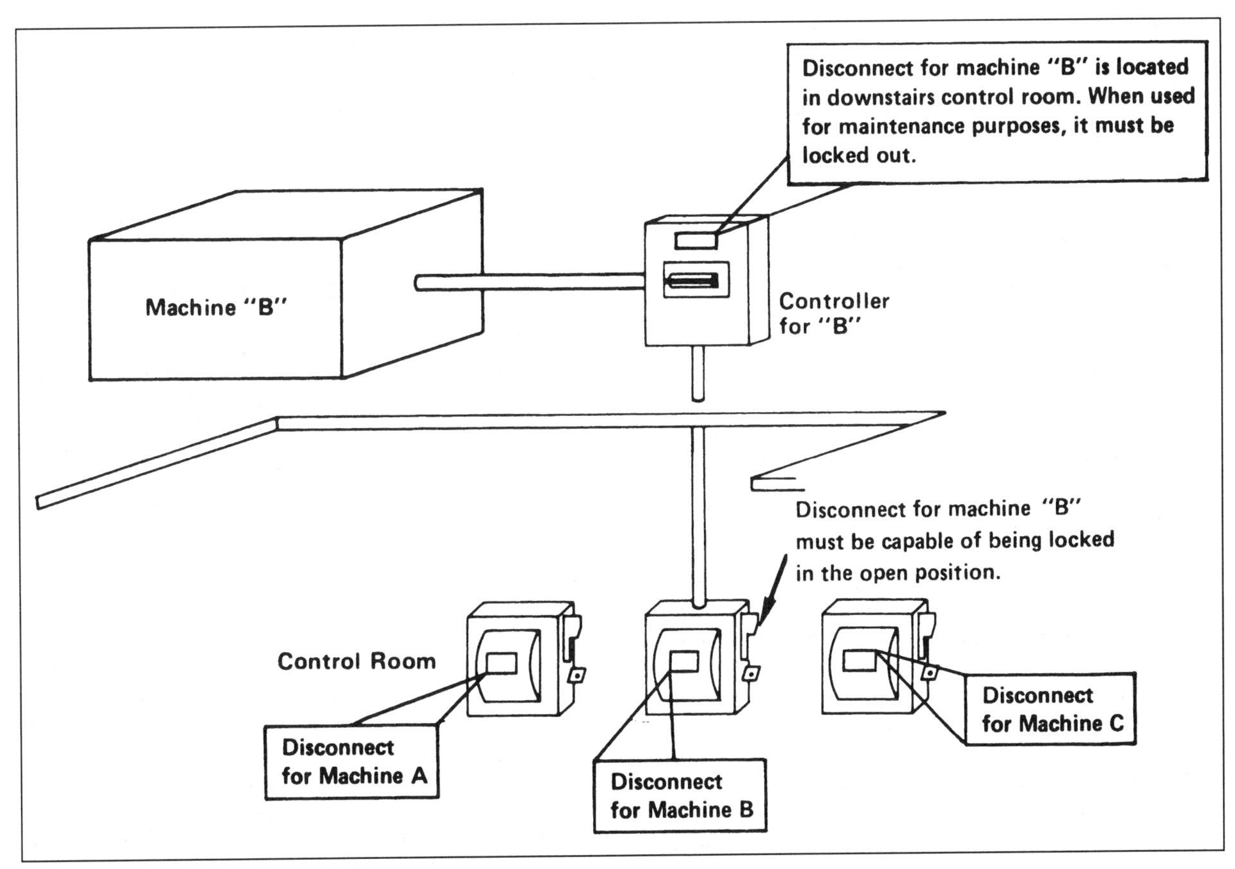

Figure 8-18. Labeling required when disconnects are out of sight.
Source: U.S. Department of Labor, Occupational Safety and Health Administration. *An Illustrated Guide to Electrical Safety,* OSHA 3073. Washington DC: OSHA, 1983, p. 79.

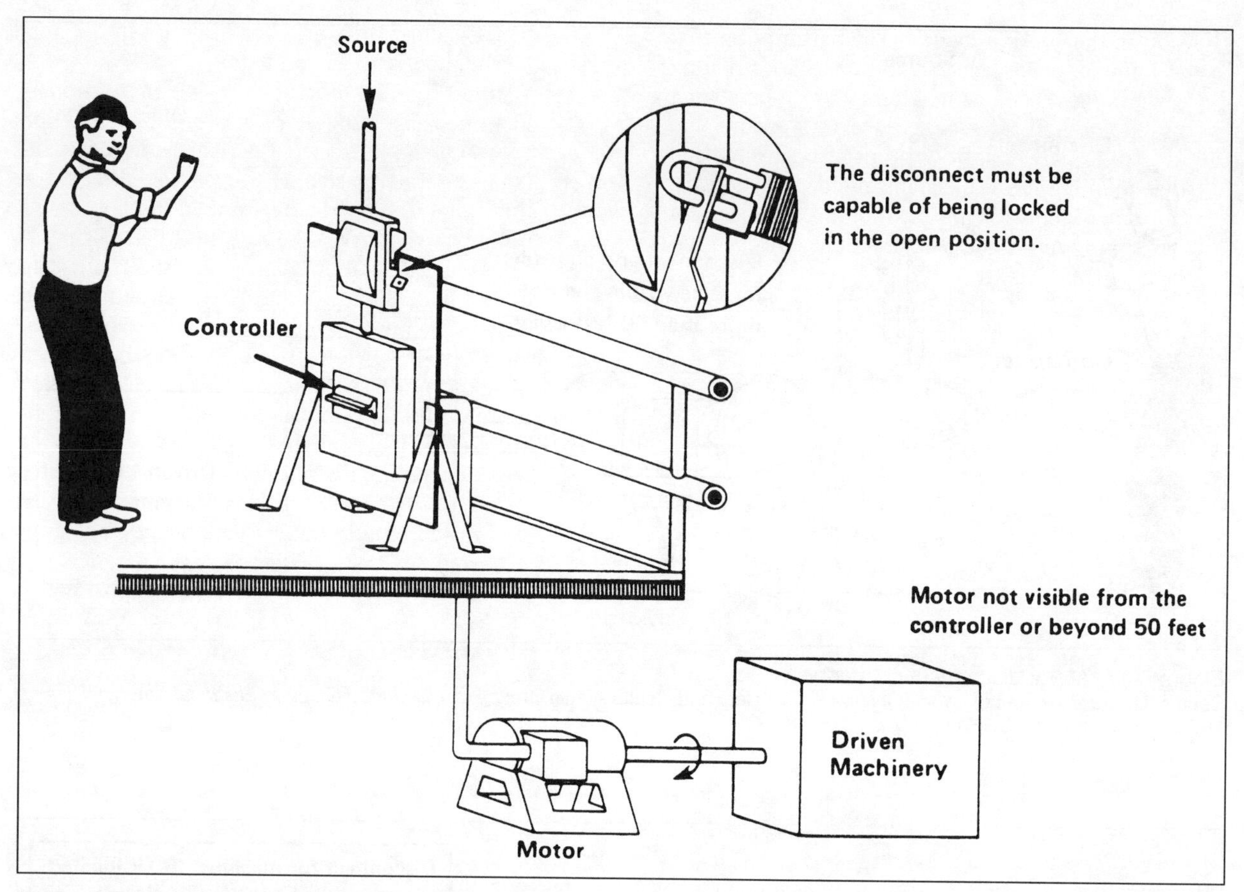

Figure 8-19. Locking controller disconnecting means.
Source: U.S. Department of Labor, Occupational Safety and Health Administration. *An Illustrated Guide to Electrical Safety,* OSHA 3073. Washington DC: OSHA, 1983, p. 80.

Batteries and banked battery systems will have to be neutralized by disconnecting the power supply leads or actuating a device that effectively isolates their stored electrical potential.

Energy may also be stored in the form of a static electrical charge. The charge (free electrons) can develop by inductive or conductive means and reach levels of thousands of volts. Charged objects will seek opportunity to discharge and achieve electrical equilibrium. Bonding and grounding normally are used to dissipate these charges before they build up and become hazardous. Transfer of materials (liquids, dusts, powders), rotation of equipment, and frictional contacts may generate static electrical charges that must be properly relieved.

Contact with stored electrical energy may range from the startling jolt when a door knob is touched to high amperage/voltage discharges with fatal consequences. Each situation must be evaluated to determine the residual electrical energy possibilities and the appropriate application of safeguards and procedures to protect the potentially exposed worker.

HYDRAULIC AND PNEUMATIC ENERGY

Hydraulic and pneumatic systems using fluids and air to magnify forces to perform work are found literally everywhere in industry. They operate across a tremendous range of forces and pressures. Their value to society is inestimable but not without price. The energy that these systems generate can be deadly if not properly managed and controlled.

Hydraulic Systems

The typical hydraulic system consists of a series of motors; pumps; accumulators/reservoirs; power, check, and relief valves; flow controls; and cylinders. The medium is a noncompressible fluid (water/oil) that is acted on to produce sizable forces/pressures. The fluids, under pressure, are transmitted through piping and other components to perform various needed actions. The piping may be steel tubing, steel braided hose, or high pressure pipe and sectionalized by appropriate manual or automatic valving. Electri-

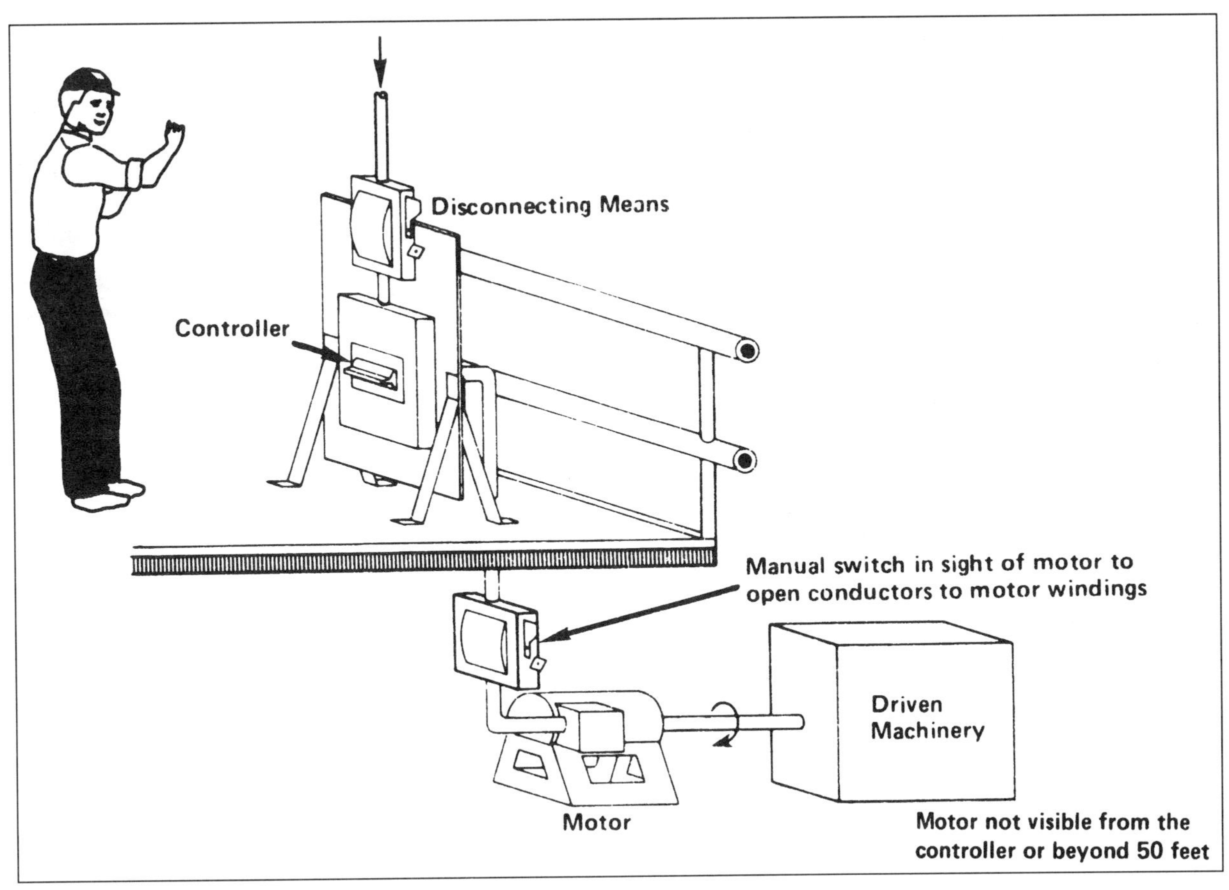

Figure 8-20. Manually operable switch within sight.
Source: U.S. Department of Labor, Occupational Safety and Health Administration. *An Illustrated Guide to Electrical Safety,* OSHA 3073. Washington DC: OSHA, 1983, p. 81.

cal motors or engines are used to drive hydraulic pumps that energize the systems.

The control of hazardous energy in hydraulic systems is accomplished by isolating the power source(s), closing the appropriate sectioning/control valves, and relieving the internal pressures in the system. The fluid will be returned to a reservoir/tank. If the system must be opened, fluids at atmospheric pressure may escape or be drained to access or change the necessary components. Pressure in the system is relieved by shifting a control valve electrically or by manual override. Accumulators, where fluids under pressure are stored, may have to be relieved dependent on the nature of the task and the design of the system and its isolating features.

A portion of a high-pressure hydraulic system is shown in Figure 8-21. Often the individual system components are not conveniently arranged or grouped. Pumps, reservoirs, and accumulators may be located in basements, whereas other gear is in the operating area. It is imperative that the complete system be understood to achieve effective isolation. In Figure 8-22 is a simple hydraulic system with an easy means of isolation available. It should be remembered that hydraulic powered equipment may tend to drift under steady-state conditions.

Figure 8-21. Complex high-pressure hydraulic system.
Source: Photo courtesy of E. Grund.

Figure 8-22. This simple hydraulic system has a convenient electrical disconnect.
Source: Photo courtesy of E. Grund.

Figure 8-23. This is a simple lockout application situation: compressed air equipment.
Source: Photo courtesy of E. Grund.

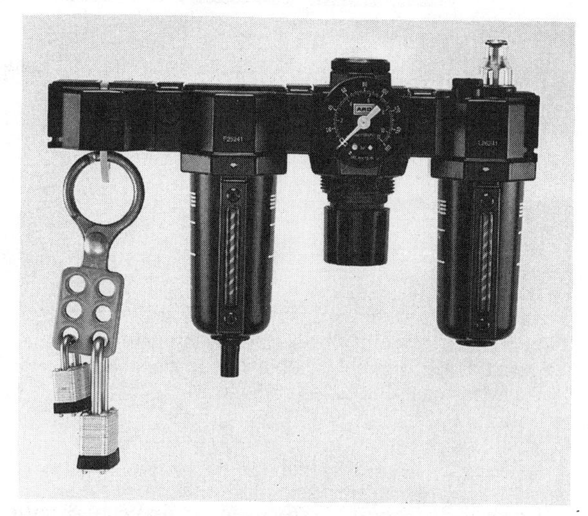

Figure 8-24. Pneumatic system: safety lock-out valve.
Source: Aro Safety Shutoff/Lockout Valve shown with Aro Module/Air 2000 © FRI. The Aro Corp., One Aro Center, Bryan, OH 43506.

Pneumatic Systems

Pneumatic systems consist of motors; reservoirs; compressors; coolers; filters, regulators; lubricators; power, check, and relief valves; silencers/mufflers; etc. The medium is air that is compressed to produce the desired force or pressure. The compressed air is transmitted through black steel pipe or tubing (e.g., steel, plastic, or copper) to perform various functions. Various control or sectioning valves are used to move or halt the internal pressure as desired. Electrical motors or engines are used to drive the compressors that pressurize the system.

Much the same as hydraulic systems, the energy in pneumatic systems is controlled by isolating the power source(s), closing the appropriate sectioning/control valves, and relieving the internal pressure. The air is typically released to the atmosphere under controlled conditions. Air under pressure cannot be relieved in such a way that it poses a danger to personnel by direct contact or by blowing dust, objects, etc. System pressure may be relieved by shifting the control valve electrically or manually by mechanically moving the valve override to exhaust at the discharge port. Surge and storage tanks or receivers may have to be relieved of pressure to safely work on the system. Do not rely on pressure gauges to convey the internal state of any closed pneumatic system.

In Figure 8-23 is a small compressed air unit with a straightforward arrangement to isolate the electrical power. Of course, piping and components would have to be relieved of pressure at the point at which work was to be performed and properly isolated.

Isolation valves with lockable features are shown on compressed air lines in Figure 8-24. In some cases, stopping the supply of air pressure or relieving it may produce a larger hazard. If air is being used as a counterbalance or restraining force on a raised component, it may be required to be left on. The decision to isolate and relieve the air pressure on any system is made only by those with full knowledge of the system and appropriate personal competence.

Both hydraulic and pneumatic systems are capable

of generating tremendous pressures under "jam" conditions. If system design does not include overpressure regulation, the pump may deliver output pressure far in excess of that required. Freeing equipment or product that is jammed can be dangerous if not done under carefully controlled conditions and sequence. The potential for jamming should be evaluated during assessment of the MEP for energy isolation. Alternative or supplementary safe practices may be required to get the MEP in a relieved or nonjam state.

When hydraulic/pneumatic systems are turned on or repressurized, valves or controls should be opened gradually to prevent sudden movements or jarring to system components.

PIPING SYSTEMS AND VALVES

As mentioned in the previous section, hydraulic and pneumatic systems involve transmission of pressure through piping networks controlled by valving. Hazardous and nonhazardous liquids and solid materials are also conveyed in piping systems, under varying pressures, which use valves for various control purposes. Valves, as energy-isolating devices, are indispensable but are not without inherent limitations. They are subject to failure or malfunction, they may leak, they are often misidentified, they may be difficult to open or close, etc. They all possess certain strengths and weaknesses based on the nature of the service and their maintenance. Knowledgeable personnel and reliable manufacturer's references/data sheets are necessary to ensure their proper use.

Piping Systems

It is often very difficult to determine what is contained in a particular pipe or set of pipes. The size and appearance of the pipe, fittings, and observable valving may provide an experienced person with what is necessary to deduce the answer. However, this is not the way that personnel should have to approach this matter. In Figure 8-25 a–b, without identification, the piping and their contents could easily be mistaken. Tracing the piping visually when possible and/or practical can assist in the determination of what components and isolating devices are involved in a particular system.

Piping and instrument drawings/diagrams (PIDs) are quite valuable for understanding the nature of the system, relationships of equipment, existence of isolating devices, etc. However, they are often inaccurate, particularly as the installation or process ages. Computer-based PID techniques may provide the long-term solution for managing the change and keeping the diagrams current. In many situations, it will be up to the individuals involved in the particular MEP isolation to gather the necessary facts about the system through a variety of means.

Identification of piping contents. Proper identification of piping contents can prove invaluable with regard to enhancing the isolation of hydraulic, pneumatic, chemical, and thermal energy. Methods of identification such as that found in ANSI A13.1-1985, *Scheme for the Identification of Piping Systems*, can be used. Color, legend, and symbols may be used in combination to assist identification (Figure 8-26). Decals,

Figure 8-25a. Piping systems illustrate the need for energy-isolating device identification.
Source: Photo courtesy of E. Grund.

Figure 8-25b. Complex piping systems present identification risks.
Source: Photo courtesy of E. Grund.

Figure 8-26. Identification of piping contents is essential.
Source: Photo courtesy W.H. Brady Co.

stenciling, sleeves, tags, and plates of a durable nature can be used to convey needed information to those individuals attempting to properly isolate a system. The markings typically are placed at junctions, valves, equipment entry and exit, intersections, terminations, etc. and periodically (75 ft [23 m]) along long runs of pipe. The markings should be placed so that they may be seen from the most logical position of personnel engaged in their use, e.g., if in an overhead pipe rack, without catwalk, the legends should be visible from an appropriate position and angle from the floor.

Valves

Energy-isolating devices have to be used with discretion and understanding of their protective limitations. The type of energy involved, nature of the intervention, the piping arrangement, location of the system, degree of hazard, position of personnel, etc. will have to be evaluated to determine the appropriate isolation practice.

The physical securing of valves to effectively lock out or isolate the system can be quite challenging when the hardware involved is considered. Valves can be operated manually via levers, wheels, T-handles, and arms or automatically by electric, pneumatic, and electropneumatic means. Nonmetallic valves in corrosive chemical service (Figure 8-27) and certain types of metallic valves (Figure 8-28) are often isolated by tagging rather than locking. Automatic valves can be isolated by interrupting the power supply and making the valve inoperable by disconnecting linkages, etc. (Figure 8-29). It must be determined that in fact the powered valve has been closed as intended and positive shutoff has occurred. Various protective appliances can be utilized to assist in the securing of valves once they have been closed for system isolation (Figure 8-30 a-b). Often custom-made portable appliances are made by the employer to secure valves in a locked position or the valves themselves adapted to be lockable.

Quick-opening valves (open with a quarter turn) may require special clamping devices or chaining that prohibits little or no movement. These valves may still present lockout difficulties when essentially no movement is required to prevent flow, unless they are designed to be secured in a totally closed position.

Each application may require a defined sequence of valve closing. The practice may require moving upstream from the equipment or work area toward the energy source(s) or the reverse. Other sequences can be necessary. What should be determined is the right sequence for each energy-isolation application. Freestyle or casual closure or opening of valves may invite error, particularly on complex systems.

Line breaking. Opening piping systems often involves danger due to failure to completely prepare for this intrusion. Pressure needs to be relieved, fluids drained or returned to reservoirs, gases vented or flared, temperatures reduced, etc. In chemical process situations, it is possible to have blockages

Figure 8-27. Plastic valving is often difficult to secure or restrain.
Source: Photo courtesy of E. Grund.

Figure 8-28. These are lever-activated metal valves with a lockable feature.
Source: Photo courtesy of E. Grund.

(corrosion, material buildup, scale, etc.) that fully or partially restrain liquids or pressures in lines or equipment. Certain parts of the system may be more prone to this type of occurrence and should be noted in procedures or employee communications. The following precautions should be considered when systems are opened to perform service/maintenance tasks:

- If system pressure cannot be released by valves or relief devices, qualified personnel under controlled conditions may have to relieve pressure at appropriate fittings (may not be possible under high-pressure, hazardous material circumstances).
- Allow time after opening for venting, drainage, off-gassing, etc.
- Loosen bolts on fittings/flanges away from the worker first; slowly untighten and be attentive for any escape of pressure/materials.
- Use shielding (rubber mat, etc.) when opening lines at joints, if possible.
- On horizontal piping, in liquid service, release tension on bottom bolts first at fittings to ascertain if any leakage/dripping will occur.
- Carefully separate flanges when frozen/stuck together; use small wedges or special tools.
- Wear protective equipment compatible with the anticipated exposure.
- Initial line-breaking activity should be done with an additional person observing or in close proximity.
- Mechanical or oxyacetylene cutting on lines/closed systems should occur only under permit circumstances.
- Move separated valves and pipe lengths as level as possible or tilt, so that the lowest open end presents no hazard if leakage or release of material occurs.
- Line separation, at elevated levels, should anticipate possible release of contents, and all affected areas should be cleared and isolated.

Figure 8-29. Automatic valving (electro-pneumatic).
Source: Photo courtesy of E. Grund.

Figure 8-30. a, (*top*) Ball-valve lockouts offer an inexpensive solution to complying with OSHA's standard; b, (*bottom*) Specialized products, such as this gate-valve lockout, provide a greater degree of security than improvised methods.
Source: Photos courtesy of Brady U.S.A. Inc., Signmark® Division.

Purging and cleaning. A variety of methods must be used to prepare or clear piping systems and process equipment for service and maintenance tasks. The methods are specific for the nature of the operation and the energies/materials involved, e.g., reactive chemicals, flammable liquids, explosive gases, toxic materials.

Steam, inert gases, boiling water, cleaners, and solvents may be used to prepare the system for work to commence. These actions may take place before isolation or after depending on the MEP situation and the scope of the intended work. Special attention should be directed at "dead ends" where process material may be trapped or not impacted properly during cleaning. Normally, lines will be cleaned/flushed/ purged before they are opened.

CHEMICAL AND THERMAL ENERGY

Chemical (corrosive, reactive, flammable, etc.) and thermal (steam, heated fluids, radiating sources, etc.) energy sources must be effectively isolated before service and maintenance personnel work on the MEP. The classic lockout/tagout of these sources may not be possible or practical. However, the concept of neutralizing, controlling, or reducing the involved energy factor to its lowest possible state is still appropriate.

Chemical Energy

Most often chemical energy exists in the materials contained in some part of a process. The equipment (e.g., vessels, reactors, columns, tanks) and the related piping (controls, valves, etc.) not only hold, react, and transfer the chemicals but are subject to its effects as well. The chemical process or system may be under such service conditions that maintenance is of an extremely high frequency. In many cases, due to the inherent hazards or characteristics of the materials, equipment spares, parallel trains, duplicate pumps, alternate piping paths, etc. are provided. This designed-in feature can expedite the work and reduce the risk involved in various maintenance tasks. The valve (sectioning, block, control, etc.), the blind (blanks, spades, slip plates, etc.), and the relief device (flare, vent, drain, bleed, etc.) are the work horses of system isolation and safeguarding. These energy-isolating devices/features are often supplemented with protective appliances (covers, chains, restraints, etc.) or proprietary/custom hardware to lock them in a secure and controlled position. The isolating devices are also frequently tagged to show their status and to warn not to operate or change their position.

A major chemical company states in one of its procedures that because valves tend to leak, two valves in series are more protective than one. They add that a double block and bleed arrangement has the greatest degree of protective potential. In Figure 8-31 a series of piping and energy-isolating device arrangements is shown with varying degree of protective reliability. In arrangement A, a blocking valve and a vented valve isolate the system under maintenance; in B a blocking valve and a blind with a vent/bleed valve in between isolates the system under maintenance; in C the classic double block and bleed arrangement is shown; in D two blocking valves in series isolate the system under maintenance; in E one blocking valve separates the maintenance work from the system in operation. Arrangements A through C would provide the greatest degree of worker protection, with B probably receiving the top preference. Arrangements D and E are less protective, with E in most cases being marginally acceptable to unacceptable.

For high pressures/temperatures, arrangement C should be supplemented by the addition of a blind between the blocking valve and the system under maintenance. For high-risk fluids, the addition of a vent and blind between the blocking valve and the system under maintenance would be appropriate.

Thermal Energy

Various sources of thermal energy are encountered in many industrial environments. Ovens, furnaces, molten materials, burners, heat of reaction, electrically heated molds, boilers, steam, process-heated water, and gaseous combustion are but a few examples of thermal energy sources that must be controlled or isolated to safely perform service/maintenance tasks.

In many cases, methods of isolation will not rely on the use of standard energy-isolating devices. For example, there is no practical way to lock/tag a 700 F (371 C) mold. Valves for fuel sources can be isolated using standard techniques. However, many of the thermal exposures can be neutralized only by removing the heating mechanism and allowing that which has gained temperature to cool over time. Many service and maintenance tasks will be necessary during operation of the MEP and be of relatively short duration. Gained thermal energy will have to be isolated using other methods. The following techniques are examples of what is often done:

- application of forced cooling methods
- use of insulating pads, shields, and barriers
- release of system heat-drain, vent, etc.
- reduce reaction rates
- use of specialized tools and equipment
- replacement/exchange of heated components
- burner throttling
- reduce MEP input
- opening MEP for heat release
- use of special personal protective equipment.

Organizations develop unique approaches to integrate the control of thermal energy sources with their

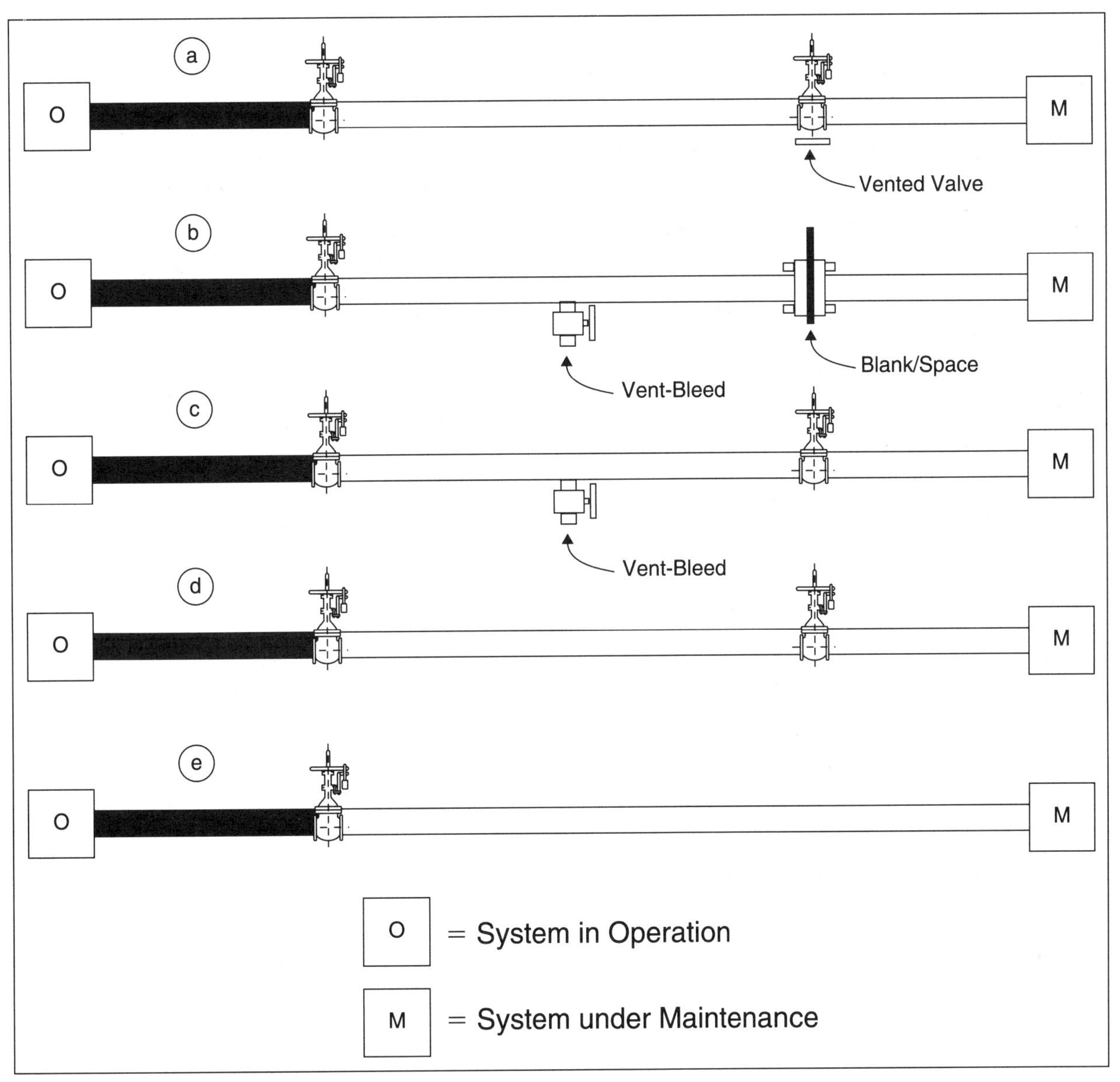

Figure 8–31. Hazardous-energy-control system.
Source: Developed by E. Grund. © National Safety Council.

general lockout/tagout procedures. A mixture of locking/tagging, placement of signs, control panel protection, posting of watchmen, etc. is used. The following actual case reveals how creative the process of controlling thermal energy can become under continuous-process conditions.

An emergency repair was necessary in a direct chill-casting pit adjacent to an aluminum-melting furnace. The repair was to occur in between casts, which took place several times during the shift. The pit was approximately 30 ft (9 m) deep with no fixed access. The furnace was normally full with 60,000 lb (27,216 kg) of molten aluminum. The pit was less than 20 ft (6 m) from the furnace tap hole. Furnaces had on rare occasion lost their refractory plugs and/or tap holes and flooded the area and pit with molten metal. The following strategy was devised to complete the repair:

- The furnace molten bath level was reduced.
- The metallic portion of the tap hole plug was welded in place.
- The mechanic was lowered into the pit on the hydraulic cylinder casting table (elevator-like arrangement).
- An emergency ladder was put in position once the casting table reached its lowest level.
- A watchman was placed at the furnace tap hole area, and an observer was assigned to the repair job.
- The repairman was provided with a safety harness and line (jib hoist and pulley available over pit).

Although a number of the precautions taken did not directly relate to hazardous energy control, the impetus for the worker/supervisor–derived plan was addressing the potential for release of the thermal (molten) energy.

GRAVITATIONAL AND MECHANICAL ENERGY

The potential energy possessed by objects suspended or elevated and by objects under compressive or tensile loads can be significant. Yet these forces are often overlooked or given limited attention in many hazardous-energy-control processes. In addition, the kinetic energy associated with various objects moving or rotating represents another force that needs to be addressed.

The force of gravity is of particular concern when objects in the workplace have been positioned or raised by other energy forms to where they pose a potential danger. Counterweights, die sets, rams, fork assemblies, crane/hoist loads, and gates may suddenly drop if not controlled. Often the movement distance is relatively small (several feet of travel) compared with the total transmission of force to objects or persons in the path of travel. Springs, cables, chains, etc. represent objects where mechanical energy is stored because of compression and tension. These forces can be released violently with self-destruction of the objects.

Halting or neutralizing the energy associated with moving equipment, materials, and objects is necessary for effective isolation. Flywheels and shafts may coast for some time even after the driving forces have been removed.

Gravitational Energy

Numerous methods are used to restrain gravitational forces when energy isolation is required. The application may involve fixed or portable restraints that are manually or automatically placed. They may range from a simple wedge to a complex electropneumatic pin insertion device. Clamps, hooks, cages, blocks, pins, props, wedges, chocks, chains, cables, locking bars, and catches are examples of the variety of simple devices in common use. The following criteria should be considered, particularly when portable restraining devices are used:

- strength: sufficient integrity and compressive/tensile performance
- stability: proper surfaces for application and prevailing conditions
- position: able to be properly placed; not subject to contact that could shift or dislodge
- condition: absence of wear, defects, damage; inspected periodically
- appearance: clean and distinctive (color)
- identification: legend, coding, etc. to define purpose and application
- design: engineering calculations, application assessments, and drawings with few exceptions.

The following two case histories reveal what may occur if this aspect of hazardous energy control is not managed properly:

- Case Number 1. A mechanic used a natural block of limestone available from the mining waste to prop up the blade of the bulldozer he was attempting to repair. He was fatally crushed when the

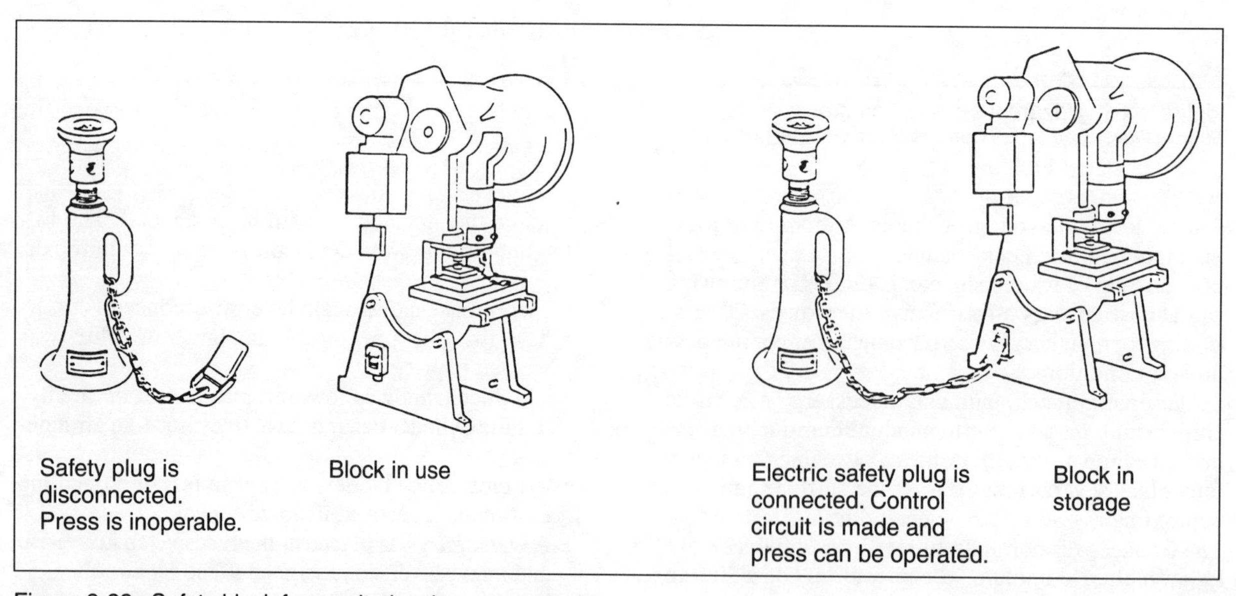

Figure 8-32. Safety block for gravitational energy restraint.
Source: Courtesy Rockford Systems Incorporated, Rockford, IL.

block broke along its natural cleavage lines, allowing the blade to fall.

- Case Number 2. A maintenance repairman was performing a service task under a large hydraulic hold-down on a continuous-process mill. The safety chain failed when hydraulic pressure was inadvertently applied by workers on another part of the unit. The hold-down force broke the links in the chain and then pinned the deceased repairman against a fixed object. The chain was common grade, and no determination of its strength was made for the installation.

In Figures 7-17a, 8-32, and 8-33 a–b, blocks, chocks, and props are shown restraining gravitational forces. It should be noted that various restraining devices such as die blocks are not intended to withstand the normal application of forces, i.e., impact of ram and die set when cycled. Personal or supervisory locks and/or tags should be applied as control measures whenever the circumstances permit. Remember, the objects are functioning as energy-isolating devices.

Figure 8-33a. Rail car wheel chock.
Source: Courtesy T&S Equipment Company.

Mechanical Energy

Momentum or forces of motion and stored energy in the form of compression or tension need to be isolated (e.g., controlled, released, stopped).

High-speed rotating equipment, flywheels, drives, etc. should be allowed to come to rest or be halted by braking systems.

Motors are available with dynamic braking features that impede the rotation of the output shaft in the shutdown mode, thus reducing the free wheeling time of connected components. Personnel should inspect the equipment to determine that all components in motion are at rest before any work is begun. Guards should not be removed during any required rundown time.

Stored mechanical energy must be approached with great care because significant forces may be imparted to the objects in compression or tension. A combination of controlled-release procedures, custom devices, and physical safeguards are used to neutralize the energy. Cable, wire rope, and chains under tension may reach the failure point if overloaded or if excessive force is applied. Caps, restraints, keepers, etc. are used to contain the energy in compressed springs (Figure 8-34 a–b) when the compression on springs is being relieved, restraints, gradual expansion, and protective devices should be used.

Equipment that is being used as a restraint or a component of a restraint (hooks, clips, chain, etc.) should be inspected periodically and when appropriate nondestructively tested. Sleeves, shields, and covers are often used as protective barriers when cable and chain are placed under tension in a restraining role.

Figure 8-33b. A supporting prop installed on a dump truck (*left*) protects against accidental lowerng of the truck bed. *Right*, The prop is in use.
Source: *Safeguarding Concepts Illustrated*, 6th ed., National Safety Council, 1993.

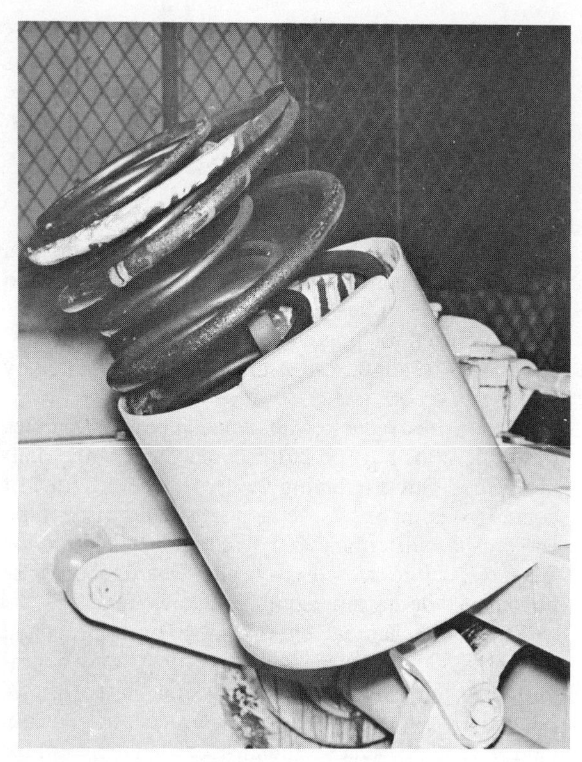

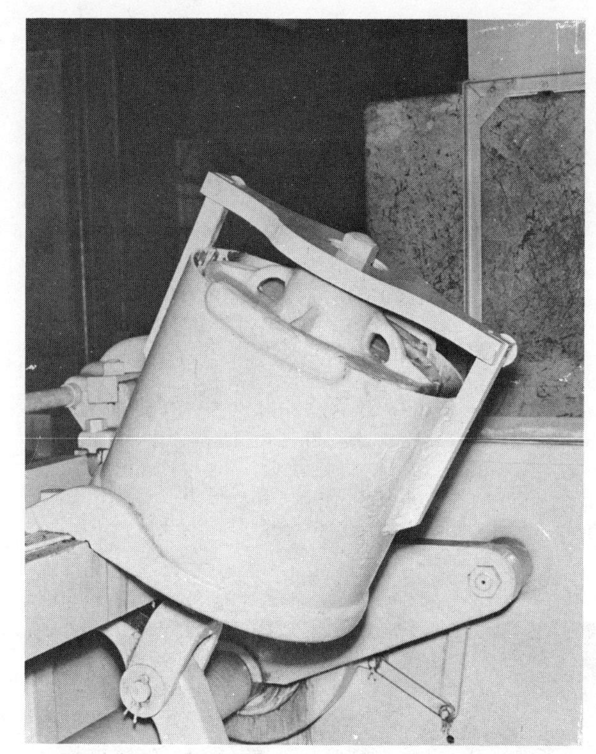

Figure 8-34. a, (*left*) What can happen if the compression spring cap of a shear brake fails. b, (*right*) How to avoid possible disastrous results by using a clamp-on cap.
Source: *Safeguarding Concepts Illustrated*, 6th ed., National Safety Council, 1993.

PROCEDURES

As mentioned in Chapter 7, System Elements, specific and alternative energy-isolation procedures are necessary in any progressive lockout/tagout system. Specific procedures will be elaborated on here; alternative procedures will be addressed in Chapter 10, Special Situations and Applications. Specific energy-isolation procedures are the blueprint for application and execution. If well prepared, they enhance the process and its prospects. If approached in a superficial way, they may become a blueprint for disaster.

Specific Procedures

Figures 7-5 and 7-6 depict possible formats for preparing specific energy-isolation procedures. There may be an inclination to emphasize the rapid creation of the written practice. This should be avoided because the real value is in the employee interaction or process that eventually generates the written procedure. The final MEP energy-isolation procedure is the culmination of what began as the energy-isolation capability survey. The intervening employee involvement and creative contribution represent the qualitative dimension of the process. The final MEP energy-isolation procedure should have these characteristics:

- all energy types/sources identified
- energy-isolating devices defined for each energy type/source
- specific isolating actions identified
- stored, trapped, residual energy relief methods defined
- required protective appliances and special safeguards defined
- verification of isolation methods emphasized
- required order/sequence of action, if applicable
- approval indicating the procedure has been field tested and effective.

The procedures should be used for authorized employee training and as a guide for execution of MEP isolation. They should be readily available in the area/department where application will take place. Posting procedures at appropriate locations can facilitate their reference and use. Job safety analyses can also be used as the procedural technique if they are prepared with energy isolation as the focus and the required content is inserted in the control portion of the format.

Alternative Procedures

Various situations are encountered where work takes place under (1) energized conditions (2) partially de-energized conditions (3) quasi-de-energized conditions. These situations should be identified, made public, and evaluated. The organizational objective should be to sanction only those alternative procedures that meet rigid safety criteria and for tasks where zero-energy conditions are not attainable for valid reasons. No tasks should be undertaken without

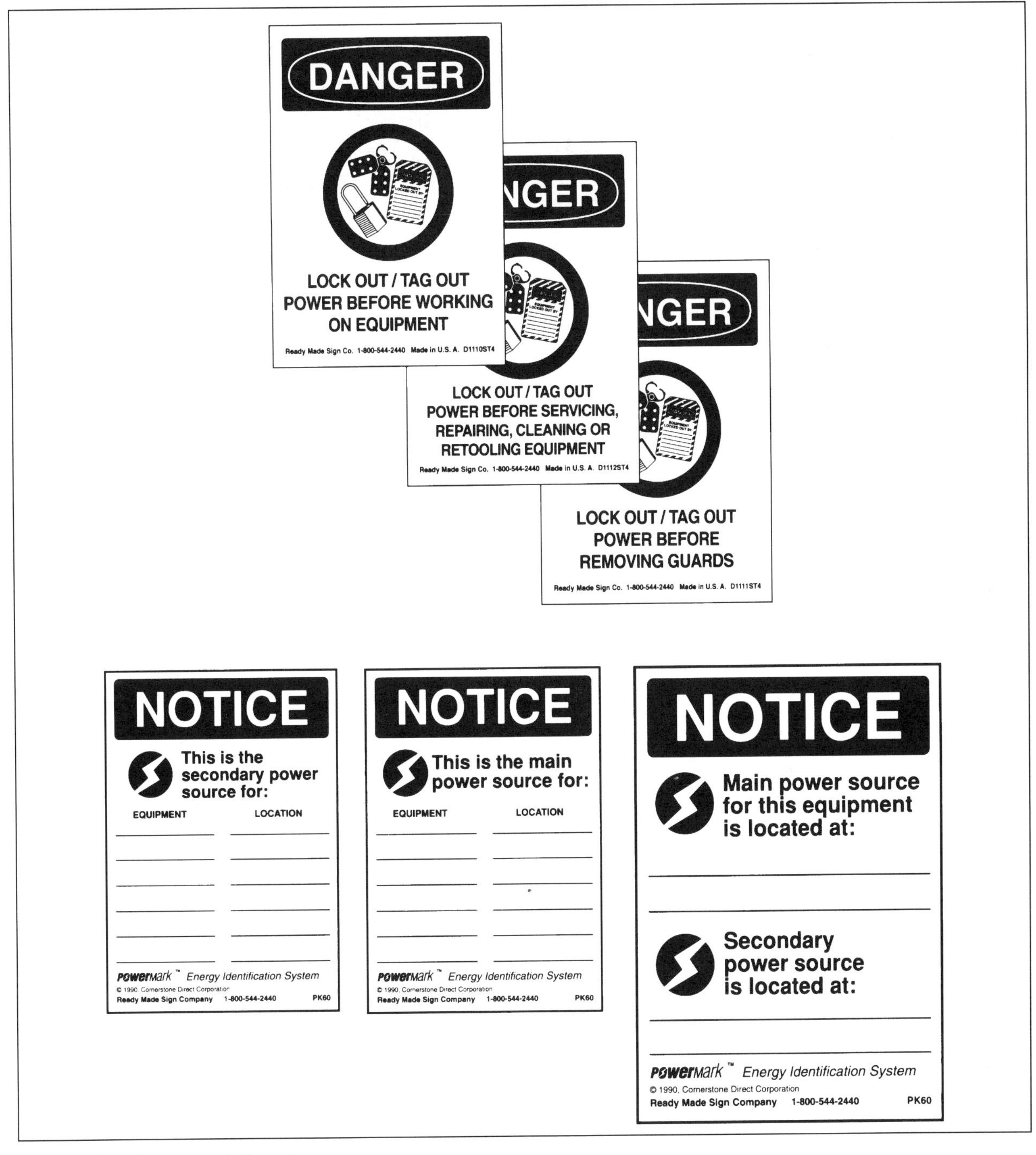

Figure 8-35. Energy-isolation signage.
Source: Courtesy Ready Made Sign Company, 480 Fillmore Ave., Tonawanda, NY 14150 (800) 544-2440.

de-energization or formal review and sanctioning of the alternative procedure to assure it is as safe as possible. See Chapter 10, Special Situations and Applications, for additional information.

Signage and graphic displays. Signs and graphic displays can be used to increase the effectiveness of lockout/tagout application. Signs may be of the general "alerting or reminder" type or actually be specific regarding some aspect of the MEP energy isolation (Figure 8-35). Signs do not usually convey enough information to replace procedures. They are supplementary in nature.

Graphic displays and procedures combine various attributes, such as visual impact, instant access, orientation, and specific direction to produce a highly effective method of communicating critical energy-isolating information. The United Auto Workers and General Motors have collaborated to produce a method

MAXIMUM # OF LOCKS REQUIRED	PLANT NAME EQUIPMENT/MACHINE NAME	LOCATION BAY X-6 BT#30723

BEFORE SERVICING THIS MACHINE, NOTIFY AFFECTED PERSONNEL

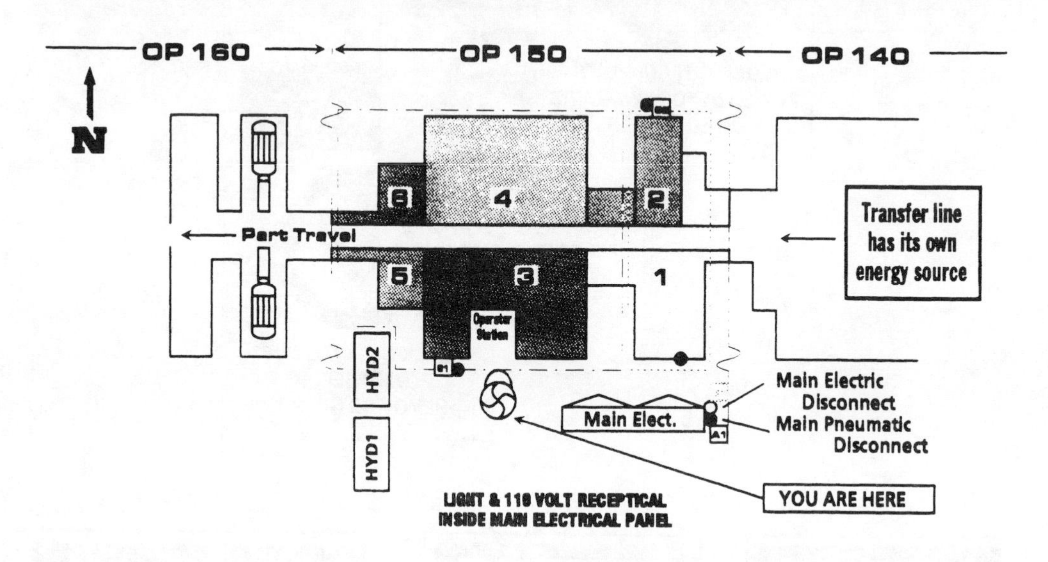

FOLLOW SHUT DOWN PROCEDURES

Energy Source	Location	Perform Action	You Must Verify
Electric	Main Electric Panel South Side of Machine	Pull down electric disconnect and lockout	Push Master Start Buttons. Check for movement
Air	A Main Air Disconnect South Side of Machine	Discharge air valve and lockout	Manually Shift Air Directional Valve. Check for Movement
Hydraulic	Main Electric Panel South Side of Machine	Pull down electric disconnect and lockout	Push Start Hydraulic Pump Button. Manually Shift Directional Valve. Check for Movement
Gravity 2 - Pins	G1 Station 3 G2 Station 2	Pin slide unit	Push Raise and Lower Head Buttons. Check for Movement

FOLLOW START UP PROCEDURES

IF LOCKOUT ENERGY CONTROL CANNOT BE PERFORMED/VERIFIED - *STOP* NOTIFY YOUR SUPERVISOR

Figure 8-36. Hazardous-energy-control system.
Source: General Motors Corporation–United Automobile Workers, *Lockout Energy Control* (Training Workbook for Authorized Employees). UAW-GM National Joint Committee on Health and Safety, 1994.

that epitomizes this graphic approach (Figure 8-36). Complete and careful evaluation of each isolation practice is done before the graphic is sanctioned for use.

Of course, signage and graphic displays do not alone assure proper execution. However, they do represent another valid step toward the control of hazardous energy.

SUMMARY

- The major components of any hazardous-energy control-system are (1) planning, (2) system elements, (3) employee involvement, (4) training, (5) execution/application, and (6) designing for safety.
- Regardless of the degree of programming and preparation, the measure of energy-isolation effectiveness is in the execution.
- An action cycle for energy isolation is offered consisting of 12 steps as follows:
  1. notifying
  2. preparing
  3. shutting down
  4. transferring
  5. assessing
  6. isolating
  7. verifying
  8. acting
  9. inspecting
  10. clearing
  11. starting up
  12. stabilizing.
- The concept of personal or individual protection is emphasized while the need for flexibility of approach is considered in complex application situations.
- The risks and the related precautions attendant to machine, equipment, and process positioning/testing, shift change, and work interruption were reviewed.
- Electrical energy isolation and control is required for safe work practices by the 29 *CFR* 1910 standard.
- Hardware issues are overhead and gantry cranes; cranes and hoists, elevated disconnects, identification of disconnecting means, remote power switching, switches, and motors, troubleshooting and testing, and stored electrical energy.
- Hydraulic and pneumatic energy system control was reviewed with emphasis on isolating the power source(s), closing appropriate sectioning/control valves, and relieving internal pressures. The requirement for pressure to remain on a system under certain circumstances and the dangers under "jam" conditions were explained.
- Valves and piping systems were covered in detail because of their pervasiveness in hydraulic, pneumatic, and chemical energy situations.
- Identification of piping systems and their contents are an important dimension of energy isolation.
- Valves, as energy-isolating devices, were addressed, covering issues such as lockability, reliability, and sequence of closing.
- Line breaking (opening closed piping systems) was reviewed, and precautions were listed for safe access.
- Purging and cleaning of piping systems in preparation for isolation were discussed with reference to reactive chemicals, flammable liquids, explosive gases, and toxic materials.
- Chemical (corrosive, reactive, flammable, etc.) and thermal (steam, heated fluids, radiating sources, etc.) energy create challenging isolation problems due to their inherent nature. For isolation of chemicals, the valve, the blind, and the relief device are the key components. Various combinations of devices are necessary to safely isolate chemical forces under varying conditions.
- Thermal energy presents unique problems in that often the gained heat is not easily eliminated under typical service and maintenance activity. A series of techniques are provided for isolation of thermal energy sources.
- Gravitational and mechanical energy are frequently overlooked in many hazardous-energy-control processes. Yet they are often involved in serious mishaps. Large forces may develop with potential for violent and disastrous release. Criteria for restraints for gravitational energy included strength, stability, position, condition, appearance, identification, and design.
- Mechanical energy was defined as the forces of motion and the stored energy in the form of compression and tension. Equipment run-down periods, braking, restraints, and specialized safeguards are proposed as methods of control for mechanical energy sources.
- The importance of specific and alternative energy-isolating procedures was emphasized. Specific procedures were identified as the blueprint for application/execution. Focus was directed at the process of development of the procedures and its qualitative value. The essential characteristics of specific energy-isolation procedures were described. The use and value of signage and graphic displays were covered and examples provided.

REFERENCES

American National Standards Institute. *Scheme for the Identification of Piping Systems,* ANSI A13.1-1985. New York: American National Standards Institute, 1985.

Boylston RP. Locking and tagging guide for industrial operations. *Professional Safety* 27, no. 8 (Aug. 1982): 21–25.

Catena D, Dietz JT, Traubert TD. Do you know where your plant pipes are? *Occupational Health and Safety Canada* 8, no. 5 (Sept–Oct. 1992):72–74; 76–78.

General Motors Corporation–United Automobile Workers National Joint Health and Safety Committee. *Lockout.* Detroit: National Health and Safety Training Publication, United Automobile Workers–General Motors Corporation Human Resource Center, 1985.

Hans M. Don't get steamed. *Safety and Health* 146, no. 2 (Aug. 1992):40–42.

Kletz TA. *Hazards in chemical system maintenance: Permits.* In Fawcett HH, Wood WS, *Safety and Accident Prevention in Chemical Operations*, 2d ed. New York: John Wiley & Sons, 1982.

National Fire Protection Association. *Standard for Electrical Safety Requirements for Employee Workplaces,* NFPA-70E. Boston: National Fire Protection Association, 1988.

National Safety Council. *Accident Prevention Manual for Business & Industry,* 10th ed. Two-volume set: *Administration & Programs* and *Engineering and Technology.* Itasca, IL: National Safety Council, 1992.

National Safety Council. *Safeguarding Concepts Illustrated,* 6th ed. Itasca, IL: National Safety Council, 1993.

Occupational Safety and Health Administration. Instruction Standard 1-7.3, *The Control of Hazardous Energy (Lockout/Tagout)—Inspection Procedures and Interpretive Guidelines,* August 11, 1990.

Occupational Safety and Health Administration. Preamble and Final Rule for *Electrical Safety-Related Work Practices* [55 *FR* 31984, August 6, 1990].

United Automobile Workers–Ford National Joint Commitee in Health and Safety. *Energy Control and Power Lockout Mini-Manual.* Detroit: United Automobile Workers–Ford National Joint Committee in Health and Safety, 1987.

9

Monitoring, Measuring, and Assessing

OVERVIEW

In 1989, in Pasadena, Texas, an explosion in a petroleum processing plant killed 23 workers and injured more than 100. An investigation of the incident revealed that better monitoring of current lockout/tagout procedures might have prevented the disaster (Wheeler, 1991). When a section of the processing equipment had become plugged, a worker had removed the affected part. This left the remaining piping connected to a live system without any warning signs. When a co-worker inadvertently opened a connecting valve, the area was flooded with explosive hydrocarbons, which triggered the deadly accident.

This incident demonstrates that simply establishing a lockout/tagout system is rarely enough. Companies must practice ongoing vigilance to ensure that procedures are appropriate to their particular work conditions and that workers are trained in how to respond to unusual or unexpected situations (e.g., report them to a supervisor rather than improvise a potentially hazardous shortcut or quick-fix.) Simply because a company operates incident free for several months or years does not really mean that all is well. It may mean that its workers are highly ingenious in finding ways to cope with day-to-day working conditions that are potentially hazardous and that are not covered by existing lockout/tagout policies and procedures. The technique employed at the petroleum plant in Pasadena, Texas, for example, may have been used several times without incident. In a very real sense, employees may be working in such a way that only time and circumstance separate them from the eventual energy-release incident.

For this reason, it is important for companies to take the initiative in evaluating their safety systems and practices. What may be viewed as acceptable day-to-day operating procedures may be unacceptable from a risk standpoint. Unless a company asks the right questions and monitors its own lockout/tagout procedures, it will never know how it is really doing in the area of controlling energy hazards. The task of monitoring, measuring, and assessing a lockout/tagout system is one of the most important functions of management. Unfortunately, it rarely receives the thought and attention it requires. The best lockout/tagout system will work only as well as it is applied, evaluated, and improved to prevent unwanted energy releases and to accommodate changes in technology, job tasks, and the work force. Today's managers must learn to view this activity as part of a company's total quality management process and not simply as a regulatory compliance "hoop-jumping" exercise. The complex nature of work in today's business environments and the sophistication of modern machinery, equipment, and processes require more stringent monitoring, measuring, and assessment programs than ever before.

Connection with Total Quality Management

The old quality adage "Expect what you inspect" was never more true than when applied to lockout/tagout systems. Total quality management (TQM) should define the requirements for an effective lockout/tagout system and monitor the resultant organizational performance against them. TQM can be defined as a "system designed to help a company succeed by continually improving a product or service" (Roughton, 1993).

TQM is not a technical issue, however. Like any quality assurance program, it is a management system issue (Roughton, 1993). Many companies have traditionally viewed quality as the improvement of the production process or the reduction of defects. However, as applied to safety and health procedures, including lockout/tagout systems, TQM requires the following:

- Commitment by the corporation to quality in all areas of organizational life, particularly safety and health. The corporate philosophy must emphasize quality as a composite entity.
- Training and education at all levels in the organization.
- Continuous improvement programs implemented throughout the organization. This includes studying process variations, e.g., lockout/tagout procedures, to identify and eliminate causes of errors.
- Measurement systems must be used to identify areas where improvement opportunities may exist.
- Communication fostered between employees and management, emphasizing employee involvement and participation in quality improvement and sharing expertise.

Quality principles to practice include employee empowerment in which employees are asked to help solve problems, improve processes, and make decisions; conformance auditing to help prevent unsafe conditions and follow-up on recommendations; and improvement efforts that include data gathering and process analysis, innovation and design, updating methods and procedures along with equipment, and removing barriers to effective performance.

TQM, particularly as it relates to safety issues, is an overall commitment to continuously improve quality performance at all levels. Workers cannot address quality issues unless they know exactly what they should be doing, how they should be doing it, and what adjustments they need to make to do the job better or more effectively. They need specific measurements by which to judge performance, e.g., number of times a lockout/tagout procedure was violated or the number of times an energy source was not isolated properly. Several questions that a TQM-related monitoring, measuring, and assessing process must answer include (Crosby, 1984):

What lockout/tagout system is needed to improve safety in our organization?
What performance standard should we use?
What measurement system do we require?
How are data integrated into the feedback system and made accessible to management?

These questions relate to the problem of how does management know it is achieving its safety goals? There must be some specific system in place that can measure performance and progress and determine if a lockout/tagout system is doing the job it was designed or intended to do. Likewise, simply inspecting a process every six months or so without generating standardized data will not tell management how it is doing over the long term. Improvements made on the spot are not likely to find their way into a safety feedback system unless there is a commitment to gathering and recording these enhancements (Northage, 1992). The answer to the question "How are we doing in terms of preventing energy-release incidents?" must be more than a qualitative, "There seem to be fewer incidents this year in the Fabricating Division." There must be some quantitative measurements to back up claims for improvement or progress toward safety goals: "According to our auditing data, we have had 43% fewer incidents in our Fabricating Division over the past six months compared to one year ago." Likewise, worker behavior needs more specific corrective guidance rather than a vague recommendation to "be more careful next time" or to do better at "following the procedures." Workers need to know exactly what to do to prevent operational error and unexpected energy release.

In addition, benchmarks and measurements used to track lockout/tagout systems can offer a range of information on an organization's work procedures, personnel, and plant safety practices that would go undetected without a strong monitoring, measuring, and assessing process. It is important to understand the nature of such a system in more detail.

Three Fundamental Tasks

The process cycle for controlling hazardous energy, introduced in Chapter 6, The Process Approach, includes the steps of monitoring and evaluating. According to the steps in the cycle, once a hazardous-energy-control system is designed and implemented, its results should be monitored or audited against the organization's goals. The results can then be evaluated with some form of measurement and assessment of progress or performance. The final step would involve applying the conclusions to shape or refine the system and create a revised design. This type of ongoing cycle where monitoring, measuring, and assessing occur over and over is critical to the prevention of hazardous-energy-release incidents.

Although some managers view monitoring, measuring, and assessing as interchangeable, they are distinctly different tasks:

- *Monitoring* has to do with the development and use of inspection and/or auditing systems. It establishes methods, techniques, and procedures through which an activity can be observed in a systematic manner. This task tells everyone involved what is expected of them and what results the company or organization is trying to achieve.
- *Measuring* is the establishment of specific yardsticks by which to evaluate performance or success. It is the process of comparing actual accomplishment against specific criteria and recording or reporting the outcome. The output or data can be channeled into a feedback system to provide "early warning" that a procedure, policy, or training program is not achieving the desired results. Measuring addresses such questions as *how often* is something being done, *who* is doing it, *where* is it being done, that it is being done properly, and *what happened* when it was done.
- *Assessing* involves interpreting, evaluating, and comparing information against stated objectives and goals. It involves qualitative judgment as to the meaning and importance of what has been observed and measured. The task should tell an organization *how well* they are doing regarding their lockout/tagout safety goals and what needs to be corrected, improved, or changed. Such information applies both to the mechanical aspects of lockout/tagout procedures and to the more elusive area of worker behavior (Tarrants, 1980).

Continuous Improvement

TQM and monitoring, measuring, and assessing are directly related to the process of continuous improvement. As nearly every manager knows, things do not improve on their own. Improvements, even in small increments, are the product of identifying and correcting defects in any system. Monitoring, measuring, and assessing are essential tools in determining what is in need of attention.

Continuous improvement requires active employee involvement in helping to detect and solve problems related to lockout/tagout systems and equipment. Training programs must alert employees to the potential limitations or shortcomings of any lockout/tagout system. Employees should inform their supervisors when conditions arise that normal procedures do not cover or that may introduce unrecognized hazards into a normally adequate system. Rather than receiving rote training in existing lockout/tagout procedures, employees can be taught to apply appropriate concepts and principles to control energy-isolation hazards. They and their fellow employees can view each situation as unique and collaborate with their supervisor in real-time problem solving. In turn, management needs to understand that they must continuously reexamine existing procedures to go beyond mere compliance and ensure that the system adequately reflects current working conditions. This task can be made easier by an effective management information system.

Feedback: Managing Information for Lockout/Tagout

In many companies, information gathered and evaluated is not always readily available to those who need it. As part of an effective monitoring, measuring, and assessing process, information must be formatted so that managers can retrieve reports on incidents, improvements, updates, or other data. The formats must be more than mere statistics, e.g., simply listing how many incidents occurred in a given time period, how many workers were injured, or how many productive hours were lost.

The data should indicate such information as during what work period most incidents occurred. For example, if more incidents occurred late at night, is there something about the 11 p.m. to 7 a.m. shift that makes employees more vulnerable to work-related injury? Perhaps there is inadequate supervision during that time, or workers take more shortcuts around the lockout/tagout procedures than do employees on other shifts.

Likewise, when corrections or improvements are made, the feedback loop should inform management about the outcome. Did the changes have the desired results? When new procedures were introduced, what happened? Did workers accept or reject them? What impact did the procedures or changes have on productivity or work processes?

Management must see to it that feedback within the information system documents the who, what, where, when, how, and why factors. It should not simply report how many incidents occurred or safety procedures were established or changed. Mere number crunching does not tell management nearly enough about how the company is doing in its efforts to improve quality and safety.

The remainder of this chapter will outline guidelines and suggestions about how to create an effective monitoring, measuring, and assessing process specifically for lockout/tagout activities.

MONITORING

Monitoring can be done through an internal, external, or custom approach. An *internal approach* includes:

- Self-assessment in which a local entity monitors its own system and procedures.
- Staff assessment in which company professional personnel from regional or corporate headquarters monitors lockout/tagout systems and safety procedures.
- Team assessment in which personnel from various facilities go to a particular location to monitor lockout/tagout activities.

An *external approach* involves hiring or using personnel from outside the company entirely:

- Professional consultants specializing in safety procedures or audits who are knowledgeable about lockout/tagout issues and activities.
- Government-consulting personnel (local, state, or federal) acting in an advisory capacity.
- Loss control personnel (insurance related) dedicated to specific task accomplishment.

A company also could devise a *custom plan* that uses a combination of these approaches in tiers timed to a regular cycle. For example, self-assessment could be done once a month, and staff, team, or external assessment could be done once every six months. Whatever approach a company decides to use, it is essential that some type of monitoring be done.

Importance of Monitoring

Both OSHA and ANSI stress the importance of establishing a monitoring system that clearly delegates responsibility for lockout/tagout procedures and mandates periodic inspections. As is stated under Section 3, Lockout/Tagout Policy and Procedure of ANSI Z244.1:

> 3.2 *Responsibilities*
>
> *3.2.1 Authority.* Compliance with this standard shall be the responsibility of the employer and the individual(s) to whom the employer delegates accountability for compliance and authority to enforce compliance.
>
> *3.2.2. Periodic Inspections.* It shall be the responsibility of the employer to verify, through periodic inspections, the organization's compliance with this standard.
>
> *3.3.3. Joint Responsibility.* The responsibility for obtaining performance in the lockout/tagout procedures shall be joint between the employer and the employee. The responsibility for compliance is that of the employer, who shall establish, communicate, train in use, and enforce procedures. The employee shall be responsible for knowing and following the established procedures. (ANSI, 1982, p. 9.)

Admittedly, the area of lockout/tagout safety and compliance is not as easily quantifiable as various aspects of the manufacturing process. Yet with a little foresight and planning, management can easily make monitoring lockout/tagout part of a proactive safety audit that can yield meaningful, quantitative results.

The problem arises when companies see no need to go beyond the minimum compliance implied in OSHA and ANSI Z244.1. They overlook the fact that their particular situation and work conditions may pose hazards that mere compliance cannot adequately address.

Components of the System

As in any other safety auditing system, there are three general components to lockout/tagout monitoring: administrative/management policies, auditing goals, and inspection/auditing procedures.

Administrative/management policies. The commitment of top management to lockout/tagout must be clearly communicated to all employees. Once this commitment is known, workers generally will be more willing to cooperate with the monitoring process (National Safety Council, 1989). Administrative policies should establish who is responsible for monitoring, whether a safety professional or an auditing team; the scope of the audit; how often the auditing is to be conducted; and how the team is to report the data they gather (National Safety Council, 1989). How often the audit should be done depends on how large the facility is, how many lockout/tagout procedures it has, how frequently they are used, and how often they must be revised or adapted in response to changes in equipment or processes.

Auditing goals. Established goals should go beyond satisfying the minimum requirements of OSHA or other lockout/tagout mandates and standards. Organizations should strive to obtain as complete a picture as possible of the actual energy-isolation situation. Goals can be established for the following:

- achieving employee training proficiency
- acquiring and maintaining the most progressive lockout/tagout equipment

- creating effective isolation procedures
- motivating employees to use the procedures
- maximizing employee involvement
- developing alternative measures for energized tasks
- maintaining high rates of lockout/tagout conformance.

These goals can be used to focus the audits on specific lockout/tagout issues and track how the system is working over time. If inspections/audits are conducted every few months, for example, an organization should be able to obtain quickly an accurate picture of what works and doesn't work on the job. There would be little guesswork and more opportunity to identify hazardous situations before they resulted in an incident.

Inspection/auditing procedures. Each organization should establish precisely how auditing will be conducted, including any questionnaires, forms, and reports that will be used. This step will need to be coordinated with employees and supervisors to ensure maximum cooperation. The auditors should seek to be thorough without disrupting production, maintenance, or repair operations any more than is reasonably necessary.

Approaches to Monitoring

There are many models of safety audits that can be adapted for use to monitor lockout/tagout systems and procedures. The important point is for the monitoring individual or team to observe workers performing actual lockout/tagout procedures and to interview employees to find out how well the systems work, i.e., actual versus plan/practice. The audit should cover three basic areas (Geller, 1992, p. 16).

Machinery, equipment, processes, and workplace conditions. What are the existing energy sources, machine/equipment design limitations, location of energy-isolating devices, layout issues, environmental factors, etc. that workers who use lockout/tagout must deal with on a daily basis?

Lockout/tagout procedures and training. Are procedures up to date? Are they accessible and available for use? When followed, do they achieve zero energy state? Are alternative procedures available?

How is training conducted, and does it appear to be effective? Do training techniques address the needs and problems of the work force? Are the training programs updated to keep abreast of changes in the workplace and in lockout/tagout equipment and procedures?

Employee behavior, perceptions, and attitudes regarding lockout/tagout. Do employees believe the procedures or equipment are adequate, cumbersome, too elaborate, ineffective, unnecessary, unrealistic? This area will give the auditing team information on how workers are actually using the equipment and procedures. If workers believe procedures for isolating energy sources are impractical, they are likely to ignore them altogether or find ways to circumvent them. This area should also uncover employee or management acts of omission, commission, or oversight regarding lockout/tagout procedures.

A closed audit loop model (Figure 9-1) provides for feedback into the process that will help achieve lockout/tagout goals (Northage, 1992). In the model, the "Questioning box" could be replaced by "measuring," and the "analyze data/prioritize box" could be replaced by "assessing." Any corrective action will be fed back into the auditing system and rechecked on the next audit. As a result, the auditing team will collect a growing body of information on real-world conditions in the plant and how changes or additions affect lockout/tagout safety procedures.

Because lockout/tagout procedures are used only in specific situations, auditing cannot be random but will have to be done on a spontaneous (breakdown) or planned (schedule) basis. Additional audits or inspections can be initiated either at the discretion of management or for such instances as when a problem either with equipment, procedures, or worker behavior does not appear to be responding to corrective action.

Training the Auditing Team

All too often the shortcoming in lockout/tagout systems lies in the fact that no one completely understands the nature of the energy or the machinery that needs to be isolated or locked out. The lockout/tagout auditing team must be trained to understand the following:

- The energy systems used. The combination of energy used (e.g., electrical, hydraulic, steam, mechanical, gravitational) and how to best reach zero energy state.
- The machinery itself. With rapid changes in technology, robotics, and computer programs, keeping up with current generation machinery can be a challenge. The auditing team should be kept up to date on changes in control, energy isolation, interconnectedness, automation, and any other aspects of the machinery that could pose a hazard in terms of lockout/tagout procedures.
- Lockout/tagout equipment and procedures. The team must be able to detect if the current equipment for lockout/tagout is appropriate for the hazards posed by the energy sources and machinery, equipment, or process. In many cases, accidents are caused by equipment limitations or deficiencies and not by any error on the part of workers. Team members should be aware of tasks done under energized or partially energized conditions.
- The work force. The auditing team must be familiar with the strengths and weaknesses of the work force. Are there problems with language barriers, cultural differences, or educational or skill differences that

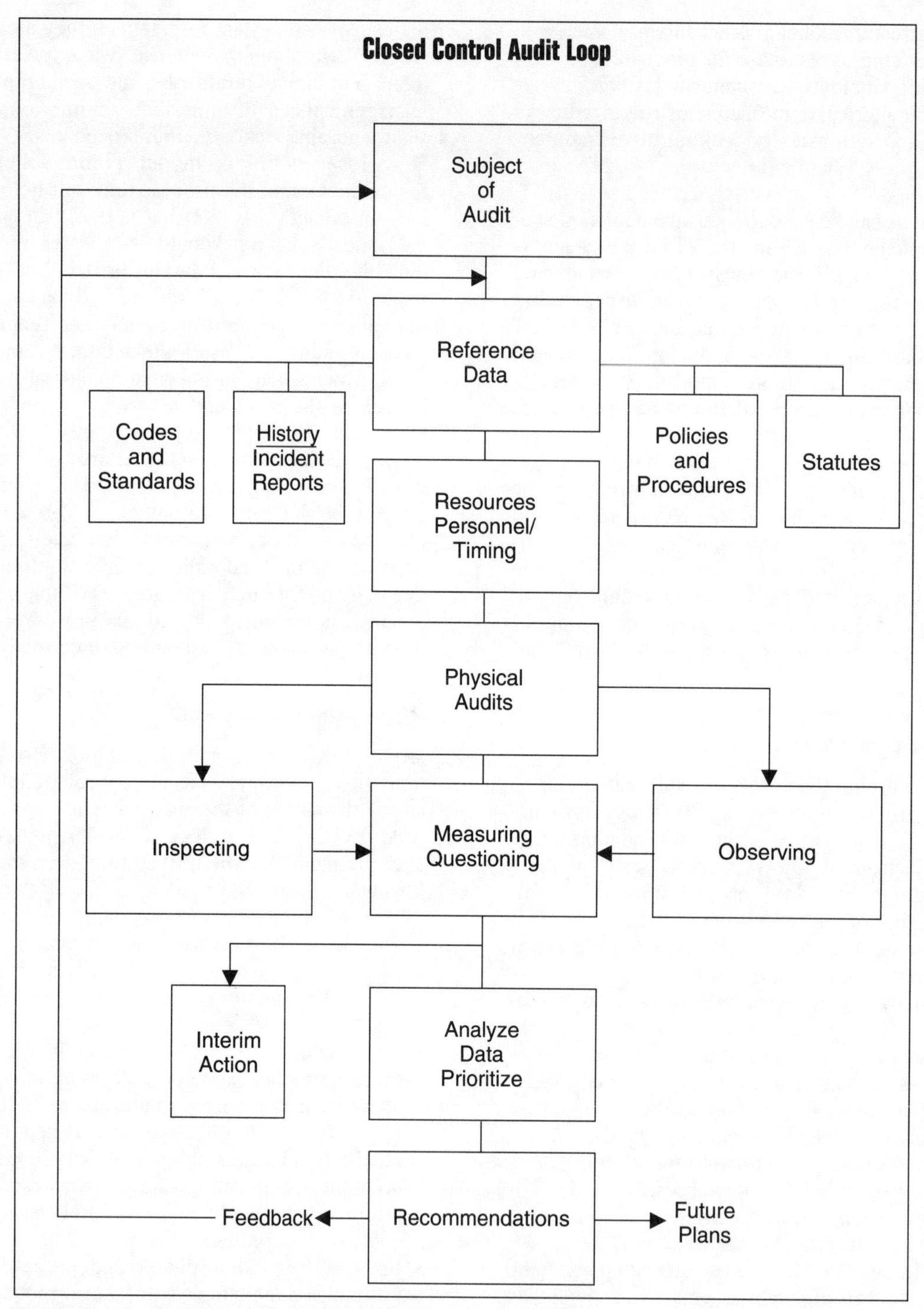

Figure 9-1. Example of audit process as a closed control loop.
Source: Northage, El. "Auditing: A Closed Loop." *Occupational Hazards,* May 1992. © National Safety Management Society.

may pose hazards in using lockout/tagout systems effectively? The team should take into account not only *when* lockout/tagout must be used but *who* will be using it during different shifts or different production or work projects.

- The work processes and environment. The team must know how and under what conditions work is actually performed. Lockout/tagout procedures on paper may be straightforward, but procedures are not performed under ideal or theoretical conditions. For example, applying lockout/tagout during weekend shut-downs can be completely different from using lockout/tagout to isolate a continuous process line in the middle of a production run.

Knowing something about all of these variables can help the team properly set up inspections and/or observations to gather data on how lockout/tagout actually works. This involves more than a perfunctory checklist or a quick observation of workers applying lockout/tagout procedures under optimal conditions.

Auditing Techniques

To fulfill the monitoring goals and objectives the organization may set, the auditing team can use a variety of techniques to obtain information (Figures 9-2 to 9-4). The major techniques include:

- *Physical inspection*. This is the most time-honored method of acquiring data and the one that is likely to yield considerable information. However, the team must know what they are looking for so that the inspection will be of real value. It is imperative that specific elements are examined rather than conducting a general overview. A skilled eye with a tight focus will generate valuable observations that can be effectively acted on.
- *Interviews*. Interviews can be a key resource for determining worker attitudes, complaints, and suggestions about lockout/tagout procedures and equipment. Interviews of operators, supervisors, and maintenance crews should be conducted.
- *Video task analysis*. Procedures can be videotaped for more detailed review and analysis at a later time. This will enable the auditing team and others to see far more than if they were just to observe an operation one time only. Videos also can be used for training to show procedures in detail, offering the ability to stop action, reverse, and repeat steps to reinforce or explain various steps in the lockout/tagout procedure.
- *Operator/employee feedback*. The team must establish some type of system in which workers can submit their concerns, observations, complaints, or suggestions at times other than formal audits or inspections. Not all important information will be gathered during interviews (Figures 9-5 and 9-6 for sample recommendation forms).
- *Computer information systems*. Data can be formatted to enable the auditing team to compare the results of audits and inspections over time. This information can help the team spot gaps in lockout/tagout procedures, flaws in the system, lapses in training, and problems with worker attitudes or behavior. A feedback system will enable these data to be entered regularly into the monitoring, measuring, and assessing process to help the organization achieve continuous improvement.

These computer systems can also track improvements in energy isolation, reductions in incidents, and decreases in injuries and downtime to show the benefits in quantitative terms. As the data base grows,

Chicago XYZ Plant
Lockout/Tagout

Self-Auditing Compliance Checklist

	Yes	No	N/A
• Does the lockout procedure require that stored energy (mechanical, hydraulic, air, etc.) be released or blocked before equipment is locked out for repairs?	___	___	___
• Are appropriate employees provided with individual keyed personal safety locks?	___	___	___
• Are employees required to keep personal control of their key(s) while they have safety locks in use?	___	___	___
• Is it required that only the employee exposed to the hazards place or remove the safety lock?	___	___	___
• Is it required that employees check the safety of the lockout by attempting a start up after making sure no one is exposed?	___	___	___
• Are employees instructed to always push the control circuit stop button prior to re-energizing the main power switch?	___	___	___
• Is there a means provided to identify any or all employees who are working on locked-out equipment by their locks or accompanying tags?	___	___	___
• Are a sufficient number of accident preventive signs or tags and safety padlocks provided for any reasonably foreseeable repair emergency?	___	___	___
• When machine operations, configuration, or size requires the operator to leave his or her control station to install parts or perform other operations, and that part of the machine could move if accidentally activated, is such element required to be separately locked or blocked out?	___	___	___
• In the event that equipment or lines cannot be shut down, locked out, and tagged, is a safe job procedure established and rigidly followed?	___	___	___

Figure 9-2. Self-auditing lockout/tagout checklist.
Source: Developed by E. Grund. © National Safety Council.

Lockout/Tagout (LO/TO) Audit

Plant ________
Date ________

Reference: OSHA 1910.147	Yes	No	Explanation
1. Does plant have local written LO/TO policy-procedure?			
2. Does plant have specific de-energization procedures for major machinery/equipment/process?			
3. Does the plant have energy-isolating devices properly identified? Indicate percent devices identified.			
4. Does the plant use lockout, tagout, or combination?			
5. Are authorized, affected, and other employees trained in LO/TO (a) general (b) specific?			
6. Does the plant LO/TO procedure require verification of isolation?			
7. If tagout in used, are tags durable with strong securing means? a. Are employees trained in limitations of tags?			
8. Has the plant conducted an application and exposure survey?			
9. Have tasks requiring "power on" been identified? Is "equivalent protection" in place and adequate?			
10. Is compliance auditing done and documented?			
11. Is there evidence of enforcement (warnings, counseling, discipline)?			

Comments:

Loss Prevention Rep. ________

Figure 9-3. Lockout/tagout compliance audit.
Source: Developed by E. Grund. © National Safety Council.

management will be able to evaluate progress against plan more effectively.

Reporting Results and Follow-up Procedures

After the audit is conducted, the team should prepare a report on their findings. Reports generally can be classified into three categories (National Safety Council, 1990):

- Emergency. Lockout/tagout equipment and/or procedures or employee training and/or behavior should be changed without delay because the current situation constitutes an immediate or imminent danger.
- Periodic or intermediate. This report covers unsatisfactory lockout/tagout equipment and/or procedures or employee behavior and/or training that should be addressed in the near future. The team should assign a specific target date for improvement and the recommended steps for achieving goals.
- Summary. This report lists all results of the audit, including improvements noted since the last audit, and current strengths and weaknesses of lockout/tagout systems.

In making recommendations for improvement, the auditing team should be guided by three rules (National Safety Council, 1990):

HAZARDOUS ENERGY ISOLATION
BEHAVIORAL PRACTICES AUDIT

Job ________________ Employees ________________

Area ________________ Date ________ Time ________

LO/TO Procedure Reference ________________

(−) Degree of Compliance (+)

Safe Practices	0	1	2	3	4	Notes
Was proper notification given?						
Was procedure/sequence followed?						
All energy sources isolated?						
Appropriate protective appliances used?						
Did isolation verification occur?						
Appropriate supplemental safeguards?						
Proper protective equipment?						
Stored energy dissipated?						
Proper work position taken?						

Comments:

Improvements:

Performance Score

Figure 9-4. Hazardous energy isolation: behavioral practices audit.
Source: Developed by E. Grund. © National Safety Council.

- Recommend correcting the cause, not just the result, of the problem. If machine design is the root cause of a problem workers are having isolating energy, the solution is not simply to prepare a better lockout/tagout procedure. It is primarily to correct the design flaw or to develop an engineering safeguard feature for the equipment.
- Report conditions even if they are outside the auditing team's official authority. Safety hazards do not respect bureaucratic boundaries. The team should be alert to other safety hazards when auditing lockout/tagout systems.
- Take action immediately as needed. If a permanent solution is going to take time (ordering new equipment, changing procedures, etc.), the team should recommend immediate temporary controls to protect workers. These can include reducing the amount of time the machine is in operation, placing a provisional safeguard on the machine, taking the machine or equipment out of production, posting warning signs, or having supervisors carefully oversee any lockout/tagout operations.

The auditing team, in some instances, may write up all three types of reports for management. The emergency report would require immediate management attention, whereas the summary report could be circulated or presented at regular safety committee meetings.

Early Detection of Trouble

As previously mentioned, the existence of well-developed, long-standing lockout/tagout systems does not necessarily signal optimum control. Further, even if energy-release-incidence experience is reasonably good, the apparent success may be only a deception. The following case history from a progressive

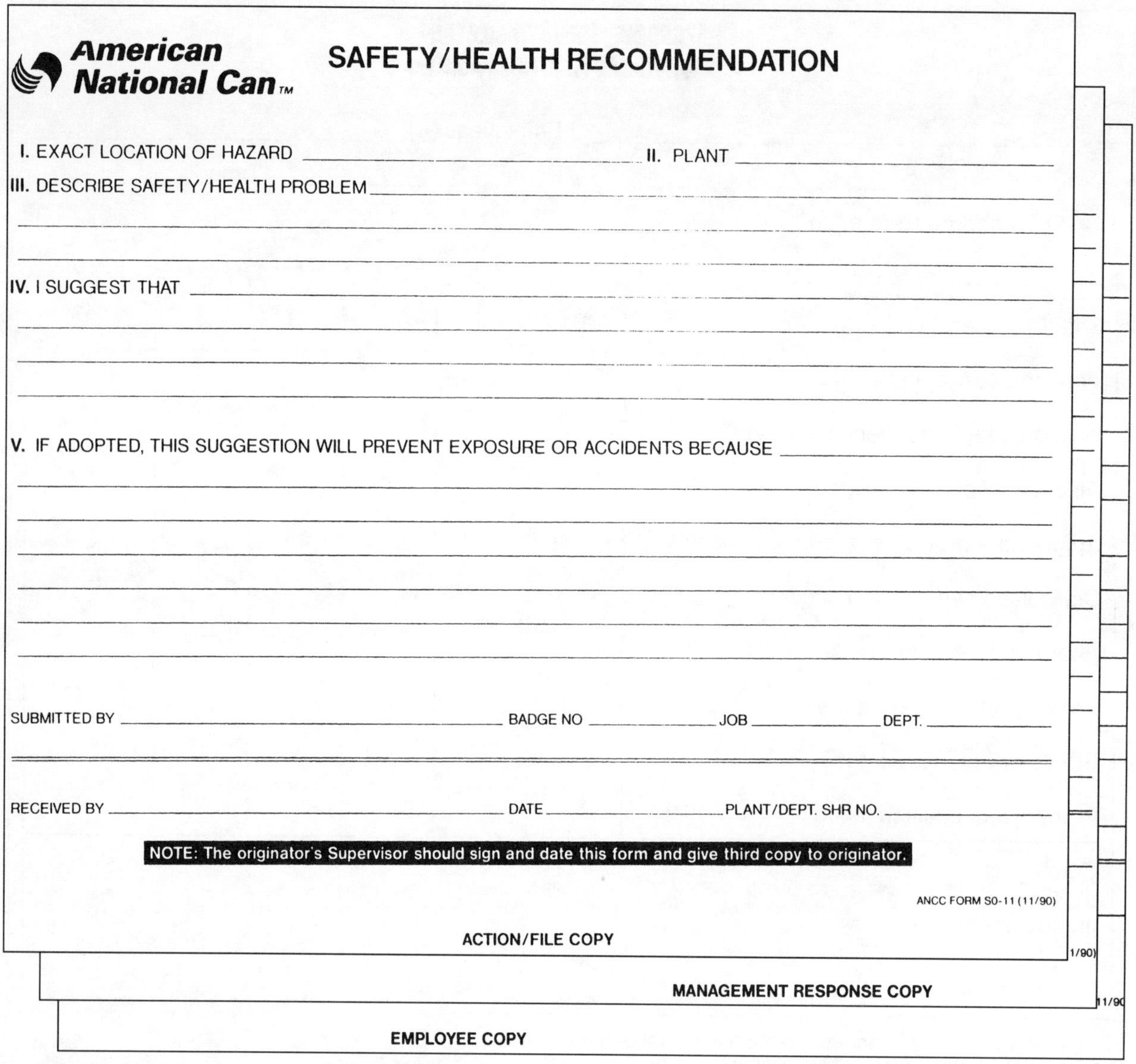

American National Can™

SAFETY/HEALTH RECOMMENDATION

I. EXACT LOCATION OF HAZARD ______________ II. PLANT ______________

III. DESCRIBE SAFETY/HEALTH PROBLEM ______________

IV. I SUGGEST THAT ______________

V. IF ADOPTED, THIS SUGGESTION WILL PREVENT EXPOSURE OR ACCIDENTS BECAUSE ______________

SUBMITTED BY ______________ BADGE NO ______ JOB ______ DEPT. ______

RECEIVED BY ______________ DATE ______ PLANT/DEPT. SHR NO. ______

NOTE: The originator's Supervisor should sign and date this form and give third copy to originator.

ANCC FORM S0-11 (11/90)

ACTION/FILE COPY

MANAGEMENT RESPONSE COPY

EMPLOYEE COPY

Figure 9-5. Employee recommendation form (*front*).
Source: Developed by E. Grund. © American National Can.

employer illustrates how trouble can exist and eventually cause tragedy if it goes undetected and untreated.

Case history. A large, hydrochemical plant employing 1,000 workers that processed and refined minerals operated 24 hours per day, 7 days per week, on a continuous annual basis. The plant infrequently shut down for process turnaround and periodically took equipment out of service while maintaining the operational output.

The bauxite ore was prepared for processing and limestone impurities were separated by using a Bradford Breaker, a cylindrical drum 22 ft (7 m) long and 14 ft (4 m) in diameter, rotating at 12 rpm. The ore was transferred by conveyor belt and gravity-fed into the Breaker. The screened/sifted ore was then moved by conveyor to storage silos before processing. The Breaker was located on the second level of a four-story building. The motor control center and control room were located on the ground level east of the Breaker building (Figure 9-7). Whenever the moisture content of the bauxite entering the Breaker exceeded 17%, the ore stuck to the drum's interior walls, eventually covering the openings in the mesh and plate, forcing operations to shut down the Breaker for cleaning. A bypass arrangement was used that permitted removal of the Breaker from operation while continuing the flow of unscreened ore into the digestion section.

During the day shift, the interior walls of the Breaker had to be cleaned of wet bauxite by maintenance personnel. However, before the end of the day shift, the bauxite buildup required another round of cleaning. The Breaker was once again scheduled for cleaning on the swing shift. A four-man crew was assigned to perform the work, and the equipment was

REVIEW AND DISPOSITION

REVIEWED BY	DATE	DISPOSITION
1.		
2.		
3.		

NOTE: Final disposition should indicate how and when originator was advised of final action.

EVALUATION

1.

2

3

YOUR SAFETY/HEALTH RECOMMENDATION HAS BEEN:

☐ FAVORABLY CONSIDERED, and corrective action:

☐ Has been taken.

☐ Will be taken by ____________ (date).

☐ BRIEFLY DESCRIBE CORRECTIVE ACTION ____________

☐ HAS MERIT, POSSIBLE FUTURE ACTION, explain ____________

☐ UNFAVORABLY CONSIDERED, because ____________

Your interest in the safety and welfare of all personnel is sincerely appreciated by Management as well as by your fellow employees. Please continue to be alert for possible hazardous conditions or practices and continue to bring them to the attention of supervision.

DATE SIGNATURE POSITION

Figure 9-6. Employee recommendation form (*back*).
Source: Developed by E. Grund. © American National Can.

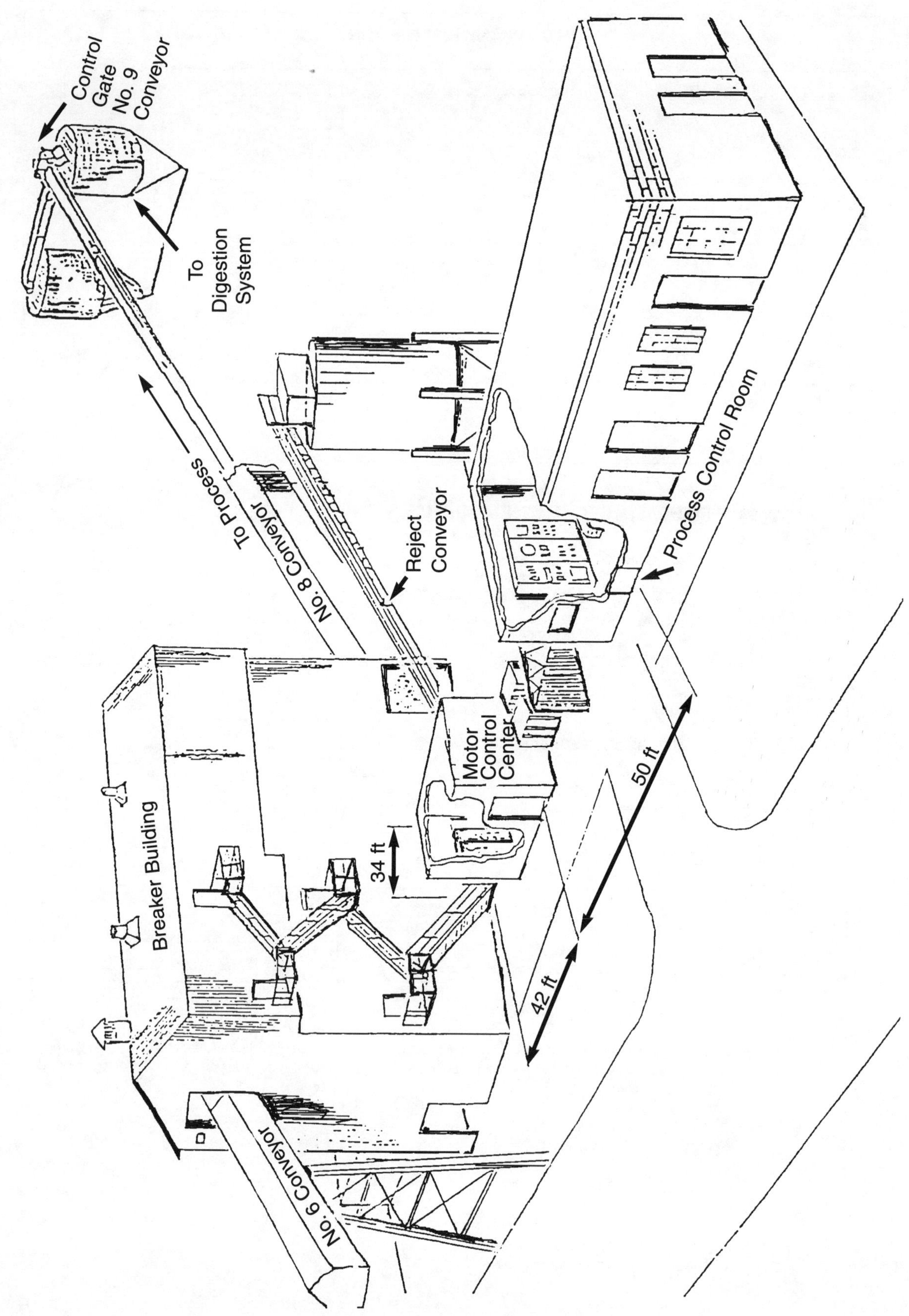

Figure 9-7. General view: hydro-chemical plant energy-release incident.
Source: Mining Safety and Health Administration.

tagged out and locked out, and the overload heaters were removed from the system's circuit breaker. The first two crewmen worked together to clean the lower half of the Breaker. The standard procedure was to work on the lower half of the Breaker, remove all tools and personnel, then request Operations to rotate the Breaker to run out loose ore so the remaining half of the drum can be cleaned. The basic procedure was as follows:

1. Bypass the screen and maintain ore levels in the surge silos.
    a. De-energize, remove overload heater, and lock out the screen's electric controls.
    b. Remove buildup from the lower screen interior surfaces.
    c. Remove equipment and personnel.
    d. Unlock, replace heaters, and re-energize screen electric controls.
    e. Rotate screen for 10 minutes to purge loosened material from the interior. Repeat (a) through (e) as required to remove remaining buildup.
    f. Remove overload heaters and the lockout of the screen.
    g. Replace feed and screening media ring.
    h. Return unit to operating department.
2. Return to routine operating mode.

At 7:00 p.m., the Breaker was rotated to allow the second two crewmen to finish the cleaning work. The electrical disconnect for the drum was tagged out first with a white Operations tag. The shift electrician secured a green Information tag and placed a red Maintenance tag with his supervisor's name on the electrical disconnect, then removed the heaters. The second two crewmen each secured a lock and blue Personal safety tag to the main disconnect and began the final phase of the cleaning work.

At 8:05 p.m., the crew removed their tools and equipment from the Breaker, and by 8:20 p.m. the maintenance work had been finished. The shift electrician was informed by his supervisor that the cleaning work was complete and that he could reinstall the three heaters on the electrical disconnect. The electrician also removed the green Information tag and the red Maintenance tag. The maintenance crew removed all Personal tags and locks. A process operator then removed the white Operations tag and placed the disconnect handle in the ON position. So far, everything had been done according to proper lockout/tagout procedures.

Unfortunately, three supervisors had decided to quickly inspect the maintenance crew's work in the Breaker and had entered the drum at the upper level without informing the workers in the motor control center. They assumed the unit was still de-energized because of additional work that needed to be done on the Breaker. There was no alarm or other indicator to signal those in the control room that someone had entered the Breaker.

A process operator inside the process control room knew that mechanics were scheduled to replace several sections of the Breaker's steel plate with wire screen later in the shift. It was necessary to run the Breaker and purge the drum of loose ore before the mechanics could do their work. At 8:30 p.m., once the electrical disconnect was activated, the operator walked to the west side of the control room and pushed the START button to run the Breaker. The supervisors were trapped inside the rotating drum.

At 8:40 p.m. another worker returned to the control room and pressed the STOP button to shut down the system. He walked up the stairway leading to the Breaker and crossed to the manway to determine if all of the ore had been cycled out of the drum. When he looked into the Breaker, the worker saw the bodies of the three supervisors. He immediately summoned help and notified his supervisor of the incident.

Several primary and contributory causes were identified for this unexpected energy-release incident:

- The supervisors failed to personally tag/lock the electrical disconnect before entering the Breaker.
- The process operator failed to check the Breaker visually before deciding to cycle the equipment.
- No prestart warning alarm existed for the Breaker but did exist for conveyor belt start-up.
- The supervisors failed to notify the operators about their intention to inspect the Breaker.

There were other distal contributory causes detected as well. However, the critical discovery was a flaw in the basic lockout/tagout system that had existed since the plant began operation years before. Supervisors were not required to use personal locks/tags for individual protection. The logic for this was deeply imbedded in the idea that supervisors were moving in and out of many jobs and that the equipment was always well isolated by the various craft personnel. Apparently this had worked quite well for years without reason for alarm.

Monitoring by qualified individuals equipped to probe and question would likely have detected this flaw in the lockout/tagout system long before this incident occurred. The tragic event provides several warnings that should be heeded:

- The release of hazardous energy exempts no one, regardless of status.
- A mechanism for detection and correction of flaws and deficiencies must be a prominent part of every lockout/tagout system.
- Auditing by serious inspectors should not be superficial, casual, or simply comforting to management.
- An absence of incidents, close calls, and complaints does not necessarily mean "all is well." Management must learn to look deeper into its

lockout/tag-out system and to ask the questions no one previously thought to ask.

MEASURING

OSHA does not specifically require measurement in its lockout/tagout inspection guidelines for compliance officers (OSHA Instruction STD 1–7.3, September 11, 1990). However, most companies consider this step essential in evaluating how effective their lockout/tagout procedures are on the job.

Measuring is an attempt to gauge performance against standards/requirements in quantifiable terms that can be used to help improve lockout/tagout systems. After the initial audit has been conducted, the company should develop yardsticks or benchmarks to establish some objective means of measuring progress toward effective control of hazardous energy. To develop an effective measurement system, management must answer such questions as (Petersen, 1980):

- What is to be measured?
- How and when is it to be measured?
- How are results to be reported?

Components of a Measurement System

As in the case of the monitoring phase, measurement systems require the commitment and backing of top management. This is a proactive task that probes into sensitive areas, and it generally will not succeed without the full support of management. Employees must be motivated to cooperate fully with the teams assigned to do the measurement work.

A measurement system generally should consist of the following (Weinstock, 1993):

- A group or organization with clearly defined responsibilities and tasks. The leader should have the authority/endorsement of upper management and be competent with regard to the assigned tasks. The tasks may change as the monitoring and measuring processes develop, but the group must start out with a clear idea of what is their objective.
- Objectives outlining what is to be measured, e.g., activities, performance, attitudes, equipment, and procedures. The team must know specifically what they are measuring. They must ask such questions as "What do we need to do? What are the issues and problems—more training? Too much pressure? Lockout/tagout procedures?"
- Measurement methods and techniques. This involves checklists, task analysis, simulations, reliability data, and comparison with other systems. Developing these instruments may take time and ingenuity on the part of management and the team doing the measurement. Outside help may be needed in developing reliable, accurate measurement techniques.
- Precise guidelines defining how and when measurement should be done. Should measurements be done by rating sheet, checklist, weighted values, strength of response? Should measurements be taken every month, every six weeks, every three months, six times a year? What sampling strategy is most representative?
- Data formats for reporting results. How should data be presented to management? In raw or summary form? In summary form with raw data appended? How often should data be reported? In weekly, monthly, or semiannual summaries? Are executive status reports required?

What Should Be Measured

Basically the areas and activities to be measured will be the same as those covered in the monitoring phase: physical conditions; lockout/tagout equipment, systems, and procedures; and human behavior. Avoid the tendency to measure what is easiest to do.

Prioritize the areas for monitoring/measuring to place more emphasis on those that are more critical to the protection of workers, e.g., administration versus energy-isolation procedures.

Sampling Techniques

Before any measurement technique can be used with accuracy to evaluate safety activities, it must be standardized and be capable of objective and continued application (Petersen, 1980). The measurement technique should meet three criteria:

- The same yardstick must be used in the same way each time a measurement is made. The yardstick cannot be used to measure lockout/tagout procedures one time and worker behavior another unless it has been specifically designed to do so. The measurement of application of lockout/tagout procedures may prove quite different than measuring individual worker behavior. Consistent application of measurements builds confidence in the accuracy and reliability of the generated data.
- What is being measured must possess certain characteristics that render it measurable. For example, how does one measure "safe behavior"? Behavior must be measured against written procedures or some other definitive standard that can be used to judge objectively whether behavior is safe or unsafe (what is the specific criteria?). Otherwise, the measurement becomes too arbitrary, and workers may feel they are being judged unfairly.
- Workers and supervisors must be able to relate their behavior and safety activities directly to the measurement standard. They must be shown that the standard is relevant to their actual work experience and conditions and that the improvements they are called on to make are possible given their circumstances.

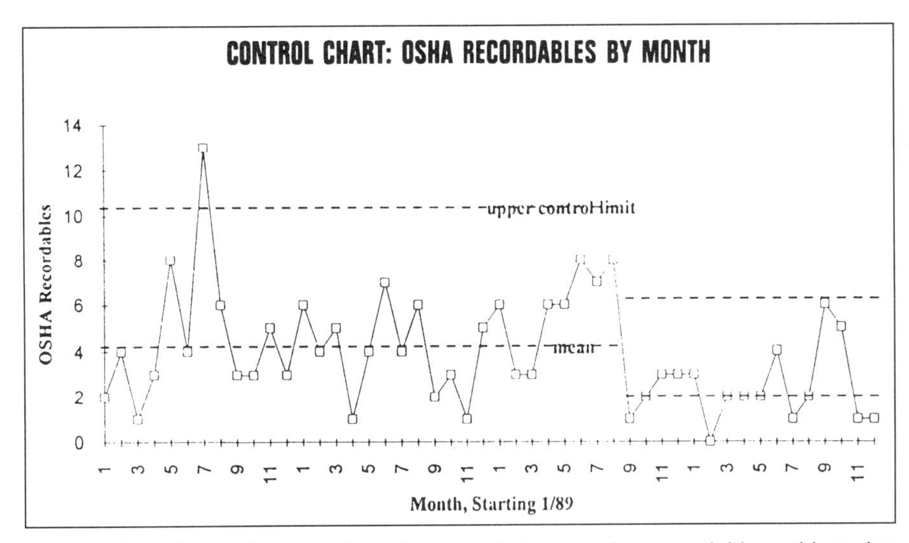

Figure 9-8. This graph is a portion of a control chart plotting recordable accidents by month, both before and after intervention, which began with supervisor/manager training in June 1991. Three months later, the accident rate dropped and remained below the initial process mean for seven consecutive months.
Source: Carder, B. "Quality Theory and the Measurement of Safety Systems." *Professional Safety,* Feb. 1994, p. 26.

The criteria and any weighted or factored values of yardsticks should be reviewed and changed to reflect progress and/or changing conditions. However, for valid comparative purposes, the team must use the same set of criteria and values throughout the organization for the duration of the assessment.

Sampling and measurement techniques include the following:

Checklists and evaluation guides. These can be developed to detect flaws or errors in key lockout/tagout steps and critical worker behaviors. Checklists are a good way to obtain a quick, overall picture of the situation being measured but are not as useful for gathering more detailed information.

Task analysis. This technique can break down lockout/tagout procedures into a series of measurable steps in which each task is analyzed for its effectiveness, safety, and necessity. This is an effective tool for determining precisely how lockout/tagout is used on the job and in what ways it can be improved.

Simulation. Computer programs now permit even small companies to do relatively sophisticated simulations of energy isolation and lockout/tagout systems application. These programs can be used to analyze the "rare event" incident, to anticipate problems, and to test various solutions before attempting them on the job. Simulators are also a good way to test trainees' knowledge once they have completed a training course. They can be used to measure performance using selected criteria with weighted values. Everyone would be evaluated using the same criteria.

Comparative analysis. The same tasks/employee performance can be evaluated at different times to identify positive or negative variance. Differentiation can quantify the impact of various intervention initiatives (before and after technique). Specific points of comparison can provide key measurements in evaluating employee performance and lockout/tagout application.

Control charts. Control charts can be used to track performance. They define upper and lower boundaries within which variation is statistically normal or expected. If accident frequency is used to measure system performance, data are plotted on a control chart (Carder, 1994). This means that frequency data are analyzed statistically to reveal changes that show significant variations in the system. This helps an organization distinguish between changes in the rate of accidents that are not significant from changes that show the system is either effective or ineffective in controlling hazards. Figure 9-8 shows a control chart recording accident frequency before and after intervention in the form of training for supervisors and managers.

Measuring Safe Behavior

Measuring worker behavior within any lockout/tagout system must be done in the context of the overall culture of the organization. Organizational culture usually is defined as the members' shared beliefs, values, expectations, norms, attitudes, values, goals, and ideologies (Petersen, 1980). From the workers' point of view, the most important elements of corporate culture are the norms and values shared by working groups, the rules by which things get done or one gets along, and the feeling or climate that exists in an organization as a result of the physical and psychological environment (Schein, 1985). These factors are a direct outgrowth of the philosophy of upper management

High-Speed Metal-Forming Machine Energy Isolation (Z.E.S.)

	Rating		
1. Preparation	U	P	A
a. Proper notification	0		
b. Clearance of machine/equipment		1	
c. Disengage metal feed			2
2. Shutdown (Controls)	U	P	A
a. Sheet feed control			2
b. Tool pack drive			2
c. Discharge (eject/conveying)		1	
3. Machine/Equipment Isolation	U	P	A
a. Disconnect breaker (440 V)			2
b. Close hydraulic supply valve (250 psi)			2
c. Close pneumatic supply valve (100 psi)	0		
4. Device Application	U	P	A
a. Placement of locks, tags, restraints			2
b. Following proper procedure		1	
5. Stored Energy Control	U	P	A
a. Insert block device-former			2
b. Relieve hydraulic pressure (return)			2
c. Bleed air pressure-relief device	0		
d. Lower discharge chute			2
6. Verify Isolation	U	P	A
a. Cycle run control (test)			2
b. Cycle feed/extract control (test)			2
c. Return control to off/neutral	0		
7. General	U	P	A
a. Utilization of safe work position		1	
b. Appropriate support equipment/safeguards			2
TOTAL	0	4	24
	28 / 4 0		

CODE U–Unacceptable (0)
P–Partial (1)
A–Acceptable (2)

Figure 9-9. Measurement tool—energy isolation performance.
Source: Developed by E. Grund. © National Safety Council.

that guides organizational policy and the rituals, language, and ceremonies established by management in the company.

Corporate culture arises in response to the problems of external adaptation and survival and to the problem of how to establish and maintain effective working relationships among group members. The hazardous-energy-control system described in Chapter 6, The Process Approach, addresses the issue of worker behavior and corporate culture (Figure 6-6). In this model, hazardous-energy-release incidents are viewed as the result of a complex set of cause and effect events with an elaborate network of contributory human action or inaction. These include disobedience, noncompliance, errors, and oversight.

Once the context of organizational culture has been established, worker behavior can be measured more accurately and fairly. As Rummler and Bache (1991) state, an organization is a system with three levels of performance: organizational, process, and job/performer levels. Measurement is necessary to get desired performance results. Without measurements, managers find it difficult to communicate to workers what is expected of them, know how workers are doing, identify problems, provide feedback, and make decisions. Workers, on the other hand, have no way of knowing what is expected of them, identifying problem or improvement areas, or recognizing when they have reached or exceeded expectations.

As mentioned in Chapter 7, System Elements, Rummler and Bache recommend the following sequence when measuring performance:

1. Identify the most significant lockout/tagout elements of the job or task.
2. Identify the critical steps or aspects and desired results for the lockout/tagout application.
3. Develop measures for each critical element.
4. Develop goals or standards for the level of performance expected of workers.

For example, a lockout/tagout procedure may have been developed for energy isolation on a high-speed, metal-forming machine. The specific procedure may involve a series of sequenced steps and discrete protective actions. A measurement tool for worker performance in using this procedure might look like the one illustrated in Figure 9-9.

Krause et al. (1990), in their text *The Behavior-based Safety Process,* detail a method for identifying and reinforcing "critical behaviors" that are essential to safe performance of the required tasks. Central to this approach is the process of measurement and feedback that enables the organization to continuously improve. Five indicators/measurements are emphasized: accident frequency, frequency of observation, exposure levels (percent safe readings), safety-related maintenance information, and involvement indicators and surveys. The key element in the general process is "behavior-based observation," which monitors what employees are doing and judges it as safe or unsafe on the basis of established critical behavioral criteria. This generates a measurement of employee behavior or performance that can be evaluated to determine organizational progress. Figures 9-10 and 9-11 depict data using this approach. See Chapter 3, Causation Analysis, for additional information regarding this behavioral approach such as Antecedent-Behavior-Consequence (ABC) Analysis (Krause et al., 1990).

Individual performance is influenced by internal factors (psychological and physiological) and external factors (situation, task, etc.) that managers must acknowledge and address. It is necessary to anticipate and appreciate the variables of human performance when measurement tools are developed. The variables are summarized in Chapter 7, System Elements, and can be refined (Figure 7-14).

On an individual level, worker behavior generally is measured on the basis of two factors: ability and motivation (Petersen, 1980).

Ability. By ability is meant the capacity to carry out tasks in an assigned manner: "Can the person do the work?" The answer to this question has to do with selection of employees and training. Is the company selecting and training employees in the best way possible to perform lockout/tagout tasks safely?

Motivation. Motivation addresses the question, "Does the employee *want* to perform the tasks?" The factors that determine motivation are complex and more difficult to measure, as shown in Figure 9-12. Behavioral influences might be considered in two components: past and current (Figure 9-13).

Past behavioral influences. These include the values and attitudes learned from family, church, school, and other important groups or figures in a person's life. These values and attitudes determine how important a person thinks it is to follow the rules, observe safety procedures, protect his or her own life, look after the welfare of others, depend on others and be depended on, value such characteristics as loyalty and honesty, obey authority figures, etc.

Current behavior influences. Although past behavioral influences can provide insights into worker actions and attitudes, current behavior influences are more relevant to monitoring, measuring, and assessing. Key current influences on worker behavior are job factors and peer pressure.

Job factors. These include:

- whether the job is important and meaningful to the employee
- the degree of employee involvement and participation in the job and in job-related procedures such as lockout/tagout
- recognition given for good performance
- the behavior theory under which management operates (i.e., are employees considered children that management must discipline and oversee? As adults who want to be involved in their jobs? As reasonable people who will respond to such motivational factors as personal growth, responsibility, initiative?)

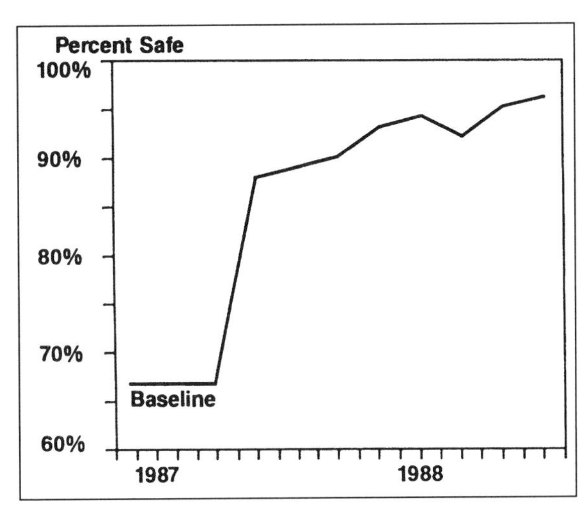

Figure 9-10. Safe performance measurement.
Source: © 1988 by Behavioral Science Technology, Inc. May not be reproduced without permission.

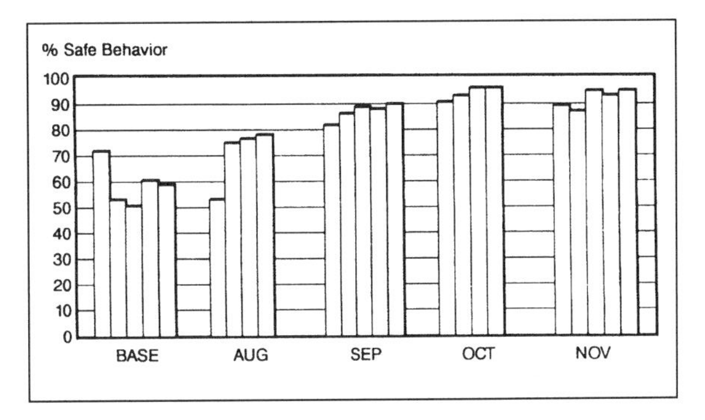

Figure 9-11. Safe behavior tracking.
Source: © 1988 by Behavioral Science Technology, Inc. May not be reproduced without permission.

The measurement team can use questionnaires to measure the influence of job factors on employees' safety behavior. How are elements of the safety program presented? How would you rate their effectiveness? What would you change in the lockout/tagout procedures? How much are employees involved in creating and refining lockout/tagout systems? How much do you feel the company cares about safety?

These and other questions will get at the basic underlying motivational factors that drive workers to adopt a safety program as their own. If a safety program is structured simply on the basis of worker satisfaction (money, status, company policies, work rules, and working conditions), workers will not feel a strong sense of involvement or "ownership" of the programs (Petersen, 1980). Such workers are likely to regard the safety program as a list of rules, posters, boring safety talks, and troublesome procedures that simply make their job harder.

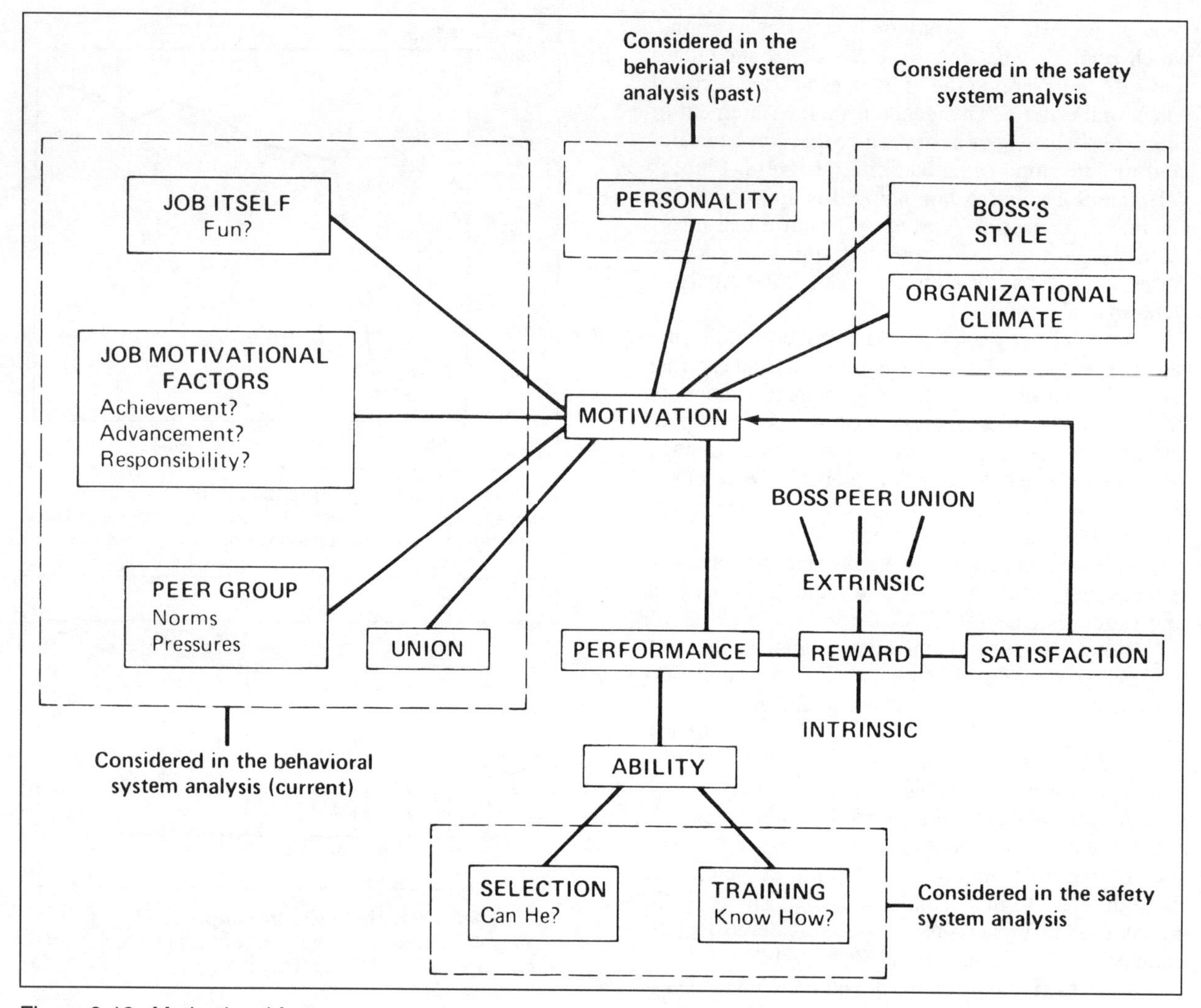

Figure 9-12. Motivational factors.
Source: Peterson, 1980, p. 85. Used with permission.

Peer group factors. Of all the motivational factors affecting workers, the influence of a peer group is the strongest (Petersen, 1980). A group is a "number of people who interact or communicate regularly and who see themselves as a unit distinct from other collections of people" (Petersen, 1980, p. 90). A group, like an individual, has several characteristics:

- A personality of its own. This is determined by the members who make up the group, whether of the same background or varied backgrounds, ages, skill levels, genders, nationalities, cultures, etc.
- Its own way of making decisions and establishing goals. If these goals are the same as those of management, ensuring compliance with safety is going to be easier. If not, compliance with and improvement of safety standards is likely to be difficult to achieve.
- A common task or purpose, which may be a design, production, or distribution process.
- Group norms or informal laws that govern the way people in the group behave. Group norms are accepted attitudes about such factors as how workers behave toward the boss, how they react to safety regulations, and how careful they are to follow work procedures.

All of these factors can be quantified and measured through interviews, questionnaires, and observation. In terms of lockout/tagout, the measurement team should look for two types of behavior in particular. The first has to do with how employees react if the procedures are inadequate. Depending on the factors just mentioned, employees may act in different ways. One employee may have enough initiative/experience to know how to fill in missing lockout/tagout steps. Or such factors as group norms may require employees to improvise their way out of problems. However, the next person may not have the same level of skill/experience. They may not know how to improvise and may work with the power not fully isolated.

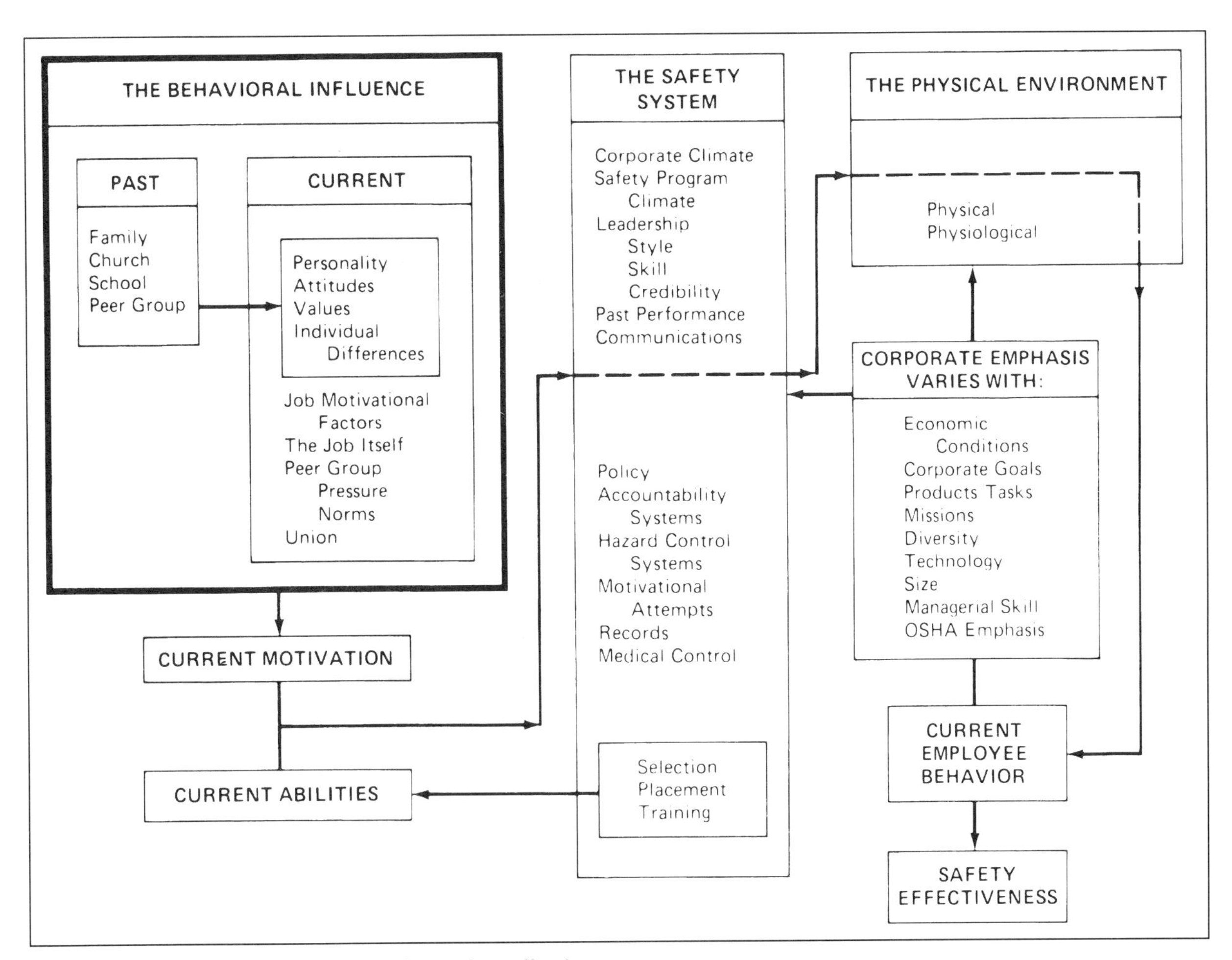

Figure 9-13. Behavioral factors affecting safety effectiveness.
Source: Peterson, 1980, p. 82. Used with permission.

The second type of behavior occurs when the written procedures are adequate but the employee does not follow them. The reasons can be varied—group norms, personal attitudes, or lack of training. The measurement team can measure actual performance against written practices to see where training, auditing, motivation, and discipline may need to be strengthened.

Taking the factors into consideration that influence behavior and measuring actual performance against written procedures should give the company a good profile of employee performance. The results should also satisfy OSHA lockout/tagout implementation and monitoring requirements.

Process measures versus results measures. Measuring employee performance and attitude focuses on *process* measures, as well as results measures (Carder, 1994). For example, process measures would be used to determine the effect of an educational program for employees. Process measures would include discipline, time management, skill development, participation, morale, and communication. These factors are fundamental to maintaining and improving safety.

Reporting Results

One way to quantify the measurement data is to construct a matrix that lists the factors measured in the audit (employee performance, lockout/tagout procedures, equipment used, etc.) and record the results of the audit against continuous improvement and TQM goals. Such a matrix would encourage the measurement of physical conditions, lockout/tagout procedures, and employee behavior against objectives. It would help management answer such questions as: "What do we need to do to improve our safety performance? What is the real problem—poor training, too much pressure to produce, inadequate lockout/tagout procedures? Where are the major problems and strengths in our program?"

The data in matrix form could be used to spot weaknesses in the organization's energy-isolation process and provide enough quantitative information to do a meaningful assessment. The emphasis should be on correcting problems and deficiencies, not on placing the blame.

ASSESSING

The monitoring function of the lockout/tagout system has generated a substantial body of opinion, perception, and fact. This information is generally of a subjective or qualitative nature, i.e., resulting from observation, impression, or value judgment. It depends in many respects on who performed the monitoring and how well they did the job or from what source the information was gathered. The process of monitoring often produces concurrent measurement of activity and corrective action directed at specific deficiencies. However, this is more of a secondary output or by-product and not the direct thrust of monitoring.

The measurement function is directed at using the input or results obtained by monitoring. These data (or information) are then translated into quantitative terms, which provides a gauge of topical performance. Trends, rates of compliance, experience statistics, attitude/perception scores, etc. are produced to establish objective values on which evaluation of performance can be based. Various categorical measurements (isolation device identification, verification of energy isolation, release of stored energy, etc.) reported in terms of time (monthly, quarterly, cumulative) and area (shift, department, unit) are available for the assessment process.

Assessment is the last but perhaps most important function in determining whether an effective lockout/tagout system is in place. Personnel who have been involved in the monitoring/measuring activity may perform the assessment, or a special group may be created to accomplish this task. During the assessment process, the qualitative and quantitative data are digested and organized for evaluation/analysis. Participants in this process should have appropriate background and experience to make valid judgments and interpretations of what the information means and its significance. They should be qualified to make an accurate diagnosis of what the data suggest are the current weaknesses or limitations of the lockout/tagout system.

The assessment team should resist the temptation to turn this part of the process into simple number crunching, apprising management of the number of errors, incidents, near misses, faulty equipment, etc. Merely saying how much is wrong is not as valuable as analyzing *what* and *why* it is wrong. The causes may not be as obvious as they may appear at first glance. A problem locking out the power supply on a computerized drill press, for example, may not be the result of negligence of the operator or maintenance personnel as it may first appear. It may really be caused by the fact that a certain level of production is expected, and following the full lockout procedure increases production downtime. Worker shortcuts may be dictated far more by management expectations than by personal risk taking or poor attitude.

Information Analysis

The available information can be analyzed in the assessment phase categorically as follows:

- Lockout/tagout procedures and equipment. Are they adequate for the organization's current processes and machinery? Do they need to be updated, improved, or modified in any way? Are they designed to meet minimum standards requirements or to address the energy-isolation problems specific to the operations of the organization?
- Effectiveness of procedures. How do the procedures work on the job? Are the number of errors, omissions, lapses, and equipment-related problems increasing, declining, or remaining the same?
- Employee performance. Is it measuring up to expectations given the organization's level of training, retraining, communication, and education efforts? What are workers' attitudes toward lockout/tagout procedures and systems? What are they doing on the job versus what they are told or trained to do?
- Compliance with standards. How does the organization's energy-control system measure up to the requirements in various applicable standards? Are the number of citations, violations, and warnings declining? Have past problems been corrected satisfactorily?
- Incident statistics. What do the numbers say about the company's lockout/tagout program? Have certain groups or production processes better or worse safety records than others? What are the reasons for these results? What can be done to improve deficient areas?
- Overall effectiveness of the lockout/tagout program. Is it meeting organizational goals of reducing injuries and fatalities, meeting TQM goals, meeting or exceeding standards requirements, and in general, ensuring a safe working environment for all affected employees?

Organizational Goals

During the process of assessing the organization's lockout/tagout system, reference should be made to the goals that were developed earlier. Is the organization on schedule? Have various milestones been reached? Have incidents been reduced? What is the state of the current lockout/tagout system compared with when we began? The answers to all of these questions and more will reveal much about lockout/tagout performance. Part of the assessment objective is to define accurately what needs to be done next to ensure that improvement of the control of hazardous energy occurs and progress is sustained.

Comparison or benchmarking activities can then be used to determine if performance is sufficient to satisfy the current requirements or goals. Often goals might be crafted based on some external definition or organizational level of excellence. There are opportunities to compare the local or internal performance with (1) past internal history, (2) competitors with industry-unique operations, (3) others with similar operations, and (4) general industrial leaders employing common practices. Benchmarking can create an objective pattern or visual image of what future state your lockout/tagout process is moving toward. It represents the destination but not necessarily the road map.

Appraisal and Reporting

Periodically those charged with the responsibility for assessing the lockout/tagout system will need to communicate with key management personnel and perhaps the organization at large. Depending on the audience, the appraisal or summary should contain the tailored synthesis of the data and information developed in the monitoring/measuring phases.

Generally, four major elements should be addressed in the summary/assessment work product. They are as follows:

1. *Situation/status.* Description or overview of current conditions/performance based on analysis of the data.
2. *Supporting data.* Results presented in terms of variance from standards, trends, rates of compliance, etc. and comparative analysis using historical references, benchmarks, etc.
3. *Conclusions.* Key observations and findings regarding lockout/tagout system effectiveness, both positive and negative outcomes.
4. *Recommendations.* Suggested action regarding correction of deficiencies and praise/recognition for individuals, groups, etc. who performed in a noteworthy manner; also mention any future considerations for enhancing the monitoring, measuring, and assessing process.

Reporting to management can occur in any number of formats from executive summary style to sectionalized notebook style. Each organization must determine the makeup of their reporting components. It may be advisable to provide different pieces of the assessment to various individuals based on their specific responsibilities or functions. It is advisable to provide feedback to the work force regarding the status and progress associated with the lockout/tagout process. Graphic treatment of the performance data is recommended to facilitate employee review and education.

Appropriate data generated during the monitoring, measuring, and assessing process should be prepared for entry into the existing information systems. The developing data base will prove useful for future reference, comparison, and report generation.

Documentation

Proper documentation of lockout/tagout activity is desirable for several reasons. First, the information can be used constructively to demonstrate the organization's proactive response to hazardous energy control. It could also be of value in support of matters such as compliance disputes, workers' compensation claims, and third-party liability incidents. Second, the data can be used to substantiate mandated lockout/tagout activity, i.e., audit of procedures, and associated performance-types statistics. Third, the documentation will provide a historical reference base to evaluate future activity against past performance.

Assessment is critical in providing quality feedback to management and ultimately to the workers themselves. The information generated is valuable for improving not only the technical aspects of the lockout/tagout system but also the human side of managerial expectations and worker performance. The better management is at monitoring, measuring, and assessing lockout/tagout performance, the more likely it is to achieve above-average results in the area of safeguarding employees from hazardous-energy-release incidents.

OSHA AND HAZARDOUS ENERGY CONTROL

Since the 1970s, OSHA has sought to provide more specific guidelines and standards for lockout/tagout and energy-control procedures. With the promulgation of the lockout/tagout standard, it now has the tool to become more vigorous in its enforcement of energy-control compliance to reduce the number of related workplace injuries and death. The agency has taken the stand that these incidents are within the power of the companies and employees to prevent if proper procedures are implemented, monitored, measured, and assessed.

Relevant Standards and Directives

In September 1989, OSHA formally placed employers on notice concerning requirements for hazardous energy control with their Final Rule on the Control of Hazardous Energy (Lockout/Tagout) (29 *CFR* 1910.147). The preamble to this standard explained the rationale behind the requirements OSHA developed and the suggested approach companies should use in implementing and monitoring their lockout/tagout systems.

Preamble: 29 *CFR* 1910.147. In the preamble, OSHA addressed the issue of the severity of the risks associated with a lapse in the implementation of the energy-control system, citing this danger as the reason for required periodic inspections. "The periodic inspection is intended to assure that the energy control procedures continue to be implemented properly, and that the employees involved are familiar with their responsibilities under those procedures." In a change from the proposed rule to the Final Rule, OSHA says of the inspector:

> The inspector, who is required to be an authorized person not involved in the energy control procedure being inspected, must be able to determine three things: First, whether the steps in the energy control procedure are being followed; second, whether the employees involved know their responsibilities under the procedure; and third, whether the procedure is adequate to provide the necessary protection, and what changes, if any, are needed.

The Final Rule also provides additional guidance regarding the inspector's duties in performing periodic inspections to obtain the needed information. Part of these duties involve reviewing each authorized employee's responsibilities under the lockout/tagout procedure. This task can be done in group meetings, as well as one-on-one interviews. Where tagout is also used, the inspector must review the responsibilities of affected employees and ensure that they understand their role in preventing accidental start-ups of the machinery being serviced. These reviews should be done on an annual basis along with the periodic inspections to ensure that employees follow and maintain proficiency in the energy-control procedure. It will also help the inspector determine whether changes are needed.

The Final Rule also contains a certification provision. It requires that a company document the operation, date of inspection, name of inspector, and names of all employees included in the inspection. This provision gives employees an opportunity to review their responsibilities and demonstrate their performance under the lockout/tagout procedure.

Although some reviewing the Final Rule suggested that the standard require employee participation in the inspections, such an addition was felt to be unnecessary. Under the standard, the employer has the obligation of ensuring proper use of the energy-control procedure. The periodic inspection is a means of ensuring compliance. If this inspection revealed flaws in the way the procedure was implemented, it was the employer who had to make changes in the procedure, retrain employees, and take other corrective steps. It was considered sufficient that the authorized inspector was also an employee and would have the necessary knowledge to evaluate the effectiveness of the procedure and report any corrective steps that needed to be taken.

In the Final Rule (29 *CFR* 1910.147), OSHA retained the periodic inspection, at least annually, to ensure that the energy-control procedure required by the standard was being followed. As stated in the Preamble:

> The employer was also required to provide effective initial training, periodic retraining, and certification of such training of employees. All the requirements in the standard were considered by OSHA to be essential to help ensure that the applicable provisions of the hazardous energy control procedure(s) are known, understood, and strictly adhered to by employees.

General Duty Clause

OSHA enforced hazardous-energy-control methods during the first 18 years of the agency's existence through use of the General Duty Clause of the Occupational Safety and Health Act. Section (5)(a)(1) of the Act requires that "each employer furnish to each of his employees employment and a place of employment which are free from recognized hazards that are causing or are likely to cause death or serious physical harm to his employees."

Because of the lack of specific federal guidelines and limited state directives on lockout/tagout, enforcement activity usually was visible only as a by-product of serious accidents and fatalities. Prosecution of these citations was usually difficult because of the absence of definitive criteria for energy-isolation procedures and methods. This changed radically with the development and issue of OSHA's lockout/tagout standard in 1989.

General and Periodic Inspection Requirements

In 29 *CFR* 1910.147 OSHA defines the employer's responsibility for establishing an energy-control program and for conducting periodic inspections to ensure its functioning effectiveness. In this standard, OSHA introduced the concept of monitoring/auditing used rarely in its earlier rule-making efforts. Its inclusion is likely connected to similar guidelines contained in the ANSI Z244.1 standard that OSHA used as a key reference.

The following general notice is provided to employers at the beginning of the standard:

> (c) *General*—(1) *Energy control program.* The employer shall establish a program consisting of an energy control procedure and employee training to ensure that before any employee performs any servicing or maintenance on a machine or equipment where the unexpected energizing, start up or release of stored energy could occur and cause injury, the machine or equipment shall be isolated, and rendered inoperative, in accordance with paragraph (c)(4) of this section.

The inspection (auditing) section was intended to provide the minimum requirements for monitoring a lockout/tagout system. The language is general, with considerable leeway for employers to develop an inspection system tailored to their particular circumstances. The focus is primarily on the monitoring or auditing function. Section (c)(6) reads as follows:

> (6) *Periodic inspection.* (i) The employer shall conduct a periodic inspection of the en-

ergy control procedure at least annually to ensure that the procedure and the requirements of this standard are being followed.

(A) The periodic inspection shall be performed by an authorized employee other than the one(s) utilizing the energy control procedure being inspected.

(B) The periodic inspection shall be designed to correct any deviations or inadequacies observed.

(C) Where lockout is used for energy control, the periodic inspection shall include a review, between the inspector and each authorized employee, of that employee's responsibilities under the energy control procedure being inspected.

(D) Where tagout is used for energy control, the periodic inspection shall include a review, between the inspector and each authorized and affected employee, of that employee's responsibilities under the energy control procedure being inspected, and the elements set forth in paragraph (c)(7)(ii) of this section.

(ii) The employer shall certify that the periodic inspections have been performed. The certification shall identify the machine or equipment on which the energy control procedure was being utilized, the date of the inspection, the employees included in the inspection, and the person performing the inspection.

A format for satisfying the requirements found in section (6) can be seen in Figure 9-14. It includes all of the specified content for inspection of both lockout and tagout application.

Compliance Directive

To effectively enforce 29 *CFR* 1910.147, OSHA issued Instruction STD 1–7.3 in September 1990. Its purpose was to establish policies and provide clarification to compliance personnel to ensure uniform enforcement of the standard. The provisions for periodic inspection (auditing) by the employer were placed in the interpretive guidance section and read as follows:

5. *Periodic Inspection by the Employer*

a At least annually, the employer shall ensure that an authorized employee other than the one(s) utilizing the energy control procedure being inspected, is required to inspect and verify the effectiveness of the company energy control procedures. These inspections shall at least provide for a demonstration of the procedures and may be implemented through random audits and planned visual observations. These inspections are intended to ensure that the energy control procedures are being properly implemented and to provide an essential check on the continued utilization of the procedures (29 *CFR* 1910.147 (c) (6) (i).

(1) When lockout is used, the employer's inspection shall include a review of the responsibilities of each authorized employee implementing the procedure with that employee. Group meetings between the authorized employee who is performing the inspection and all authorized employees who implement the procedure would constitute compliance with this requirement.

(2) When tagout is used, the employer shall conduct this review with each affected and authorized employee.

(3) Energy control procedures used less frequently than once a year need be inspected only when used.

b. The periodic inspection must provide for and ensure effective correction of identified deficiencies (29 *CFR* 1910.147 (c)(6) (ii).

c. The employer is required to certify that the prescribed periodic inspections have been performed (29 *CFR* 1910.147 (c)(6)(ii).

Process Safety Management Standard

In early 1991, OSHA promulgated 29 *CFR* 1910.119, the Process Safety Management of Highly Hazardous Chemicals Standard. It is mentioned here because of the inclusion of auditing (monitoring) as a key element that mirrors the inspection requirement of 29 *CFR* 1910.147. The emerging pattern suggests that OSHA believes that monitoring (auditing/inspecting) is a critical feature in prevention programming and that employers should be held accountable for its use.

Section (o) of 1910.119 provides insight as to OSHA's view regarding auditing specific to the Process Safety Management standard and in general. It reads as follows:

(o) *Compliance Audits.* (1) Employers shall certify that they have evaluated compliance with the provisions of this section at least every three years to verify that the procedures and practices developed under the standard are adequate and are being followed.

(2) The compliance audit shall be conducted by at least one person knowledgeable in the process.

(3) A report of the findings of the audit shall be developed.

(4) The employer shall promptly determine and document an appropriate response to each of the findings of the compliance

Lockout/Tagout (LO/TO)
Compliance Audit

Plant ______________________________ Date ______________

Machine/Equipment/Process ______________________________

Energy Isolation Procedure (I.D.) ______________________________

Employees Involved (name/I.D.) ______________________________

____

____

Deficiencies Detected ______________________________

____

____

Corrections Required (who/when) ______________________________

____

____

Review of Procedural Responsibilities:

Lockout ☐ Authorized Employees Tagout ☐ Authorized Employees
☐ Affected Employees
☐ Tag Limitations

Auditor(s) ______________________________

Other Observations ______________________________

____

____

____

Reviewed by ______________________ Date ______________

Figure 9-14. Inspection of lockout/tagout application.
Source: Developed by E. Grund. © National Safety Council.

audit and document that deficiencies have been corrected.

(5) Employers shall retain the two (2) most recent compliance audit reports.

Audit Commentary

Appendix C of the 29 *CFR* 1910.119 standard provides further nonmandatory guidelines for employers in setting up an auditing/monitoring process. The main points of these guidelines are as follows:

1. Employers need to select a trained individual or assemble a trained team of people to audit the process safety management system and program. Selection of qualified, knowledgeable team members is critical to the success of the audit. A small process or plant may need only one knowledgeable person to conduct the audit. The audit should be conducted or led by someone knowledgeable in audit techniques and who is impartial toward the facility.
2. The essential elements of an audit program include planning, staffing, conducting the audit, evaluation and corrective action, follow-up, and documentation.
3. Planning in advance is essential to the success of the auditing program. Each employer should establish the format, staffing, scheduling, and verifica-

tion methods before conducting the audit. If a checklist can be properly designed, it could serve as the verification sheet to ensure that no requirements of the standard are omitted. It would also identify the elements that require evaluation or correction and be used for follow-up.
4. Team members should be chosen for their experience, knowledge, and training. For a large, highly complex process, it may be necessary to include members with expertise in process engineering and design, process chemistry, instrumentation and computer controls, electrical hazards and classifications, safety and health disciplines, etc.
5. An effective audit includes a review of relevant documents and process safety information, inspection of the physical facilities, and interviews with all levels of plant personnel. Using the audit procedure and checklist developed in the preparation stage, the audit team can systematically analyze compliance with the provisions of the standard and with any other corporate policies that are relevant.
6. Auditors should select as part of their preplanning a sample size large enough to give a degree of confidence that the audit reflects the level of compliance with the standard. The audit team can document areas that require corrective action. Such action provides a record of the audit procedures and findings and serves as a data baseline for future audits.
7. Corrective action includes not only addressing identified deficiencies but also planning, follow-up, and documentation. This stage usually begins with management review of the audit findings to determine which correction actions are appropriate. All actions taken should be documented and reasons given for what was done and why.
8. To control the corrective action process, employers should consider using a tracking system. This system might include periodic status reports shared with affected levels of management, specific reports such as completion of an engineering study, and a final implementation report to provide closure for audit findings. The tracking system provides the employer with some idea of how the correction action process is going and also provides documentation required to verify that deficiencies identified in the audit were appropriately corrected.

Inspection Guidelines

In addition to the interpretive guidance in OSHA's Instruction STD 1–7.3 regarding periodic lockout/tagout inspection by employers, information regarding the conduct of compliance inspections is provided for compliance safety and health officers (CSHOs). Section H, Inspection Guidelines, begins with the statement that "the standard incorporates performance requirements which allow employers flexibility in developing lockout/tagout programs suitable for their particular facilities."

The lockout/tagout provisions of this standard are for the protection of general industry workers while performing servicing and maintenance functions and augment the safeguards specified in Subparts O, S, and other applicable portions of 29 *CFR* 1910. Companies are expected to meet the requirements contained in these directives as *minimum* compliance with OSHA regulations.

Section H, Inspection Guidelines, reads as follows:

1. The compliance officer shall determine whether servicing and maintenance operations are performed by the employees. If so, the compliance officer shall further determine whether the servicing and maintenance operations are covered by 29 *CFR* 1910.147 or by the requirements or employee safeguarding specified by other standards as discussed in I.1 (of this instruction).
2. Evaluations of compliance with 29 *CFR* 1910.147 shall be conducted during all general industry inspections within the scope of the standard in accordance with the *Field Operations Manual*, Chapter III, D.7. and 8., "Additional Information to Supplement Records Review." The review of records shall include special attention to injuries related to maintenance and servicing operations.
3. The compliance officer shall evaluate the employer's compliance with the specific requirements of the standard. The following guidance provides a general framework to assist the compliance officer during inspections:
   a. Ask the employer for any hazard analysis or *other basis* on which the program related to the standard was developed. Although this is not a specific requirement of the standard, such information *when provided* will aid in determining the adequacy of the program. It should be noted that the absence of a hazard analysis does not indicate non-compliance with the standard.
   b. Ask the employer for the documentation including: procedures for the control of hazardous energy including shutdown, equipment isolation, lockout/tagout application, release of stored energy, verification of isolation; certification of periodic inspections; and certification of training. The documented procedure must identify

the specific types of energy to be controlled and, in instances where a common procedure is to be used, the specific equipment covered by the common location. The identification of the energy to be controlled may be by magnitude and type of energy. Note the exception to documentation requirements at paragraph 1910.147 (c) (4) (i), "Note." The employer need not document the required procedure for a particular machine or equipment when all eight (8) elements listed in the "Note" exist.

c. Evaluate the employer's training programs for "authorized," "affected," and "other" employees. Interview a representative sampling of selected employees as a part of this evaluation (29 *CFR* 1910.147 (c)(7)(i)). (1) Verify that the training of authorized employees includes:
   (a) Recognition of hazardous energy
   (b) Type and magnitude of energy found in the workplace
   (c) The means and methods of isolating and/or controlling the energy; and
   (d) The means of verification of effective energy control, and the purpose of the procedures to be used.

(2) Verify that affected employees have been instructed in the purpose and use of the energy control procedures.

(3) Verify that all other employees who may be affected by the energy control procedures are instructed about the procedure and the prohibition relating to attempts to restart or reenergize such machines or equipment.

(4) When the employer's procedures permit the use of tagout, the training of authorized, affected, and other employees shall include the provisions of 29 *CFR* 1910.147 (c)(7)(ii) and (d)(4)(iii).

d. Evaluate the employer's manner of enforcing the program (29 *CFR* 1910.147(c)(4)(ii)).

4. In the event that deficiencies are identified by following the guidelines in H.3. of this instruction, the compliance officer shall evaluate the employer's compliance with specific requirements of the standard, with particular attention to the interpretive guidance provided in section I. and to the following:

a. Evaluate compliance with the requirements for periodic inspection of procedures.

b. Ensure that the person performing the periodic inspection is an authorized employee other than the one(s) utilizing the procedure being inspected.

c. Evaluate compliance with retraining requirements which result from the periodic inspection of procedures and practices, or from changes in equipment/processes.

d. Evaluate the employer's procedures for assessment, and correction of deviations or inadequacies identified during periodic inspections of the energy control procedure.

e. Identify the procedures for release from lockout/tagout, including:

(1) Replacement of safeguards, machine or equipment inspection, and removal of nonessential tools and equipment;
(2) Safe positioning of employees;
(3) Removal of lockout/tagout device(s); and
(4) Notification of affected employees that servicing and maintenance is completed.

f. Ensure that when *group lockout or tagout* is used, it affords a level of protection equivalent to individual lockout or tagout as amplified in I.7. through I.9. of this instruction.

5. The lockout/tagout standard is a performance standard; therefore, additional guidance is provided in Appendix C of this instruction to assist in effective implementation by employers and for uniform enforcement by OSHA field staff.

The remainder of the instruction provides guidance on the scope of 29 *CFR* 1910.147, lockout/tagout procedures, employees and training, periodic inspection, equipment and group lockout/tagout, compliance with group lockout/tagout, compliance of outside personnel, and classification of violations of 1910.147.

OSHA Enforcement Posture

Since issuance of the lockout/tagout standard (29 *CFR* 1910.147), OSHA has become very assertive in its enforcement. This emphasis resulted in large part from its internal Instruction Standard 1-7.3, which in Section H.2. states that "evaluation of compliance with 29 *CFR* 1910.147 *shall be* conducted during *all* general industry inspections." This places the standard, along with a few others such as Hazard Communication (1910.1200), in a unique enforcement status—priority attention by directive.

The results of this priority attention are revealed when the most frequently cited OSHA general industry standards violations are examined for fiscal 1991 to 1993 (Figure 9-15). Lockout/tagout, which functions in tandem with the machine guarding standards, would rank second if the two were combined. This should place employers on notice that deficiencies in compliance with 29 *CFR* 1910.147 will likely be detected and viewed quite seriously! The ongoing incidence and severity of lockout/tagout accidents, which often draw OSHA inspection attention, also contribute to the enforcement emphasis.

In Instruction STD 1-7.3, OSHA provides compliance personnel with guidance on how to classify violations for lockout/tagout deficiencies. They are found in Section J as follows:

> J. *Classification of Violations*
> 1. A deficiency in the employer's energy control program and/or procedure that could contribute to a potential exposure capable of producing serious physical harm or death shall be cited as a serious violation.
> 2. The failure to train "authorized," "affected," and "other" employees as required for their respective classifications should normally be cited as a serious violation.
> 3. Paperwork deficiencies in lockout/tagout programs where effective lockout/tagout work procedures are in place shall be cited as other-than-serious.

From a practical standpoint, very few violations of the lockout/tagout standard (1910.147) will be viewed as "other than serious." Therefore, infractions deemed "serious" potentially carry a maximum penalty of $70,000 per instance. Normally there would be downward adjustments for mitigating factors such as good faith, size of establishment, and past violation history. In spite of these adjustments, employers can anticipate significant fines depending on the inspection circumstances and the findings of the compliance officer.

Enforcement Activity

In slightly more than three years, OSHA has cited employers for over 12,000 violations of the lockout/tagout standard (29 *CFR* 1910.147). In addition, states with jurisdiction for their own OSHA programs have issued thousands of similar violations of the standard. There appears to be no downward trend in the activity even though the specific standard is now entering its fifth year of existence and enforcement. The following examples of agency enforcement action taken from public sources illustrate the nature of lockout/tagout violations and the associated fines:

Rank	Subject	Violations
1	Hazard Communication	48,827
2	Posting/OSHA Log	23,916
3	Machine/Power Transmission Guarding	23,357
4	Lockout/Tagout	12,735
5	General Duty	5,110

Figure 9-15. General industry most frequently cited OSHA standards, fiscal 1991–1993.
Source: Developed by E. Grund. © National Safety Council.

- Packaging company. Failure to establish an energy-control program and provide employee training. Proposed penalty $560 (Ohio, 1993).
- Ceramics company. Employee death during operation of a computer-controlled lathe; failure to provide protective safeguard (interlock) to prevent accidental start-up. Proposed penalty $70,000 (Colorado, 1991).
- Paper company. Miscellaneous lockout/tagout and guarding violations; 41 alleged repeat violations and 45 alleged serious violations. Proposed penalty $803,000 (Mississippi, 1992).
- Glass company. Failure to use specific procedures for machine lockout/tagout and to de-energize when making adjustments; serious violation. Proposed penalty $5,000 (California, 1993).
- Metals company. Willful failure to establish an energy-control program and to implement lockout/tagout procedures during maintenance; employee death as a result of machine cycling unexpectedly. Proposed penalty $20,000 (Connecticut, 1990).
- Automotive company. Fifty-seven alleged willful violations involving training, retraining, and inadequate procedures for energy isolation; maintenance employee death when pneumatic lift accidentally started. Proposed penalty $2,800,000 (Oklahoma, 1991).
- Tire company. Ninety-eight alleged willful violations involving failure to establish specific lockout/tagout procedures, failure to provide locks and other devices, failure to conduct inspections, and failure to properly train all employees at risk; employee death during servicing of tire assembly machine. Proposed penalty $7,500,000 (Oklahoma, 1994).

Elements of the 1910.147 standard that appear to be most frequently cited are as follows:

> (c)(1) energy control program
> (c)(4) energy control procedures
> (c)(6) periodic inspection
> (c)(7) employee training.

OSHA contends that compliance with the standard is well within the capability of every employer and its employees and that effective lockout/tagout implementation is expected.

SUMMARY

- Simply establishing a lockout/tagout program is rarely enough. Organizations must ensure that procedures are appropriate to their particular working conditions and that workers are properly trained to respond to unusual, as well as routine, situations. For this reason, organizations must develop effective monitoring, measuring, and assessing programs to evaluate their lockout/tagout systems.
- These programs must be viewed as part of the organization's safety culture and of total quality management. This requires a commitment by top management, training and education at all employee levels, strong employee participation, continuous improvement programs, and measurement of performance.
- Monitoring has to do with the development and use of inspection and/or auditing systems. Measuring is the establishment of specific yardsticks by which to evaluate performance or success. Assessing involves interpreting, evaluating, and comparing information against stated objectives and goals.
- Data from these three processes must be formatted so that it is readily available to those who need it. This will ensure proper feedback and corrective action regarding all lockout/tagout systems in the organization. This information will enable management to determine how well corrections or improvements are working.
- Monitoring can be done through internal or external approaches. Both OSHA and ANSI stress the importance of establishing a monitoring system that clearly delegates responsibility for lockout/tagout procedures and mandates periodic inspections. Components of monitoring include administrative/management policies, auditing goals, and inspection/auditing procedures.
- Monitoring should cover three basic areas: machinery, equipment, processes, and workplace conditions; lockout/tagout procedures and training; and employee behavior, perception, and attitudes regarding lockout/tagout. The auditing team must be properly trained and knowledgeable in the areas they are inspecting, including the energy systems, equipment and machinery, lockout/tagout procedures and equipment, the work force, and work processes and environment.
- Auditing techniques include physical inspection, interviews, video task analysis, operator/employee feedback, and computer information systems. The final report should be classified into three categories: emergency concerns, periodic or immediate concerns, and summary. The auditing team should be guided by three rules: correct the cause, not simply the result; report conditions outside, as well as within, the auditing team's authority; and take action immediately as needed.
- OSHA does not specifically require measurement, but most organizations consider this step essential in evaluating their lockout/tagout procedures. Measurement systems usually consist of a measurement group, specific objectives about what is to be measured, measurement methods and guidelines, and data formats for reporting results.
- Measurement techniques should meet three criteria: consistent yardsticks that are used the same way each time; measurable performance, behavior, or tasks; and measurement standards that workers can relate directly to their behavior and safety activities. Sampling and measurement techniques include checklists, task analysis, simulation, comparative analysis, and control charts.
- Measuring safe behavior must be done in the context of the organization's culture and philosophy. Measuring performance should be done in sequence beginning with the most significant lockout/tagout steps, identifying the critical steps and desired results, developing measures for each critical element, and developing goals or standards for performance levels. Worker behavior is influenced by past behavior, current behavior, job factors, and peer group factors.
- Data from measuring can be reported in matrix form. This would encourage the measurement of physical conditions, lockout/tagout procedures, and employee behavior against objectives and help to identify what corrective steps were required.
- Assessing is the last but perhaps most important function in determining whether an effective lockout/tagout system is in place. During assessment, quantitative and qualitative data are analyzed and evaluated. This is not mere number crunching but a process that determines not only *what* is wrong but *why* it is wrong.
- The assessment team analyzes lockout/tagout procedures, employee performance, compliance with standards, incident statistics, and overall effectiveness of lockout/tagout programs. These data can be compared with organizational goals established at the beginning of the monitoring process. The summary report to top management addresses four elements: situation/status, supporting data, conclusions, and recommendations.
- Documentation of assessment information is critical to prove proactive responses and compliance with standards, provide quality feedback, and establish a historical reference base for future evaluations.
- OSHA has sought to provide more specific guidelines and requirements for lockout/tagout and energy-control procedures in its Final Rule for 29 *CFR* 1910.147 (lockout/tagout standard). The preamble to the standard explains the rationale behind OSHA

requirements and its suggested monitoring and evaluating approach. The Final Rule for 29 *CFR* 1910.147 provides guidance for the inspector's duties in performing periodic inspections and contains a certification provision requiring documentation of an inspection.

- To enforce the lockout/tagout standard effectively, OSHA issued Instruction STD 1–7.3 in September 1990. Its purpose was to establish policies and provide clarification to compliance personnel to ensure uniform enforcement of the standard. The Process Safety Management of Highly Hazardous Chemicals standard also included an auditing element that underscores OSHA's belief that monitoring is a critical feature in accident prevention programming.
- The process safety standard (29 *CFR* 1910.119) provides nonmandatory guidelines for auditing that are of value in designing the monitoring, measuring, and assessing elements of any lockout/tagout system. In addition, Instruction Standard 1–7.3 identifies the OSHA regulatory emphasis regarding auditing of lockout/tagout specifically and the standard in general.
- OSHA has been vigorously enforcing the lockout/tagout standard with over 12,000 violations cited. Most violations of the standard are classed as serious so employers can anticipate significant fines, depending on the inspection circumstances and the findings of the compliance officer. OSHA states that compliance is well within the capabilities of every employer and organization and that effective lockout/tagout implementation is expected.

REFERENCES

American National Standards Institute. *Safety Requirements for the Lock Out/Tag Out of Energy Sources*, ANSI Z244.1-1982. New York: American National Standards Institute, 1982.

Carder B. Quality theory and the measurement of safety systems. *Professional Safety* 39, no. 2 (Feb. 1994):23–28.

Creech B. *Five Pillars of TQM*. New York: Dutton, 1994.

Crosby PB. *Quality Without Tears*. New York: McGraw-Hill, 1984.

Durban TJ. Safety and quality at PPG Industries. *Occupational Hazards* 50, no. 5 (May 1993): 79–82.

Enforcement. *Occupational Safety & Health Reporter*, April 15, 1992.

Geller E. Three ways to evaluate safety. *ISNH* 26, no. 12 (Dec. 1992):16.

Krause TR, Hidley JH, Hudson SJ. *The Behavior-based Safety Process*. New York: Van Nostrand Reinhold, 1990.

National Safety Council. *Product Safety: Management Guidelines*. Chicago: National Safety Council, 1989.

National Safety Council. *Public Employee Safety and Health Management*. Chicago: National Safety Council, 1990.

News briefs. *Occupational Health & Safety News*, June 15, 1990, p 7.

Northage EI. Auditing: A closed control loop. *Occupational Hazards* 54, no. 5 (May 1992): 87–90.

Petersen D. *Analyzing Safety Performance*. New York: Garland STMP Press, 1980.

Roughton J. TQM: Integrating a total quality management system into safety and health programs. *Professional Safety* 38, no. 6 (June 1993):32–37.

Rummler GA, Brache AP. *Improving Performance—How to Manage the White Space on the Organization Chart*. San Francisco: Jossey-Bass Publishers, 1991.

Schein EH. *Organizational Culutre and Leadership*. San Francisco: Jossey-Bass, 1985.

Tarrants WE. *The Measurement of Safety Performance*. New York: Garland Safety Management Series, 1980.

U.S. Department of Labor. OSHA Instruction STD 1–7.3, September 11, 1990.

U.S. Department of Labor. 29 *CFR* 1910.147: Control of Hazardous Energy Sources (Lockout/Tagout); Final Rule, *Federal Register*, September 1, 1989.

U.S. Department of Labor. 29 *CFR* 1910.119: Process Safety Management of Highly Hazardous Chemicals. *Federal Register*, vol 57, no. 36, February 24, 1991, p. 6407.

U.S. Department of Labor. 29 *CFR* 1910.147: The Control of Hazardous Energy (Lockout/Tagout)—Inspection Procedures and Interpretive Guidance, *Federal Register*, September 1990.

Weinstock MP. Environmental auditing: A measure of safety. *Occupational Hazards* 55, no. 5 (May 1993):73–77.

Wheeler KF. Comply or pay: OSHA getting tough about lockout/tagout enforcement. *Occupational Health & Safety* 60, no. 11 (Nov. 1991):24–26, 51.

10

Special Situations and Applications

The control of hazardous energy is often viewed in an oversimplified, unidimensional manner. Consensus practices and regulatory requirements focus on methods designed for relatively simple machinery, minimal energy involvement, and straightforward energy-control application. Even where performance criteria exist, people seem to deny the existence of a host of special situations or circumstances that warrant advanced control tactics applied under the most flexible of conditions. Although such circumstances are often treated as excursions from the norm by the regulator, they are in fact quite normal for the environment in which they exist. Regulations, by their design, should not impede or obstruct the development of options and tactics for the control of hazardous energy in complex situations. Principles and the integrity of control methods should be promoted and protected while still allowing the necessary flexibility to address energy hazards as they are encountered.

Special situations for energy isolation may arise because of the environment (underground, outdoors, elevated, etc.); the number and nature of personnel; machine, equipment, and process complexity; energy dimension (types, magnitude, delivery structure, etc.); degree of automation/mechanization; process risks (explosions, toxins, radiation, etc.); duration of energy isolation; unique operational characteristics, and so on. These circumstances may dictate approaches that are not considered mainstream or "pure vanilla" but nevertheless have evolved or have been developed to achieve the same objective: the effective control of hazardous energy. The litmus test is this: Are they soundly conceived, and do they meet the spirit and intent of the energy-control regulations and concepts?

In this chapter, the following special situations/applications for hazardous energy isolation are discussed: complex isolation of automated systems and manufacturing processes, alternative procedures for hazardous energy control, high-risk energy scenarios, the role of guarding and interlocking, robotics, host-contractor interaction, and other energy-isolation issues. These subjects do not represent all of the unusual or special circumstances under which hazardous energy needs to be effectively controlled. They do, however, illustrate the breadth of opportunity for unique and innovative hazardous-energy-control techniques.

COMPLEX ISOLATION

No definition for complex isolation is found in any of the consensus or regulatory standards. The subject of complex isolation is usually addressed by treatment of certain inherent characteristics such as shift change, group lockout, and high voltage. However, it is more constructive to deal with this situation frontally. For example, the following definition might be used:

> *Complex isolation:* Any situation in which machine, equipment, or process energy requires special control methods due to the nature of the hazards and the variables influencing effective isolation.

Hazards and variables can be the number of involved personnel, the type and quantity of energy-isolating devices, special risks and treatments, nature and duration of the work, environmental factors, etc.

The identification of a hazardous-energy-control situation as a complex isolation would serve to signal those concerned (regulators, auditors, investigators, and users) that the application will likely employ a number of techniques that depart from traditional methods. The focus could then legitimately shift from how the situation differs from the conventional and the problems it presents to how to accomplish what is desired.

Employers would need to identify each situation where complex isolation would be applied and prepare the appropriate intervention tactic. The burden for the employer would be the requirement to demonstrate that the totality of all employed technology, methods, and devices was capable of producing the desired effect: effective energy isolation during the accomplishment of the prescribed work.

Complex isolation tactics may be used routinely in such industries as petroleum refining, chemical processing, power generation/distribution, mineral smelting and refining, vehicle/ship/airplane production, aerospace, nuclear development/application, iron/steel production, and so on. Many other industries using advanced manufacturing systems also would qualify for energy isolation at the complex level.

As complexity of the process increases, a greater need develops for elements in the isolation approach that reveal the total picture. It is therefore common to use various visual methods such as process schematics, flow diagrams, pictorial drawings, process control graphics, computer graphics, and computer-generated procedures to guide those personnel involved in system lockout/tagout. Figures 10-1 through 10-5 illustrate various visual methods used in complex energy-isolation applications.

During the public hearings and written comment period on the lockout/tagout standard (29 *CFR* 1910.147), various individuals and industries that made use of complex isolation procedures (petroleum, chemical, electric power, etc.) made persuasive arguments regarding their unique demands and approach to energy isolation. A number of commentors stated that OSHA should not try to "force fit" a machinery standard to process systems and piping networks. In addition, others were adamantly in favor of retaining their method, which involved the use of a work permit or work authorization system with provisions for use of blinds, disconnection of pipes, and post-isolating cleaning and testing. Another contributor stated that a somewhat different procedure was required for safe performance of process system maintenance such as: (1) deactivation, (2) content removal, (3) isolation, (4) decontamination, (5) restraining, (6) verification, (7) control, and (8) communication. OSHA concluded that the work authorization procedures used by process industries were consistent with the procedures set forth in the proposed rule. Further, the agency felt that the utilization of the lockout/tagout standard in process type industries was not a force fit but a logical tailoring of the procedures to a different type of equipment.

However, in the Final Rule, 29 *CFR* 1910.147, *The Control of Hazardous Energy*, most of what is specifically said about complex isolation is presented in section (f), additional requirements. In less than one page this standard addresses testing and positioning of machines/equipment, outside personnel (contractors), group lockout or tagout, and shift or personnel changes. Group lockout/tagout or complex isolation will therefore be addressed here in greater detail because of its prevalence and criticality.

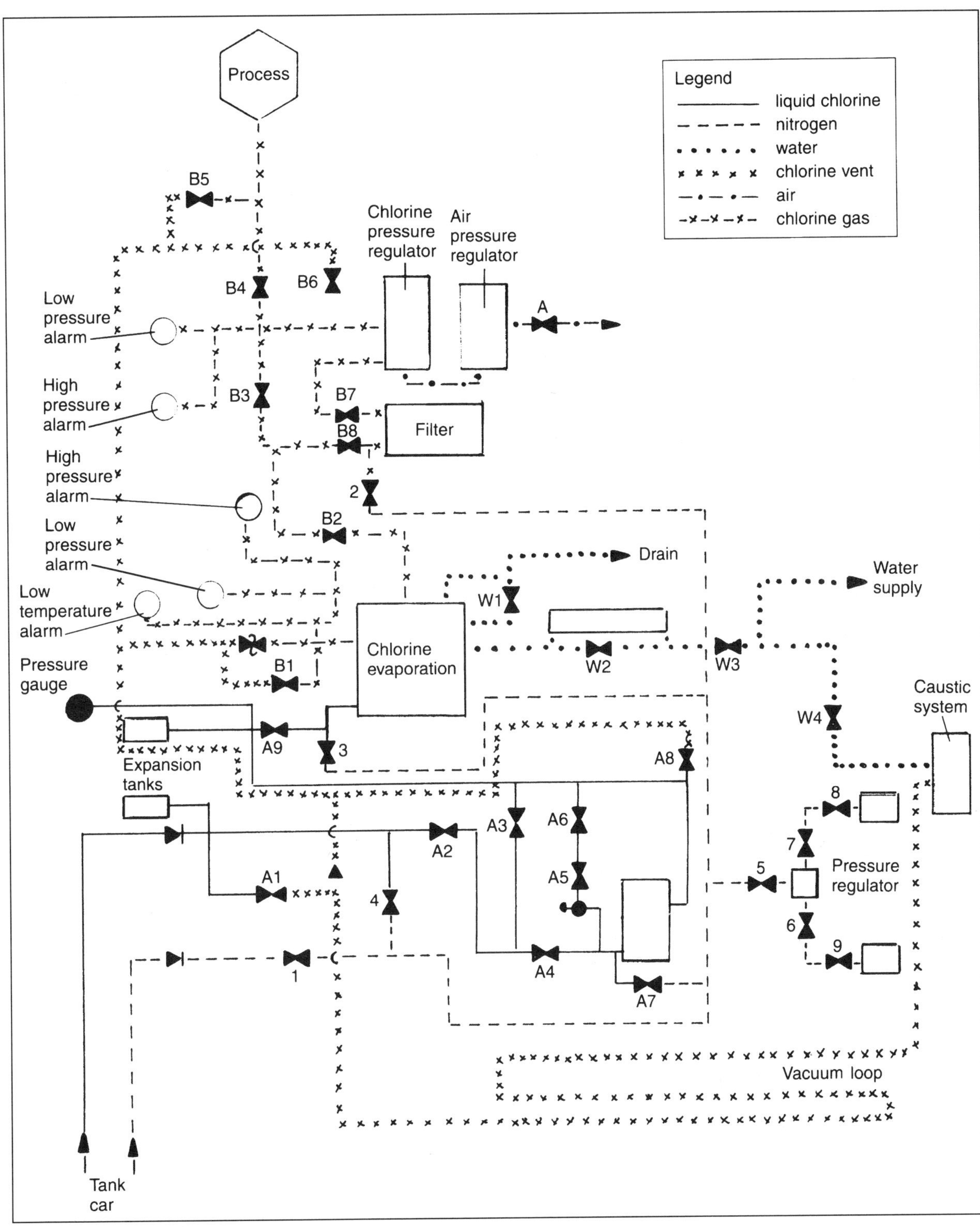

Figure 10-1. Process schematic for energy isolation—hazardous materials.
Source: Developed by E. Grund. © National Safety Council.

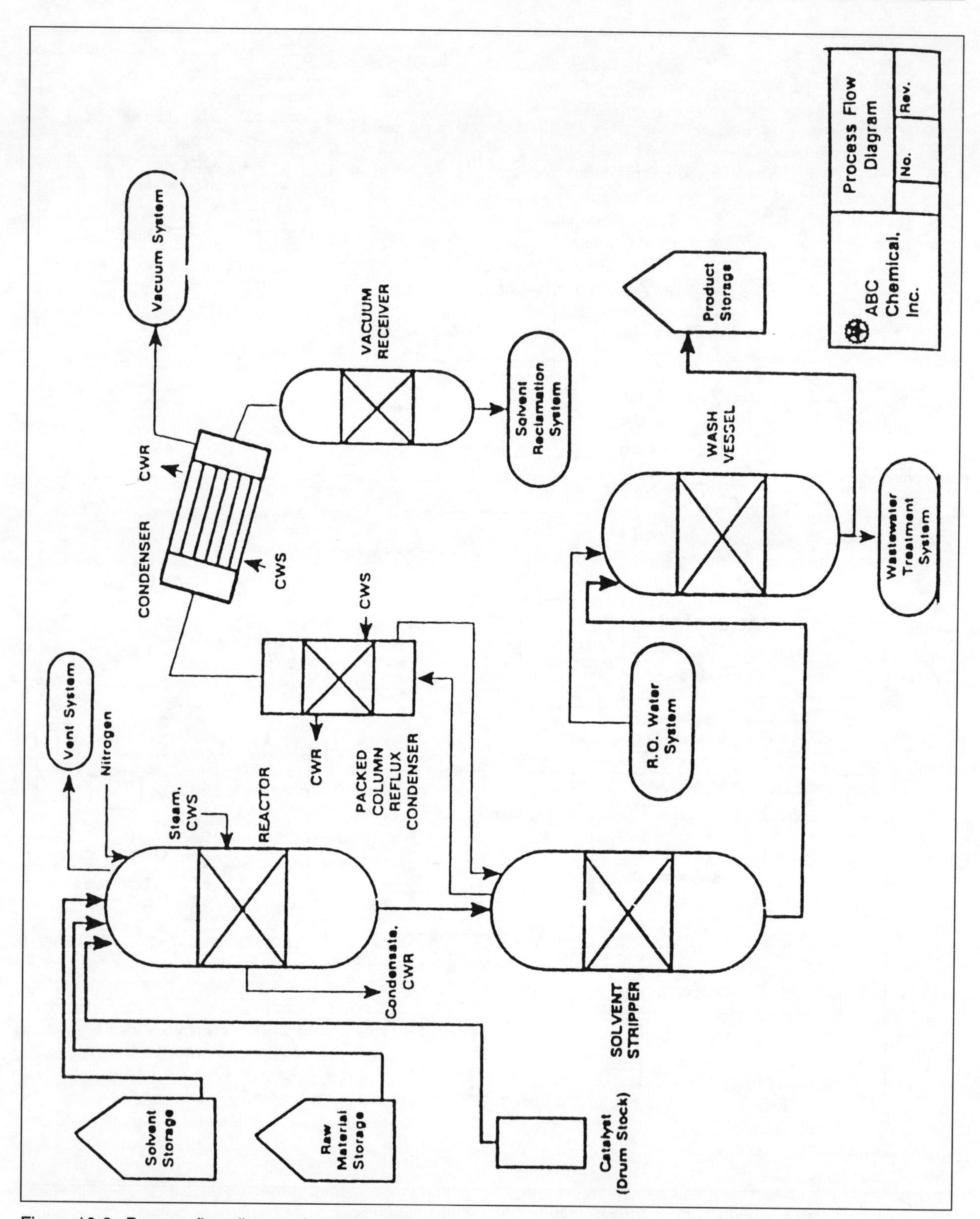

Figure 10-2. Process flow diagram for energy isolation.
Source: OSHA, 29 *CFR* 1910.119, Process Safety Management of Highly Hazardous Chemicals.

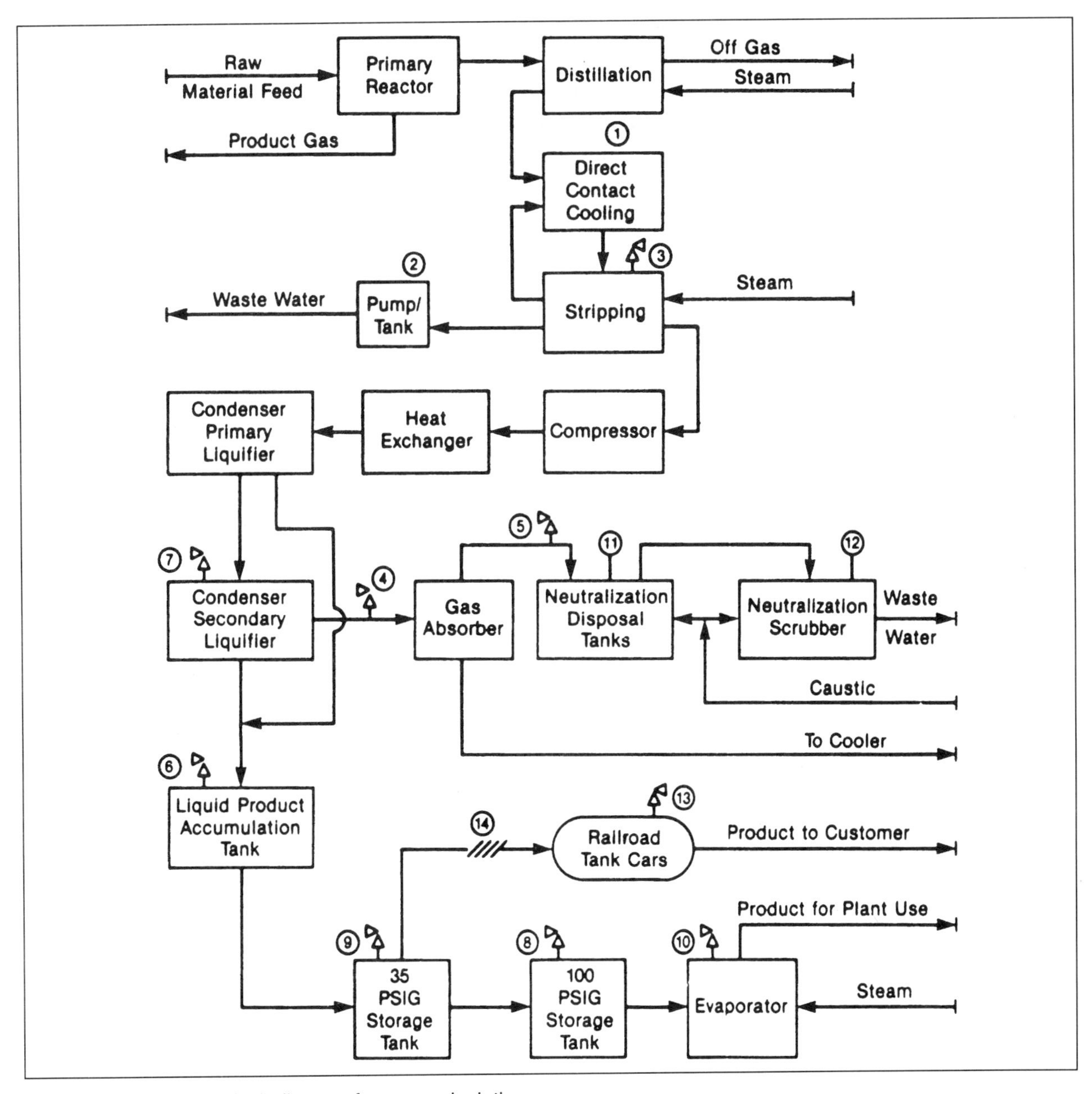

Figure 10-3. Process block diagram for energy isolation.
Source: OSHA, 29 *CFR* 1910.119, Process Safety Management of Highly Hazardous Chemicals.

GROUP LOCKOUT/TAGOUT

OSHA's appreciation for and recognition of complex isolation is not apparent when reviewing the content of 29 *CFR* 1910.147. The preamble to the final standard and instruction standard (administrative guidance for compliance personal) 1-7.3 provide much more insight as to OSHA's understanding of the complexity of many existing energy-isolation practices. The American Petroleum Institute, Chemical Manufacturer's Association, Edison Electric Institute, and others contributed during the public hearings and written comment period to ensure that OSHA fully understood the nature of and need for complex isolation procedures.

Standard Requirements

In the preamble to 29 *CFR* 1910.147, OSHA explained *group lockout* as the performance of servicing or maintenance activities using a group lockout device, by more than one employee, with an authorized employee directly responsible for the performance of the overall servicing or maintenance. This explanation, which is not an official definition, covers a huge amount of

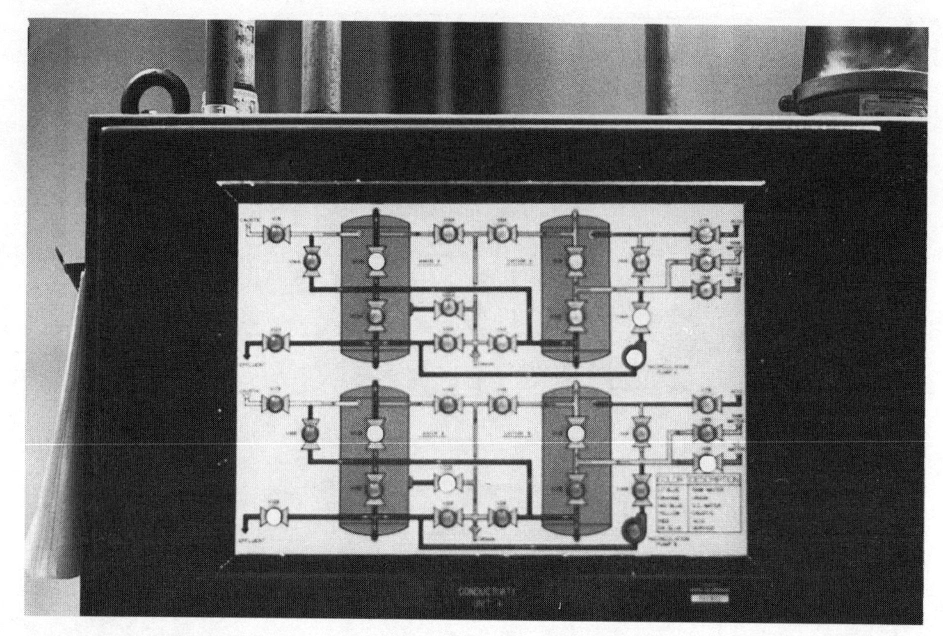

Figure 10-4. Process control display—energy-isolation reference.
Source: Photo courtesy E. Grund.

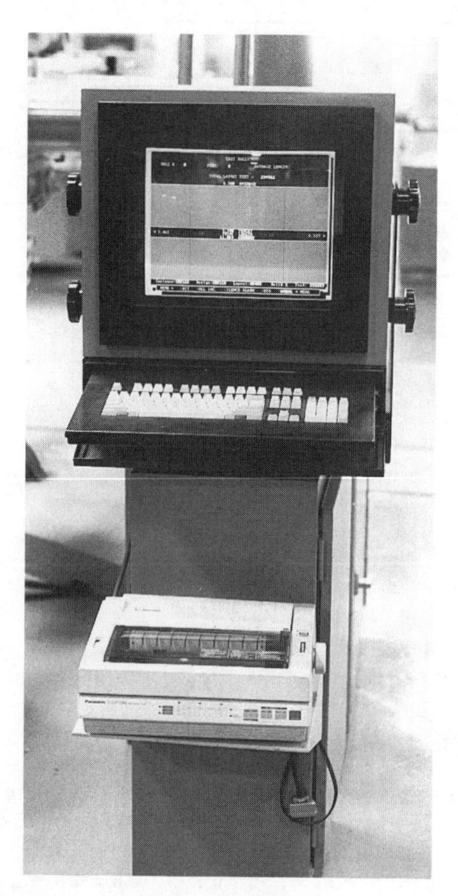

Figure 10-5. This is an example of a computer used to display applicable energy-isolation procedures in complex processes.
Source: Photo courtesy of E. Grund.

energy-isolation work in almost every industry in the manufacturing sector. At this point, OSHA has not apparently deemed it necessary to define formally the term *group lockout/tagout*. The final standard addresses the issue in section (f) (3) as follows:

(3) *Group lockout or tagout.*

(i) When servicing and/or maintenance is performed by a crew, craft, department or other group, they shall use a procedure which affords the employees a level of protection equivalent to that provided by the implementation of a personal lockout or tagout device.

(ii) Group lockout or tagout devices shall be used in accordance with the procedures required by paragraph (c)(4) of this section including, but not necessarily limited to, the following specific requirements:

(a) Primary responsibility is vested in an authorized employee for a set number of employees working under the protection of a group lockout or tagout device (such as an operations lock);

(b) Provision for the authorized employee to ascertain the exposure status of individual group members with regard to the lockout or tagout of the machine or equipment; and

(c) When more than one crew, craft, department, etc. is involved, assignment of overall job-associated lockout or tagout control responsibility to an authorized employee designated to coordinate affected work forces and ensure continuity of protection; and

(d) Each authorized employee shall affix a personal lockout or tagout device to the group lockout device, group lockbox, or comparable mechanism when he or she begins work, and shall remove those devices when he or she stops working on the machine or equipment being serviced or maintained.

In the standard preamble, after reexamination of the public comment, OSHA concluded that an additional element was necessary for the safety of servicing and maintenance employees: Each employee in the group needs to be able to affix his/her personal lock/tag device as part of the group lockout. This is reinforced in sections (3)(i) and (3)(ii)(d) of the final standard. OSHA provided the following five reasons for justifying the final requirement:

1. Placement of a personal lock/tag device would enable that employee to have a degree of control over his/her own protection, rather than having to depend completely upon other people.
2. Use of a personal device will enable each servicing employee to verify that the equipment has been properly de-energized in accordance with the energy-control procedure and to affix his/her device to indicate that verification.
3. Presence of an employee's lock/tag device will inform all other persons, including the other servicing employees and supervisors, that the employee is still working on the equipment.
4. As long as the device remains attached, the authorized person in charge of the group lockout or tagout knows that the job is not completed and that it is not safe to re-energize the equipment.
5. The servicing employee will continue to be protected by the presence of his/her device until he/she removes it.

OSHA stated that "it was convinced that the use of individual lockout or tagout system devices to supplement the group lockout device is necessary for the safety of the servicing and maintenance employees." In section (f)(3)(ii)(b), OSHA further requires the authorized employee to determine the exposure status of individual crew members and to take appropriate measures to control or limit that exposure. This provision as seen by OSHA requires at least the following steps:

1. Verification of shutdown and isolation of the equipment or process before allowing a crew member to place a personal lockout or tagout device on an energy-isolation device or on a lockout box, board, or cabinet.
2. Ensuring that all employees in the crew have completed their assignments; removed their lockout and/or tagout devices from the energy-isolating device, the box lid, or other device used; and are in the clear before turning the equipment or process over to the operation personnel or simply turning the machine or equipment on.
3. Providing the necessary coordinating procedures for ensuring the safe transfer of lockout or tagout control devices between other groups and work shifts.

Definitions

OSHA, in its Instruction Standard 1-7.3, dated September 11, 1990, provides further insight as to how group lockout/tagout must function. The standard details inspection procedures and provides interpretive guidance for compliance personnel to ensure uniform enforcement, and its Appendix C contains further information relating to group lockout/tagout application. The following terms relating to group applications are defined:

Primary Authorized Employee is the authorized employee who exercises overall responsibility for adherence to the company lockout/tagout procedure. (See 29 *CFR* 1910.147 [f] [3] [ii] [a].)

Principal Authorized Employee is an authorized employee who oversees or leads a group of servicing/maintenance workers (e.g., plumbers, carpenters, electricians, metal workers, mechanics).

Job-Lock is a device used to ensure the continuity of energy isolation during multi-shift operation. It is placed upon a lockbox. A key to the job-lock is controlled by each assigned primary authorized employee from each shift.

Job-Tag with Tab is a special tag for tagout of energy isolating devices during group lockout/tagout procedures. The tab of the tag is removed for insertion into the lockbox. The company procedure would require that the tagout job-tag cannot be removed until the tab is rejoined to it.

Master Lockbox is the lockbox into which all keys and tabs from the lockout or tagout devices securing the machine or equipment are inserted and which would be secured by a "job-lock" during multi-shift operations.

Satellite Lockbox is a secondary lockbox or lockboxes to which each authorized employee affixes his/her personal lock or tag.

Master Tag is a document used as an administrative control and accountability device. This device is normally controlled by the operations department personnel and is a personal tagout device if each employee personally signs on and signs off on it and if the tag clearly identifies each authorized employee who is being protected by it.

Work Permit is a control document which authorizes specific tasks and procedures to be accomplished.

Group Procedural Structure

In Appendix C, Instruction Standard 1-7.3, OSHA outlines a procedural structure that could be used under group lockout tagout circumstances:

- A *primary* authorized employee would be designated to exercise primary responsibility for implementation and coordination of the lockout/tagout of hazardous energy sources.
- The primary authorized employee would coordinate with equipment operators before and after completion of servicing and maintenance operations which require lockout tagout.
- A verification system would be implemented to ensure the continued isolation and de-energization of hazardous energy sources.
- Each authorized employee would be assured of his/her right to verify individually that the hazardous energy has been isolated and/or de-energized.
- When more than one crew, craft, department, and so on is involved, each separate group of servicing/maintenance personnel would be accounted for by a *principal* authorized employee from each group. Each principal employee is responsible to the primary authorized employee for maintaining accountability of each worker in that specific group in conformance with the company procedure. No person may sign on or off for another person or attach or remove another person's lockout/tagout device unless the provisions of the exception to 29 *CFR* 1910.147(e)(3) are met.

Group Lockout/Tagout (Lockbox Procedures)

In Appendix C, OSHA describes various methods of lockout/tagout where lockboxes are used as the controlling technique. Three variations are presented as follows:

Type B. Under a lockbox procedure, a lock or job-tag with tab is placed upon each energy isolation device after de-energization. The key(s) and removed tab(s) are then placed into a lockbox. Each authorized employee assigned to the job then affixes his/her personal lock or tag to the lockbox. As a member of a group, each assigned authorized employee verifies that all hazardous energy has been rendered safe. The lockout/tagout devices cannot be removed or the energy isolating device turned on until the appropriate key or tag is matched to its lock or tag.

Type C. After each energy isolating device is locked/tagged out and the keys/tabs placed into a master lockbox, each servicing/maintenance group principal authorized employee places his/her personal lock or tag upon the master lockbox. Then each principal authorized employee inserts his/her key into a satellite lockbox to which each authorized employee in that specific group affixes his/her personal lock or tag. As a member of a group, each assigned authorized employee verifies that all hazardous energy has been rendered safe. Only after the servicing/maintenance functions of the specific subgroup have been concluded and the personal locks or tags of the respective employees have been removed from the satellite lockbox, can the principal authorized employee remove his/her lock from the master lockbox.

Type D. During operations to be conducted over one shift (or even many days or weeks) a system such as described here might be used. Single locks/tags are affixed upon a lockbox by each authorized employee as described at Type B or Type C above. The master lockbox is first secured with a job-lock before subsequent locks by the principal authorized employees are put in place on the master lockbox. The job-lock may have multiple keys if they are in the sole possession of the various primary authorized employees (one on each shift). As a member of a group, each assigned authorized employee verifies that all hazardous energy has been rendered safe. In this manner, the security provisions of the energy control system are maintained across shift changes while permitting reenergization of the equipment at any appropriate time or shift.

Group lockout/tagout using lockbox techniques usually involve (1) pyramiding control methods; (2) company or general locks/tags for isolating devices; (3) multilevel boxes or cabinets; (4) master/job locks/tags; (5) personal cards, tabs, etc.; and (6) multiple lockout devices. In Figures 10-6 and 10-7, lockout boards (complete setups for complex isolation) are shown that incorporate these elements.

Pyramiding control methods are often used in complex isolation situations. The approach can be used with locks, tags, or a combination of the two. Personnel security is managed by establishing a hierarchy of control in which no single individual has the authority or ability to move/remove any energy-isolation device without appropriate procedural approval. Figure 10-8

Figure 10-6. Standard lockout board.
Source: Photo courtesy Kaiser Aluminum Trentwood Works, Spokane, WA.

Figure 10-7. See-through lockout board.
Source: Photo courtesy Kaiser Aluminum Trentwood Works, Spokane, WA.

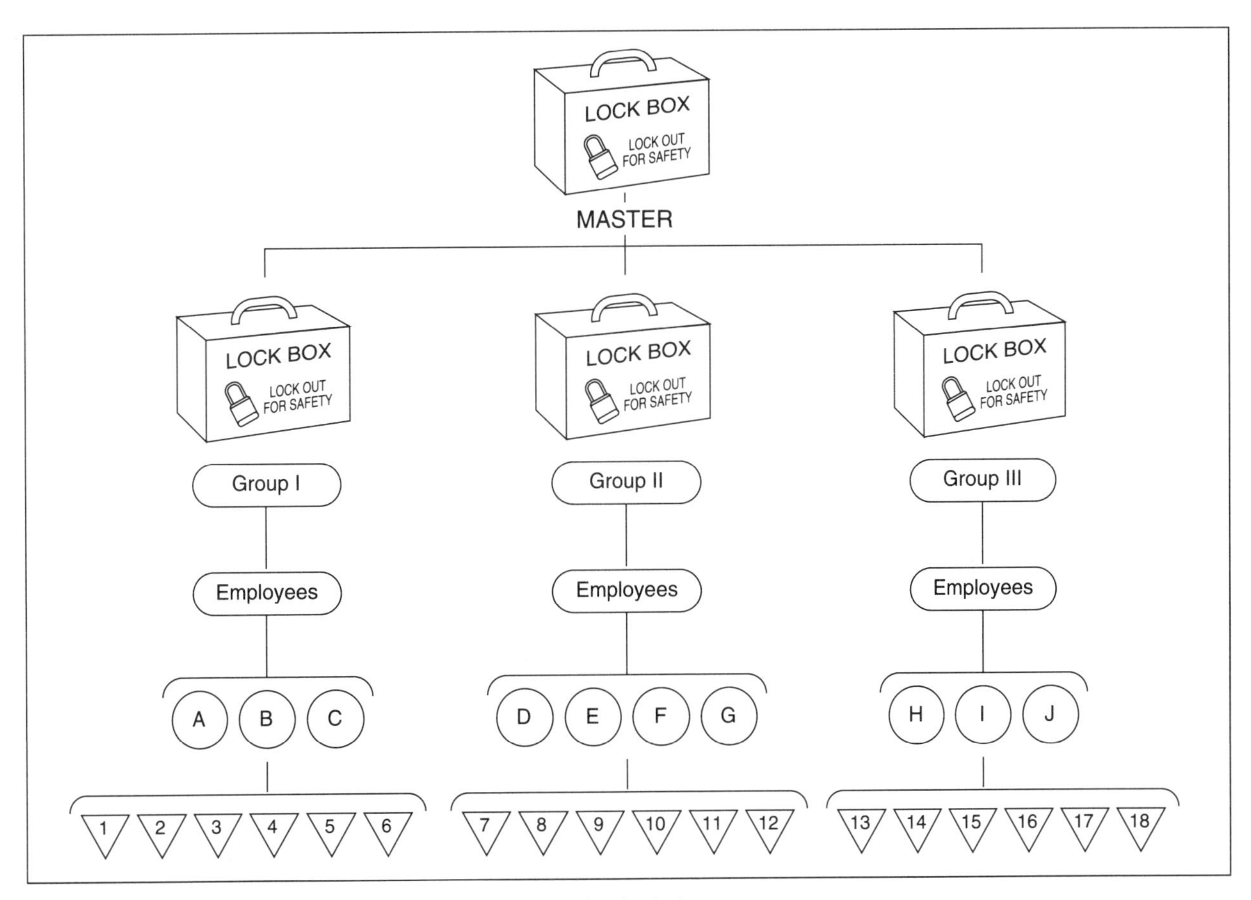

Figure 10-8. Lockout/tagout security pyramid for complex isolation.
Source: Developed by E. Grund. © National Safety Council.

illustrates a scheme for establishing such a control structure. The following explanation applies:

General: Three groups of employees are involved in the energy-isolation activity in order to perform service/maintenance. A total of 10 employees (A–J) will be involved. Eighteen energy-isolating devices (1–18) will need to be locked/tagged by the employees.

Step 1: Employees (A–C) in Group I will operate energy-isolating devices (1–6) and lock/tag using company/equipment type locks/tags.

Step 2: Keys/tag stubs are placed in Group I lockbox from company/equipment type lock/tags. Employees (A–C) place personal locks/tags on multiple lockout device on lockbox. Principal authorized employee (A) collects personal keys/stubs from employees (A–C).

Step 3: Principal authorized employee (A) takes personal keys/stubs from employees (A–C) and deposits them in master job lockbox.

Step 4: Groups II and III follow the same pattern in steps 1 through 3 above.

Step 5: All keys/stubs securing group's lockboxes and representing employees (A–J) are now deposited in the master job lockbox.

Step 6: Each principal authorized employee (A, D, and H) for each group places a group lock/tag on the master lockbox and retains the key/stubs.

Step 7: The primary authorized employee places a job lock/tag on the multiple lockout device on the master lockbox and retains the key/stub. Outcome: All keys/stubs for energy-isolating devices are secured in group lockboxes. All employee personal keys/stubs are secured in master lockbox. Keys/stubs for master lockbox are in possession of three principal authorized and one primary authorized employee(s).

Note: A number of variations are possible in this type of control pyramid to achieve the end result, that is, personal security for all potentially exposed personnel.

Group Lockout/Tagout (Permit Systems)

OSHA Instruction Standard 1-7.3 (1990) offers guidance regarding energy isolation where machine, equipment, and process complexity demand more exhaustive control methods. Petroleum, chemical, steel, and industrial gas production, among others, involve operations in which adaptation and modification of normal group lockout/tagout procedures is needed in order to ensure the safety of employees performing service and maintenance activities. To provide greater worker safety through implementation of a more feasible system and to accommodate the special constraints of the standard's requirement for ensuring employees a level of protection equivalent to that provided by the use of a personal lockout or tagout device, an alternative procedure may be implemented if company documentation justifies it. Lockout/tagout, blanking, blocking, and similar practices are often supplemented in these situations by the use of work permits and a system of continuous worker accountability. In evaluating whether the equipment being serviced or maintained is so complex as to necessitate a departure from the normal group lockout/tagout procedures to the use of an alternative procedure, the following factors (which often occur simultaneously) are among those that must be evaluated: the physical size and extent of the equipment being serviced/maintained, the relative inaccessibility of the energy-isolating devices, the number of employees performing the servicing/maintenance, the number of energy-isolating devices to be locked/tagged out, and the interdependence and interrelationship of the components in the system or between different systems.

After the machinery, equipment, or process is shut down and the involved hazardous energy has been controlled, maintenance, servicing, and operating personnel must verify that the isolation of the equipment is effective. The workers may walk through the affected work area to verify isolation. If there is a potential for the release or reaccumulation of hazardous energy, verification of isolation must be continued. The servicing/maintenance workers may further verify the effectiveness of the isolation by attempting procedures that are used in the work (for example, using a bleeder valve to verify depressurization, flange-breaking techniques, and so on). Throughout the maintenance and/or servicing activity, operations personnel normally maintain control of the equipment. The use of the work permit or master tag system (with each employee personally signing on and off the job to ensure continual employee accountability and control), combined with verification of hazardous energy control, work procedures, and walk-through, is an acceptable approach to compliance with the group lockout/tagout and shift transfer provisions of the standard.

The following scenario illustrates the control method that may be used during isolation of hazardous energy found in complex processes:

1. Complex process equipment scheduled for servicing/maintenance operations is generally identified by plant supervision. Plant supervision issue specific work orders regarding the operations to be performed.
2. In most instances where complex process equipment is to be serviced or maintained, the process equipment operators can be expected to conduct the shutdown procedure. This is due to their

in-depth knowledge of the equipment and the need to conduct the shutdown procedure in a safe, efficient, and specific sequence.

3. The operations personnel normally prepare the equipment for lockout/tagout as they proceed and identify the locations for blanks, blocks, and so on by placing operations locks and/or tags on the equipment. The operations personnel can be expected to isolate the hazardous energy and drain and flush fluids from the process equipment following a standard procedure or a specific work permit procedure.
4. Upon completion of shutdown, the operations personnel review the intended job with the servicing and maintenance crew(s) and ensure their full comprehension of the energy controls necessary to conduct the servicing and maintenance safely. During or immediately after the review of the job, the servicing and maintenance crews install locks, tags, and/or special isolating devices at previously identified equipment locations following the specified work permit procedure.
5. Line openings necessary for the isolation of the equipment would normally be permitted only by special work permits issued by operations personnel. (Such line openings should be monitored by operations personnel as an added safety measure.)
6. All of the previous steps should be documented by a master system of accountability and retained at the primary equipment control station for the duration of the job. The master system of accountability may manifest itself as a master tag that is subsequently signed by all of the maintenance/servicing workers *if* they fully comprehend the details of the job and the energy-isolation devices actuated or put in place.
7. After the system has been rendered safe, the authorized employees verify energy controls.
8. Specific work functions are controlled by work permits issued for each shift. Every day, each authorized employee assigned must sign in on the work permit at the time of arrival to the job and sign out at departure. Signature, date, and time for sign-in and sign-out are recorded and retained by the applicable crew supervisor who, upon completion of the permit requirements, returns the permit to the operations supervisor. Work permits can extend beyond a single shift and may subsequently be the responsibility of several supervisors.
9. Upon completion of the tasks required by the work permit, the authorized employees' names are signed off the master tag by their supervisor once all employees have signed off the work permit. The work permit is then attached to the master tag. (Accountability of exposed workers is maintained.)
10. As the work is completed by the various crews, the work permits and the accountability of personnel are reconciled jointly by the primary authorized employee and the operations supervisor.
11. During the progress of the work, inspection audits are conducted.
12. Upon completion of all work, the equipment is returned to the operations personnel after the maintenance and servicing crews have removed their locks, tags, and/or special isolating devices following the company procedure.
13. At this time, all authorized employees who were assigned to the tasks are again accounted for and verified to be clear from the equipment area.
14. After the completion of the servicing/maintenance work, operations personnel remove the tags originally placed to identify energy isolation.
15. Operations personnel then begin checkout, verification, and testing of the equipment prior to being returned to production service.

T. A. Kletz, writing on the basic principles of a permit system in Chapter 36 of *Safety and Accident Prevention in Chemical Operations,* makes the following statements (Fawcett & Wood, 1982):

> Many of the raw materials, intermediates, and products handled by the chemical industry are hazardous. They may be flammable or explosive, toxic or corrosive; they may be handled hot or under pressure; they may possess all these properties. Therefore, before any equipment is opened up for inspection, repair, or modification, all hazardous materials should, if possible, be removed. If this is not possible then the workers carrying out the inspection, repair, or modification must be told of the hazards that remain and told what precautions they should take.
>
> These objectives are usually achieved by issuing a permit-to-work or clearance certificate. The permit or clearance:
>
> 1. Provides a checklist for the worker preparing the equipment and this reduces the chance that any part of the procedure will be missed.
> 2. Informs the workers carrying out the repair, modifications, or inspection of the hazards that are present and the precautions that should be taken.
>
> Many accidents have occurred because the permit system was not satisfactory or because the system, though satisfactory, was not followed. Errors in the preparation of equipment for maintenance are one of the commonest causes of serious accidents in the chemical and allied industries.

He describes a clearance certificate used by a British company with three major parts as follows:

Part A—Preparation: checklist of potential hazards; need and methods for physical equipment/process isolation; precautions taken; hot work/entry permits required; radioactive source control; electrical isolation requirements; preparation completion section.
Part B—Operation: jobs to be done, tag numbers, equipment rotation check, personal protective equipment required, safeguards for all hazards identified in Part A above.
Part C—Transfer and Acceptance of Responsibility: transfer from-to [between operations and maintenance]; sign on and off provisions; job completion data; return transfer; equipment status checks.

Permits or clearance certificates are usually issued by a lead operator or other person in a supervisory position. In each facility or part thereof, only one person at a time should be allowed to issue permits. Whoever issues permits must understand the purpose and working of the permit system and have some knowledge of incidents that have occurred because the system was not followed.

Permits should be accepted by the worker who is going to do the job. When several workers will be doing a job, the permit should be accepted by the one in charge. In some companies, it is the practice for a supervisor to accept all permits. If this system is followed, the permit should be shown to the worker or workers who will do the job. A convenient way of doing this is to display the permit on the job.

When many permits have to be issued, for example, at a shutdown lasting several days or if new construction is taking place alongside an operating facility, then a day process supervisor should be appointed with special responsibility for issuing permits and for liaison with maintenance or construction.

All permits should cease to be valid after a stated period, usually one shift or one day but never more than one week (one shift when hot work or entry is involved), and must then be handed back or renewed. The period of validity should be stated on the permit. Permits should be issued only when the maintenance team is ready to start the job. It is bad practice to issue permits for jobs that will be done "if we have some spare time."

The following special considerations should be addressed in typical permit or clearance systems:

- energy/hazardous materials isolation
- purging and testing
- equipment/process identification (verification)
- changes of intention
- disarming of protective systems
- relief devices, vents, bleeds, etc.
- modification action/documentation
- confined space/vessel entry
- hot work
- radiation
- piping networks (content/area effects)
- contractor interaction
- line/equipment opening
- flammables/explosives
- inert/toxic gases
- corrosive materials
- trapped energy/materials
- personal protective equipment.

Occidental Chemical Corporation's facility in Castle Hayne, North Carolina, uses a *lock, tag, and try* energy-isolation procedure. Oxychem uses the *custodian* concept, in which each machine, equipment, or process (MEP) has one individual who is responsible for ensuring that their MEP is operating properly and that appropriate isolation procedures are followed. When servicing or maintenance is needed, the custodian places a multiple lockout device and a special, extra-long shackle lock(s) on the energy-isolation devices. The custodian is the first person to secure his/her padlock and the last person to remove the lock from the energy-isolating device. Oxychem uses six distinct groups of color-coded locks to differentiate various departments.

Because jobs frequently extend beyond a single shift, a master key system is used. A single department master key is in the possession of the custodian who is responsible for ensuring the job proceeds with appropriate energy isolation. All padlocks are uniquely numbered and used in conjunction with tags (Germain, 1990).

Zeneca's Bayport, Texas, site uses a unique safe-work-file approach to ensure effective communication regarding equipment status and proper isolation. The safe work file is required when:

- an isolation must be made to ensure the safety of a job task
- the isolation must remain for the duration of the job task.

A safe work file is initiated to record all isolations made. The site's lockout/tagout policy is used in conjunction with the safe work permit (Figures 10-9 through 10-11). The equipment coordinator and shift or area supervisor is responsible to ensure that a safe work file is initiated and that the work order number, equipment number, and job description are entered on the file. The general process proceeds as follows:

1. A competent person (such as an area process technician) isolates the appropriate piece of equipment according to the site lockout/tagout procedure.
2. The competent person then signs the "Isolation Complete—Job Ready" line on the front of the safe work file, after ensuring that the isolations are adequate and completed.

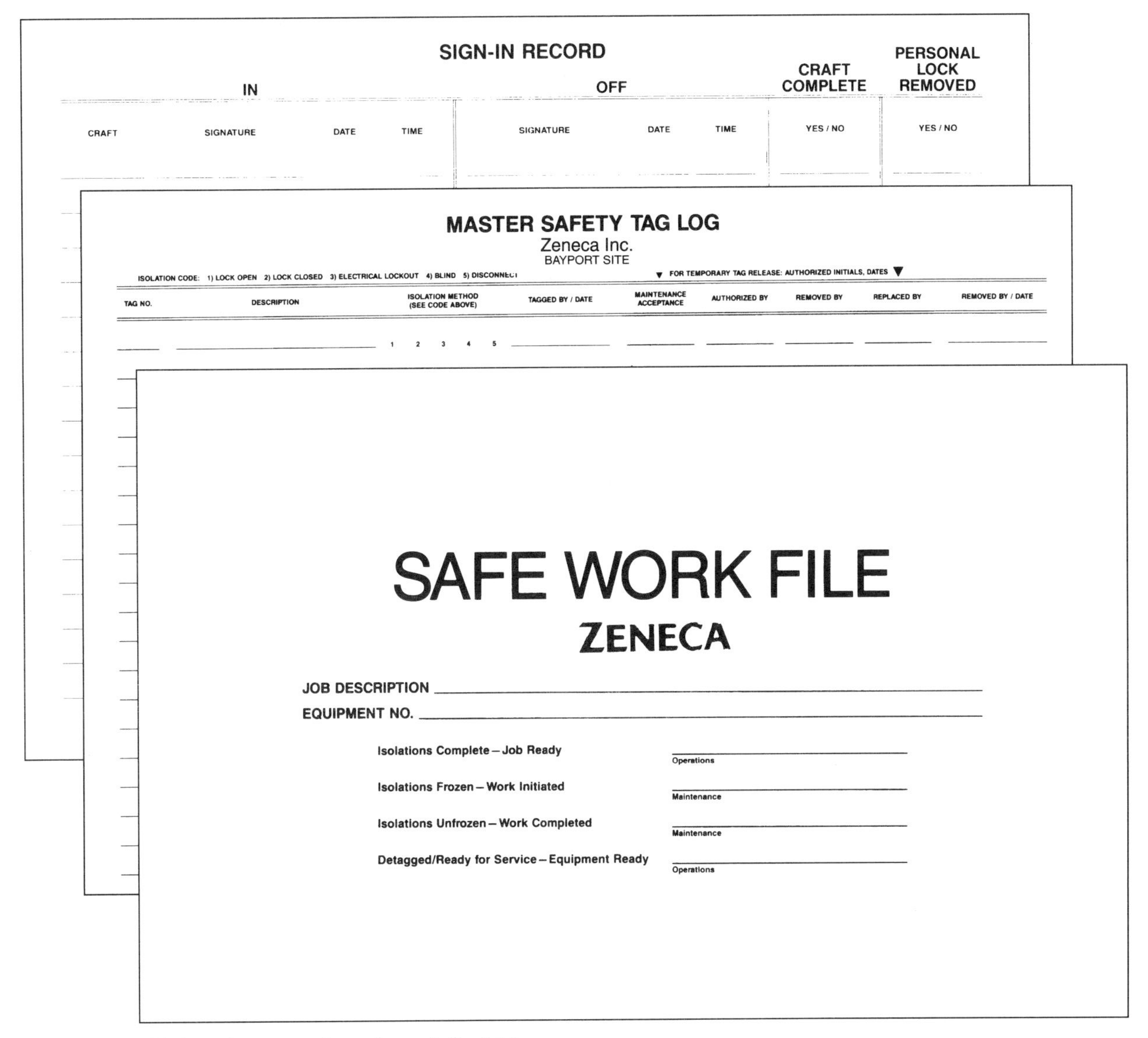

SIGN-IN RECORD

IN				OFF			CRAFT COMPLETE	PERSONAL LOCK REMOVED
CRAFT	SIGNATURE	DATE	TIME	SIGNATURE	DATE	TIME	YES / NO	YES / NO

MASTER SAFETY TAG LOG
Zeneca Inc.
BAYPORT SITE

ISOLATION CODE: 1) LOCK OPEN 2) LOCK CLOSED 3) ELECTRICAL LOCKOUT 4) BLIND 5) DISCONNECT

▼ FOR TEMPORARY TAG RELEASE: AUTHORIZED INITIALS, DATES ▼

TAG NO.	DESCRIPTION	ISOLATION METHOD (SEE CODE ABOVE)	TAGGED BY / DATE	MAINTENANCE ACCEPTANCE	AUTHORIZED BY	REMOVED BY	REPLACED BY	REMOVED BY / DATE
		1 2 3 4 5						

SAFE WORK FILE
ZENECA

JOB DESCRIPTION ____________

EQUIPMENT NO. ____________

Isolations Complete – Job Ready ____________ Operations

Isolations Frozen – Work Initiated ____________ Maintenance

Isolations Unfrozen – Work Completed ____________ Maintenance

Detagged/Ready for Service – Equipment Ready ____________ Operations

Figure 10-9. Various features of a safe work file folder.
Source: Courtesy of Zeneca Inc., Bayport, TX.

3. The first maintenance person to initiate work will sign the “Isolations Frozen—Work Initiated” line on the front of the file. This signature freezes all isolation associated with this job and cannot be unfrozen (changed) unless a temporary tag release is authorized or job is completed.
4. The maintenance person signing on the front of the file shall also sign the “Sign In/Record” section on the back of the file. All maintenance personnel who work on this piece of equipment must sign it. The “Sign Off” section will be used
   - at the end of the job
   - at the end of the day
   - when personnel are removed from that job to work on another job. Maintenance personnel will then complete the “Craft Complete” and the “Personal Lock Removed” columns.
5. The maintenance person that completes the job task must clear the file by signing the “Isolations Unfrozen—Work Complete” line on the front of the file. This line will only be signed after ensuring that all involved with this job have completed their tasks. This can be accomplished by reviewing the “Sign-in Record” and ensuring that no other craftsman locks are on the electrical isolation point. If any doubts remain, do not sign off the file until verification with the foreman of the craftmen involved with this file.
6. The competent person will then remove all isolations associated with this job according to the lockout/tagout procedure and sign the “Detagged/Ready for Service-Equipment Ready” line on the front of the file. The temporary tag release section is to be used whenever an isolation temporarily needs to be

Zeneca Inc.
FORMCO NO. AM 9950 (REV. 4/92)

SAFE WORK PERMIT
BAYPORT SITE

FORM DISTRIBUTION:
(1) Original - To Be Posted At Job Site
(2) Duplicate - Equipment Foreman

SECTION I ☐ I & E ☐ Process ☐ Maint. ☐ Contractor

Requested By: ____________ Date ________ For Hours ______ a.m./p.m. to ______ a.m./p.m.

Location of Equipment ____________ Equipment No. ________

Description of Job Task ____________

TYPES OF WORK

Class "A" ☐ Hot Work (Welding or Cutting)
☐ Other ________
☐ Chemicals ________

Class "B" ☐ Hand and Power Tools Capable of Producing Sparks
☐ Excavation with Power Equipment
☐ Electrical Hot Work
☐ Chemicals ________
☐ Other ________

Class "C" ☐ Hand Tools
☐ DC Powered tools or Equipment
☐ Chemicals ________
☐ Other ________

SECTION II a. Authorized person must check each line or cross out whole section if not applicable.
Fill Out Completely b. Technician Completing task (Preparation) Must initial each section under "Operations Completed."

Preparations and Tests	To Be Made YES	To Be Made NO	Operations Completed
Empty	☐	☐	______
Vent	☐	☐	______
Purge	☐	☐	______
Blind	☐	☐	______
Tag Out/Positive Isolations	☐	☐	______
Electrical Lockout	☐	☐	______
Safe Work File Initiated	☐	☐	______
Wash	☐	☐	______
Steam	☐	☐	______
Cool	☐	☐	______
Explosive Gas Test (LEL)	☐	☐	______
Other	☐	☐	______

Precautions/Safety Equipment Required	Required YES	Required NO	Maintenance Accepts
Locate Emergency Equiment	☐	☐	______
Safety Harness	☐	☐	______
Face Shield	☐	☐	______
Respirator	☐	☐	______
Chemical Cartridge Type ______	☐	☐	______
SCBA	☐	☐	______
Bottled Breathing Air (EBA)	☐	☐	______
PVC Gloves	☐	☐	______
Other Gloves	☐	☐	______
Acid Suit	☐	☐	______
Hearing Protection	☐	☐	______
Fire Watch Required	☐	☐	______
Barricade Area	☐	☐	______
Safety Instructions	☐	☐	______
Chemical Hazard Instructions	☐	☐	______

Supplementary Instructions: ____________

SECTION III

A. ____________ Approval Signature

B. ____________ Maintenance Signature

CLASS "A' ONLY
C. ____________ Final Approval

D. ____________ Outside Contractor

____________ On-Site Contact Person

Figure 10-10. Safe work permit (*front*).
Source: Courtesy of Zeneca Inc., Bayport, TX.

SWP PROCEDURE—APPENDIX 1

SWP CLASSIFICATIONS AND APPROVAL LEVELS

CLASS "A":

Anhydrous Ammonia lines that can not be decontaminated
Caustic/sludge — above 60°C
Chlorine lines
Dimer/Oxidizer slurry
HCL
Methyl Chloride
Nitrogen — pressure 150# and above
Organics — Above 60°C
Peroxides
PP796 (Emetic)
Steam — pressure over 35#
Sodium — all except railcar changeout
Sodium Ammonia
Sulfuric Acid
All hot work — spark producing, open flame, electric welding, or above the ignition point of the chemical.

APPROVAL LEVEL: Site Manager
Designated Representative

CLASS "B":

Anhydrous Ammonia — vapor side only
Bipyridyl — above 60°C
Caustic/sludge lines — Ambient temperature
Methylene Chloride
Organics — below 60°C
Paraquat lines
Sodium railcar changeout
Steam — up to 35#
Valeric Acid
354 — Filming amines
Any combination of chemicals in an undecontaminated line not listed in Class "A"
Tools: Hand & Powered tools capable of producing spark
Drill
Saw
Air Hammer
Impact Wrench (electric)
Cold cutting
Excavation with power equipment

APPROVAL LEVEL: Plant Manager
Designated Representative
Temporary Representative

CLASS "C":

Other chemical on plant not listed in Class "A" or Class "B"
or DC powered tools and equipment
Portable battery driven tools and equipment
Hand tools

APPROVAL LEVEL: Designated Representative
Qualified Technician/Craftsman

Figure 10-11. Safe work permit (*back*).
Source: Courtesy of Zeneca, Inc., Bayport, TX.

changed (for example, re-energizing a motor to verify rotation prior to completing the job). The procedure outlined in the lockout/tagout policy shall be followed.

7. All safe work files will be retained for a minimum of 18 months. The originator of the safe work file will be responsible for its retention.

The unique characteristics of each operational setting will determine the approach to the isolation of hazardous energy. Complex isolation exists far more frequently than is generally appreciated. However, even under situations deemed to require complex isolation tactics, a gradient of approach exists that is defined by such job nomenclature as *small to large, short to long, repair to rebuild,* and *on-stream to turnaround.* In Figures 10-12 and 10-13, a complex isolation procedural example and sketch (faulty pump) is provided in which four individuals are working with five energy-isolation devices. In Figures 10-14 and 10-15, a complex isolation procedural example and sketch (fractionation tower) is shown involving five groups, 50–60 workers, and over 100 energy-isolation devices. Procedural creativity is a must when considering the proper approach to energy isolation. The objective is to adhere to the principle of protecting individuals at risk of exposure to energy-release incidents without being trapped or restricted by a preconceived generic notion of lockout/tagout.

CONTRACTORS

Contractors represent a challenging safety issue for employers who engage them to perform various services. OSHA, recognizing the need to respond to the less-than-satisfactory safety experience of contractors, has included them by reference in four of their most far-reaching regulatory actions. The *Hazard Communication, Lockout/Tagout, Process Safety, and Confined Space* standards all attempt to address the relationships between employers (hosts) and contracting personnel or companies with regard to safety.

A *contractor*, for our purposes, can be defined as an independent individual, group, or company that has been engaged by a business/entity (host, property owner, etc.) to perform various services according to some contractual (written) agreement. The engaging business/entity defines the nature of the services (normally involving the use of external personnel) to be performed and all of the associated requirements. The parties negotiate and eventually agree on all matters of interest relating to the performance of the services. The host then attempts to ensure that the contractor delivers what has been agreed upon without jeopardizing the liability (if any) of the engaging party.

The possible number of arrangements for contracting is limited only by the imagination of the involved parties. The nature of the arrangement or relationship between parties will determine in large part how safety is addressed and to what degree liability is borne by either party. If an incident or accident occurs affecting either or both parties to the contract, case law, factual findings, and the like will eventually determine who is liable for what.

The following relationships between host and contractor represent the most common arrangements:

1. Full-time or continuous employment on host's site (supervised by contracting/management personnel)
2. Full-time or continuous employment on host's site (supervised by host)
3. On-demand hiring (frequent use of regular personnel) supervised by contractor or host
4. On-demand hiring (infrequent use of independent agency or hiring hall personnel) supervised by contractor or host
5. Any version of arrangements 1–4 in which contractor employees are comingled or work together with the host's employees
6. Arrangement 4 where services to be performed are deemed to be unique or high risk, for instance, stack work, demolition, aerial lifting, high-pressure/voltage, diving, and well fire control.

The key variables—who supervises, knowledge of the host's premises and processes, degree of utilization, past experience with host, and specialized skills—will determine how safety is approached to protect all involved parties. It is generally believed that when the employer (host) provides supervision (directly or indirectly) and/or mixes the work groups, a greater standard of care is assumed by the employer for the safety of *all* involved personnel. This may or may not be desirable depending on the circumstances.

Typically, the host/employer has no direct control over the actions of the contractor's employees. Influence is exerted indirectly through contractor supervision or the supplier (contract administrator, owner, general manager, etc.). This ability to influence provides the mechanism to achieve the following:

- contractor compliance with host's safety provisions/requirements
- contractor compliance with state and federal regulatory requirements
- correction of deficiencies (behavioral or physical conditions)
- contractor's control over individual employees who fail to perform as expected or who might jeopardize the host employee's safety
- contractor guidance under circumstances where host has specialized expertise necessary for safe performance of required services

The specific requirements for conforming to the contractor provisions in OSHA's *Lockout/Tagout* standard are found in 29 *CFR* 1910.147(f)(2) as follows:

LOCKOUT/TAGOUT
(Complex Isolation - Small Job)

Job: Removal of faulty pump and re-installation following servicing and/or maintenance

Isolating Devices: Two block valves, two blinds, one electrical circuit breaker

Involved Personnel: Process operator, three mechanics

STEP 1. The process operator turns off the electrical switch to shut down the pump.

STEP 2. The process operator closes the two block valves and tags them with a white "DO NOT OPEN" tag and applies a multiple lockout device and process lock(s). Process attaches a pink "INSTALL BLIND HERE" tag to each of the two blinding points.

STEP 3. The process operator locks out the electrical circuit breaker with a specially designed hasp. This hasp can accept a number of separate locks. The process operator installs a process lock on the hasp.

STEP 4. The process operator clears and drains the pump.

STEP 5. Some time later (perhaps hours or days), a mechanical crew of three mechanical employees will arrive to "pull" the pump so that it can be taken away for servicing and maintenance. The process operator must first issue a permit to mechanical. This permit requires both process and mechanical to make a safety tour of the jobsite. During this safety tour, mechanical verifies that the valves have been closed and tagged and process also identifies for mechanical each location to be blinded. Mechanical is given a written "BLINDING LIST" on which the location of each blind to be installed is recorded. Both process and mechanical sign the permit at the jobsite, and a copy of the permit is posted in a conspicuous place at the jobsite. Process cannot open the valves while the mechanical permit is active.

STEP 6. Mechanical applies a mechanical lock to the electrical lockout hasp. At each of the blinding locations, mechanical unbolts the flanges, inserts the blind, and rebolts the flanges locking the blind in place. Each blind has a metal piece which extends beyond the flange providing visual observation that a blind is in place. Mechanical attaches tagout devices to the blinds.

STEP 7. Mechanical pulls the pump and sends it for repair, removes the mechanical lock from the electrical circuit hasp, signs and returns the permit (with the completed blinding list) to process, and leaves the jobsite.

STEP 8. Some time later (perhaps several shifts or days), a completely different mechanical crew of three mechanical employees will arrive to reinstall the repaired pump. Process and mechanical will conduct a jobsite safety tour-process identifies each location to be unblinded, and process issues a new permit with the blinding list to mechanical. Once the permit is signed by mechanical, the valves and blinds may not be moved by process.

STEP 9. Mechanical places their lock on the electrical circuit hasp, reinstalls the pump, removes the blinds, removes their tags and electrical lock, signs and returns the permit, and leaves the jobsite.

STEP 10. Process removes their tags from valves and starts up the pump.

Figure 10-12. Procedural example for complex isolation (small job).
Source: OSHA Draft Instruction Standard CPL 2.85–1989.

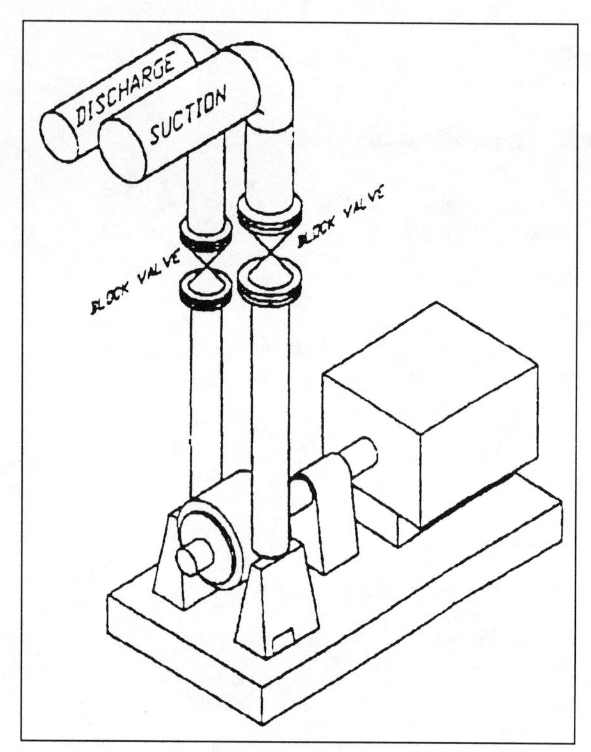

Figure 10-13. Process pump, simplified sketch.
Source: OSHA Draft Instruction Standard CPL 2.85–1989.

> Outside personnel (contractors, etc.)
>
> (i) Whenever outside service personnel are to be engaged in activities covered by the scope and application of this standard, the on-site and outside employer shall inform each other of respective lockout or tagout procedures.
>
> (ii) The on-site employer shall ensure that his/her personnel understand and comply with restrictions and prohibitions of the outside employer's energy control procedure.

These requirements provide very little guidance to employers beyond the mutual informing of both parties regarding their respective lockout/tagout procedures. In the lockout/tagout standard preamble, OSHA states that the proper utilization of these provisions, when they are understood and agreed upon, is a way to prevent misunderstandings by either plant employees or outside service personnel regarding the use of lockout/tagout procedures in general and with regard to the use of specific lockout/tagout devices that are selected for a particular application.

Other OSHA standards, notably 29 *CFR* 1910.110, *Process Safety Management of Highly Hazardous Chemicals*, and 29 *CFR* 1910.146, *Permit-Required Confined Spaces,* convey further insight as to regulations related to employer-contractor safety relationships.

The process safety management standard covers the employer-contractor situation in 29 *CFR* 1910.119(h) contractors as follows:

> (1) Application.
>
> This paragraph applies to contractors performing maintenance or repair, turn around or major renovation, or specialty work on or adjacent to a covered process. It does not apply to contractors providing incidental services that do not influence process safety, such as janitorial work, food and drink services, laundry, delivery or related services.
>
> (2) Employer Responsibility.
>
> (i) The employer, when selecting a contractor, shall evaluate data regarding the contractor's safety performance and program.
>
> (ii) The employer shall inform contractors of known potential fire, explosion or toxic release hazards related to required work and the process.
>
> (iii) The employer shall explain to contract employers the applicable provisions of the Emergency Action Plan required by paragraph (n) of this section.
>
> (iv) The employer shall develop and implement safe work practices consistent with paragraph (f)(4) of this section, to control the entrance, presence and exit of contract employers and contract employees in covered process areas.
>
> (v) The employer shall periodically evaluate the performance of contract employees in fulfilling their obligations as specified in paragraph (h)(3) of this section.
>
> (vi) The employer shall maintain a contract employee injury and illness log related to the contractor's work in the process areas.
>
> (3) Contract Employer Responsibilities.
>
> (i) The contract employer shall assure that each contract employee is trained in work practices necessary to safely perform his/her job.
>
> (ii) The contract employer shall assure that each contract employee is instructed in the known potential fire, explosion or toxic release hazards related to the job and the process, and the applicable provisions of the Emergency Action Plan.
>
> (iii) The contract employer shall document that each contract employee has received and understood training required by this paragraph. The contract employer shall prepare a record that contains the identity of the contract employee, date of training and means used to verify that the employee understood the training.
>
> (iv) The contract employer shall assure that each contract employee follows the safety rules of the facility including the safe

LOCKOUT/TAGOUT
(Complex Isolation - Intermediate Job)

JOB: Extensive repairs to a distillation tower in a refinery

Isolating Devices: 65 block/control valves, 30 small diameter pipe locations require plugs, 35 large diameter pipes require blinds; almost all devices at elevated locations.

Involved Personnel: Five mechanical contractors, employing 50-60 workers, using crews of 2-10 workers, working two eight-hour shifts per day for one month.

Other: Contractors will follow hosts lockout/tagout procedure and require a permit from process operations.

STEP 1: The process operator with the assistance of process employees from other locations closes the 65 valves and tags each with a white "DO NOT OPEN" tag. Process attaches a pink "INSTALL BLIND HERE" tag to each of the 35 blinding locations.

STEP 2: Process clears and drains the tower.

STEP 3: Process inserts a plug at each of the 30 plugging locations. In each instance, process is working in close physical proximity to the energy isolating valve which was closed and tagged in Step 1. Process includes each plug on a master blinding list.

It will take 2-3 days to complete Steps 1, 2, and 3. At this point, the tower has been shut down and is ready for the mechanical contractors.

STEP 4: A mechanical contractor erects scaffolding on the outside of the tower. In doing this work, the contractor is not engaged in servicing or maintenance of the tower and is no more dependent on the closed valves and plugs than the physical integrity of the pipes themselves.

STEP 5: A mechanical contract crew will install the 35 blinds. Process issues a permit to each crew before work begins and includes a blinding list identifying the blinds to be installed. During the jobsite safety survey for each permit, the blinding locations are pointed out to mechanical. Each permit is posted at the jobsite. When a crew completes installing and tagging blinds, the permit is signed and returned with the blinding list. Process includes each new set of installed blinds on the master blinding list for the tower repair.

Figure 10-14. Procedural example for complex isolation (intermediate job). (*continued*)
Source: OSHA Draft Instruction Standard CPL 2.85–1989.

work practices required by paragraph (f)(4) of this section.

(v) The contract employer shall advise the employer of any unique hazards presented by the contract employer's work or of any hazards found by the contract employer's work.

The permit-required confined spaces standard addressed the employer-contractor responsibilities in much the same manner in 29 *CFR* 1910.146(c)(8) and (9):

(8) When an employer (host employer) arranges to have employees of another employer (contractor) perform work that involves permit space entry, the host employer shall:

(i) Inform the contractor that the workplace contains permit spaces and that permit space entry is allowed only through compliance with a permit space program meeting the requirements of this section.

(ii) Apprise the contractor of the elements,

STEP 6: Before beginning any servicing and maintenance work, each mechanical crew must receive a permit from process. Crews arrive to open manways to provide access inside the tower. They are followed by crews that erect scaffolding inside the tower. Additional mechanical crews enter and perform the actual servicing and maintenance tasks. There are also inspection crews to monitor the work done. Any or all of this work can be ongoing at any time under separate permits. Finally, when the work inside the tower is completed, crews close the manways. Each permit must be signed and returned when the authorized work is finished. Process cannot unblind a blind while a permit is active on the tower. Over the 30 days required to complete these activities, 300 total permits may be issued.

STEP 7: Now that the actual servicing and/or maintenance of the tower has been completed, mechanical contract crews arrive to unblind the tower. Before these crews arrive, process attaches a pink "REMOVE THIS BLIND" tag to each location to be unblinded. Each crew is issued a permit by process with a blinding list identifying the location of the blinds that crew will remove. Process conducts a jobsite safety tour and points out the location of these blinds. The combination of pink locator tags, permit blind list and jobsite safety tour singularly identify to mechanical the exact location of each blind to be removed. Each mechanical crew removes its assigned blinds and then signs and returns the permit with its blinding list to process. Process indicates that these blinds have been pulled on the master blinding list.

STEP 8: When all the mechanical-installed blinds have been removed, the mechanical aspects of the work will have been completed. Now process begins the work of starting-up the tower. Process removes the 30 plugs and indicates on the master blinding list that all blinds have been removed. Finally, process untags and opens the 65 valves.

STEP 9: To the extent not already removed, a mechanical contractor would remove the scaffolding from the exterior of the tower.

STEP 10: Finally, process starts up the tower.

Figure 10-14. (*concluded*)

including the hazards identified and the host employer's experience with the space, that make the space in question a permit space.

(iii) Apprise the contractor of any precautions or procedures that the host employer has implemented for the protection of employees in or near permit spaces where contract personnel will be working.

(iv) Coordinate entry operations with the contractor, when both host employer personnel and contractor personnel will be working in or near permit spaces, as required by paragraph (d)(11) of this section.

(v) Debrief the contractor at the conclusion of the entry operations regarding the permit space program followed and regarding any hazards confronted or created in permit spaces during entry operations.

(9) In addition to complying with the permit space requirements that apply to all employers, each contractor who is retained to perform permit space entry operations shall:

(i) Obtain any available information regarding permit space hazards and entry operations from the host employer.

(ii) Coordinate entry operations with the host employer, when both host employer personnel and contractor personnel will be working in or near permit spaces as required by paragraph (d)(11) of this section.

(iii) Inform the host employer of the permit space program that the contractor will follow and of any hazards confronted or created in permit spaces, either through a debriefing or during the entry operation.

OSHA, in its Field Operations Manual for compliance personnel (chapter V(F) Issuing Citations—Special Circumstances) identifies policy for dealing with multi-employer worksites as follows:

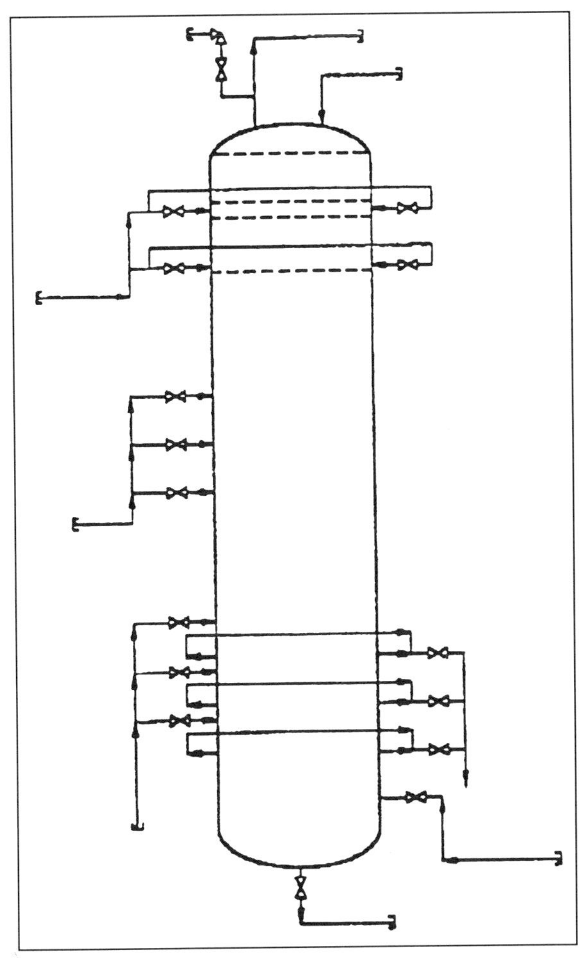

Figure 10.15. Fractionation in tower, simplified sketch. Source: OSHA Draft Instruction Standard CPL 2.85–1989.

2. *Multi-employer Worksites.* On multi-employer worksites, both construction and non-construction, citations normally shall be issued to employers whose employees are exposed to hazards (the exposing employer).

a. Additionally, the following employers normally shall be cited, whether or not their own employees are exposed:

(1) The employer who actually creates the hazard (the creating employer).

(2) The employer who is responsible, by contract or through actual practice, for safety and health conditions on the worksite; i.e., the employer who has the authority for ensuring that the hazardous condition is corrected (the controlling employer).

(3) The employer who has the responsibility for actually correcting the hazard (the correcting employer).

b. It must be shown that each employer to be cited has knowledge of the hazardous condition or could have had such knowledge with the exercise of reasonable diligence.

c. Prior to issuing citations to an exposing employer, it must first be determined whether the available facts indicate that the employer has a legitimate defense to the citation as set forth below:

(1) The employer did not create the hazard.

(2) The employer did not have the responsibility or the authority to have the hazard corrected.

(3) The employer did not have the ability to correct or remove the hazard.

(4) The employer can demonstrate that the creating, the controlling and/or the correcting employers, as appropriate, have been specifically notified of the hazards to which his/her employees are exposed.

5) The employer has instructed his/her employees to recognize the hazard and, where necessary, informed them how to avoid the dangers associated with it when the hazard was known or with the exercise of reasonable diligence could have been known.

(a) Where feasible, an exposing employer must have taken appropriate alternative means of protecting employees from the hazard.

(b) When extreme circumstances justify it, the exposing employer shall have removed his/her employees from the job to avoid citation.

d. If an exposing employer meets all the conditions in F.2.c., that employer shall not be cited. If all employers on a worksite with employees exposed to a hazard meet these conditions, then the citation shall be issued only to the employers who are in the best position to correct the hazard or to ensure its correction. In such circumstances the controlling employer and/or the hazard-creating employer shall be cited even though no employees of those employers are exposed to the violative condition. (See, however, F.2.e.)

e. In the case of general duty clause violations, only employer(s) whose own employees are exposed to the violation may by cited.

V. Gallagher, in "Liability, OSHA, and the Safety of Outside Contractors," comments as follows regarding the tenuous relationship between employer and contractor for site safety:

OSHA would examine contract documents and/or actual practices between parties to determine whether the employer has the right to control or has the *authority for ensuring* that

> a hazardous condition is corrected. Often, the contract between an employer and outside contractor requires that the contractor comply with OSHA standards. Often, the employer also has the right to stop work for contract violations. . . . If the employer assumes more responsibility to manage the safety of an outside contractor, OSHA will likely perceive that the employer has "authority to ensure" the correction of the hazard. Conversely, if a hands-off policy is adopted, the employer will be less likely to be cited by OSHA—but more likely to have unsafe work being performed by an outside contractor who may decide to cut safety costs. With no one watching, the contractor may take chances (Gallagher, 1993).

The Stanford University Department of Civil Engineering, conducting safety research for the Business Roundtable Construction Industry, identified the following methods used by employers who successfully dealt with contractors to manage safety:

- requiring permits for hazardous activities
- requiring contractors to designate safety to someone on site
- providing the contractor with safety guidelines to follow
- discussing safety at owner/contractor meetings
- conducting safety audits for the contractor during construction
- requiring immediate reporting of contractor accidents
- stressing safety as a part of the contract prior to engagement
- investigating contractor accidents
- maintaining accident statistics of contractor safety performance
- conducting regular safety inspections
- setting goals for construction safety
- factoring in safety to prequalify a contract bid
- establishing a construction safety management department to monitor contractor safety
- including safety guidelines in the contract.

In addition, Stanford's research provided three major indicators that could be used to compare and select safe contractors (Figure 10-16). With regard to lockout/tagout, a contractor lockout/tagout capability inquiry can prove useful in determining whether the contractor will be competent in controlling hazardous energy during the execution of the contract (Figure 10-17). The use of the inquiry does not guarantee a safe employer/contractor relationship but takes the process another step in the right direction.

The National Safety Council's *Accident Prevention Manual for Business & Industry*, 10th edition, called for a more proactive and assertive approach to safety management of outside contractors. Various key elements were identified: prejob safety conferences, evaluation of the contractor's safety program, preplanned safety procedures, correction of safety deficiencies, periodic inspections/audits, safety meetings, and others. The NSC's outline for a prejob safety conference can be used as a guide for conducting such a meeting between employer/contractor (Figure 10-18).

In J. Dean's article "Plant/Contractor Safety Partnerships," the following basic program for contractor safety is described:

- The purchasing department must work with the safety department to evaluate the safety performance of contract bidders.
- The legal department must understand the various safety requirements as supplied by the safety department and line management, then incorporate them into an actual contract.
- Line management must be willing to promote and support actions by a "third entity," composed of a composite team of company/contractor representatives that actually perform the work, and reinforce that commitment at all levels of the plant organization.
- Engineering, maintenance, and other internal groups must be open and honest about the facility's capabilities, willing to accept nontraditional ways of doing things, supportive of the partnership effort, and dedicated to improving and guiding the relationship.
- Safety personnel must be aware of and enforce the agreed-upon safety practices as the job progresses.
- The contractor must have the ability and the motivation to conform to the facility's culture, to expand and adapt elements of its organization to interface with those of the facility, and to anticipate its idiosyncrasies. The contractor must also be willing to assume initiatives beyond what is the norm in contractor's past roles.

Dean goes on to say that the partnering relationship begins with a well-conceived and mutually agreed-upon contract between parties. Early project involvement is essential. After the project scope, control, safety, quality, and partnering behavior are defined, the following four-stage action plan can be initiated:

Step 1. Evaluate contractors. Planning for safety starts even before the contract is awarded. Purchasing should use the past safety record and safety policies of a project contractor as an important criteria in evaluating all proposals.

Step 2. Prejob safety review. Before the job actually begins, the safety provisions of the contract are turned into a safety checklist through joint discussions between the plant owner and the contractor. This list provides the specific details of all safety programs.

Step 3. Ensure safety on the job. Line supervisors and the safety department must manage and reinforce the safety procedures as the work progresses.

Comparing and Selecting Safe Contractors

One way to determine the difference between contractors regarding safety values is to use indicators that resulted from research performed by Stanford University, Dept. of Civil Engineering, for the Business Roundtable Construction Industry. The three indicators, listed in order of importance, are:

1] PAST SAFETY RECORD. Examine the contractor's workers' compensation and OSHA experience. Workers' compensation experience is reflected in the experience modification rate [EMR] which is the ratio of actual losses to expected losses over a three-year period. It reflects the average loss experience for the previous three years and is a good indicator of a contractor's past safety performance and for comparing contractors who perform similar work.

a] *Experience Modification Rate*

$$\text{EMR} = \frac{\text{Actual Losses}}{\text{Expected Losses}}$$

EMR for construction contractors ranges from 0.3 to 2.0. It is not uncommon for contractors in the same industry to have significantly different EMRs.

b] *OSHA Incidence Rate.* Two OSHA incidence rates can be calculated from data furnished by the bidder.

The first relates to frequency:

$$\frac{\text{No. of Injuries and Illnesses} \times 200{,}000}{\text{Total Hours Worked by All Employees During Period Covered}}$$

An OSHA severity rate can be calculated as follows:

$$\frac{\text{No. of Lost Work Days} \times 200{,}000}{\text{Total Hours Worked by All Employees During Period Covered}}$$

2] MANAGEMENT SAFETY ACCOUNTABILITY. Accountability is a key element in managing a safety program. If managers cannot "get in trouble" for poor safety performance, the program will likely fail. Individual performance is a key element in a successful management program. The Business Roundtable suggested evaluating performance based on the following information:

a] The recipients of accident reports and frequency distribution of reports [field superintendent, vice president of construction, firm president].
b] Frequency of safety meetings for field supervisors.
c] Frequency of project safety inspections and the degree to which they include project and field superintendents.
d] Compilation method for accident records and the frequency of reporting. Those contractors that subtotal accidents by superintendent and foreman, rather than by company, have a more detailed accountability system.
e] Compilation method for accident costs and the frequency of reporting. Again, greater accountability comes from a more detailed system that measures project accident costs of superintendents and foremen.

3] FORMAL SAFETY PROGRAM. Components of a contractor's safety program found to be associated with better safety performance are:

a] Orientation of new workers and foreman.
b] Frequency of toolbox meetings.
c] Existence of a written safety program.

Figure 10-16. Safe contractor selection.
Source: Gallagher VA. "Liability, OSHA, and the Selection of Outside Contractors." *Professional Safety* (Jan. 1993):20–32.

Step 4. Postjob safety review. When the job is completed, representatives from both companies should meet and analyze the results of the safety programs (Dean, 1993).

A consistent ingredient in any progressive company's safety structure is a defined approach to employer/contractor relationships. For example, Mobil's Joliet refinery requires prospective contractors to complete a safety questionnaire that covers workers' compensation–insurance experience. Contractors must provide a copy of their written safety and health program, including provisions for hazard communication, respiratory protection, hearing conservation, confined space entry, and fire training.

Mobil requires written statements attesting that the contractor will comply with applicable federal and local regulations and Mobil's corporate safety policies. The contractor must provide Mobil with copies of material safety data sheets for any and all substances or commodities the contractor might bring to or use on Mobil property. Contractors, in turn, are entitled to copies of all refinery material safety data sheets.

"Contractors must demonstrate a commitment to safety through an excellent work-safety record and a

CONTRACTOR
LOCKOUT/TAGOUT CAPABILITY INQUIRY

Facility: ______________________ Date: ____________

Nature of Work: ______________________

Energy Isolation Requirements: ______________________

Applicable LO/TO Procedure (Local): ______________________

Inquiry Elements	Yes	No	Comment
Does contractor have a LO/TO program?			
Does contractor exhibit LO/TO competence, i.e., experience, equipment, procedures, trained personnel, etc.?			
Is contractor aware of state/federal LO/TO standard requirements?			
Does contractor confirm that assigned employees are trained in LO/TO procedures?			
Has company/contractor informed each other of their respective LO/TO procedures?			
Has contractor experienced LO/TO OSHA violations or injury incidents on previous jobs?			

Company Authority ______________________ Contractor/Agent ______________________

Title ______________________ Title ______________________

Other Information: ______________________

Figure 10-17. Contractor lockout capability inquiry.
Source: Developed by E. Grund. © National Safety Council.

comprehensive written safety-and-health program to be considered for work in the refinery," Mobil forewarns (Britt, 1993).

The use of outside contractors is often an emotional issue, particularly in facilities where unions represent the work force. "Contracting out" is often identified in union-management labor contracts as an issue that demands some degree of discussion/negotiation. Typically, notification is called for, with further dialogue depending upon the nature of the job, skills involved, lay-off status, and so on. If relations are strained at the outset, required coordination for the control of hazardous energy may be difficult to achieve. The tension that may result from any dispute over outside contractors may be a detriment to effective lockout/tagout application.

The controversial John Gray Institute Study, funded by OSHA, examined the use of contract labor in the U.S. petrochemical industry and its relationship to workplace safety. The study was commissioned as a result of catastrophic accidents and fatalities within the industry and to assist OSHA in its deliberations on the process safety management standard. Robert E. Wages, vice-president of the Oil, Chemical, and Atomic Workers Union and National Steering Committee member for the study, testified that researchers found "dominant patterns indicating contract employees receive less training, sustain more injury, have

Sample Outline for Preconstruction Safety Conference

A. Preconference activities
   1. Define the conference's purpose
      a. Evaluation of proposed program
      b. Discussion of job organization and operating procedures
      c. Preplanning the work and agreeing to a means for applying standard procedures
   2. Notify all parties
   3. Evaluate proposed program
   4. Decide on conference facilities
   5. Determine meeting attendance
   6. Determine how conference will be recorded (minutes of meetings)

B. Agenda for the conference
   1. Orientation
      a. Explain why the program is necessary
      b. Advantages in terms of economy and efficiency
      c. Prescribed safety standards
      d. Review
         (1) Accident prevention agreements
         (2) General conditions of specifications on safety
         (3) Special conditions of specifications on safety
         (4) Lockout/tagout
         (5) Hazard communication
      e. Other requirements—local, state or provincial, and federal
      f. Supervision
         (1) Organization of, at project site
         (2) Functions of personnel at the site
         (3) Responsibilities and accountability
         (4) Delegated authorities
         (5) Relations regarding enforcement and discipline
   2. Discussion of proposed program
      a. Plans about layout of temporary construction, site, buildings, etc.
      b. Action taken toward planning and coordinating activities between different operations and crafts
      c. Access to work areas
      d. Safety indoctrination and safety education
      e. Delegation of safety responsibilities to supervisors
      f. Integration of safety into operating methods and procedures
      g. Housekeeping program
      h. Safety factors in job-built appurtenances
      i. Traffic control and parking facilities
      j. Fire protection
      k. Lighting, ventilation, protective apparel, and medical care
      l. Safe operating conditions and maintenance of equipment
   3. Discussion of follow-up procedures
      a. Methods for meeting objectives
      b. Plans for periodic readjustment of safety objectives
      c. Handling safety deficiencies
      d. Arrangements for additional meetings and periodic staff meetings
      e. Following up on agreements made in the preconstruction meeting

Figure 10-18. Employer/contractor prejob safety conference.
Source: National Safety Council, *Accident Prevention Manual for Business and Industry: Engineering and Technology,* 10th ed. (Chicago: National Safety Council, 1992).

higher turnover rates, work under less comprehensive safety and health protection policies, and are routinely segregated from labor, management, and employee involvement systems governing permanent employees." OSHA, because of research and interpretation problems, did not use the study specifically as the basis for the process safety management standard, 29 *CFR* 1910.119. Despite the debate over what precisely can be concluded from the study, sufficient reason exists to question how well contractor safety is being addressed by employers.

A host of specific lockout/tagout issues must be resolved between employers/contractors prior to work commencement. The degree to which the contract employees will interface with host personnel will greatly determine the various roles of individuals, actuation of energy-isolating devices, and application of available lockout/tagout practices/procedures. The following items represent a sample of questions that may need to be answered before work proceeds:

- Will a common lockout/tagout technique be used?
- Will host employees, contractor employees, or some combination thereof actuate/install energy-isolation devices?
- Who will manage the shift-change personnel activity?
- Can lockbox procedures be used where employer/contractor personnel are intermixed?
- Who will be responsible for work permit or authorization paperwork?
- Will company equipment (restraints, barriers, tags, etc.) be used by the contractor?

- Will contractors follow written company lockout/tagout procedures?
- Who will coordinate energy-isolation activities between parties?
- What method will be used to resolve conflicts involving energy isolation?
- How will the contractor work area be isolated from other operations?
- Are contractors authorized to cycle equipment? What and to what degree?
- How will compliance with lockout/tagout be monitored and enforced?

Employer/contractor lockout/tagout effectiveness will undoubtedly be a subset of the overall approach and treatment of safety in general. Reason for concern certainly exists when contractors do not exhibit general safety competence or are not willing or capable of documenting their safety awareness/proficiency.

MACHINERY GUARDING/INTERLOCKS

OSHA has prominently addressed the relationship between machinery guarding and lockout/tagout. In 29 *CFR* 1910.147 (a) (2) (ii) application, it states that normal production operations are not covered by the standard (see Subpart O, Machinery and Machine Guarding). Servicing and/or maintenance that takes place during normal production operations is covered by the lockout/tagout standard only when the following occurs:

> (a) An employee is required to remove or bypass a guard or other safety device: or
> (b) An employee is required to place any part of his or her body into an area on a machine or piece of equipment where work is actually performed upon the material being processed (point of operation) or where an associated danger zone exists during a machine operating cycle.
> **Note:** *Exception to paragraph (a) (2) (ii):* Minor tool changes and adjustments, and other minor servicing activities, which take place during normal production operations, are not covered by this standard if they are routine, repetitive, and, and integral to the use of the equipment for production, provided that the work is performed using alternative measures which provide effective protection (See Subpart O of this Part).

As early as January 1980, the National Institute of Occupational Safety and Health (NIOSH) published its *Request for Information: Lockout and Interlock Systems and Devices* and provided OSHA with the responses. In late 1989, in the preamble to OSHA's Final Rule, 29 *CFR* 1910.147, *Control of Hazardous Energy*, the relationship between machinery, guarding, and lockout/tagout was elaborated further:

> In evaluating servicing performed during normal production operations, the first question to be asked is whether employees must bypass guards or otherwise expose themselves to the potential unexpected release of hazardous energy. If no such exposure will occur, either because of the method in which the work is performed or because special tools, techniques, or other additional protection is provided, the lockout or tagout requirements of this standard apply. However, if the servicing operation is routine, repetitive and must be performed as part of the production process, it is obvious that lockout or tagout cannot be performed, because these procedures would prevent the machine from economically being used in production, OSHA will continue to treat these operations as being covered by the general machine guarding requirements of subpart O. The employer must provide appropriate guarding to protect employees from points of operation, nippoints, and other areas of the equipment where the employees might be endangered. The use of alternative protective methods to keep employees' bodies out of danger zones, such as specially designed servicing tools, remote oilers, and the like, would meet this requirement.

The logic diagram in Figure 10-19 illustrates the decision-making network for determining whether lockout/tagout or alternative measures providing effective protection will be used. Where the employer can demonstrate that guarding and safeguards are in place and that no exposure exists to hazardous energy, employees can perform the task without further protective action (such as visual inspection of machine cycling).

Guarding

Various methods are used to safeguard machinery point-of-operation and power-transmission hazards. In 29 *CFR* in 1910.212 *General Requirements for All Machines,* and 29 *CFR* 1910.212 *Mechanical Power-Transmission Apparatus*, OSHA provides extensive detail regarding the proper methods of safeguarding. Enclosures, fixed or interlocked barriers, automatic feed/discharge, two-hand trip devices, restraints and pull away devices, presence-sensing devices, type A/B gates, and so on are some of the safeguards that may be used (Figure 10-20). Each method has advantages and disadvantages that influence in varying degree the activities associated with hazardous energy isolation (for example, a solid total enclosure of a flywheel may not permit observance of the run-down or at-rest condition.).

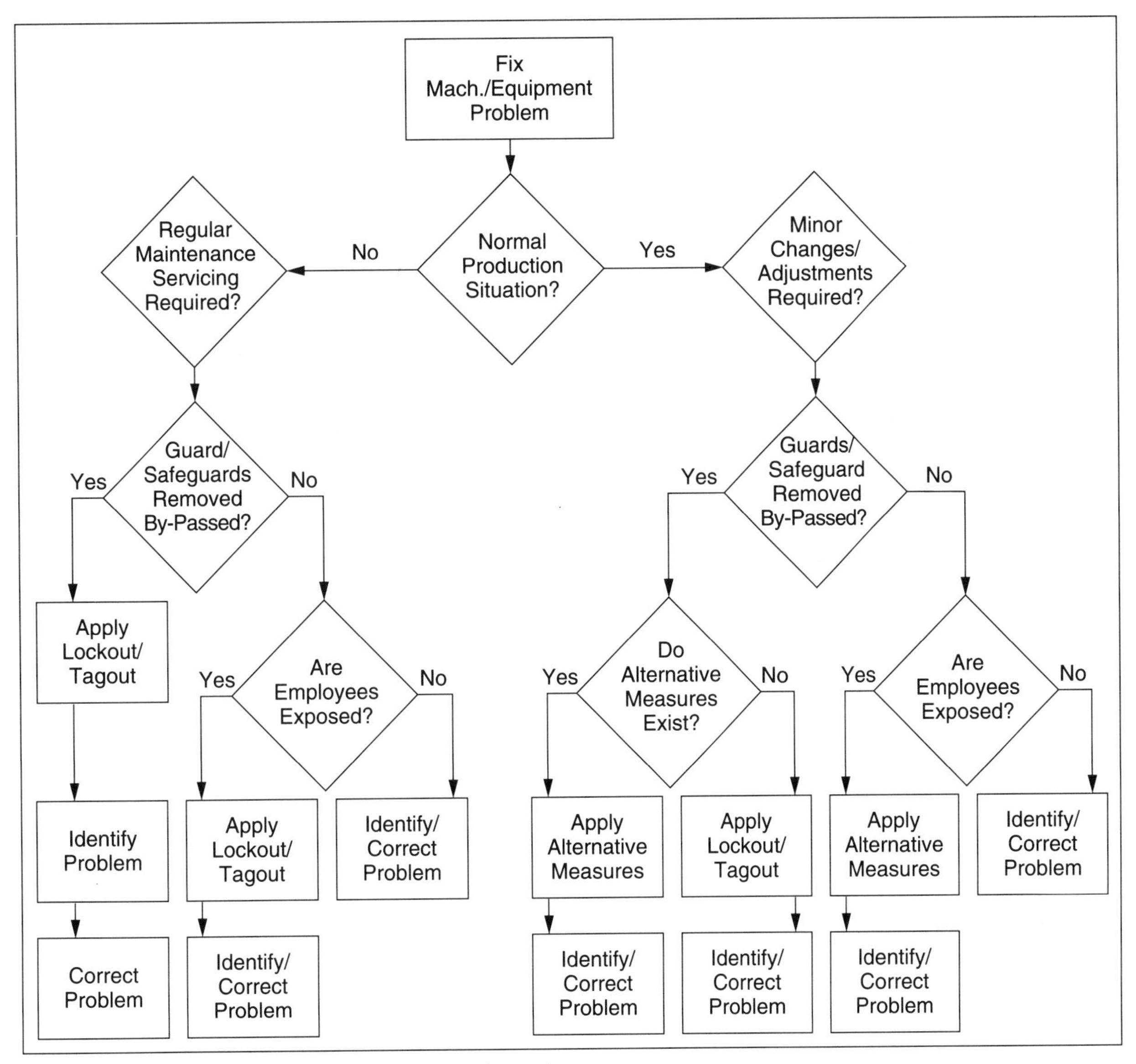

Figure 10-19. Logic diagram—lockout/tagout or alternative measures.
Source: Developed by E. Grund. © National Safety Council.

The U. S. Bureau of Labor Statistics conducted a survey of arm, hand, and finger amputations occurring between December 1980 and May 1981. Twenty-three states participated in the survey, in which 861 cases were reported for analysis. All those surveyed were blue-collar workers, with 60% employed in manufacturing. Approximately 70% of the cases involved fixed machinery/power saws. Sixty-two percent of those injured on fixed machinery were operating the equipment, whereas unjamming, cleaning, repairing, setting up, and servicing accounted for the balance of the cases. Six of the seven most severe injuries were related to contact with moving machinery. Three cases involved employees setting up or unjamming equipment that was mistakenly activated by co-workers. One worker was staging a piece of lumber to be cut when a co-worker activated the saw. The injured worker was in such a position that a hold-down bar trapped his hand, allowing the saw to amputate it. The second was unjamming a piece of cutting equipment and asked his co-worker to run the machine backward to help clear it. The co-worker activated the wrong switch and the machine ran forward, pulling the injured worker's hand into cutting range. The injured worker attributed the accident to poor machine design because the forward and reverse controls were not clearly marked. The third worker was unjamming a piece of farm equipment when a co-worker unexpectedly turned on the machine. Another worker, who lost part of an arm, was caught in moving machinery while measuring it for replacement parts. He stated that he had complained to his supervisor that the equipment was still running but was ordered to perform this task because the equipment was "running slow enough to get the measurements."

Methods of Machine Safeguarding

There are many ways to safeguard machinery. The type of operation, the size or shape of stock, the method of handling, the physical layout of the work area, the type of material, and production requirements or limitations will help to determine the appropriate safeguarding method for the individual machine.

As a general rule, power-transmission apparatus is best protected by fixed guards that enclose the danger area. For hazards at the point of operation, where moving parts actually perform work on stock, several kinds of safeguarding are possible. One must always choose the most effective and practical means available.

We can group safeguards under five general classifications:

1. Guards
   - A. Fixed
   - B. Interlocked
   - C. Adjustable
   - D. Self-adjusting
2. Devices
   - A. Presence Sensing
     - (1) Photoelectrical (optical)
     - (2) Radiofrequency (capacitance)
     - (3) Electromechanical
   - B. Pullback
   - C. Restraint
   - D. Safety Controls
     - (1) Safety trip control
       - (a) Pressure-sensitive body bar
       - (b) Safety triprod
       - (c) Safety tripwire cable
     - (2) Two-hand control
     - (3) Two-hand trip
   - E. Gates
     - (1) Interlocked
     - (2) Other
3. Location/Distance
4. Potential Feeding and Ejection Methods to Improve Safety for the Operator
   - A. Automatic feed
   - B. Semi-automatic feed
   - C. Automatic ejection
   - D. Semi-automatic ejection
   - E. Robot
5. Miscellaneous Aids
   - A. Awareness barriers
   - B. Miscellaneous protective shields
   - C. Hand-feeding tools and holding fixtures

Figure 10-20. Machine safeguarding methods.
Source: National Safety Council, *Safeguarding Concepts Illustratred,* 6th ed. Itasca, IL: National Safety Council, 1993.

The conditions or events table resulting from the survey, reproduced here as Table 10-A, provides an interesting perspective as to what precipitated these incidents. It provides clues as to where and to what degree various failure modes are involved in machinery-related incidents. Note that safeguards were a factor in 27% of the cases and that hazardous-energy-control issues appear to be represented in 39%.

Maintenance employees account for a significant portion of the amputations because their responsibilities include emergency repairs as well as routine setups, servicing, and cleaning of equipment. Their work involves access to hazardous areas on equipment, areas that are routinely guarded during normal production operation.

Due to the nature of the machine maintenance jobs, awareness of safe working procedures is imperative to reduce the exposure of workers' hands to injury. A review of accident data indicates that there is a need to remind mechanics of their exposure to inherent hazards during their work routine. Major causes of severe hand injuries include

- blindly reaching into a machine to locate and diagnose potential problems through sense of touch
- contacting moving parts when correcting production problems with equipment operating
- failure to de-energize equipment being worked on that may start automatically
- failure to tag out or lock out equipment being worked on
- failure to use safety blocks when placing hands in die areas
- using inappropriate or defective hand tools to do the job
- failure to communicate with co-workers on actions to be taken in diagnosing and correcting production problems.

Supervisors need to observe the work habits of maintenance mechanics when equipment is down so that proper work procedures can be emphasized to these key employees.

Generally, machine/equipment safeguards should meet the following minimum requirements: They should prevent contact, be durable and secure, protect from falling objects, create no new hazards, create no interference, be user friendly, and facilitate maintenance. The machine/equipment safeguarding survey shown in Figure 10-21 can be used to identify the status of guarding and to generate an improvement schedule. In one company with over 10,000 employees in 50 locations, 1,600 deficiencies relating to machine guarding were identified during the period of the survey.

The state of any facility's guarding/safeguards is a dynamic one, under the constant pressure of change. Without periodic formal survey, it is practically impossible to identify to what degree this protective feature is functioning. The following example illustrates this point:

Table 10-A. Conditions or Events Contributing to Injury: Arm, Hand, or Finger Amputations, Selected States, December 1980–May 1981

Condition or event	Workers	Percent
Indicate any conditions or events which you feel led to your injury.		
Total	861	(¹)
Did not realize hand was in hazardous area	246	29
In a hurry	212	25
Tool or machinery not equipped with safeguard (such as a barrier guard)	147	17
Work material shifted position or broke	146	17
Misjudged time or distance needed to avoid injury	138	16
Co-worker did something that caused your injury	110	13
Tool or machinery accidentally activated	110	13
Tool or equipment was in bad condition (for example: Dull blade)	106	12
Little or no instructions given on how to do task	94	11
Attention not fully on task	93	11
Tool, machinery, or work material shifted position or slipped	85	10
Not looking at hand	74	9
Hand slipped	72	8
Gloves, clothing, jewelry, or watch got caught in the equipment	72	8
Recent change in work routine or procedures	70	8
Tool or machinery had been altered or modified at the job site (for example: Barrier guard removed)	70	8
Unfamiliar with tool or equipment used	67	8
View of hand blocked by part of machine or other object	60	7
Tool or machinery broke or malfunctioned	53	6
Lost balance, slipped, or fell	50	6
Hand/finger(s) pulled into machine by work material	44	5
Upset or under stress	44	5
Tool or machinery continued to run after being shut off (coasting)	43	5
Other factors contributed to injury	35	4
Using wrong type of tool or equipment for job	32	4
Machine power not turned off	30	3
Reacted to loud noise or other distraction	28	3
Accidentally hit foot pedal	28	3
Tired or bored	22	3
Task not being done according to instructions	21	2
Cleaning tool, cloth, or rag got caught in the equipment	21	2
Safeguard failed	20	2
Visibility poor due to inadequate lighting, dust, or glare	18	2
No contributing factors indicated	27	3

¹ Because more than one response is possible, the sum of the responses and percentages may not equal the total. Percentages are calculated by dividing each response by the total number of persons who answered the question.

NOTE: Due to rounding, percentages may not add to 100. See appendix A for types of injuries included in the survey.

Source: Bureau of Labor Statistics.

A drive guard had been provided on a piece of machinery. Maintenance personnel believed it was difficult to remove for equipment access. They cut a 4"x 6" opening in the guard to access the sprocket/belt intersect for observation and lubrication. The opening existed for several years without incident. An operation employee dropped an

Machine/ Equipment Safeguard Survey

Division ______________ Plant______________ Dept./ Area______________

Survey Team ______________ Survey Date______________

Machine/Equipment	Safeguarding Status	Action Needed/Schedule	Responsibility

Note: Review machinery/equipment for exposures to point of operation, power transmission, reciprocating, rotating, transverse, cutting, punching, and shearing hazards; identify guards which are poorly designed, damaged, inadequate, not in place, unstable, insecure, function poorly, need interlocking, unpainted, etc., *or* exposures which are in need of guarding.

Figure 10-21. Format for machine/equiment safeguarding survey.
Source: Developed by E. Grund. © National Safety Council.

object which entered the opening but was visible and reachable. While attempting to retrieve the object, the workers hand was caught in the unnoticed nip point amputating his finger tip.

Interlocks

Interlocks, more specifically safety interlocks, elicit a wide variety of reaction and responses from those who attempt to use them or have to deal with them in an application situation. It is even difficult to precisely define what is meant by the term *interlock*. *Merriam Webster's Collegiate Dictionary*, 10th edition, defines *interlock* as "to connect so that the motion or operation of any part is constrained by another." Telemecanique, in its November 1988 technical publication, *Safety Interlocking in the Automated Factory*, defines *safety interlock* as "a device that ensures the operator's protection where intervention is required on a machine or mechanism that can put the operator's health or safety at risk." Why address interlocks at all as a dimension of the control of hazardous energy? Safety interlocking plays an important ancillary role in achieving the objective of safeguarding employees who are potentially exposed to the release of hazardous energy. Although the use of interlocking is pervasive, the abuse of interlocking is almost as universal.

Seim and Beutler, in *Occupational Health Safety*, had this to say about interlocks:

> Safety interlocks, complimenting the main power lockout/tagout system, add another dimension of protection to operators and service personnel. An interlock is an electro-

mechanical arrangement in which the operation of one control or mechanism automatically brings about or prevents the operation of another. When installed as an ancillary lockout/tagout device at the point of access, the interlock switch opens an electrical circuit, when the machine guard is opened, breaking the circuit of power [Figure 10-22]. The interlock can mechanically prevent the opening of a safety barrier/guard until it is safe to do so. An effective auxiliary lockout/tagout interlock device heightens safety in four ways:

- It prevents access until the hazards represented by stored/residual energy have dissipated.
- It serves as a secondary point of electrical power isolation via a positive opening set of switch contacts.
- It prevents the inadvertent restoration of power by mechanically preventing the closing of the machine guard until the mechanical lockout/tagout (padlock) has been removed.
- Its mechanical design prevents it from easily being bypassed or defeated (Seim & Beutler, 1993).

Primary/Secondary Protection

It is important to point out that interlocks function as a *secondary* means of protection for personnel. Many uninformed individuals view the interlock as a primary form of protection and use an interlocked safety guard as an isolating device rather than a line voltage main disconnect where both are present. The interlock may cause workers to overvalue it and rely too greatly on its protective capability. Its convenience and simplicity often offer an attractive alternative to identifying and isolating at a primary disconnect. The following incident illustrates the folly of such a practice:

A repairman opened the interlocked cover type guard on a production machine to correct a problem being experienced with the tooling. The console stop control was within ten feet and the feed power electrical disconnect was on the wall nearby. Neither of these two isolating components were used by the repairman. While working on the tooling the machine cycled, mangling his hand. The interlock had failed to danger and automatic system control had determined that no fault or open circuit existed; therefore, the signal to run was sent and the machine was in a mode to function. There were numerous contributing factors in this incident but the one that was most prominent related to the repairman's conditioned behavior. He had become accustomed to using the in-

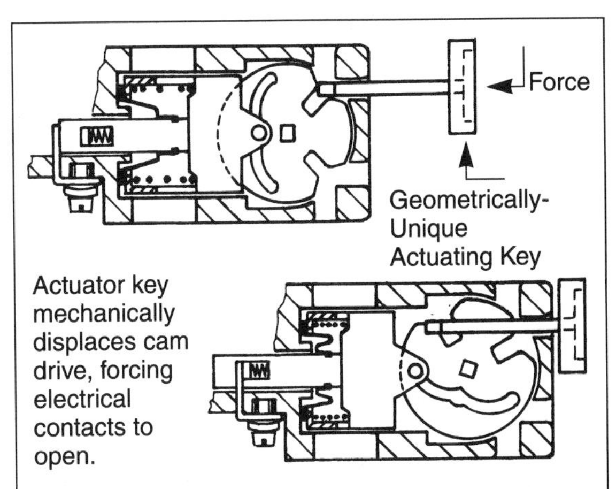

Because positive-opening, tamper-resistant safety interlocks directly control the circuit of power on a machine, they augment lockout/tagout applications where movable machine guards are used.

An electrical switching mechanism, an interlock, designed with positive-break contacts, should be mounted to the movable machine guard—such as an access door, protective grating, equipment hood/cover or work-area barrier.

Tamper resistant. A geometrically unique actuator key (see figure) mates with the interlock's electrical switching mechanism when the machine guard is closed. This key prevents an operator from easily bypassing or defeating the mechanism, just as a door lock prevents entry without the properly matched key.

In the closed position, the electrical contacts are "closed" to provide electrical circuit continuity and permit the machine/equipment to operate; the equipment power is thus safely "enabled" to work.

Positive-opening. When the machine guard is opened to perform set-up, service, or maintenance, the actuator key mechanically rotates the cam mechanism within the switch body—*forcing* the electrical contacts to open. This positive-opening mechanism breaks the electrical circuit, ensuring the interruption of electrical power once the machine guard is opened.

Applying a padlock lockout or a tagout notice to this interlock switch serves to prevent inadvertent closure and restoration of power. Upon key removal, the safety interlock can be fitted with a padlock-lockout accessory. The positive-opening mechanism of the interlock physically blocks the key-actuator opening to ensure the electrical contacts cannot be reclosed (enabling power to the equipment).

Figure 10-22. Safety interlock switch with positive operation.
Source: Courtesy K.A. Schmersal GmbH & Co.

terlocked cover as his protective shelter and on this occasion it failed!

Interlocking can be of great value when applied properly. However, any presence sensor can detect the proper position of a guard and signal when it is out of place. As well, most of the so-called safety interlocks currently in use in the United States are magnetic, proximity, photoelectric, or limit switches. These sensors do not necessarily ensure safety.

Standards

International Safety Standards (IEC 204, EN 292, and EN 954) and the European Union Machinery Safety directive now require the use of positive-opening safety interlock switches (fail-safe type). In the United States, no standard for the common-use limit, proximity, and photoelectric sensors requires them to be fail-safe. Only light barriers used to detect presence near machinery and equipment are mentioned in this regard.

The new IEC 947-5 standard, *Control Circuit Devices and Switching Elements*, recommends a new symbol to be used with the standard contact symbol to reflect the positive-opening feature.

Most of the European mechanical control switches are limit switches, which incorporate the positive-opening feature. The U.S. market today is principally oriented toward slow-break, slow-make contacts for safety purposes. These contacts follow the principles of positive-opening contacts, although they have inherent limitations because of their slow speed.

U.S. standards are expected to begin reflecting the advancements shown in European standards relating to safety switches. ANSI B11.19 (1990) safeguarding reference for the B11 Machine Tool Standard Series and the proposed UL 491 standard for *Power-Operated Machine Controls and Systems* are currently the most progressive U.S. references.

The British Standard Code for *Safety of Machinery,* BS 5304:1988, section nine, provides an outstanding treatment of the considerations for interlocking. The subject is extensively covered in over 50 pages of text and drawings that served as the source for much of this section.

Guarding/Interlock Systems

An interlock connects the machinery/equipment guard with the control or power system. The guard and interlock should be designed and installed so that

- the interlock prevents the machine from operating by interrupting the power until the guard is closed
- the guard remains in place protecting the worker from the associated hazard(s), or opening the guard ensures there is no hazard encountered during access.

Interruption of the power medium itself may be adequate to control the hazard before access occurs. However, where the hazard cannot be eliminated immediately by power interruption, the interlocking system must include guard locking and/or a braking system. The four most common energy types used in interlocking systems are electrical, mechanical, hydraulic, and pneumatic. The choice of energy medium will depend on the application circumstances and the assessment of advantages and disadvantages.

Two primary means of interlocking ensure that power is interrupted when a guard is opened:

1. *Power interlocking*: device directly interrupts power
   - Guard-inhibited: interlock device has to be moved to OFF position before guard can be opened
   - Guard-operated: interlock device is operated by guard movement
2. Control interlocking: device indirectly interrupts power through a control system (may include guard-operated or -inhibited types)
   - dual-control with cross-monitoring (with or without self-checking)
   - dual-control without cross-monitoring
   - single-control interlocking.

A hazard may still exist after power interruption due to release of stored energy, machinery momentum, etc. Under these conditions, systems should be designed to include a device to:

- cause the hazard to be neutralized (braking)
- prevent the guard from opening until the danger has passed (guard locking).

The choice of which interlocking system to use should be made after assessing the following criteria:

- frequency of exposure
- probability of negative consequence
- severity of employee injury anticipated
- resources necessary to minimize the risk.

See Figure 10-23 for interlock selection guidance.

Interlock Failures

When interlocks are inappropriately used as primary protection or used in conjunction with alternative protective measures, their functioning effectiveness is critical. The system design objective should be to prevent failure to danger. *Failure to danger* can be defined as any failure of the machinery, related safeguards, control circuits, or power supply that creates an unsafe condition. The reliability or integrity of the interlock system is a function of its inherent design and upkeep.

Interlock system integrity can be improved by correct installation, circuit overload protection, high-quality components, avoidance of grounding potential, and minimizing failure to danger and defeatability.

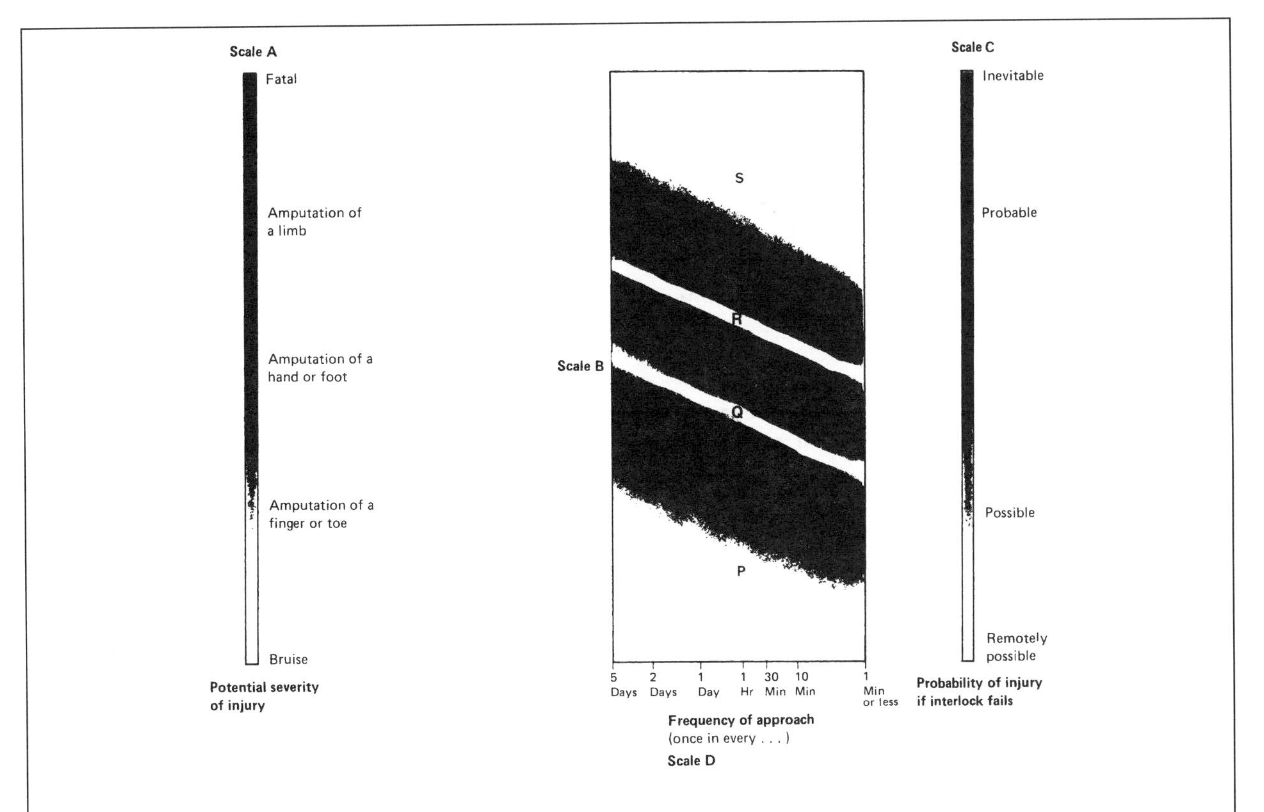

Appendix B. Selection of interlocking system

B.1 This appendix is intended to provide a graphical method of carrying out the selection process outlined in 9.6. While it provides a more formal approach, it should be recognized that it has limitations due to the lack of data available.

B.2 Having assessed the severity and probability of injury if the interlocking system fails (see section four), select the equivalent points on scales A and C

This assessment will be influenced by:

(a) the method of working the machinery;

(b) the nature of the hazard;

(c) the need to approach it.

Any operating histories and accident records for the machinery should also be taken into account.

B.3 Using a straightedge, connect these points to obtain an intersection point *y* on scale B.

B.4 Select a point *x* on scale D equivalent to the expected frequency of approach.

B.5 Project a horizontal line from the point *y* and a vertical line from the point *x*. The area in which these lines intersect gives an initial indication of the outlay, in time, money and effort, that should be spent on the interlocking system to reduce the risk of injury to an acceptable level. If the intersection falls in area P, the indication is that the outlay will be sufficient to provide for a single control system of interlocking. Intersections falling above area P will indicate that a greater outlay is warranted allowing provision for higher integrity interlocking systems as follows:

Area Q: dual-control system interlocking without cross-monitoring;

Area R: guard operated power interlocking;

Area S: either dual-control system interlocking with cross-monitoring or guard inhibited power interlocking.

The intersection may fall into one of the shaded portions of areas P, Q, R or S. Where this occurs the merits and cost of using higher quality components, to improve the integrity of the interlocking system provided for below the shaded portion, should be compared with the merits and cost of providing a system of interlocking provided for above the shaded portion.

Before the final selection is made the merits of an interlocking system requiring an initially higher capital outlay but with lower inspection and maintenance costs should be compared with an initially cheaper system requiring higher inspection and maintenance costs.

Figure 10-23. Interlocking selection guidance.
Source: British Standards Institution, *Safety of Machinery*, BS 5304, Section 9:1988. Extracts from BS 5304:1988 are reproduced with permission of BSI. Complete copies can be obtained by post from BSI sales, Linford Wood, Milton Keynes, MK14 6LE.

Electrical components should possess good corrosion and vibration resistance, electromagnetic interference resistance, durability, duty rating, reliability, and longevity.

Power supply failures occur more frequently than failure of system components (relays, contactors, valves, solenoids, springs, brakes, etc.); therefore, the components that rely on the power supply for their function should be installed so that power interruption produces a failure to safety in the system. Components and interconnections usually fail with one or more conditions present, e.g., open or closed circuit, on or off, engaged or disengaged, etc. Therefore their mode of installation should be carefully examined.

The most commonly occurring component failures in any interlocking system are as follows:

- failure, interruption, or variation of energy supply
- open circuits in electrical system
- disconnection/rupture of hoses or piping
- mechanical failure
- electrical environment malfunction
- malfunction due to vibration
- power medium contamination malfunction
- earth faults (grounding)
- brake failure
- single component failures causing changes or loss of function
- cross-connection faults.

Interlocking Switches

Various types of switches are used for interfacing guard operation with an ON/OFF electrical control signal. Switches should be selected only after manufacturers have determined them suitable for specific safety applications. The following types are in common use:

- cam-operated position switches
- tongue-operated switches
- captive-key switches
- trapped-key control of electrical switches
- inductive proximity switches
- magnetic switches
- diode links
- manually operated delay shotbolts
- solenoid-operated bolts.

Captive-key or trapped-key control interlock systems are not to be confused with the common lockbox key method used in crew or multiple-employee lockout/tagout applications. Some conceptual similarities exist, but the lockbox/crew tactics might be more aptly described as secured-key or pyramid-key systems to avoid confusion.

Generally, in a captive-key switch, a key is secured to the movable part of the guard, and a combined lock-and-switch unit is attached to the stationary member.

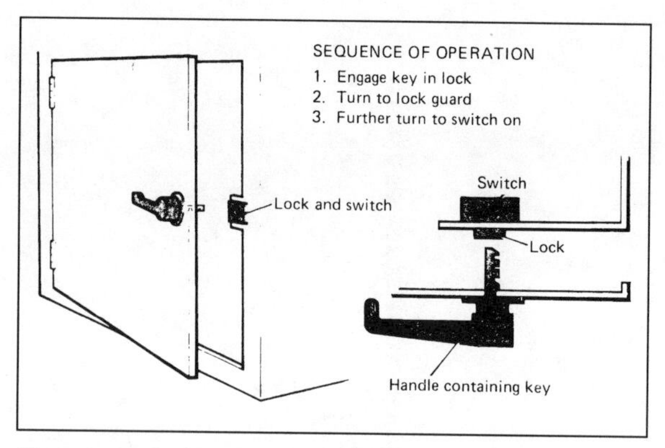

Figure 10-24. Captive-key switch.
Source: British Standards Institution, *Safety of Machinery*, BS 5304:1988. Reprinted with permission; see Figure 10-23.

To open the guard, the key is moved, putting the switch to the OFF position and releases the key from the lock allowing the guard to open (Figure 10-24). A time or remote delay feature, using an integral electromechanical bolt, can be incorporated to ensure that the key cannot be turned to the GUARD OPEN position until the machine has come to rest.

In a trapped-key control system (key exchange), the guard lock and locking switch are separate components. The removable key is either trapped in the guard lock or the switch lock. The guard lock is designed so that the key can only be released when the guard is closed and locked. The key can then be moved to the switch lock. Closing the switch traps the key so that it cannot be removed with the switch on. Where there is more than one source of power, control, and guards, a key exchange box is used where isolation keys can be transferred to and locked in place. This allows the access key(s), of a different configuration, to be released for movement to the guard lock(s) (Figure 10-25a–b). Where process sequence is critical, the transferable key is locked in and exchanged for a different one at each stage.

Interlock Defeatability

F. B. Hall, a developer of the concept of zero mechanical state, provides special insight on the subject of safety interlocks in "Safety Interlocks—The Dark Side." He believes that the perfect interlock is more myth than fact. The article describes a manufacturer who holds two patents, one for a safety-interlocked cover and the other for the defeat of the safety-interlocked cover. The ANSI Z241.1 standard recognizes the circumstantial need for defeat of interlocks in paragraph 4.1.7.5, "Defeating protective devices," and in paragraph 4.1.7.4, "Trouble shooting with power on." Very few workers in U.S. manufacturing would be able to say they have never observed interlock switches tapped over, jammed with objects, unbolted and paired together, bypassed, disarmed, or jumpered. In some cases there may have been a legitimate reason

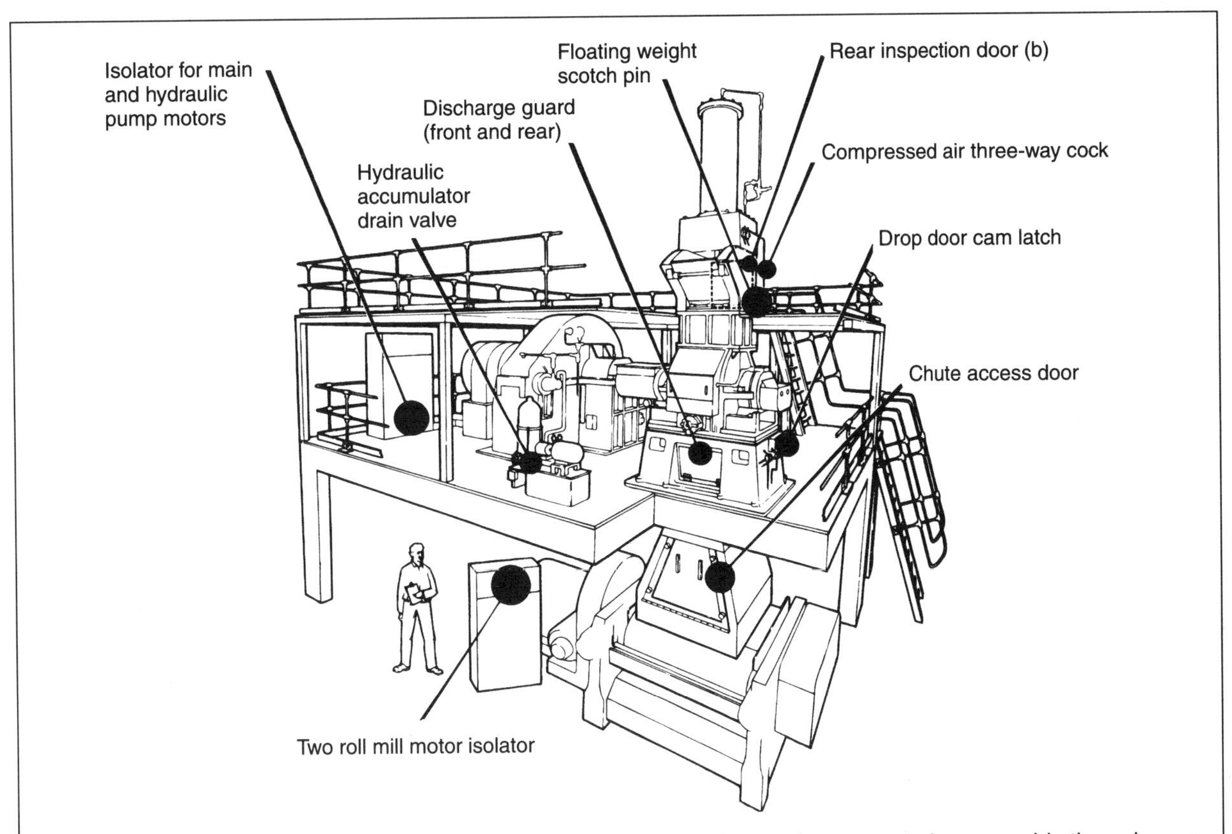

At an internal mixing machine, used in the processing of rubber, key exchange control may provide the only practicable means of preventing access to a number of widely separated danger areas. The principle employed is that all sources of power are isolated, and all stored energy is dissipated, before access is possible.

Interlocking is achieved by means of a main key exchange box, shown in (b), into which all isolating keys have to be inserted before any access key can be released. When a guard is open, an access key is trapped in it; likewise, when a part controlled by an isolating key is live, the key remains trapped in it. For example, before the rear inspection door, shown in (a), can be unlocked and opened, the compressed air supply has to be locked shut; with both ends of the floating weight cylinder vented to atmosphere and the floating weight supported on a scotch pin which is locked in position. In addition, the isolator for the main and hydraulic pump motors has to be locked open and the rotors have to be at rest. This latter condition is achieved by a time delay unit which will release an isolating key for the main exchange box only after sufficient time has elapsed for the motor to come to rest.

Access to the other dangerous parts is obtained by a similar process.

Figure 10-25a. Rubber process mixing machine—trapped-key control interlocking system.
Source: Courtesy Castell Interlocks, Inc.

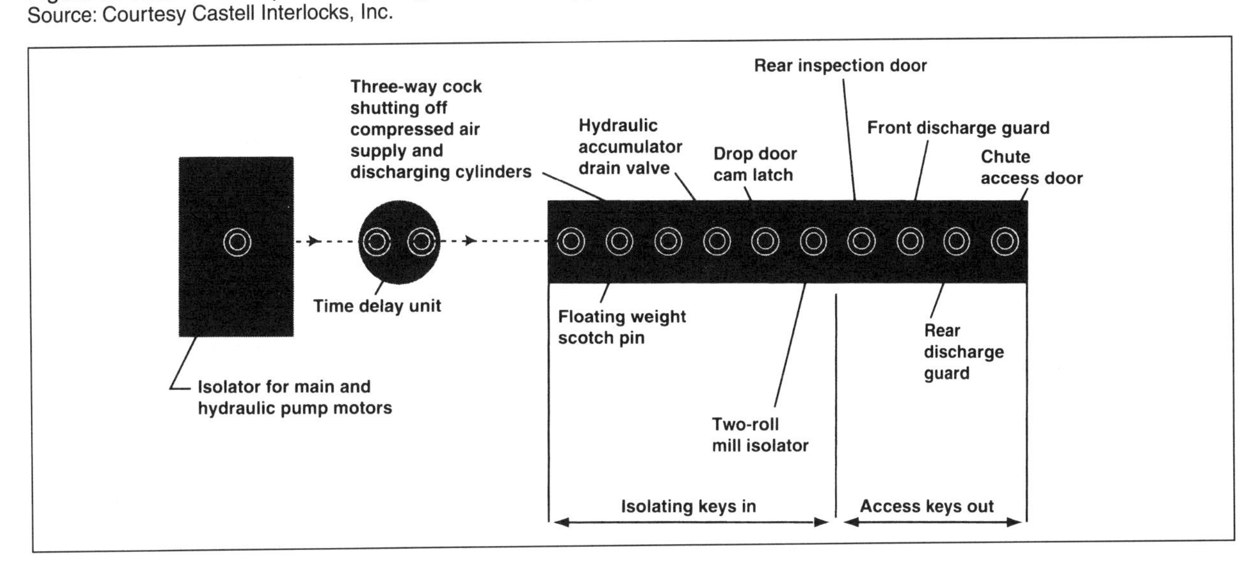

Figure 10-25b. Main key exhange box.
Source: Courtesy Castell Interlocks, Inc.

for such actions, but far too frequently they are symptomatic of something wrong that urgently needs to be addressed by management and employees alike.

Some of the ways that an electrical interlock can be defeated or may fail to function are as follows:

1. Interlock shorted or grounded:
   - in the switch itself
   - in the wiring to the switch
   - in the enclosure of the control cabinet
   - by capacitance. (Possible with long runs and light burden. Can be marginal and erratic.)
2. Interlock falsely actuated:
   - mechanical actuator stuck, bent, or broken
   - internal mechanism stuck, bent, or broken
   - substituted magnet or metal mass
   - actuator tied, wedged, or otherwise restrained
   - foreign field from magnetically held tool
   - shock, impact, or vibration
   - stray magnetic fields on proximity device.
3. Active bypassing:
   - trapped interlock key released by duplicate key
   - trapped interlock key released by gimmick
   - electrical troubleshooting bypassing interlock
   - mechanical (manual) actuation of magnetic device (for example, starter or relay operated manually or proximity switch by magnet or steel substitute)
   - manual overrides on pneumatic or hydraulic valves
   - starter (contactor wedged closed) drive started and stopped with disconnect switch.

The following partial list of reasons for defeat or failure of interlocks illustrates the scope of the problem and draws attention to issues that must be appropriately resolved:

- removal for repair, maintenance, or inspection
- removal to check, test, or replace
- removal of guard or interlocked member to allow increased production or convenience of access
- interlock device broken or requiring repair, with no replacement units available; becomes a "temporarily permanent" condition
- permanent removal or defeat of interlock to avoid trouble or expense of repairing it
- contamination of electrical interlock by environmental conductive dust
- contamination of electrical interlock by conductive liquids
- vibration teasing contacts—causing them to weld
- deterioration of interlock seals, causing sticking
- short circuiting caused by
  —Mechanical crushing of conduit, fittings
  —Abrading of insulation on wires
  —Deterioration of insulation by overload currents
  —Deterioration of insulation from external heat (Hall, 1992).

Various classification systems are in use to describe the resistance to defeat or effectiveness of safety hardware under field conditions. Designers and engineers should be aware of and anticipate various scenarios that might compromise the performance of devices, such as interlocks, restraints, and so on that are intended for employee protection. Installation of a mechanical plunger interlock for critical safety purposes may be totally irresponsible due to its limitations. The following safety defeatability index from the ASSE *Dictionary of Terms Used in the Safety Profession* illustrates this thought process:

> **Safety defeatability index.** A system for use by engineering designers and product safety analysts in assessing the susceptibility of machinery, equipment, and products—under reasonably foreseeable conditions of service—(a) to malfunction; (b) to inadvertent or nonpurposeful misuse or nonuse; and (c) to intentional misuse, abuse, bypass, disablement, alteration, etc. The system facilitates selection and specification of superior or highest-order safety hardware and/or software, in the sense of its invulnerability to being operated within the safety features functioning as intended. The ranked or rated list of modes or conditions employed by owners/operators to defeat the safety features of machinery, equipment and products is as follows:
>
> - Class 1: user changes or does not follow recommended (intended) operating procedure, thus defeating safety features.
> - Class 2: user employs simple tools and/or contrivances to defeat safety features.
> - Class 3: user employs special or complex tools or devices to defeat safety features.
> - Class 4: user redesigns, remanufactures, or rebuilds in order to defeat safety features
> - Class 5: user disrupts or damages safety features, requiring repair in order to restore originally intended function.
> - Class 6: user irreversibly destroys product and/or safety features, thus rendering them incapable of being reset, restored, or restarted.

Interlock systems and their individual components can be evaluated in terms of their reliability and susceptibility to defeat. All available components are not equal in terms of their inherent safety. Final decisions regarding which type of system individual components, arrangement, and method of application to choose should be sensitive to limitations and the resultant risks.

F. B. Hall provides this caveat in summary:

> Every safety interlock brings to an interlock application its own risks which tend to offset the intended safety. The balancing of

those risks against the safety afforded must always be considered in the ultimate decision of whether or not, under all the circumstances of the intended application, the safety device should be used at all. Often the risks outweigh the safety benefits (Hall, 1992).

ALTERNATIVE PROCEDURES

In Chapter 7, System Elements, an energized tasks exposure survey was introduced for purposes of identifying all work activities that take place on facility machines, equipment, and processes under energized conditions with some degree of employee exposure. Ideally, all activities would be performed under zero energy state conditions. However, to achieve this ideal state in all circumstances presents some very real difficulties for many industries.

OSHA, perhaps reluctantly, attempted to recognize the practical aspects of the zero energy state objective and its potential impact. If total energy isolation was imperative in all cases and under all circumstances, the consequences would be monumental. OSHA did two things of note in its rulemaking for lockout/tagout.

First, the agency, in 29 *CFR* 1910.147 (a)(2)(ii)(note), attempted to provide some flexibility by introducing the exception for minor activities that are routine, repetitive, and integral to the use of production equipment. At the same time, the phrase *alternative measures which provide effective protection* was born as the companion proviso. This action ostensibly acknowledges the existence of circumstances under which certain production tasks would not be accomplished under zero energy conditions.

Second, because of input from the American Petroleum Institute and its members, OSHA excluded from coverage hot-tap operations involving transmission and distribution systems for substances such as gas, steam, water, or petroleum products when they are performed on pressurized pipelines. Hot-tapping involves the cutting and welding of equipment (pipelines, vessels, tanks, etc.) under pressure in order to make repairs or modifications without interruption of service. Special equipment and procedures are used by internal maintenance personnel or outside contractors. The American Petroleum Institute's publication 2201, *Procedures for Welding or Hot Tapping on Equipment Containing Flammables* (1978), was referenced by OSHA as representing an acceptable procedure.

OSHA's intent in proposing this exception from the requirements of the standard was to allow a particular type of work (hot-tapping) in a limited number of cases (that is, when continuity of service is essential and shutdown is impractical) while providing for an acceptable level of safety for employees. Without this exception to the requirements of this standard, a hot-tap operation could not be conducted.

OSHA's treatment of the exceptions for incidental production activities and hot-tapping indirectly establishes precedent for numerous additional exceptions that would likely have to be viewed as legitimate, such as, electrical troubleshooting, electrolytic cell processes, high-tension electrical transmission, molten materials flow control, robot training, and so on. Each industry undoubtedly can identify a host of activities that are not currently done under zero energy conditions and in which the relative potential hazard or risk to the employee is somewhat greater.

The challenge for employers who believe certain activities should take place under less than zero energy state conditions is substantial. The following must be demonstrated:

- The activity requires that it be done under energized conditions because:
  —zero energy state conditions will have excessive impact on production and customers
  —it is more hazardous to de-energize
  —there is no technological alternative.
- No changes are possible that would reduce the overall risks encountered under the energized method.
- Alternative methods have been developed involving the personnel performing the work.
- Alternative methods are capable of delivering a level of safety reasonably equivalent to that provided by machine guarding and/or energy-isolation practices.
- Employees are trained, knowledgeable, skilled, and have the ability to execute the alternative measures.
- The local or general experience with the alternative measures substantiates their effectiveness.

The employer's obligation, therefore, is to provide, as the British term it, a safe system of work. The priorities are clear:

- Secure the machine, equipment, or process to preclude exposure to energy releases.
- Attempt to convert as many tasks done under partial or total energization to the lockout/tagout state or mode.
- Design alternative measures that are reasonable, capable of a high degree of safety, and effective.

What are legitimate alternative measures that provide effective protection and minimize the risk that lies between the polar points—energized/de-energized? The assessment and endorsement of these alternatives may be quantifiable but most often will be done subjectively. If employers have a task that they believe, for various reasons, must be done under full or partially energized conditions, they must be able to document the assessment process that was used to determine that the *resultant protection is effective*.

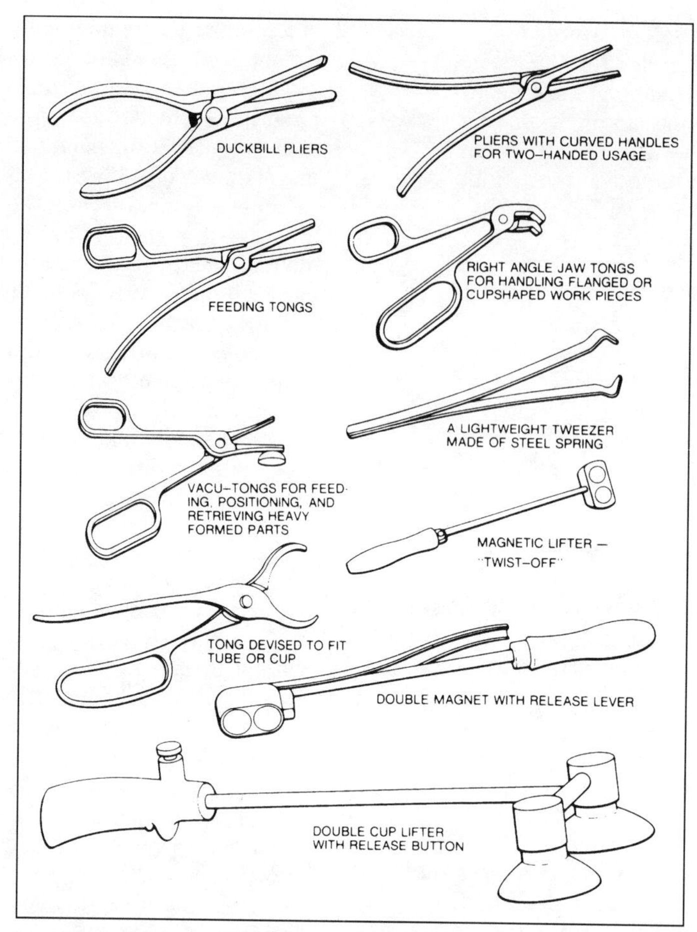

Figure 10-26. Custom manual holding tools.
Source: National Safety Council, *Safeguarding Concepts Illustrated,* 6th ed. (Itasca, IL: National Safety Council, 1993).

For instance, a metals fabricator may believe that employees should be able to clean sheet-processing rolls under power to get them clean enough to make high-quality product. The objective might be sound, but cleaning under powered conditions is unacceptable. What are possible alternatives to complete isolation and lockout/tagout? First, the employer should be able to clearly define why the task cannot be done under zero energy conditions. If this case can be legitimately made for substantial reasons, the employer may then and only then be in a position to consider alternatives. The employer must then establish the alternative measure with, ideally, consensual concurrence from those who supervise and those who perform the task.

Review and approval of the alternative should also be undertaken by someone like a safety manager or department manager. Some of the options that might be considered individually or collectively are as follows:

- equipment placed in jog-only mode
- roll access limited to out-running side
- interlock-trapped key locking of controls
- cleaning device/tool devised for manual use
- cleaning fixture designed—permanent hardware
- pressure mat interrupt for quarter-roll rotation cleaning/cycle
- lockout of redundant control circuitry.

Other options might not require total energy isolation. Effective training of personnel when using alternative measures should receive special emphasis.

The following general techniques/protective measures are often employed as safeguards when tasks must be performed without complete energy isolation:

- custom manual holding tools (Figure 10-26)
- tool extensions
- shielding (for example, Plexiglas internal shields on electrical cabinet exposed contacts)
- clearance distances
- specialized control circuitry (lockable)
- blocks, pins, wedges
- restraining devices
- protective gear (mats, insulating blankets, etc.)
- personal protective equipment
- interlock devices
- jog- or inch-only modes
- interrupt modes (protective systems—audible

alarms/delayed start)
- manipulators (powered)
- feed diversion/obstruction devices
- use of security personnel
- specialized training
- nonconductive tools, etc.

All consensual alternative measures should be in written form that precisely defines what actions/safeguards are to be used to achieve effective employee protection. The alternative measures should be companion pieces to the specific MEP energy-isolation procedures. As a result, each MEP should have written procedures covering all associated tasks done under energized circumstances as well as the specific procedure for complete isolation (zero energy state).

AUTOMATED SYSTEMS/ROBOTICS

Advanced machinery systems, automated programmable assembly, and robotics are designed to relieve workers of many stressful tasks and manual handling activities. The anticipated health and safety benefits provide reason for optimism. However, with these general gains may come setbacks in certain areas; the control of hazardous energy may present increasing difficulties unless appropriate emphasis and priority are applied. System complexity, computer interfacing processes, rapidly changing technology, reliability deficiencies, experiential voids, technical skill demands, etc. may be the ingredients that trigger unexpected energy-release incidents. Factory workers of the future may be stretched to the mental limit after being provided the necessary physical relief. Although the outlook is favorable overall, management needs to ensure that no aspects of this progress endanger employee safety.

Automated Systems

As manufacturing processes become more automated and capital intensive, various influences emerge that have a direct bearing on service and maintenance tasks. Continuous process lines, integrated factory cells, progressive assembly stations, etc. all have common characteristics that create energy-isolation dilemmas. From a service or maintenance worker's viewpoint, consider the following:

- Downtime costs are stratospheric.
- Troubleshooting can be demoralizing.
- Equipment interrelationships are complex.
- System controls demand special skills and time to think.
- Quality of product requirements are extraordinary.
- Tolerances and performance specification are tight.
- Numerous forms of energy are present.
- Energy isolation is often complicated.
- Everyone tends to wait and watch.

If just-in-time (JIT) manufacturing procedures are in place, added stress may be generated because of lack of surge capacity, shorter production runs, reduced inventory, and shorter delivery times, which exacerbate production interruptions. This could create a climate in which shortcuts are taken with regard to energy-isolation requirements.

Under these circumstances, it is imperative that breakdowns be minimized or eliminated. Equally important for operators and technicians is the reduction of interruptive tasks such as cleaning, adjusting, and setting up. If greater run time is achieved, worker exposure to hazardous energy release will be correspondingly reduced. Computerized maintenance management systems (CMMS) are now being used to enhance preventive maintenance (performance of scheduled tasks) and predictive maintenance (intervention by process monitoring). These systems speed up the process of maintenance while upgrading the quality and outcome of the activity. The old adage "If it ain't broke, don't fix it" has seen its day.

Total productive maintenance (TPM), a Japanese conceptual approach aimed at eliminating breakdowns and increasing worker productivity, is also gaining widespread recognition. Prevention of breakdown is addressed beginning with the improvement of equipment development and design. Five basic conditions for factory equipment of the future are stressed (development, reliability, economics, availability, and maintainability). In highly automated operations, safety inherent in equipment design is critical to successful control of hazardous energy.

Automated manufacturing systems using computer-controlled machines, mechanized transfer/transport technology (automated guided vehicles), robotics, multi-axis transfer press developments, and so on represent the wave of the future (Figure 10-27). PLC (programmable logic controller) controls have enabled users to manage systems with diagnostic, fault, tutorial, and instructional feedback capability. This on-board automated ability can reduce risks to service and maintenance personnel when troubleshooting and making decisions

Figure 10-27. Automated guided transfer vehicle.
Source: Photo courtesy of E. Grund.

regarding corrective action. In many cases, the machinery system is self-regulating, thus eliminating the need for frequent operator intervention. It is not beyond reason to expect equipment in the near future to unjam, clean, or adjust itself.

The advancement of control systems using PLCs exclusively or in conjunction with relays, and the benefits that are derived are not totally without cost or risk. Electronic control systems are subject to noise effects (electromagnetic and electrostatic interference) that can generate spurious signals. A broad variety of methods exist to combat this technological Achilles heel that involve grounding, shielding, and the like. This situation has great import for personnel servicing and maintaining equipment who expect a certain control response for their protection.

Järvinen and Karwowski, in a study involving 103 accident cases that was completed in 1993 and entitled "Accidents in Advanced Manufacturing Systems," reported that

> manufacturing automation has eliminated some dangerous tasks and traditional risks of injury, but new types of risks have appeared. Automated manufacturing systems are usually comprised of several computer controlled machines assembled into an interacting system. It is difficult for the operators to understand the system as a whole and to perceive the hazards. During normal operation there is seldom a need for direct intervention of the process, but when unusual events or production stoppages occur, the operators may face unfamiliar situations with limited time to decide the proper actions. According to previous research, most accidents in automated manufacturing take place during production disturbances (Järvinen & Karwowski, 1993).

Most of the accidents (about 52%) occurred within the context of stand-alone automated equipment, followed by flexible manufacturing systems/cells (14%). Typically, the accidents involved materials-handling equipment such as conveyors, automated guided vehicles (AGV), shuttle carts, and automated storage and retrieval systems (ASRS). Operators were injured most frequently (67%), with maintenance/repair personnel following (20%). The activity of the injured workers and main defects in safeguarding were defined as minor (part of operating cycle to keep machinery running) and major (preparing the systems for use or responding to significant disruptions) interventions. Slightly more than half of the incidents occurred when the machine had stopped but had not been isolated. In this condition, the equipment was capable of being unexpectedly restarted, for example, by a control system fault or inadvertent activation of a sensor. About one-fifth of the incidents involved death or amputation.

The factors identified as principle causes were as follows:

- improper procedures followed (present in 44% of the incidents)
- human error (38%)
- incompatible controls (25%)
- lack of awareness (20%)
- inadequate training (16%).

Energy-isolation consideration for the safety of personnel working on automated systems should be first addressed in the conceptual/design stage. If the design incorporates safeguards for prevention of unexpected energy release, service and maintenance tasks will be performed with less risk. The following energy-isolation features will benefit worker safety:

- fail-to-safety control components
- energy-isolating devices located for convenient rapid access
- interlocking features with high resistance to defeat
- energy-isolating devices made prominent by color and target discs, identification tags/labels, contrast techniques, etc.
- graphic/pictorial lockout/tagout procedures displayed in area
- self-relieving devices for stored energy where possible
- physical restraints designed for the specific application
- delayed start control and warning alarms
- emergency stop controls
- lockable energy-isolating devices not requiring supplemental hardware
- safeguard monitoring controls/circuits
- alerting devices (flashers, panel lights, etc.) to indicate service/maintenance in progress
- interlocked pinning safeguards capable of resisting applied loads
- automatic or exposure-free lubrication design
- process material/reservoirs supplied fixed systems or using low-risk resupply features, etc.

Automated production requires a formal approach to assessing operational risk. Unique hazards involving complex control systems, interdependent components, powered interface panels, software programs, and so on demand first-class technical review. Hazard analyses, fault-tree analysis, what-if methodology, simulation, and hazard and operability studies will become common tools for protecting the service/maintenance worker from the release of hazardous energy.

Robotics

The first industrial robot went into operation in 1961 at General Motors' facility in Trenton, New Jersey.

However, the term *robot* had been introduced in 1920 and the first robotlike equipment, Televox, was built in 1927 by Westinghouse.

An industrial robot includes a system of moving parts (physical structure), the arm (articulated member), the wrist (end part of arm), the gripper/tool (end effector), and the hand (part of the gripper with jaws for grasping). The control system varies from a simple electromechanical system to sophisticated computer control (see Figure 10-28). Robots can be divided into the following groups:

- limited sequence robots (fixed and variable)
- point-to-point (playback robot)
- continuous path control (real-time teaching)
- intelligent robots (have sensory perception and programmed decision-making capabilities).

On a worldwide basis, an estimated 30,000 industrial robots were in use in 1981. In 1990, the robot population had grown to 400,000 in the principal industrialized countries. By the year 2000, industrial robots are expected to number in the millions. This exponential growth may prove to be good news for the elimination of labor-intensive tasks and the soft-tissue injuries (sprains, strains, pains) that go with them. However, the potential bad news is the rise in severe traumatic injuries due to workers and robots interacting less than harmoniously.

Definitions. The Robotic Industries Association of the United States defines an industrial robot and industrial robot system as follows:

- *Industrial robot*: a reprogrammable, multifunctional manipulator designed to move material, parts, tools, or specialized devices through variable programmed motions for the performance of a variety of tasks.
- *Industrial robot system*: a system that includes industrial robots, the end effectors, industrial equipment, and the devices and sensors required for the robot to be taught or programmed or for the robots to perform the intended automatic operations as well as the communication interfaces required for interlocking, sequencing, or monitoring the robots.

The Japanese definition of a robot uses much broader language:

- *Robot:* an all-purpose machine equipped with a memory device and a terminal, capable of rotation and of replacing human labor by automatic performance of movements.

Classification and application. The Japan Industrial Robot Association (JIRA) places robots in six different classifications as follows: (1) manual manipulator, (2) fixed sequence robot, (3) variable sequence robot, (4) playback robot, (5) numerically coded robot, and (6) intelligent robot.

Figure 10-28. Multiaccess industrial robot.
Source: *Accident Prevention Manual for Business & Industry: Engineering & Technology*, 10th ed. (Itasca, IL: National Safety Council, 1992).

The applications for industrial robots can be categorized into the following three groups:

- workpiece gripped by robot—handling or transport of workpieces, loading/unloading equipment, and manipulation of materials
- tool handled by robot—metal working, joining of materials, surface treatment, testing, marking, stamping, and packaging
- assembly—various products that require robot features such as tactile/optical sensors, versatile gripping devices, fast computer software, special kinematic joints, and sophisticated feeding devices.

Robot Systems

Industrial robots are powered electrically (up to 440 V, 60 Hz, 30 amp), pneumatically (80 psi; 5.4 atm) and/or hydraulically (1,500 psi; 102 atm). Several types of robot control systems exist: servo-controlled point-to-point, non–servo-controlled point-to-point, and servo-controlled continuous path. Usually a robot has three major axes of motion: vertical, horizontal, and swing or rotation. In addition, pitch, yaw, and roll parameters are also possible.

The area or space within which a robot functions is known as the *envelope*. Several important distinctions must be made regarding the envelope as follows:

- *Work envelope*. The greatest area a robot can reach (width, height, and depth); the volume of space defined by the maximum reach.

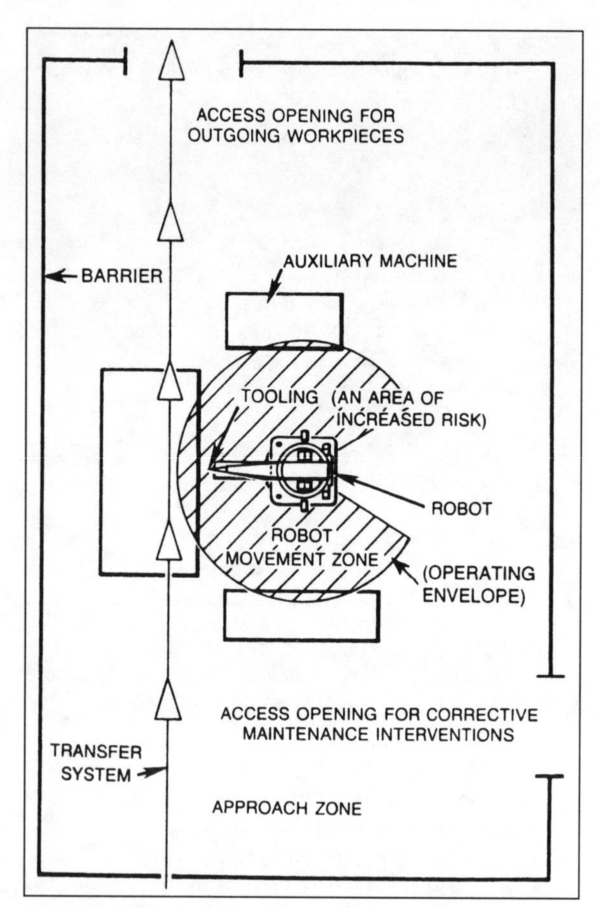

Figure 10-29. Robot movement zone.
Source: NIOSH, *Safe Maintenance Guide for Robotic Workstations,* Publication No. 88-108. Cincinnati, OH.

- *Restricted envelope.* The maximum distance a robot can move in all directions after a limiting device is in place.
- *Operating envelope.* That part of the work envelope used by the robot during automatic operation; it may change without warning due to programming or to satisfy operational requirements.

Overlapping envelopes may exist when robots are required to reach/enter another robot's operating envelope. The extended rear of the robot must be considered when defining the envelope (Figure 10-29).

Rules of Robotics

The impact on humans of the introduction of the "steel-collar worker" has been addressed in many ways. Y. Yokomizo, Y. Hasegawa, and A. Komotsubara devised the following rules of robotics:

> *Rule 1*: Robots must be made and used for the purpose of contributing to the well-being and development of people.
> *Rule 2*: Robots must not replace people on jobs people prefer to do themselves, but must replace people on jobs that they do not wish to do or that they believe to be hazardous.
> *Rule 3*: Robots must follow the command of people so that they do not psychologically and physically oppress people.
> *Rule 4*: Robots must follow the command of people so that they do not harm other people, only damaging themselves.
> *Rule 5*: If robots are to replace people in certain jobs, the prior approval of the people affected must be obtained.
> *Rule 6*: Robots must be made so that they can be easily operated by people and they can readily perform the role of assistants to people.
> *Rule 7*: As soon as robots finish their assigned tasks, they must leave the area so that they do not interfere with people and other robots. (Yokomizo et al, 1987)

It is evident rule number four is being violated with growing frequency. Shall we blame the robots for not complying or their designers, maintainers, and users? The first fatal robot-related accident involved a maintenance worker in Osaka, Japan, in 1981. By 1987, the number of deaths associated with industrial robots in Japan had risen to 10. The Ministry of Labour listed the causes as "operational error of victim (4) and spontaneous start of robot (6)."

Accident Experience

The Japanese Prefectural Labour Standards Office conducted a survey in 1982 covering 190 workplaces with 4,341 industrial robots in use. In this study, 300 "robot-related problems" were reported and analyzed by N. Sugimoto (Table 10-B). He developed a fault-tree analysis of robot incidents arising from unprogrammed operations (Figure 10-30). The unexpected operations were attributed to operators in 38% of the incidents and to the robot system in 62% (Sugimoto, 1987).

K. Lauck of General Motors, in *Safeguarding of Industrial Robots*, reports a series of robot-related incidents that reveal the danger to personnel when hazards are uncontrolled. Several of these are presented to provide a qualitative perspective to the discussion of experience:

- A robot was used to unload a mechanical power press and place the stamped-out discs on a trolley conveyor. An employee crossed into the work envelope to pick up some discs that had fallen from the conveyor. His sleeve became tangled on the robot arm, and he was dragged into the press. At that time another employee had observed the situation and opened an interlocked gate causing the robot to stop.
- A machine repairman climbed over the standard barrier into the robot work envelope to investigate

Table 10-B. Causes of Industrial Robot-Related Incidents (300)

Classification of problem	Percentage distribution	Main description of problem—percentage by class	Distribution of 85 problems due to unexpected operation by percentage
Failure of electrical systems (controllers, etc.)	52.2	Control circuits, 30.0; internal and sensors, 11.1; external poor contact, 34.0; wire breakage of cable, etc., 15.7; noise, 4.6; others, 4.6	23.9
Defect of mechanical system	8.5	Maladjustment	5.3
Failure or defect of drive system (including actuator)	7.1	Servo-valves, 47.6; solenoid valves, packings, etc., 42.9; others, 95.	6.19
Mistake in interlocked operation with peripheral machinery	3.7		8.0
Abnormal air pressure, oil pressure, or voltage	1.71		0.9
Release or fall of workpiece	7.51	Handling mistakes, 90.9; power failure, 9.1	16.8
Human error, etc.	18.4	Operating mistakes, 29.6; inadvertent approach to robot, 22.0; hazardous start operation, 14.8; program mistakes, 18.5; call address, 7.4; others, 7.7	38.1
Others	0.88		0.81

Source: N. Sugimoto, "Subjects and Problems of Robot Safety Technology," *Occupational Health and Safety in Automation and Robotics,* ed. K. Noro (London: Taylor and Francis, 1987). Reproduced with permission.

an air leak. He was squeezed between the back end of a robot and a solenoid panel. Serious injury was prevented because abnormal resistance was detected by over-current to the robot servo-motors, and the program shut off the robot.

- A maintenance engineer sustained a neck injury when he was in the work envelope of a robot observing its operation. He failed to realize that he was also in the work envelope of another robot and was blind-sided by that robot.
- A maintenance repairman was servicing the welding electrodes on a robot. He was not aware that a product model change altered the operational envelope of an adjacent robot and that he was now standing in that envelope. He was struck by that robot and suffered a serious laceration to the back.
- The hydraulic system to a robot was turned on in the morning. Because of a failure in the servo-valve, the robot arm swung at full speed until it smashed into an adjacent robot. It was not possible to depress the emergency stop button in time to prevent collision.
- Four people were in the work envelope of a robot, with the robot in teach mode and slow speed. Because of a fault in a controller card, the robot suddenly swung to its end stop at high speed. There was a software fix that would have shut the robot down rather than allow full operation. However, the fix had not been communicated to the user (Lauck, 1986).

Järvinen and Karwowski's recent (1993) study of accidents in advanced manufacturing systems included 20 cases involving industrial robots. Nine of the cases involved stand-alone robots, and 11 involved robots that were part of a manufacturing system. About 75% of the robot incidents involved parts handling. Operators were involved in 65% of the cases studied. Production disturbance activities (clearing a blockage, adjustment, and fault finding and correction) accounted for 50% of the robot accidents. The manufacturing equipment was in the automatic mode in 75% of the incidents. Fifteen of the incidents occurred during a programmed movement. In 8 of 20 cases, the safeguard was either removed or defeated. Two fatalities were included in the robot incidents. Improper procedures, incompatible workplace layout, and incompatible controls were contributing factors in one-third of the accidents (Järvinen & Karwowski, 1993).

Hazards and Risk Situations

A problem that pervades the conventional versus robotic machinery spectrum is the inherent differences in their operation or behavior. Conventional machinery, while capable of high speeds, is usually repetitive and predictable; the danger zones are visible and stationary. Sequences and motion paths are cyclic and patterned. Machinery does not tend to dwell on idle

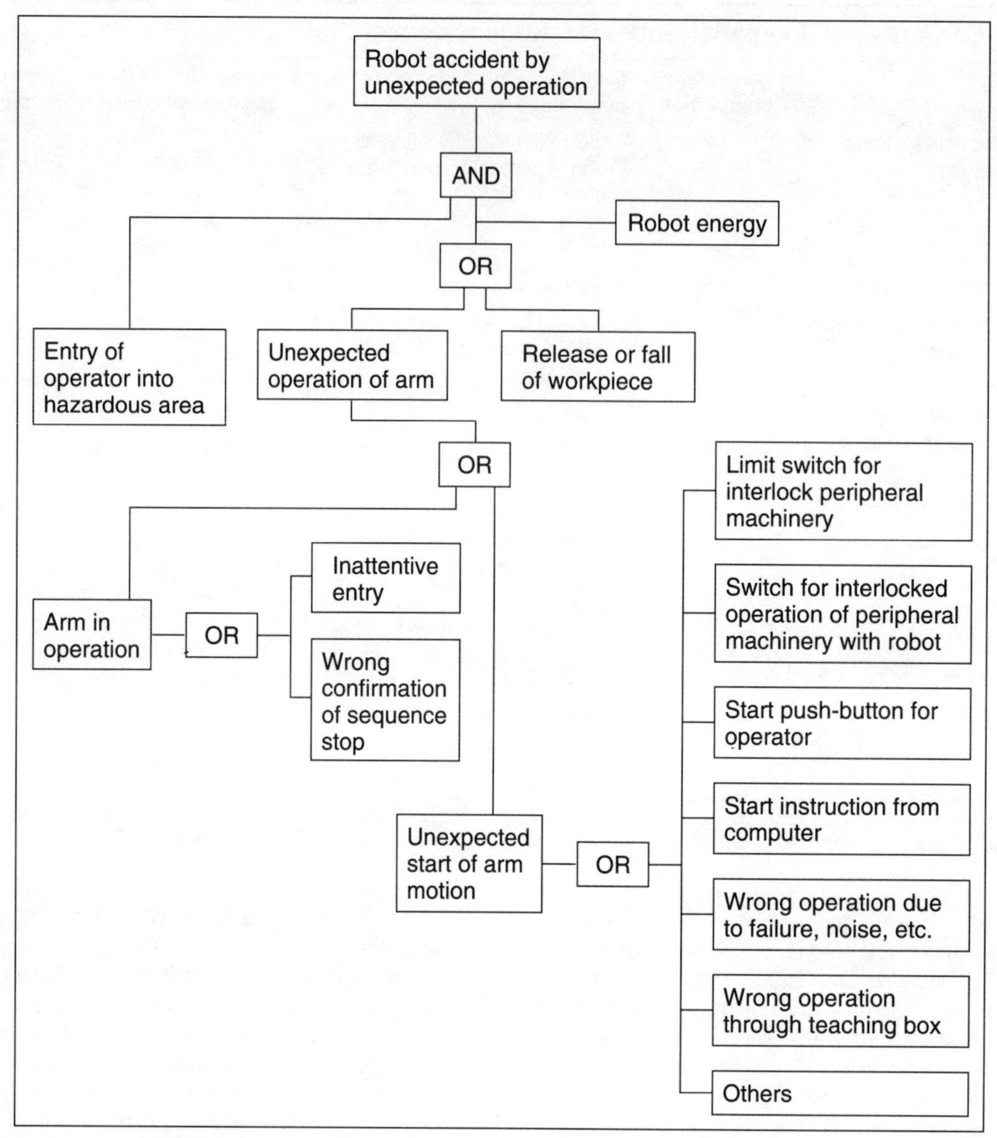

Figure 10-30. Fault tree analysis.
Source: N. Sugimoto, "Subjects and Problems of Robot Safety Technology," *Occupational Health and Safety in Automation and Robotics,* ed. K. Noro (London: Taylor and Francis, 1987). Reproduced with permission.

for varying lengths of time. Conventional machines normally will not operate without the action of an operator or a routine input signal.

Conversely, robots may present danger at any point they may reach. If motionless, they may merely be internally processing, and they may suddenly return to operation. Previous motion paths may be abruptly changed. Their three-dimensional movement may be unpredictable and therefore lead to inaccurate judgments of safe approach distance. Robots possess characteristics such as freedom of movement, speed, momentum, range, and cognitive programming capability. Robots, even when performing as intended, always have the potential for surprise.

Various types of accidents are associated with robotic operations, with the caught-between type being most prevalent. The following types of accidents and examples are based on the general experience to date:

- Individuals are caught between moving and stationary objects or two moving objects (entrapment).
- Individuals are struck by moving objects (for example, a robot drops or throws an object, a worker comes in the path of a robot, a robot malfunctions, etc.).
- Energy is released—electrical, mechanical, hydraulic, pneumatic, gravitational and stored (for example, workers are exposed to trapped pressures in lines, hoses, tanks, accumulators; energy failure–induced falls of objects or gravitational movement of components; energized electrical circuits; valve, switch, relay malfunction).

- Individuals are caught in or on an object (pinch points/projections).
- Individuals move toward an object and make contact with it (robots and auxiliaries).
- Individuals come in contact with end effectors such as hot welding components, tooling not in, chemical applicators, etc.

Robotic operations involve a host of general and unique hazards (physical conditions, system flaws, and behavioral factors). The following illustrate some of the possibilities:

- unauthorized access in the restricted zone
- failure to follow procedures
- intentional breach of safeguards
- system control errors
- power interruptions
- unawareness of stored energy
- safety sensor failures
- unguarded dangerous motion
- excessive equipment inertia
- application program logic errors
- improper robot-mounting installation
- inadequate training or skill level
- mechanical failures
- layout limitations
- cluttered restricted zone
- overly complex power-up sequences.

Various situations involving risk to personnel are encountered in robotic operations. Activities such as robot rebuilding or remanufacture, equipment installation, start-up, electrical-mechanical repair, preventive maintenance, programming, debugging, process adjustments, and material movement possess potential for employee injury if hazards are not effectively controlled.

Maintenance and Repair

Safety Requirements for Industrial Robots and Robot Systems (ANSI/RIA 15.06-1992, section 6.8) provides guidance for safeguarding maintenance and repair personnel. A number of its requirements have direct bearing on the control of hazardous energy.

> 6.8 Safeguarding maintenance and repair personnel
> 6.8.1 The user shall ensure that personnel who perform maintenance or repair on robots or robot systems are trained in accordance with clause 9.
> 6.8.2 Personnel that maintain and repair robot systems shall be safeguarded from injury due to hazardous motion.
> 6.8.3 A procedure shall be followed that includes lockout/tagout of sources of power and releasing or blocking of potentially hazardous stored energy.
> 6.8.4 When a lockout/tagout procedure is not used, alternate safeguards or safeguarding procedures shall be established and used to prevent injury.
> 6.8.4.1 Prior to entering the restricted envelope (space) while drive power is on, the following procedure shall be performed:
> a. The robot system shall be visually inspected to determine if any conditions exist that are likely to cause malfunctions;
> b. If pendant controls are to be used, the motion controls, emergency stop, and enabling device shall be function tested prior to such use to ensure their proper operation;
> c. If any damage or malfunctioning is found, correction shall be completed and retested before personnel enter the restricted envelope (space).
> 6.8.4.2 Personnel performing maintenance or repair tasks within the restricted envelope (space) when drive power is available shall have total control of the robot or robot system. This shall be accomplished by the following:
> a. Control of the robot shall be removed from the automatic mode;
> b. Maintenance and repair personnel shall have single point of control of the robot system;
> c. Movement of other equipment in a robot system shall be under the control of the person in the restricted envelope (space) if such movement would present a hazard;
> d. All robot system emergency stop devices shall remain functional;
> e. To restore automatic operation the following shall be required:
> 1. Exiting of the restricted envelope (space);
> 2. Restoring of safeguards required for automatic operation;
> 3. Initiating of deliberate start-up procedure.
> 6.8.4.3. Additional safeguarding methods may be provided as follows:
> a. Certain maintenance tasks can be performed without exposing personnel to trapping or pinching points by placing the robot arm in a predetermined position;
> b. The utilization of devices such as blocks/pins can prevent potentially hazardous motion of the robots and robot motion of the robots and robot systems;
> c. If a second person is stationed at the robot control panel, this person shall be prepared to respond properly to the potential hazards associated with the robot system.
> 6.8.5. If, during maintenance or repair, it becomes necessary to bypass any safeguards, alternative safeguarding shall be provided

and the bypass method shall be identified. The bypassed safeguards shall be returned to their original effectiveness when the maintenance task is complete.

See Figure 10-31 and Table 10-C regarding safeguards and their selection. Many of the safeguard options are interfaced with the hazardous-energy-control procedures associated with service/maintenance tasks.

In 1989, the International Labour Organization published *Safety in the Use of Industrial Robots*. Chapter 4 contains an extensive table of 211 safety-related items taken from the relevant robotic documents (laws, standards, regulations, etc.) of France, Germany, Japan, the former USSR, the United Kingdom, and the United States. The following requirements are of particular relevance to hazardous energy control:

1. A means should exist of preventing hazards arising from
   - the malfunctioning of robot components
   - opening of grippers and dropping of a load
   - release of sources of energy (including lockout fixtures)
   - electrical interference (conducted, electrostatic, electromagnetic, radio, lightning, etc.)
   - unintentional operation of actuating controls (power/motion).
2. Linked robots should be capable of operating individually (e.g., during repair work).
3. All robots should have one or more emergency stops that, without fail, immediately stop all movements.
4. The robot systems should stop automatically and safely and remain stationary until restarted when

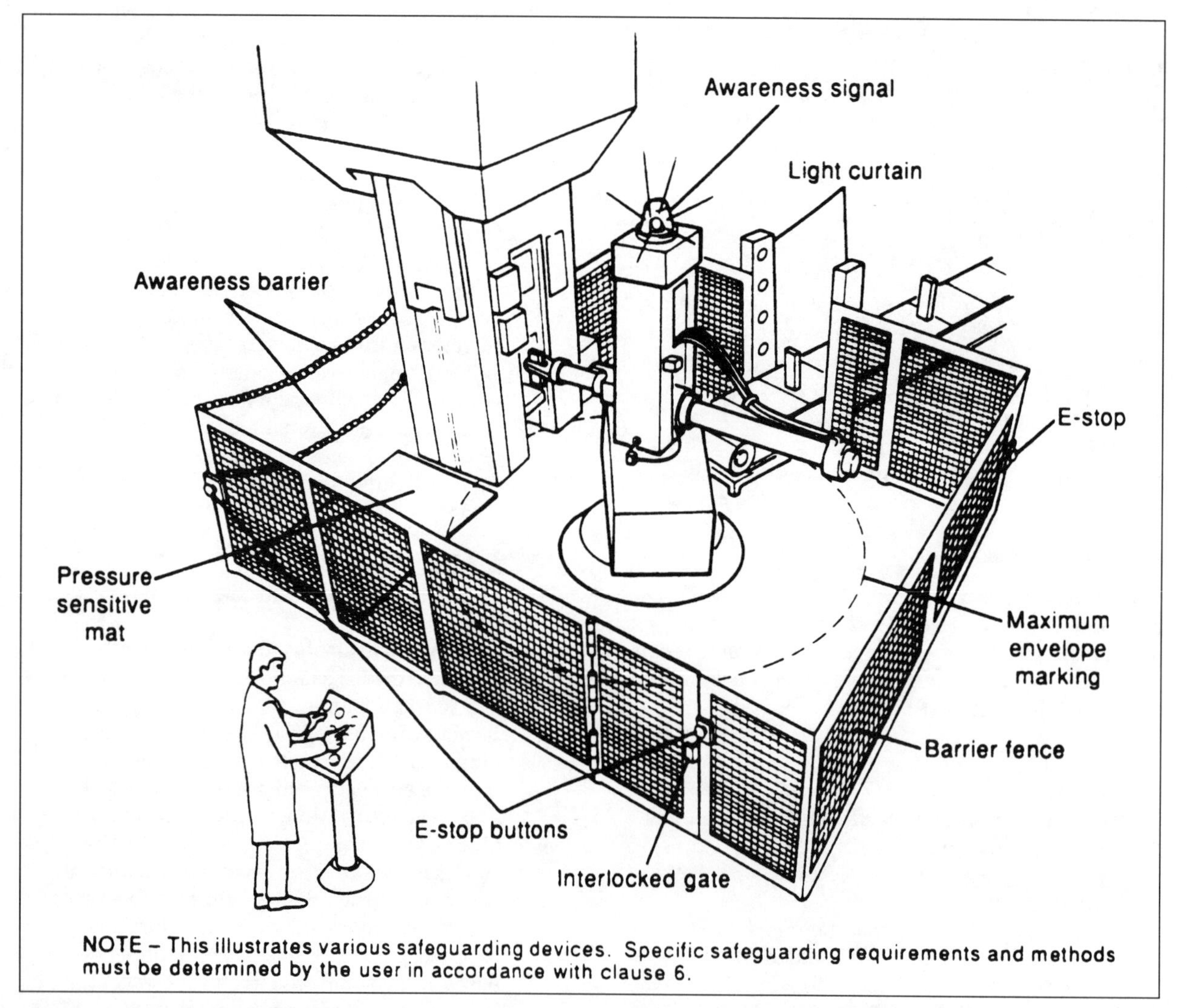

Figure 10-31. Composite illustration of robotic safeguarding devices.
Source: This material is reproduced with permission from the American National Standard *Safety Requirements for Industrial Robots and Robot Systems,* © 1992 by the American National Standards Institute. Copies of this standard may be purchased from the American National Standards Institute, 11 West 42nd Street, New York, NY 10036.

Table 10-C. Safeguard Selection

Safeguard	Action	Advantages	Limitations
EMERGENCY STOP button	Stops robot when activated. (required on all robots)	• Immediate shutdown of a robot or a robot system. • Requires little maintenance. Intended for emergencies only.	May cause damage to hydraulic robot if stopped in work cycle.
Deadman switch	Stops robot when activated (required on all robots)	• Immediate shutdown of robot. • Requires little maintenance.	
Gates with interlocking devices	Provides a barrier.	• Can be tailored to a specific system. • Can provide maximum protection. • Can be prefabricated in-house.	May require maintenance. May interfere with visibility. May be easy to bypass.
Light curtains/light screens	Stops robot when beam is broken.	• Allows access to work area without removing fixed guards. • Unobstructed vision. • Can be custom-sized to specific system.	May require adjustment and maintenance.
Locks (see facility and UAW-GM lockout procedures)	Prevents robot power and/or robot system power from being activated.	• Requires little maintenance. • Provides maximum protection.	
Mechanical steel pins	Restricts robot's movement.	• Can be prefabricated in-house. • Requires little maintenance.	May cause trapping points if not properly designed either on the robot or arm.
Pressure sensitive mats, plates, strips	Stops robot when activated.	• Unobstructed vision. • Allows access to work area without removing fixed guards. • Can be custom-sized for specific system.	May cause tripping hazard if not maintained. May be easy to bypass.
Screenguard	Provides a barrier.	• Can be tailored to a specific system. • Can be prefabricated in-house. • Can provide maximum protection. • Requires little maintenance.	May interfere with visibility. May require removal for maintenance or repair of robot. Causes hazards if not replaced. May cause trapping points if not properly designed.
Trip wire	Stops robot when activated.	• Can be prefabricated in-house. • Requires little maintenance.	May require removal for maintenance or repair of robot. May require special fixtures to hold work. May cause tripping hazards.
Interlocks	Shuts off or disengages energy and prevents start-up of robot and/or associated machinery.	• Allows access to work area without removing fixed guards.	May be easy to bypass or disengage.

Source: United Autoworkers—General Motors, *Robotic Safety* (Detroit: UAW-GM National Joint Committee on Health and Safety, 1987).

- a change in oil, air pressure, or voltage could result in faulty operation
- the power supply is interrupted or cut off
- related equipment has developed a fault
- an abnormality exists in the control system
- any part moves beyond the danger area.

5. If possible, a robot should stop automatically if the end effector makes contact with a person.
6. The robot should use a reduced speed (and preferably torque) when switched to teaching mode.
7. The operation of the robot should allow teaching, inspection, and other tasks to be done easily and safely.
8. It should be possible to release safely residual pressure in the pneumatic system.
9. Barrier guards for maintenance and cleaning should have a ground clearance no greater than 5.9 in. (150 mm).
10. All electrical connectors that could cause hazardous movements if mismatched should be uniquely keyed and labeled.
11. The shutdown of associated equipment should not result in hazardous motion of the robot.
12. When teaching, inspecting, repairing, adjusting, cleaning, or lubricating expose the worker, establish and enforce rules relating to the following:
    - starting up and operating the robot
    - speed of manipulator during teaching
    - signalling between operators
    - emergency procedures
    - restart after emergency stop
    - operator positioning
    - actions for inappropriate response or faulty operation
    - service/maintenance practices
    - trapped-key exchange practices (breakers, fuses, valves)
    - reduced speeds and special regular speed precautions.
13. During repairs, any parts capable of movement that could cause danger should be secured mechanically.
14. Control panel precautions to prevent inadvertent operation should include locking a cover over the panel, locking out individual controls, hanging AT WORK signs, etc.
15. Stored energy systems such as accumulators, springs, counterweights, and flywheels should be fail-safe if possible.
16. To diagnose a malfunction, maintenance personnel may need to approach the robot under energized conditions. These personnel should be protected by physical restraints and other methods such as permits, safe systems of work, pressure mats, electronic safety devices, etc.
17. Before beginning work, warning signs should be posted. When detaching counterweights or driving mechanisms, vertical sliding members must be restrained. Do not depend on control system monitors for zero energy condition verification. Use other dependable methods.
18. Some energy hazards should be isolated by means of primary devices and not by secondary control circuitry. Once isolation is complete, means should be taken (preferably lockout) to ensure that no intentional or unintentional activation can occur.

The following general guidelines for protecting service and maintenance personnel supplement or reinforce those previously presented. They are derived from a variety of sources and experience.

1. Obtain and follow robot manufacturer's instructions for care, service, and repair.
2. Be aware of changes, modifications, and upgrades of robot systems.
3. Access manuals, procedures, instructions, drawings, schematics, etc. routinely.
4. When service, repair, troubleshooting requires entry into robot envelope,
    - choose position in clear
    - mentally select escape route
    - determine robot home position and avoid.
5. When using the teach pendant, plan for possible unexpected movement due to energy loss or control error (position to avoid impact or entrapment).
6. When lockout/tagout is not possible or safeguards are bypassed, use alternate effective protection:
    - emergency stop
    - robot slow speed
    - second person at control panel
    - blocking devices
    - control over robot and associated equipment when in envelope.

Standards Activity

Underwriters Laboratories (UL) is expected to publish its *Industrial Robot Safety* standard UL 1740 during 1994. The standard will cover industrial robots and associated equipment intended for installation in accord with the National Electric Code and robotic systems intended for indoor use applications such as welding, parts assembly, parts transfer, inspection, loading, die casting, deburring, and sealant dispensing. The document will cover equipment construction and safety performance. UL has also developed standards for robotic safety-related software (UL 1988) and tests for safety-related controls employing solid-state devices (UL 991).

ELECTRICAL ENERGY (HIGH-VOLTAGE)

High voltages, pressures, and temperatures create circumstances where personnel will be exposed to significant forces if an unwanted energy-release incident should occur. Electrical forces reaching 750,000 V, pressures over 10,000 psi (680 atm), and temperatures exceeding 2,000 F (1,093 C) demand special procedures and safeguards to adequately protect personnel. High-voltage issues will be discussed in this section as they relate to energy isolation and protective practices. High-pressure and temperature safety techniques and procedures can be found in specialized references or equipment manufacturer's data and publications.

High voltage, as defined in article 710 of the National Electric Code, is more than 600 V, nominal. This definition is commonly used in general U.S. industrial establishments, although voltages up to 2,300 V may still be viewed as low. However, these voltages are dwarfed when considering the electrical generation and distribution industry's operational environment.

Many industrial plants have substations, rectifier stations, emergency power systems, and the like that involve substantial voltages. In some cases, companies are actively operating power generation and distribution networks within their business sphere. Occasionally, surplus power is sold and introduced to community utility grids for public consumption. It is not uncommon for facility electricians to deal routinely with voltages in the 440- to 2,300-V range (Figure 10-32). These large forces and their safe management is somewhat tempered by the fact that relatively small voltage and current levels can be lethal. Approximately one-half of all electrocutions involve 600 V or less.

The control of high-voltage hazardous energy, in concept, is essentially the same as controlling risk from low voltages. However, the numerous practices, safeguards, and methods used vary greatly in addressing the unique environmental requirements. In addition, standards and regulations exist that often differentiate between the control of low- and high-voltage risks.

Experience

In chapter 2, NIOSH's Fatal Accident Circumstances and Epidemiology (FACE) project was referenced, which reported on 201 electrocution incidents occurring during an eight-year period ending in 1990. In this study, two-thirds of the cases involved high voltage (>600 V). (See Figure 2-11 for the distribution by voltage.) In 48% of the incidents, workers made direct contact with an energized power line or equipment.

In a similar study, Fatal Occupational Electrical Injuries in Virginia, 196 deaths were reported as having occurred between 1977 and 1985. Most accidental electrocutions resulted from power line contact (53%) and machine or tool usage or repair (22%). Approximately 60% of the deaths involved high voltages. Only 3% of the deaths were associated with extremely high voltages (125 KV–750 KV). The annual mortality rates per 100,000 workers were 10.0 for utilities, 5.9 for mining, 3.9 for construction, 2.9 for agriculture, and 0.9 for manufacturing.

Figure 10-32. Testing a high-voltage circuit.
Source: Courtesy Gulf Publishing Co., Video Publishing.

Two major strategies were proposed in the Virginia study for preventing occupational electrical fatalities:

- *Visual enhancement of power lines*: streamers, bright markers, guided aerial maneuvering by a designated lookout worker, and protective sleeving
- *Protective measures*: electrical grounding, ground-fault interrupters, and machine/tool maintenance.

Of critical importance was the institution of passive measures (such as double insulated tools, underground power lines, and so on) and extensive electrical safety education and training (Jones et al, 1991).

OSHA engaged the Eastern Research Group, Inc., (ERG) to prepare a study entitled "Preparation of an Economic Impact Study for the Proposed OSHA Regulation Covering Electric Power Generation, Transmission, and Distribution—June 1986." The study found more accidents associated with transmission and distribution (power lines) than with substations or power generation (installations). ERG reported a potential annual benefit of 60 fatalities and 1,633 lost-workday injuries prevented at a first-year cost of $40 million ($22 million for subsequent years).

Regulations

Many standards have direct bearing on the safe performance of high-voltage work. In 1972, OSHA adopted regulations applying to the construction of power transmission and distribution lines and equipment in subpart V of part 1926. The National Electrical Safety Code (ANSI-C2) provides broad guidance on preventing electric shock. The 29 *CFR* 1910.268 standard, pertaining to the telecommunications industry, is also relevant.

In 1989, OSHA published a proposed standard on electric power generation, transmission, and distribution work and on electrical protective equipment. The Edison Electric Institute (EEI) and International Brotherhood of Electrical Workers (IBEW) developed a similar negotiated draft standard that was submitted for consideration. On January 31, 1994, the final rules were promulgated for 29 *CFR* 1910.269 *Electric Power Generation, Transmission and Distribution* and 29 *CFR* 1910.137 *Electrical Protective Equipment.* In 29 *CFR* 1910.269, OSHA expanded the original proposal's scope to include not only electric utilities but also equivalent installations of industrial establishments. In 29 *CFR* 1910.137, *Electrical Protective Equipment*, requirements for personnel protection in the form of insulated gloves, rubber matting, insulation blankets and hoods, insulation line hose and sleeves are specified (Figure 10-33).

Figure 10-33. Use of electrical protective equipment.
Source: Coutesy Gulf Publishing Co., Video Publishing.

The NFPA 70E standard for *Electrical Safety Requirements for Employee Workplaces* (1988) and *CFR* 1910.331–.335, *Electrical Safety-Related Work Practices*, both contain general lockout/tagout procedures, overhead line clearances and practices, and test equipment provisions that are applicable to high voltage.

Electrical Power Generation, Transmission, and Distribution

In this new standard, 29 *CFR* 1910.269, sections (d) lockout-tagout, (l) working on or near exposed energy parts; (m) de-energizing lines and equipment; (n) grounding; (o) testing; and (q) overhead lines are particularly informative regarding protecting personnel from energy transfer/release incidents.

Section (d), lockout-tagout, is much the same as the generic 29 *CFR* 1910.147, although the utilities industry is primarily a user of tagging systems. The agency has determined that compliance with 1910.147 will satisfy the requirements found in 1910.269 (d). OSHA stated in the 1910.269 preamble that

> the industry's (utility) tagging systems generally provide protection equivalent to that obtained by the use of a lockout program. However, the employer who uses a tagging system must demonstrate that it will provide full employee protection. For the employer to demonstrate that a tagging program is as protective as a lockout for a lockable piece of equipment, additional elements that bridge the gap between the two will have to be shown.

In section (m), de-energizing lines and equipment, OSHA addresses the special hazards associated with electrical transmission and distribution systems. Because these lines are located outdoors, they may be subject to re-energization by means other than what is normally expected, such as

- lightning strikes
- other cogeneration sources
- induced voltages
- contact with other energized sources.

In section (m)(2) and (m)(3), the following provisions for de-energizing lines and equipment for employee protection are found:

> (2) general
>
> (i) If an employee must depend on others to operate switches to de-energize lines or equipment on which the employee is to work or if the employee must secure special authorization before operating such switches, all of the requirements of paragraph (m)(3) shall be observed, in the order given, before work is begun.

(ii) If an employee other than the system operator is in sole charge of the lines or equipment and of the means of disconnection, that employee shall also comply with all of the requirements of paragraph (m)(3) of this section, in the order given, taking the place of the system operator as necessary.

(iii) If an employee is working alone and the means of disconnection are accessible and visible to the employee, the requirements of paragraphs (m)(3)(ii), (m)(3)(v), (m)(3)(vi), and (m)(3)(vii), and (m)(3)(xi) of this section shall be observed. However, tags required by these provisions need not be used.

(iv) Any disconnecting means that are accessible to persons outside the employer's control (for example, the general public) shall be rendered inoperable while they are open for the purpose of protecting employees.

(3) *Deenergizing lines and equipment.*

(i) A designated employee shall make a request of the system operator to have the particular section of line or equipment deenergized. The designated employee becomes the employee in charge (as this term is used in paragraph (m)(3) of this section) and is responsible for the clearance.

(ii) All switches, disconnectors, jumpers, taps, and other means through which electric energy may be supplied to the particular lines and equipment to be energized shall be opened. Such means shall be rendered inoperable, as design permits, and appropriately tagged to indicate that employees are at work.

(iii) Automatically and remotely controlled switches that could cause the opened disconnection means to close shall also be tagged at the point of control and shall be rendered inoperable, if design permits.

(iv) Tags shall prohibit operation of the disconnection means and shall indicate that employees are at work.

(v) After the applicable requirements in paragraphs (m)(3)(i) through (m)(3)(iv) of this section have been followed and the employee in charge of the work has been given a clearance by the system operator, the employee in charge of the work has been given a clearance by the system operator, the employee in charge shall verify by test that the lines and equipment to be worked are deenergized.

(vi) Protective grounds shall be installed as required by paragraph (n) of this section.

(vii) After the applicable requirements in paragraphs (m)(3)(i) through (m)(3)(vi) of this section have been followed, the lines and equipment involved may be worked as deenergized.

(viii) If two or more independent crews will be working on the same lines or equipment, each crew shall independently comply with the requirements in paragraph (m)(3) of this section.

(ix) To transfer the clearance, the employee in charge (or, in case of forced absence, the employee's supervisor) shall inform the system operator; employees in the crew shall be informed of the transfer; and the new employee in charge shall be responsible for the clearance.

(x) To release a clearance, the employee in charge shall:

(A) Notify employees under his or her direction that the clearance is to be released;

(B) Determine that all employees in the crew are clear of the lines and equipment;

(C) Determine that all protective grounds installed by the crew have been removed; and

(D) Report this information to the system operator, thus releasing the clearance

(xi) Only after all protective grounds have been removed, all protective tags have been removed from points of disconnection, all crews working on the lines or equipment have released their clearances, and equipment, may action be initiated to reenergize the lines or equipment.

(xii) The identity of the person requesting tag removal shall be the same as that of the employee requesting placement, unless responsibility has been transferred under paragraph (m)(3)(ix) of this section.

Electrical Protective Equipment

OSHA revised its existing requirements for electrical protective equipment (29 *CFR* 1910.137) on January 31, 1994. The agency chose to use performance language when defining provisions for use and care of the following protective equipment: rubber insulating gloves, rubber insulating matting, rubber insulating blankets, rubber insulating covers, rubber insulating line hose, and rubber insulating sleeves.

The revision maintains the protection presently afforded to employees by the referenced ANSI/ASTM standards. All relevant requirements are now placed in the text of the regulations, making referral to reference documents unnecessary.

Electrical Safety-Related Work Practices

Technically, OSHA's generic lockout/tagout standard found in 29 *CFR* 1910.147 Subpart J excludes "exposure to electrical hazards from work on, near, or with conductors or equipment in electric utilization installations." OSHA, by plan, addressed this exclusion by incorporating electrical lockout/tagout requirements in 29 *CFR* 1910.331–.335 Subpart S Electrical Safety-Related Work Practices. Specifically, energy-isolation direction is found in section .333, Selection and use of work practices. However, from a practical standpoint 29 *CFR* 1910.147 and 1910.333 are essentially the same and present no conflicts or special features that would complicate general lockout/tagout compliance. OSHA notes in section .333(b)(2) "lockout and tagging" that compliance with 1910.147(c) through (f) will generally satisfy the electrical requirements for energy isolation.

OSHA Instruction Standard 1-16.7 provides inspection procedures and interpretive guidelines for compliance officers evaluating employer's safe practices relating to electricity. The 29 *CFR* 1910.331 standard excludes work by qualified persons on or with electric generation, transmission, and distribution installations (utilities or comparable situation). *Qualified persons* are defined as those familiar with the construction and operation of the equipment and the hazards involved. Training of qualified persons in high-voltage practices, clearances, and use of protective equipment is either stated or implied.

In 29 *CFR* 1910.333, "Selection and use of work practices," guidance for energized and de-energized work is found. Section (c) contains the following requirements:

> *(c) Working on or near exposed energized parts.*
>
> (1) Application. This paragraph applies to work performed on exposed live parts (involving either direct contact or contact by means of tools or materials) or near enough to them for employees to be exposed to any hazard they present.
>
> (2) Work on energized equipment. Only qualified persons may work on electric parts or equipment that have not been deenergized under the procedures of paragraph (b) of this section. Such persons shall be capable of working safely on energized circuits and shall be familiar with the proper use of special precautionary techniques, personal protective equipment, insulating and shielding materials, and insulated tools.
>
> (3) Overhead lines. If work is to be performed near overhead lines, the lines shall be deenergized and grounded, or other protective measures shall be provided before work is started. If the lines are to be deenergized, arrangements shall be made with the person or organization that operates or controls the electric circuits involved to deenergize and ground them. If protective measures such as guarding, isolating, or insulation are provided these precautions shall prevent employees from contacting such lines directly with any part of their body or indirectly through conductive materials, tools, or equipment.
>
> *(i) Unqualified persons.*
>
> (a) When an unqualified person is working in an elevated position near overhead lines, the location shall be such that the person and the longest conduction object he or she may contact cannot come closer to any unguarded, energized overhead line than the following distances:
>
> (1) For voltages to ground 50 KV or below-10 ft (305 cm);
>
> (2) For voltages to ground over 50KV-10 ft (305 cm) plus 4 in. (10 cm) for every 10 KV over 50 KV.
>
> (b) When an unqualified person is working on the ground in the vicinity of overhead lines, the person may not bring any conductive object closer to unguarded, energized overhead lines than the distances given in paragraph (c)(3)(i)(a) of this section.
>
> NOTE: For voltages normally encountered with overhead power lines, objects which do not have an insulating rating for the voltage involved are considered to be conductive.
>
> *(ii) Qualified persons.* When a qualified person is working in the vicinity of overhead lines, whether in an elevated position or on the ground, the person may not approach or take any conductive object without an approved insulating handle closer to exposed energized parts than shown in [Table 10-D] unless:
>
> (a) The person is insulated from the energized part (gloves, with sleeves if necessary, rated for the voltage involved are considered to be insulation of the person from the energized part on which work is performed), or
>
> (b) The energized part is insulated both from all other conductive objects at a different potential and from the person, or

Table 10-D. Clearance Distances

Voltage range (phase to phase)	Minimum approach distance
300 V and less	Avoid contact.
Over 300 V, not over 750 V.	1 ft 0 in. (30.5 cm)
Over 750 V, not over 2 KV.	1 ft 6 in. (46 cm)
Over 2 KV, not over 15 KV.	2 ft 0 in. (61 cm)
Over 15 KV, not over 37 KV.	3 ft 0 in. (91 cm)
Over 37 KV, not over 87.5 KV.	3 ft 6 in. (107 cm)
Over 87.5 KV, not over 121 KV.	4 ft 0 in. (122 cm)
Over 121 KV, not over 140 KV.	4 ft 6 in. (137 cm)

Source: OSHA 1910.333 Selection and Use of Work Practices.

(c) The person is insulated from all conductive objects at a potential different from that of the energized part.

In section (b)(iv)(b), qualified persons using test equipment to determine that circuit elements (equipment) are de-energized in situations involving more than 600 V must check the equipment for proper operation immediately before and after each test.

Energy-Isolation Practices

Various practices are used to protect personnel from exposure to or contact with high voltages such as maintaining proper clearance distances, use of specialized protective equipment, use of insulating personal protective equipment, application of grounding techniques, and use of lockout/tagout procedures. The following cases illustrate what happens when these protective practices/techniques are used improperly or not at all:

- A plant electrician opened and incorrectly wired a pole-mounted 2,400-V cutout and started disconnecting conductors on the load side. One conductor was still energized because the cutout was bypassed, and he was electrocuted.
- An electrician attempted to replace the main 4,160-V circuit breaker in a control panel without turning off the power supply. Open exposed conductors shorted together, and the flash burned his eyes.
- A station operator died from electrical burns when he accidentally grounded an energized circuit in a generating plant. He had just tested two buses and found one de-energized and the other energized at 12 KV. In a moment of apparent inattention, he clamped the ground connection onto the hot bus.
- An electronics technician troubleshooting on a power supply opened the circuit and reached in to disconnect a test lead without waiting for an 8-KV capacitor to be drained of its charge. He suffered a severe shock and burns when he touched a terminal on the capacitor.
- A mechanic was troubleshooting with a screwdriver in a radar transmitter to detect transformer hum. He received a severe shock when he accidentally contacted an energized 20-KV terminal.

In "Electrical Safety Clearance Procedures," E. Palko, electrical engineering editor of *Plant Engineering Magazine*, defines *clearance* as taking measures to ensure that equipment is de-energized and reinforcing those measures with formal safeguards against altering that de-energized status for as long as clearance is required. A *clearance system* (sometimes called a lockout/tagout system) is a formal policy that defines procedures to be observed in securing safety clearances. An effective system defines three distinct classes of work:

- no formal clearance required
- only nominal, but effective, clearance required
- rigid clearance, secured by a formal clearance permit, required.

Two distinct types of clearance tags are suggested—one to be used when the equipment status is not to be altered under any circumstances and another to be used when the equipment must be operated periodically during the work (Figures 10-34 and 10-35). Palko also recommends a clearance permit system for electrical work

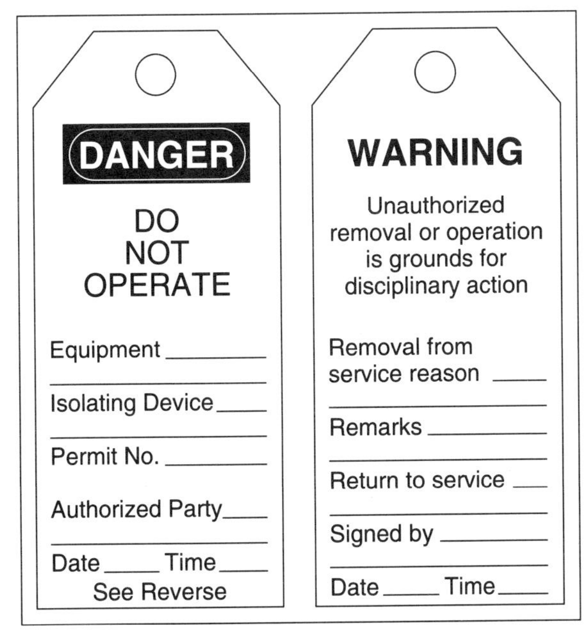

Figure 10-34. Electrical warning tag used in clearance system.
Source: Developed by E. Grund. © National Safety Council.

Figure 10-35. Electrical warning tag used in clearance system (intermittent use).
Source: Developed by E. Grund. © National Safety Council.

performed on medium- and high-voltage circuits and in confined spaces/equipment.

Industrial Approach to High Voltage

Various manufacturing companies have developed special internal practices for high-voltage hazards. Often these practices are in addition to the regular lockout/tagout procedures that are in use. Typically, they are solely in the domain of the electrical craftsman! These high-voltage practices frequently preceded any consequential use of generic lockout/tagout procedures. Most likely they were spawned by the utility industry and found their way into industrial use.

The LTV Steel Company and its predecessor organizations have used high-voltage rules for years. As early as 1934, General Safety Order #19—High Voltage Rules—was in effect; the most recent issue, Works Safety Order #23 (1990), is their offspring. The following rules excerpted from the document are of particular relevance to energy-isolation practices:

> Rule 2.2 *Working on high voltage circuits or equipment.*
>
> Anyone who is required to work on high voltage circuits or equipment (>600 volts) must get in touch with the Electrical Supervisor. The Electrical Supervisor will be responsible for clearing the circuit and shall see that all proper precautions have been taken for the safety of the workers.
>
> Rule 3.2 *Opening breakers and disconnects*
>
> Before employees work on high voltage circuits or equipment, breakers must be opened and disconnects must be opened at both ends of the circuit or equipment and properly protected with lock and tag, or block and tag, to prevent anyone from throwing same in while employees are working on circuits or equipment.
>
> 3.5 A check must be made to be sure that no potential is left on the line with a fluorescent "glow stick" immediately after breakers or disconnects are opened and before grounding or further work proceeds.
>
> Rule 7.3 *Exposure to high voltage circuits*
>
> In order to prevent accidents through misunderstanding, operators in charge of equipment receiving unwritten messages or instructions shall repeat them back to the sender. Persons issuing such unwritten messages or instructions shall require the receiver to repeat back such instructions. The sender and receiver of such messages shall secure the full name and title of the other party.
>
> Rule 10.1 *High voltage cutout fuses*
>
> No plug-type high voltage cutout fuses (primary cutouts) shall be changed while power is on. This also applies to high voltage cartridge-type fuses wherever possible. In cases where this cannot be done, the following rule shall be observed: "No primary cutout of the plug type is to be handled without the use of a plug puller. A switch hook is to be used in handling the expulsion types. The plug puller and switch hooks to be fitted with treated wood handles not less than 3½ feet in length approved for voltage up to 7500 volts; the length of the handle for higher voltages to be in accordance with the voltage of the circuit."

OTHER ISSUES

Often, conventional energy-isolation practices must be altered or supplemented with special safeguards, unique requirements, or specialized training. Each industry, type of technology, or process may operate under lockout/tagout practices that have evolved over years of application and operational experience. This is not to say that long-standing energy-isolation practices are flawless and deserve to be left undisturbed. Many of these practices represent compromises that were made at the time because of circumstances, absence of technology, managerial misconception, design limitations, lack of knowledge, etc.

Lockout/Tagout Device Removal

In 29 *CFR* 1910.147, OSHA requires the individual who placed his/her lockout/tagout device(s) to remove them upon completion of the protected task. This principle is sound and can be applied even in group/crew situations, albeit in another manner. However, some circumstances preclude this from being done: an employee becomes ill, forgets and goes home, or is injured on the job. OSHA provides an exception for legitimate circumstances that many progressive employers have been using for years in some form or another.

OSHA permits the removal of lockout/tagout devices when the original applier is not available (for legitimate reasons) by or under the direction of the employer. Specific procedures and training for this exception must be developed, documented, and incorporated into the employer's energy-control program. The procedure must provide a level of safety equivalent to the removal of the device by the authorized employee who applied it. Section (e) (3) states that as a minimum the employer's procedure shall contain the following elements:

> (i) Verification by the employer that the authorized employee who applied the device is not at the facility;
>
> (ii) Making all reasonable efforts to contact the authorized employee to inform him/her

that the lockout/tagout device has been removed; and
(iii) Ensuring that the authorized employee has this knowledge before resuming work at that facility.

Many companies have provisions that require the employee to return to work, if possible, when the reason for nonremoval was a memory lapse. Under these circumstances, various disciplinary measures may be instituted because of the serious implications of such failures. Generally, employers use the following process for removing lockout/tagout devices under exceptional circumstances:

- Identify the abandoned devices and their owners.
- Search/check for the whereabouts of their owners.
- Verify, if possible, departure from company premises.
- Attempt to make contact with owner by phone.
- If successful, have owner return or authorize removal.
- If unsuccessful, survey work area for clearance.
- Supervision/employee representative remove lock/tag.
- Return MEP to operational status.
- Prepare lockout/tagout device removal report for owner's supervisor and management.
- Counsel employee upon return to work.

Machine, Equipment, Process Start-up

A number of safety considerations must be kept in mind even when MEP is re-started after service, repair, or testing/positioning. In section (f)(1), OSHA describes the special requirements for machines/equipment when lockout/tagout devices must be temporarily removed to test or position the MEP. They are as follows:

(i) clear the MEP for tools and materials
(ii) remove personnel from the MEP
(iii) remove the lockout/tagout devices
(iv) energize and proceed with test/positioning
(v) de-energize all systems and reapply energy control measures.

This is a logical, sequential approach that is easier to comprehend and follow where the MEP is relatively simple. It becomes much more challenging as complexity increases. In some cases, a partial re-energization involving MEP components, sections, or energies may be possible. The focus is to determine what is to be accomplished and what potential exposure may be produced for the workers. Various safeguards may be used during this period to provide the necessary protection.

In situations such as conveyor systems, long process lines (painting, rolling, coating of metals), extrusion, blown-film production, liquid transfer, vacuum systems, hazard protective systems, and large mining equipment, warning must be given prior to movement related to the MEP. Often a preoperational alarm (audible/visual) is activated that alerts personnel that start-up is imminent. In addition, control circuits prevent immediate response by timed or delaying features. These safeguards allow personnel who are far from the control work area to have time to clear the MEP. These individuals may or may not have been involved with the service/repair activities. Further, a number of remote or secondary emergency controls often exist that would allow exposed personnel to interrupt the start-up if they were endangered—for example, conveyors could be equipped with prestart alarms and safety lanyards.

Machine, Equipment, Process Out-of-Service

Some circumstances require the MEP or some portion thereof to be taken out of service. This occurs frequently as the result of maintenance activity that has resulted in the removal of components for bench or shop repair, a hold status while parts are procured, temporary isolation of unsafe components, isolation of energy sources to protect an inoperable MEP, and so on. Under these conditions, a system is needed to protect personnel and property while not detracting from regular energy-isolation practices. In many cases, employers establish tagging/locking methods that are unique to out-of-service applications. The following features are typically incorporated:

- Tags and locks are different (shape, color, size, etc.).
- Information describing what is out of service and why is provided.
- Protection is provided as required (barriers, blanking, bypasses, guarding, etc.).
- Logs are maintained in the area identifying what is out of service and the expected return-to-service date.
- Control systems (manual and electronic) are developed to manage the out-of-service MEP inventory.
- Tags and locks are of the supervisory/departmental (security) type, with no personal protection status; they are prohibited from being used for any regular energy-isolation purposes.
- Authorization-to-work, permit, and clearance systems often incorporate out-of-service provisions as part of the process.

A well-designed MEP out-of-service element should not detract from mainstream energy-isolation procedures but should support their integrity while providing necessary supplemental protection.

Optimizing Manufacturing Performance

Manufacturing organizations are currently doing everything possible to remain competitive by optimizing their performance. Although this is a worthy and necessary endeavor, reason for elevated concern, if not alarm, exists with regard to its ramifications for safety. More specifically, what possible impact on the control of hazardous energy could result from enhancing manufacturing performance?

Many industries have gained renewed interest in cycle-time reduction in manufacturing processes. The intense interest is based on drawing down unit costs by reducing the nonproductive/operating times experienced in various operations, for example, changing press dies for a different product, changing inks/coatings for a new run, and adjusting tooling for different product specifications. Dr. Shigeo Shingo's recent text, *A Revolution in Manufacturing*, shows how to reduce changeovers by an average of 98%! He invented the SMED (single-minute exchange of die) system for Toyota as the centerpiece for just-in-time (JIT) manufacturing. As one might expect, these initiatives have great appeal for managers attempting to improve their business performance. Two schools of thought have developed on how "quick change" will be accomplished: One is that technology holds the key to performance optimization, whereas the other believes that labor is the crucial factor. In either case, the risks related to the control of hazardous energy appear to be increasing. The technological track seems to offer more promise for employee exposure reduction if safety (energy isolation) has been introduced as a design consideration.

The area of greatest concern is the misguided efforts by some organizations to reduce cycle time via labor-based initiatives. Working smarter may be desirable, but working faster may prove to be a disaster. Too often, workers are being engaged in stopwatch races or crew competitions in which safety has been ignored or compromised. Cycle-time reduction with a corresponding increase in risk should be totally unacceptable. These quick changes may produce unreasonable pressure on employees to opt for an energized status rather than a zero energy state for task accomplishment. Management's increasing interest in this area suggests the need for careful safety appraisal.

Although many claims of success for quick change are being offered, little or no data are available regarding its impact on the safety of involved tasks. Even more necessary is an assessment of whether employee exposure to the release of hazardous energy increases or decreases as a result of such initiatives.

SUMMARY

- Complex isolation of automated/advanced manufacturing systems, work under energized conditions, high-risk energy scenarios, guarding/interlocking technology, robotic energy-control issues, and host-contractor interaction are representative of situations requiring special control application.
- Complex isolation of the MEP can be viewed as any situation requiring special control methods because of the nature of the hazards and the variables (number of personnel, type and quantity of energy-isolating devices, nature and duration of the work, special risks, environmental factors, etc.) influencing effective lockout/tagout. As MEP complexity increases, it is common to use various visual methods such as process schematics, flow diagrams, drawings, process control graphics, computer graphics/simulation, and computer-generated procedures to facilitate system lockout/tagout.
- OSHA has determined that group lockout/tagout needs to provide a level of protection equivalent to that of personal lockout/tagout. The agency concluded that servicing/maintenance employees needed to be able to affix their lock/tag device as part of any group application. In its Instruction Standard 1-7.3, OSHA provides additional lockout/tagout definitions specific to group energy-isolation scenarios as well as procedural approaches involving lockbox techniques and permit systems. Clearance, safe-work-file and authorization-to-work systems utilized by various companies for complex lockout/tagout procedural approaches are reviewed as drafted by OSHA.
- In OSHA's *Process Safety Management* and *Permit-Required Confined Space* standards, the duties of each party are defined, and OSHA's compliance posture for multi-employer worksites is covered. Key research findings by Stanford University's Department of Civil Engineering on methods used by employers to successfully deal with contractors and the National Safety Council's prejob safety conference outline are presented. A lockout/tagout capability inquiry form is shown, which can be used to assess contractor competence. A list of various lockout/tagout issues needing resolution between contractors/employers is presented to stimulate needed review of this subject.
- The relationship between OSHA's 1910.147 lockout/tagout and 1910.212 machine guarding standards is reviewed with discussion of the exceptions to lockout/tagout application. U.S. Bureau of Labor Statistics data are reported regarding arm, hand, and finger amputation and the causal connection to machinery guarding isolation.
- Safety interlocks are discussed in detail because of their role in safeguarding machinery and as supplemental (secondary) energy-isolation devices. A description of interlock function and relevant international standards is provided. Power and control interlocking are explained, and the British Standard Institute's method for interlock selection is presented. Interlock failure and safe design features are reviewed. Captive-key and trapped-key control systems used in guarding applications are reviewed with graphic explanation of how they work. Interlock defeatability is addressed by describing various failure modes and warning about their limitations. A

safety defeatability index is shown that ranks the performance of safety hardware/software according to the degree of difficulty involved in its circumvention or defeat. F. B. Hall (1992) warns that often the risks outweigh the safety benefits provided by interlock devices.

- Alternative procedures used when partial or full energization of the MEP is necessary during service/maintenance tasks are discussed. OSHA recognizes this exception situation in its hot-tapping and incidental production activities exclusions. The phrase "alternative measures which provide effective protection" describes OSHA's posture with regard to activity not occurring under full lockout/tagout control. Six criteria are provided for demonstrating that alternative measures are warranted. Various methods and safeguards are provided that are often used singly or collectively as alternatives to a zero energy state.
- Automated systems and robotics are examined with regard to their special demands for energy isolation and the protection of personnel. Factors influencing the successful isolation of complex MEP are given from the maintenance worker's viewpoint. Various modern initiatives such as just-in-time and computerized maintenance systems are addressed. Five basic conditions for future factory equipment are highlighted (development, reliability, economics, availability, and maintainability).
- Järvinen & Karwowski's 1993 study, "Accidents in Advanced Manufacturing Systems" reveals that the principal causes of accidents are improper procedures, human error, incompatible controls, lack of awareness, and inadequate training. Design features of equipment that will enhance energy control are presented.
- Robotic safety, with emphasis on hazardous energy and service/repair, is covered broadly. Robot definitions, classification, and applications are presented. Robot systems, the envelope, and rules of robotics are discussed. Robot accident experience studies are reviewed and reveal various failure modes; for example, production disturbance activities accounted for 50% of the robot accidents in the Jarvinen and Karwowski survey. Types of robot accidents and hazards are presented to increase awareness of what needs to be controlled. The ANSI/RIA Standard for Robot Safety is quoted for safeguarding maintenance and repair personnel. The ILO publication, *Safety in the Use of Industrial Robots*, is excerpted, revealing requirements of special relevance to hazardous energy control. Robot installation safeguards and their selection are treated generally. General guidelines for protecting service and maintenance personnel are offered.
- High-voltage electrical activity relevant to lockout/tagout and clearance procedures are presented. Electrical accident experience is reviewed to identify key factors that should be controlled. Safeguards, regulatory requirements, and employer work practices are examined.
- Lockout/tagout device removal by someone other than the applier, MEP start-up, and MEP out-of-service issues are reviewed with suggestions for improving the activity. A warning is voiced regarding initiatives to reduce manufacturing cycle time that do not address or are insensitive to worker safety.

REFERENCES

American National Standards Institute. *Safety Requirements for Industrial Robots and Robot Systems*, ANSI/RIA 15.06. New York: American National Standards Institute, 1992.

American Petroleum Institute. *Procedures for Welding or Hot Tapping on Equipment Containing Flammables,* No. 2201. Washington DC: API, 1978.

American Society of Safety Engineers. *Dictionary of Terms Used in the Safety Profession,* 3d. ed. Des Plaines, IL:1988.

Bohme H. *The Dream Factory*. Frankfurt, Germany: Lufthansa Bordbuch, 1993.

British Standards Institution. *Safety of Machinery*, BS 5304, Section 9. London: HMSO, 1988.

Britt P. Owners own up to contractor safety. *Safety & Health* 148, no. 6 (Dec. 1993):44–48.

Dean JR. Plant/contractor safety partnerships. *Plant Services Magazine* (July 1993).

Gallagher VA. Liability, OSHA and the safety of outside contractors. *Professional Safety* 38, no. 1 (Jan. 1993):26–33.

Germain A. Occidental uses safety lockout system to protect employees. *Chemical Processing* 53, no. 9 (July 1990):76, 81.

Hall FB. Safety interlocks—The dark side. *Triodyne Inc.*, 7, no. 3 (June 1992).

International Labor Organization. *Safety in the Use of Industrial Robots*, No. 60. Geneva, Switzerland: 1989.

International Organization for Standardization. *Manipulating Industrial Robots—Safety*, ISO/10218. Geneva, Switzerland: 1992.

Järvinen J, Karwowski W. Accidents in Advanced Manufacturing Systems. Paper delivered at National Robot Safety Conference, Center for Industrial Ergonomics, University of Louisville, 1993.

Keltz TA. Hazards in chemical system maintenance: Permits. In *Safety and Accident Prevention in Chemical Operations*, 2d ed., edited by Fawcett H, Wood WS. New York: John Wiley & Sons, Inc., 1982.

Lauck KE. *Safeguarding of Industrial Robots: Working Safely with Industrial Robots*. Dearborn, MI: RIA, 1986.

Levitt RE, Samuelson NM. *Improving Construction Safety Performance: The User's Role*, Technical Report #260. Stanford, CA: Department of Civil Engineering, Stanford University, 1981.

LTV Steel. *High Voltage Rules*, LTV Steel Works Safety Order No. 23. Cleveland, OH: LTV Steel, June 1990.

National Institute of Occupational Safety & Health. Etherton JR. *Safe Maintenance Guidelines C-88-2193 for Robotic Workstations*, Publication No. 88-108. Cincinnati, OH: Occupational Safety & Health. 1988.

National Safety Council. *Accident Prevention Manual for Business & Industry: Engineering & Technology*, 10th ed. Itasca, IL: National Safety Council, 1992.

National Safety Council. *Accident Prevention Manual for Industrial Operations: Engineering and Technology*, 9th ed. Chicago: National Safety Council, 1988.

National Safety Council. *Safeguarding Concepts Illustrated*, 6th ed. Itasca, IL: National Safety Council, 1993.

OSHA Field Operations Manual, Issuing Citations—Special Circumstances, Chapter V, 1989.

OSHA Instruction Standard 1-7.3, *Control of Hazardous Energy—Inspection Procedures and Interpretive Guidance*, 1990.

Palko E. Electrical safety clearance procedures. *Plant Engineering* (Feb. 1977):106–111.

Scheel PD. Robotics in industry: A safety and health perspective. *Professional Safety*, no. 3 (Mar. 1993): 28–32.

Seim B, Beutler WB. Ancillary safety interlocks enhance conventional lockout/tagout procedures. *Occupational Health & Safety* 62, no. 10 (Oct. 1993):47–50.

Strubhar PM (ed.). *Working Safely with Industrial Robots*. Dearborn, MI: Robotics International of SME and Robotics Industries Association, 1986.

Sugimoto N. Subjects and problems of robot safety technology. In *Occupational Health and Safety in Automation & Robotics*, ed. K Noru. London: Taylor & Francis, 1987.

Telemecanique. *Safety Interlocking in the Automated Factory*. Westminster, MD: Technical Publication, 1988.

29 *CFR* 1910.147. *Control of Hazardous Energy (Lockout/Tagout)*. Final Rule, 1989.

29 *CFR* 1910.269. *Electric Power Generation, Transmission, and Distribution*. Final Rule, 1994.

29 *CFR* 1910.137. *Electrical Protective Equipment*. Final Rule, 1994.

29 *CFR* 1910.331–.335. *Electrical Safety-Related Work Practices*. Final Rule, 1990.

29 *CFR* 1910.146. *Permit-Required Confined Spaces*. Final Rule, 1993.

29 *CFR* 1910.119. *Process Safety Management of Highly Hazardous Chemicals*. 1991.

United Autoworkers/General Motors. *Robotic Safety*. Detroit: Human Resource Center, 1987.

U.S. Bureau of Labor Statistics. *Work-Related Hand Injuries and Upper Extremity Amputations*, Bulletin 2160. Washington DC: U.S. GPO, 1982.

Yokomizo Y, Hasegawa Y, Komotsubara A. Problems of and industrial medicine measures for the introduction of robots. In *Occupational Health and Safety in Automation and Robotics*, ed. K. Noro. London: Taylor & Francis, 1987.

11

Beyond Compliance

OVERVIEW

After more than 50 years of experimentation in controlling hazardous energy release, it is natural to ask: Where do we stand now? What can we be doing better? What more should we be doing now and what more should we do in the future to safeguard workers?

At present, although workplace fatalities and injuries have been reduced, we are often operating in the old paradigm and thinking in ways that were shaped by the industrial revolution. Yet, the working environment is now vastly different in most industries. It has changed significantly even from the 1950s to the 1970s when most of the lockout/tagout policies, procedures, and standards were developed. All indications are that it will continue to change rapidly well into the next century, bringing new technologies and methods of work that we can only dream of now.

Even today, despite all the efforts of employers, unions, associations, and government, workers are still seriously injured or die on the job through unexpected energy-release incidents. What then are the missing ingredients? Do employees and management refuse to take hazardous energy control seriously or give it the proper priority? Is there reason to believe that the risks of new working conditions will be adequately controlled by regulations and enforcement forged out of the old paradigm? Without consequential change, is there any likelihood of making real progress?

There is a prevailing presumption that if an employer develops a lockout/tagout program meeting the requirements of the 29 *CFR* 1910.147 standard that compliance will be achieved and energy-release incidents will be more or less eliminated. This is not true for a number of reasons:

- 29 *CFR* 1910.147 represents only minimum safety criteria; employers need to go beyond minimalism.
- This OSHA standard is basically a performance style standard, that is, the details of the energy-control process must be developed by the employer.
- The standard does not cover all critical lockout/tagout issues, such as equipment design, alternative methods for energized tasks, enforcement strategies, and quality of employee training.

- Employer safety policy and philosophy affect the implementation and overall effectiveness of all programs.
- Methods to reinforce safe behavior must be in place to condition lockout/tagout response (positive strokes).
- No basic understanding of the scope and magnitude of tasks exists where a "zero energy state" is not currently practical to achieve.
- Continuous competition for the time of management and employees who must attend to a variety of important issues, for example, quality and statistical process control, compliance with clean air and water regulations, confined space entry procedures, hazardous waste disposal, cost reduction, and downsizing.
- Safety-related disciplinary "quagmire," for example, an ever-increasing number of rules, employee fears of retaliation for pointing out safety lapses, competing supervisory demands, arduous grievance procedures, and monitoring limitations.
- Management/employee perceptions regarding lockout/tagout risks, that is, because a serious energy-release incident rarely occurs, all of these safety precautions must be overkill and therefore not really necessary.
- In many companies, less line supervision and greater worker autonomy can result in varying compliance with safety procedures.
- New technology or technical changes that workers are currently adapting to or whose complexity is likely to result initially in more errors and incidents (that is, the classic learning curve).

Furthermore, looking at the Hazardous-Energy-Control System shown in Figure 6-1, 40 subelements are identified in a prevention-oriented approach. The 29 *CFR* 1910.147 standard directly or indirectly addresses only about 35% of these for purposes of compliance.

Since January 1990, the effective date of the OSHA lockout/tagout standard, much has been happening in general industry regarding the installation or upgrading of preventive practices. Even progressive employers with mature lockout/tagout programs (over 20 years of use) may still not be in effective compliance with all aspects of the standard. Areas that are likely to receive the lowest compliance scores are the following:

1. written lockout/tagout enforcement policy and documentation
2. specific procedure development for various machines, equipment, and processes
3. alternative procedures (energized work, partial power, etc.) or "equivalent to" practices
4. training in specific lockout/tagout tasks
5. inspection (auditing) of lockout/tagout application for determining effectiveness of the procedures
6. energy-isolating device availability and lockable feature for new/modified equipment (after January 2, 1990).

Employers/employees who have used lockout/tagout procedures for years may still not realize everything that must be done or why. Unfortunately, they may have few resources (safety professionals, experienced engineers/maintenance specialists with lockout/tagout experience, etc.) to construct a lockout/tagout system using state-of-the-art techniques. They may be inclined simply to implement what they believe is necessary for compliance or continue to use what they have acquired from others without understanding its limitations.

Shortly after the lockout/tagout standard was promulgated, a number of court challenges were launched by trade associations, unions, and employers. Although the U.S. Circuit Court of Appeals remanded the standard to OSHA to respond on three issues, it refused to stay the standard. One issue, "applicability of the standard to all industries without appropriate justification," may encourage employers in lower-risk businesses to delay implementing lockout/tagout programs in the hope that the Court will force OSHA to exclude them from coverage.

In documenting the need for a lockout/tagout standard, OSHA indicated that the rule would affect 1.7 million businesses and 39 million workers in general industry. It would not apply to construction; agriculture; oil well and gas well drilling and servicing; electric utilities performing power generation, transmission, and distribution; mining; and federal agencies/establishments.

OSHA estimated that the lockout/tagout standard would prevent 122 deaths, 60,000 disabling injuries, 28,400 lost workdays, and 31,900 restricted workdays each year. The cost of compliance would be about $214 million the first year and $135 million annually thereafter. The 29 *CFR* 1910.147 standard was viewed as a major federal rule. There is good reason to believe that the cost was simply underestimated and the benefit overestimated at this point. There has been no federal attempt to assess the impact of the standard with respect to actual cost/benefit or effectiveness during the more than three years of its existence. According to the Bureau of Labor Statistics, general occupational injury/illness rates for U.S. manufacturing during 1989 to 1991 are virtually unchanged. For example, data from the UAW would seem to suggest that no consequential improvement has occurred in reducing lockout/tagout deaths per year even though the union data base reflects principally progressive high-profile companies with mature, comprehensive lockout/tagout programs. This data may suggest that employers will need to reach "beyond compliance" to produce the desired results and that the real causes of energy-release incidents run much deeper than the essential provisions of 29 *CFR* 1910.147 can cover.

Incremental Gain or Quantum Leap?

If this is the case, what must be done to change the paradigms of the past? Since the turn of the century, energy generation and consumption have spiraled, automation and mechanization have expanded exponentially, the pace of work has multiplied in terms of speed and output, new technology arrives routinely, and the rate of change has accelerated to the point where even industry experts struggle to keep up.

Yet despite this torrent of growth and commensurate risk, incremental gains have been made with respect to hazardous-energy-release incident occurrence. Companies actually are doing a relatively better job protecting workers today than they were 50 years ago. The questions are: What is good enough? Can they do better? How well can they do given the rate of growth and change? Are they trapped into making small gains, offset by large changes, or can they make the quantum leap to a new level of safety and performance? These questions need to be addressed by all employers and their workers when determining their hazardous-energy-control goals and the actions needed to achieve them.

Apathy or Advocates

Management can cling to a time-worn security blanket view that their lockout/tagout program is all right—the problem is that the employees refuse to comply with it. Or management may embark on an unbiased total reassessment of their program. Either way still leaves unanswered the question, What are the real performance inhibitors and how can they be neutralized or eliminated?

The prevention of "energy-release incidents" will require an unprecedented commitment by manager and employee alike to change. The use of a systems approach, total involvement method, and process style will be required for a good start. Understanding the causes of energy-release incidents will help in the design of prevention initiatives that effectively address the *root* causes (for example, unsafe machine design) and not merely the symptoms (operators experiencing hand injuries). Companies must develop a supportive safety culture to create the environment for effective resolution of lockout/tagout issues. Lockout/tagout must have top management priority and be established as a critical agenda item. Finally, lockout/tagout needs a high-level sponsor, champion, or advocate to raise it from the "world of words" to the "arena of action."

Prospects: The Triad

In addition to adoption of the new patterns just mentioned to move beyond compliance, three critical areas that have been neglected in regard to existing lockout/tagout programs or approaches need to be addressed: (1) designing for safety, (2) managing change, and (3) continuous improvement of process and system (see Chapters 6, The Process Approach, and 7, System Elements). Designing for safety has been universally underutilized with respect to solving lockout/tagout problems. The second and third items have received considerable management attention but usually from a financial, manufacturing, or quality perspective and not from an emphasis on hazardous energy control.

As a result, this chapter will focus on safety issues in corporate culture, designing for safety, management of change, and the work of continuous improvement from the viewpoint of improving safety in the workplace. If the first 250 years of the industrial revolution have produced astonishing changes in the way we work and the types of work we are able to perform, the next stage—the post–industrial revolution—is likely to bring about even more profound changes in work life. Everyone involved will be needed to improve hazardous-energy-control efforts in an increasingly complex, technological environment. It is an environment that is becoming the global rule rather than the exception.

SAFETY AND THE CORPORATE CULTURE

An organization's culture is a combination of its values, beliefs, mission, goals, performance measures, and sense of responsibility to its employees, customers, and community. This culture determines what is tolerable in terms of risk and safety in an organization. It also determines the type and success of its safety program—from initial hazard and risk assessment to designing for safety to the development and implementation of safety programs. Without the commitment and routine involvement of management in safety matters, no program will be effective in the long run.

This fact will be even more true in the future as organizations undergo radical change in the 1990s and beyond. The question is, Where does safety appear on the list of management priorities? Companies are looking for ways to reduce operating expenses and worker and management payrolls, to downsize their organizations and eliminate marginal divisions or subsidiaries, and to reengineer their operations to become more efficient and profitable.

In this environment, what will the board of directors or senior management decide are acceptable standards of risk or safety performance? Whatever they decide will significantly impact the attitudes and performance of their work force. This essential fact is why establishing safety as a high priority in the corporate culture remains a key factor in implementing the lockout/tagout triad of designing for safety, managing change, and continuous improvement of processes and systems.

Characteristics of Safety-Minded Organizations

As discussed in Chapter 7, System Elements, safety initiatives do not flourish in an organization with a negative safety climate. The complex elements of the hazardous-energy-control (lockout/tagout) system—design, environment, employees, management, and technology—can be implemented and sustained only in a cooperative culture (Figure 7-1). The corporate culture that places safety among its highest priorities generally has the following characteristics:

- *Management is involved and backs up words with actions.* Too often in some corporate cultures, management pays only lip service to safety policies and programs while placing responsibility for safety on the employees' shoulders. Yet the actions that management takes influence thousands of other decisions made in creating the total work environment, which includes such areas as the level of risk accepted, compliance with regulations, design and engineering standards, funds for hazard control, safety procedures and equipment, accident investigation, and emergency planning.

  In a corporate culture where safety is a major concern, top management are involved in a comprehensive way. Senior executives and managers serve as the role models for the rest of the organization. The company publishes its commitment to safety in such statements as: "We will inform employees and the public about any workplace hazards or community dangers," "We will comply with and enforce all occupational health and safety regulations and laws in each of our plants and worksites," and "We will review all operations regularly to ensure they meet or exceed legal requirements for worker health and safety."
- *Management establishes accountability for performance in safety, health, and environmental (SHE) affairs.* To ensure an effective accountability system, top management must establish agreed-on safety goals with all management staff. There are clearly stated rewards for goals that are met and equally clear consequences for goals not met. In this manner, every executive and manager from the CEO to the line supervisor becomes accountable.
- *The company has a strong SHE organization and staff.* One clear indication of an organization's cultural commitment to safety is the quality of its personnel. The safety, health, and environment professional should have a prominent place in the organization's hierarchy and be among the top decision makers. In addition, the professional staff should be perceived by workers as being unbiased and credible, and their work should be regarded as critical to achieving overall organizational goals.
- *Effective communication and information systems facilitate safety efforts at all levels.* These systems enable the organization to maintain a vital flow of information on safety-related areas such as hazard control, employee training, prevention techniques, safety performance, and action schedules. Such information is given high priority within the company and taken seriously by everyone in the organization. Effective communication and information systems also help an organization cope with rapid changes in technology, procedures, or regulations.
- *Design and engineering considerations for processes, equipment, and facilities is a matter of primary importance.* The SHE staff is involved at the beginning in decisions regarding design and engineering specifications. Too often in current organizations, safety professionals are left to deal with the hazards of equipment/machinery or work processes after delivery or installation. Involving the SHE staff early in design decisions can save companies considerable expense; it is often far more costly to retrofit existing equipment than to include safety provisions in the original design.
- *Preventive maintenance is given a high priority.* Maintenance is one of the more visible indications to workers of management's commitment to safety. If machines, equipment, and worksites are kept in good repair, the message is clear: Management is willing to put money and resources into the business to keep hazards to a minimum and prevent worker injury. A strong preventive maintenance program can produce its own momentum. A machine that is well cared for encourages its operators to report problems quickly and to be more responsible in how the machine is used.
- *Safety training is consistently high quality and ongoing.* In a safety-minded corporate culture, safety training begins with senior management and includes all levels down to line supervisors and workers. Senior management, who must serve as role models for other employees, are taught to recognize or evaluate the risks of their business and to learn hazard prevention and control techniques. Training needs are anticipated when there are plans for new or altered facilities, processes, or technology.

  At the supervisory and worker level, training must be well planned, continuous, and geared for results. Supervisors set the example and tone for the workers, communicating management intention and ensuring safety training and procedures are followed. Workers must know what management expects of them and the level of safety they must achieve and continually seek to exceed.

  However, the company does not make the mistake of considering safety training the main method of preventing incidents. Rather, it is used to support a total approach of safety design, hazard recognition, and other aspects of the organization's safety program.
- *There is a high level of employee involvement in all aspects of safety.* Like training, employee involvement is a critical component of successful safety pro-

grams. A corporate culture that encourages employee participation generally creates a workplace in which workers are conditioned to think routinely about safety. Such an organization includes employees in such safety activities as design suggestions, hazard identification, proposing solutions to safety problems, developing safety interest activities and promotions, and training programs. The more workers have contributed to a safety program, the more investment they will have in making sure that it works.

- *The design and content of the safety/health/environment program are well thought out, effectively implemented, and routinely updated.* The elements of the program focus on the anticipation, identification, assessment, and control of hazards and their consequences. Issues that are addressed include:
  - machine/equipment/process safety design
  - purchasing/procurement
  - ergonomics/human factors
  - occupational health hazards
  - environmental pollutants
  - hazardous materials/wastes
  - safety/health/environment capital budgeting
  - employee education and training
  - monitoring and measuring systems
  - communication and information processes
  - emergency planning and response
  - medical services and surveillance.

Corporate culture must be recognized by safety professionals and others as a powerful factor that influences the ability of an organization to improve workplace safety. A group cannot go beyond mere compliance or create a superior safety program and record without top management demonstrating, through their commitment, that they intend to achieve it. The organization will always be judged by its accomplishments, not its intentions.

Consequently, senior executives must take a stronger, more active role in overseeing the health, safety, and working environment of their organization. The major responsibility and accountability for safety must be on the shoulders of those who make the major decisions and shared by those who must do the actual work in the business.

Corporate Culture and OSHA

OSHA is also recognizing the importance of developing a corporate culture that makes safety an integral part of the corporate mission and operations. To encourage this trend, OSHA has adopted several programs and guidelines, including the *Voluntary Protection Program* (1982) and the *Safety and Health Program Management Guidelines* (1989). The updated Voluntary Protection Program (VPP) states (*Federal Register*, 1988):

> Requirements for VPP participation are based on comprehensive management systems with active employee involvement to prevent and control the potential safety and health hazards of the site. Companies which qualify generally view OSHA standards as a minimum level of safety and health performance and set their own stringent standards where necessary. . . .
>
> OSHA has long recognized that compliance with its standards cannot of itself accomplish all the goals established by the Act. The standards . . . will never cover all unsafe activities and conditions.
>
> The purpose of the Voluntary Protection Program (VPP) is to emphasize the importance of, the improvement of, and recognition of excellence in employer-provided, site-specific occupational safety and health programs. These programs are comprised of management systems for preventing or controlling occupational hazards. . . .
>
> When employers apply for and achieve approval for participation in the VPP, they are removed from programmed inspection lists.

Based on the superior safety record of workplaces qualifying for the VPP, OSHA developed its *Safety and Health Program Management Guidelines* in 1989. The four major elements in the *Guidelines* reflect the emphasis on creating the appropriate corporate culture.

1. Management commitment and employee involvement are complementary.
2. Worksite analysis is critical to identify existing and potential hazards.
3. Hazard prevention and control involve effective design of the job site and work as well as safe work practices and safety equipment.
4. Safety and health training addresses the safety and health responsibilities of all personnel.

Each element is then followed by a series of recommended actions. It is interesting to note that employee involvement is included in the first element. In OSHA's Final Rule for *Process Safety Management of Highly Hazardous Chemicals*, 29 *CFR* 1910.119, employee participation was placed at the beginning of the rule, emphasizing its importance. It also stressed the employers' obligation to consult with employees and their representatives. Section (c) Employee participation states the following (*Federal Register*, 1992):

> (1) Employers shall develop a written plan of action regarding the implementation of the employee participation required by this paragraph.
>
> (2) Employers shall consult with employees and their representatives on the conduct and development of process hazards analyses and on the development of the other elements of process safety management in this standard.

(3) Employers shall provide to employees and their representatives access to process hazard analyses and to all other information required to be developed under this standard.

The same four elements contained in the *Guidelines* form the basis for several OSHA initiatives that seek to move occupational safety and health a step further. These include *OSHA's Ergonomics Program Management Guidelines for Meatpacking Plants* (1990) and *OSHA's Final Rule for Process Safety Management of Highly Hazardous Chemicals*.

One of the most important areas to address in improving safety performance is designing safety into the worksite and jobs. This step should be part of an organization's improvement process. However, it is likely to occur only in a corporate culture in which safety is considered a major priority.

DESIGN FOR SAFETY

Lockout/tagout standards, procedures, and practices are generally conceived of and implemented under normal conditions. For example, the supervisor informs an operator that maintenance workers will do a routine repair of his or her machine on Tuesday morning. But what happens when conditions are not normal, when you cannot achieve zero energy state under the equipment's present design, or when to de-energize creates substantial problems? How then do employees proceed?

Generally people are hurt not because of the way that energy-isolation devices are secured but because of a host of issues related to design limitations, unanticipated circumstances, procedural flaws, training deficiencies, and lack of enforcement. Because one cannot write a procedure for every unusual situation that may arise, continuously monitor workers, or shut down completely for every task, the issue of designing safety into machinery, equipment, and work processes becomes paramount. The best promise for the future may be offered by designing for safety before machine/equipment manufacture or process installation. Under this system, the ideal hierarchy of safety control would look like the following:

1. Design to prevent or minimize anticipated hazards.
2. Eliminate hazards or substitute less hazardous materials and conditions.
3. Control or correct the hazards.
4. Warn/alert workers to hazards.
5. Train workers to handle or avoid hazards.
6. Use personal protective equipment.

Currently tactics 4, 5, and 6 are generally the more widely used approaches to safety, but a growing number of safety professionals and industry experts are urging manufacturers and employers to focus their attention on the design element. In addition to reducing risk, the concept of safety in the design process has been shown to:

- increase worker productivity
- improve quality
- improve people and processing flexibility
- facilitate and enhance operating performance
- reduce costs
- reduce hazards in service and maintenance activities
- achieve effective upstream environmental control.

Some organizations, for example, the General Motors Corporation, have developed an action process to make design-in safety a routine part of the manufacturing process. Their system has five steps (*ProAct*, 1992):

1. Identify opportunities for designing in safety elements and features.
2. Analyze the "critical few" target opportunities using risk-exposure analysis.
3. Plan which opportunities are the most feasible to accomplish.
4. Implement the design controls and safeguards.
5. Evaluate their effectiveness.

This five-step approach or a process like it can help to make safety in design a routine part of an organization's operations.

Supply-Side Safety

Designing safety features into new machinery or processes, procuring or creating safer materials, and upgrading current technology can be termed "supply-side safety." This approach refers to preoperational steps taken by the employer to ensure that safety is built "in" not added "on" to the process. It anticipates the problems of human/machine/environment integration and optimizes the design for safety. It also considers the nature and implications of materials used in the process. As proactive tools, anticipation and prevention can stop trouble before it starts, as more organizations are emphasizing to their top management (ProAct, 1992).

There are three concepts behind supply-side safety. The first is that the safety of an operation can be determined before the workers, processes, and plant and equipment are assembled at the worksite to produce a particular product. The more that safety is built into the work, the fewer hazards and risks people will have to cope with on the job.

Second, the farther upstream that improvements for safety can be made, the more cost-effective they are likely to be. It is usually cheaper to build in safety the first time than to retrofit machinery or processes later.

Third, management systems, employee training, and behavioral conditioning are only part of the safety

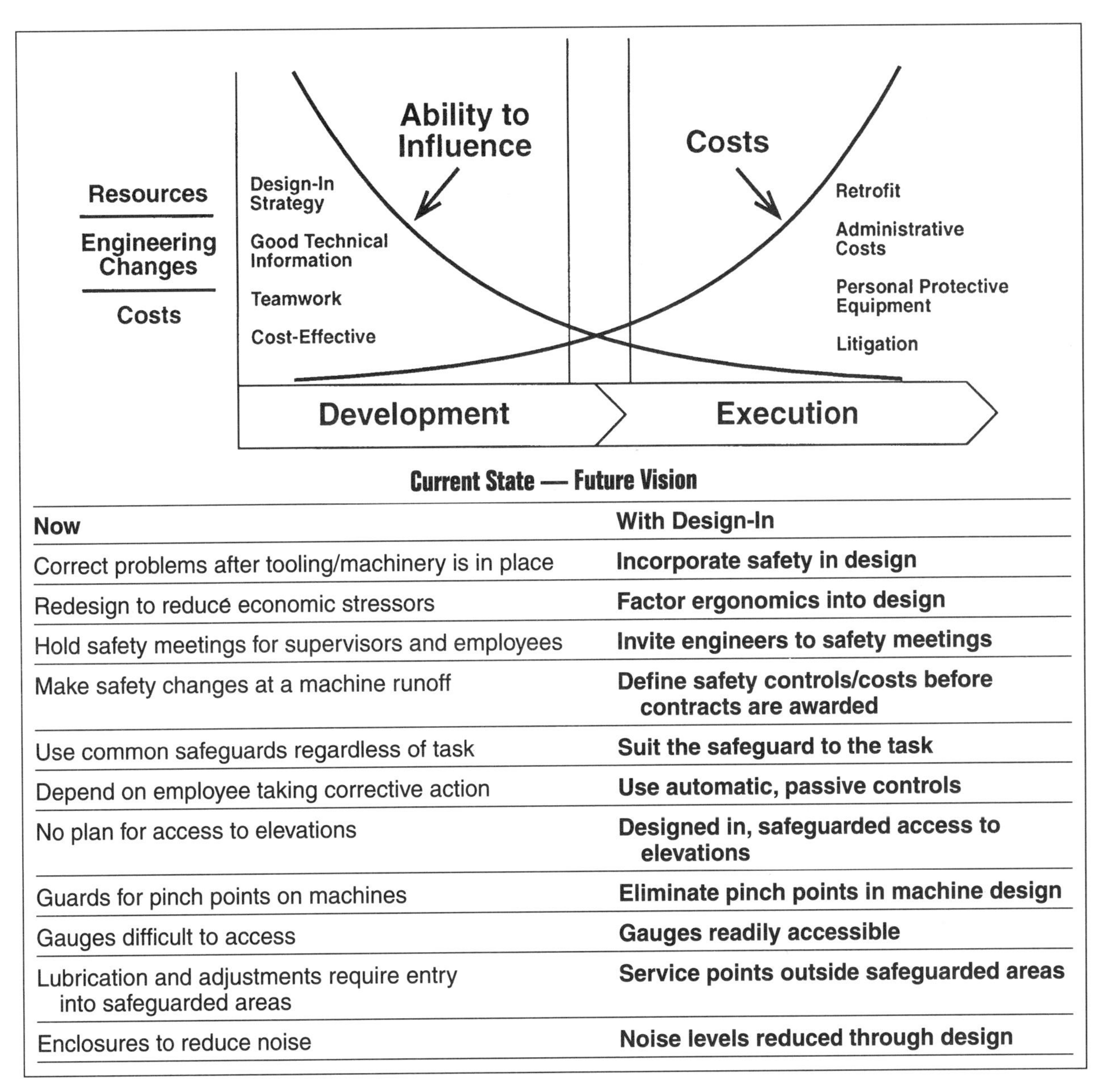

Current State — Future Vision

Now	With Design-In
Correct problems after tooling/machinery is in place	**Incorporate safety in design**
Redesign to reduce economic stressors	**Factor ergonomics into design**
Hold safety meetings for supervisors and employees	**Invite engineers to safety meetings**
Make safety changes at a machine runoff	**Define safety controls/costs before contracts are awarded**
Use common safeguards regardless of task	**Suit the safeguard to the task**
Depend on employee taking corrective action	**Use automatic, passive controls**
No plan for access to elevations	**Designed in, safeguarded access to elevations**
Guards for pinch points on machines	**Eliminate pinch points in machine design**
Gauges difficult to access	**Gauges readily accessible**
Lubrication and adjustments require entry into safeguarded areas	**Service points outside safeguarded areas**
Enclosures to reduce noise	**Noise levels reduced through design**

Figure 11–1. Design-in cost-effective safety.
Source: Courtesy of General Motors.

equation and by themselves will not be enough to move workplace safety substantially forward. Training or behavioral conditioning, no matter how well done, will improve safety *only if the problem is a deficiency in the skill and knowledge of the worker.* If the problem is a hardware flaw, it should be addressed at the design or engineering stage and not through employee instruction and training.

Supply-side safety requires a future vision of the organization compared with the current state. For example, the current state–future vision contrast for a manufacturing company might look like that in Figure 11-1. The figure also shows how design-in safety is a more cost-effective way to approach safety issues.

In designing for safety, everyone—safety professionals, managers, and workers—has an opportunity to give advice and counsel in this area. Some of the efforts occurring now include:

- applying quality assurance principles to safety practices to produce safer work procedures, better hazard control, and more accurate incident reporting
- adding the concepts of hazard anticipation, elimination, mitigation, or control to the ideas on which concurrent, or simultaneous, engineering are based
- increasing importance of OSHA's requirements for hazard analysis in "planning new facilities, processes, materials, and equipment"

- adapting ergonomic/human factor concepts (designing work to fit the worker) to focus company design efforts on the workplace and work methods.

Identifying the Hazards

One of the first steps in safety design is to conduct an analysis of the machinery, equipment, or process to identify the hazards that exist or can be anticipated. Several excellent hazard-analysis models, for example, gross hazard analysis, hazard and operability studies (hazop), fault-tree analyses, safety analysis tables, and failure mode and effects analyses (FMEA) have been developed over the years to make this task easier and more accurate. Specifications and criteria for a new machine or work process or an updated version of an old machine or process should be based on a detailed hazard analysis or hazard-analysis audit. OSHA, through some of its newer standards, also is requiring hazard analyses for modified, planned, and new facilities, processes, materials, and equipment.

Some firms, for example, Matthew Hall Engineering in England, have established a three-phase safety audit procedure specifically tailored to help develop and evaluate safety in design (*Occupational Safety and Health*, 1988). Phase 1 is Design Safety Management in which engineering, process, and environmental control managers and staff are asked to fill out three questionnaires. These identify normal and atypical hazards in the workplace associated with a particular equipment or process. Phase 2 is Detailed Design Safety Appraisal, in which all engineering disciplines working on a contract fill out questionnaires developed to promote safety-oriented thinking throughout the design process. Phase 3 is Operability and Information Appraisal, which evaluates the flow of technical and safety-related information to the operator of a new machine or process.

This safety audit provides a means of examining design safety in detail, providing evidence that safety procedures and standards are being implemented and creating documentation to show the reasons behind safety design decisions.

Hazard analysis for safety design must include more than simply considering what risks exist when the machine is operating normally according to specifications (Figure 11-2). In today's working environment, with its complex technology, automation of

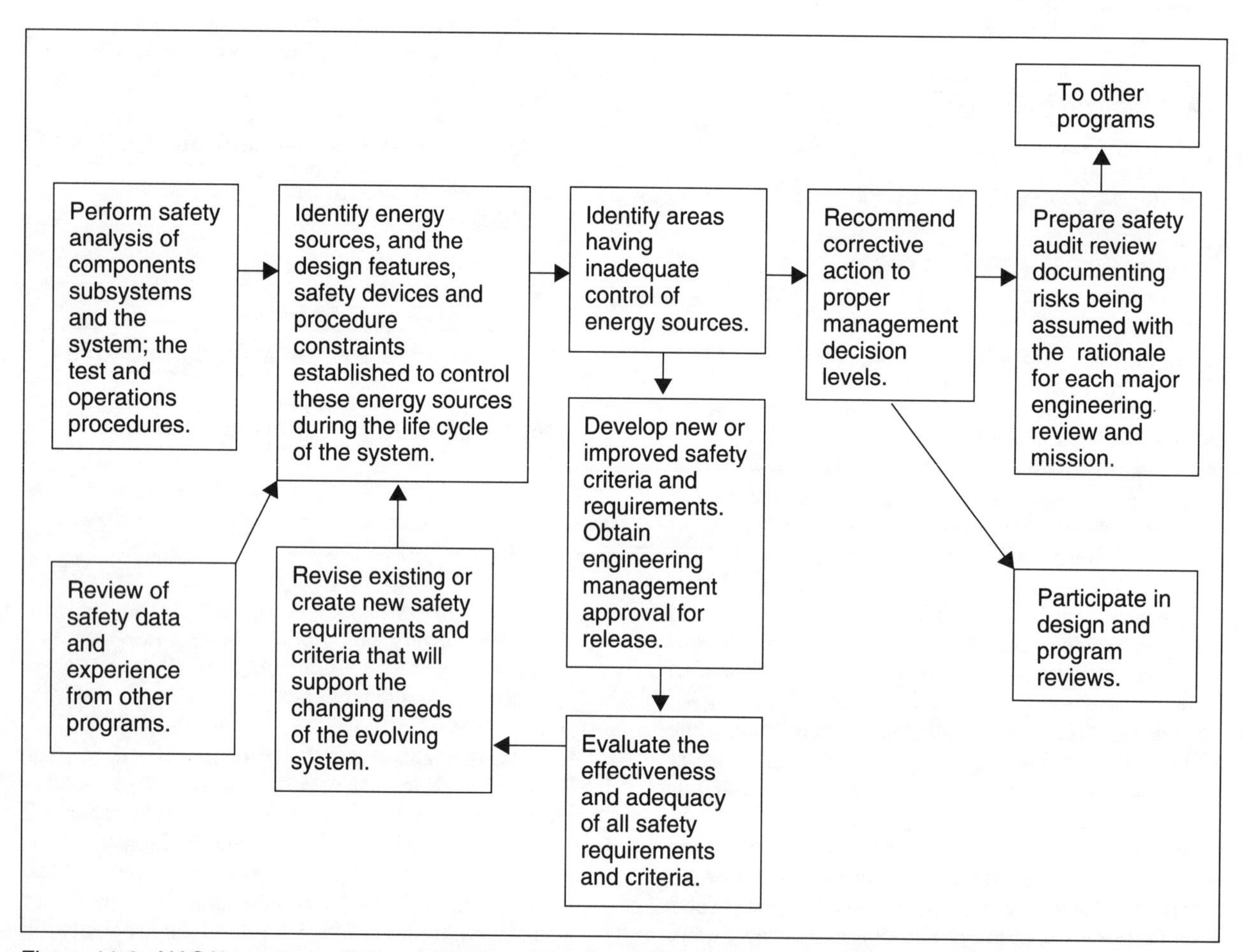

Figure 11-2. NASA's system safety activities functional flow, idealized functions for new projects.
Source: Johnson WG. *MORT Safety Assurance Systems.* New York: Marcel Dekker, 1980, p. 258. Reprinted by courtesy of Marcel Dekker, Inc.

many tasks, and interdependent systems, many machines or processes may operate under a wide variety of normal and abnormal conditions. As a result, workers may be exposed to or must deal with many hazards in which one or more of the following situations exist:

- Some degree of energy must be present to accomplish work (test, setup, or troubleshoot equipment).
- Isolation of energy is impractical because of the nature of the task, its frequency, the ratio of task time to isolation time, etc.
- Power is used to free jammed material in combination with other techniques.
- Design limitations dictate less than required isolation practices (multiple machines on one feed circuit without individual disconnects).
- Machine-guarding systems are bypassed to perform tests, inspection, etc.
- Equipment performance/product quality, etc., is being tuned (timing, sequencing, registration, flow).
- Computer/programmed systems require backup/alternate power supplies.
- Power supply is uninterruptable (hospital life support systems, power grids, etc.).
- Work is performed live by design (power distribution tasks, electrolytic cell processes, hot-tapping, etc.).
- Training/teaching/programming with energized robots and automated systems is created.

In addition to these situations, workers face other high-risk conditions for which little or no hazard analysis has been done to reduce the risks to a minimum. Management may believe that abnormal operating situations do not occur often; therefore, these "exceptions" do not need to be taken into consideration when written safety procedures are developed. The skilled worker, it is believed, should know how to protect himself or herself during these "rare" breakdowns. Or the machinery may have fundamental flaws that result in the operator having to service the machine in a manner that was never intended in the original concept and design.

This leads to a type of "catch-22," or trap, for the worker. With no input into the design of a machine or piece of equipment, the worker is nevertheless called on to solve operating problems with the machinery often by putting himself or herself at risk. By coping with process limitations and maintaining output, however, the worker makes it appear that the machine is operating acceptably. If the worker is injured, management may regard the incident as due to "worker failure to follow directions or carelessness," when, in fact, the real cause is the machine design. Because the worker has been blamed for the incident, no one addresses the machine's design flaws during the development of the next generation. The worker is then presented with a "new, higher-output version" of the old machine with the same or added safety difficulties.

These areas represent a significant opportunity for design improvements in existing machinery or processes and in new or replacement machines, equipment, or methods. The policy of involving various groups of people—safety professionals, engineers, scientists, workers, and management—in identifying hazards and developing specifications for safety design is part of what is known as concurrent, or simultaneous, engineering.

Concurrent Engineering

Concurrent engineering means that from the first step, manufacturing, production, quality control, safety, purchasing, finance, and marketing concerns all are examined simultaneously at the conceptual/design stage. The principal objective of the approach is to develop a systematic method for designers to anticipate, evaluate, and eliminate or control hazards and risks before design and engineering work begins. It may be thought of as holistic engineering.

When a traditional or sequential engineering approach is used, the employer, safety manager, and workers are often confronted with a design that is virtually complete. Any problems identified at this stage or later in the production stage are increasingly costly to correct. In contrast, in concurrent engineering all the groups who use or are involved with the proposed machinery, equipment, or process are included in discussions from the outset. This means that safety concerns will influence the hundreds of decisions that must be made when a design is developed or upgraded.

From a systems-engineering perspective, every project has a generalized life cycle. It can be modeled in five phases: concept, design, development, operation, and disposal. The duration and activity of each phase depends on the particular system being engineered. In addition, phase overlap or transition may vary in terms of time. Activities from more than one phase often are performed simultaneously.

1. *Concept phase*. The concept phase starts when a need for a system is identified. It includes all those activities that lead to a decision to design a system to satisfy the need. The system task is defined during this period, and the feasibility of accomplishing the task is evaluated. Alternative approaches to system design are established, and trade-off studies are made to select the best alternative. In addition, cost and schedule estimates are made for the balance of the life cycle.
2. *Design phase*. During the design phase, the concept of the system is refined, and details are established regarding the form that the system will take. Decisions are made concerning the specific equipment that will be included in the system, and system task procedures are prescribed. The capabilities and limitations of each system component are evaluated. Design details are worked out for

the components so that they perform as a unit. Provisions are also made to assure that the entire system is capable of operating in the intended environment. Modifications for improving the system may be made to the resulting design throughout the development and operation phases.

3. *Development phase*. During the development phase, the design information is used to prepare the system for operation. Development activities include purchase or manufacture of system machinery, construction, installation, and the assignment and training of operating personnel. Evaluations may be conducted during this phase to measure system effectiveness and other system criteria.
4. *Operation phase*. The system performs its intended task during the operation phase. Periodic performance reviews and maintenance are usually conducted to assure that a high level of system effectiveness is maintained.
5. *Disposal phase*. Once the system has satisfied its purpose completely and is no longer useful, the disposal phase is entered. This period includes all of those activities necessary to scrap, salvage, abandon, disassemble, phase out, replace, destruct, or otherwise dispose of the system.

A project life cycle—safety through design—model in Figure 11-3. depicts concurrent engineering methodology for incipient hazard elimination or control.

Generally, there are two major situations where opportunity exists to design-in safety. First, when new machinery, equipment, processes or plants are installed/constructed (green field sites). Second, when existing hardware and facilities are upgraded, expanded, or modified. Often the enhancements to existing machinery and process are substantial in terms of change and/or cost. The very frequency of general improvements to machinery, equipment or process, when compared to green field projects, make it an ideal candidate for enhancing safety through design. Unfortunately, this opportunity is often wasted because organizations have failed to integrate safety into their business processes. In contrast, concurrent engineering methodology can capitalize on and maximize any occasion where physical assets are to be changed.

A more productive alternative to the traditional engineering approach is to use concurrent engineering as a basis for design. In this method the following type of information is used:

- *Experience of the past*. This is the internal and external history of the item or process from the manufacturer's and user's point of view. What do manufacturer and user records show about test results, defects, failure rates, operating parameters? What have users been saying about the item or process to the manufacturer or employer?

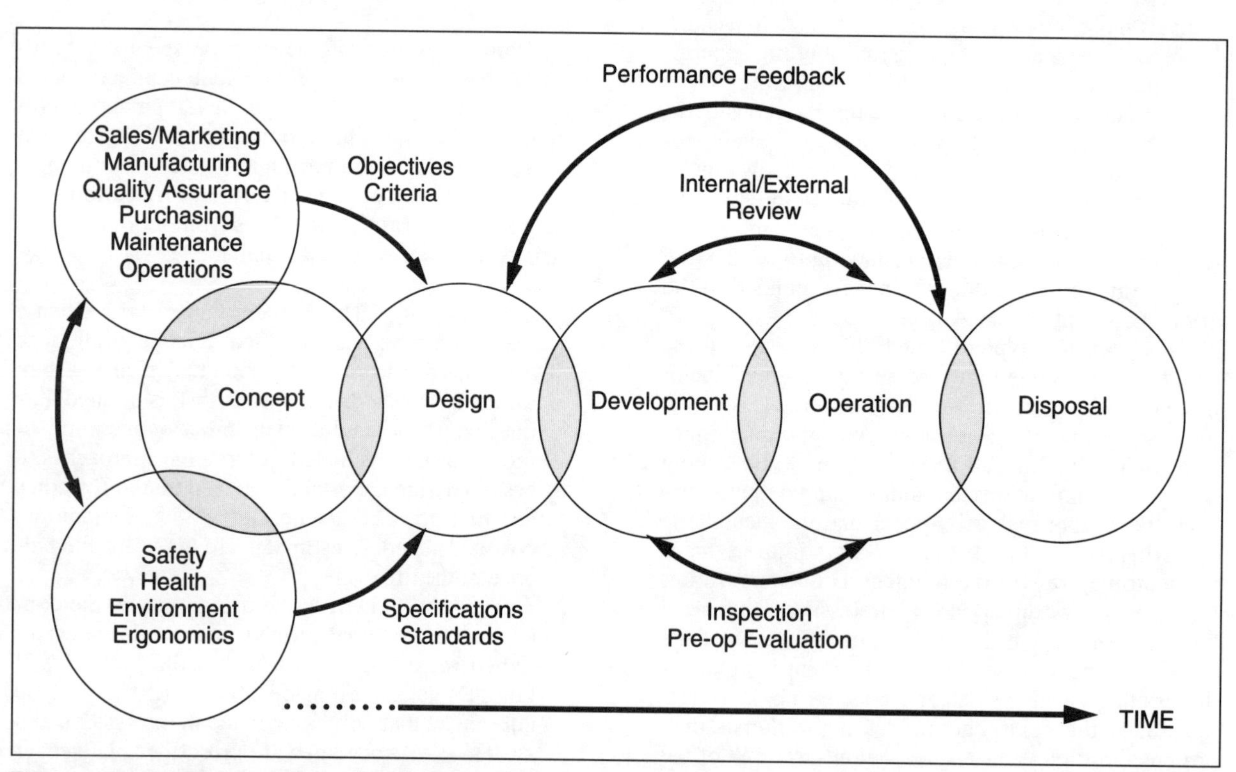

Figure 11-3. Project life cycle—safety through design.
Source: Developed by E. Grund. © National Safety Council.

- *Criteria and specifications for the new item or process.* What safety concerns have been noted? For example, if workers experience problems with a machine operating at 500 rpm and the new design calls for an increase in speed to 700 rpm, what compounded safety concerns should designers address?
- *Input from operators.* How does the item or process work on the job? Generally designers may convene a panel of operators and question them about how well the item or process runs when everything is working as it should, but what happens when things go wrong?
- *Input from maintenance workers, safety professionals, supervisors, engineers.* These are the people who must intervene when something breaks down or accidents occur. The information they have regarding frequency of breakdowns, types of problems that arise, length of downtime, and other

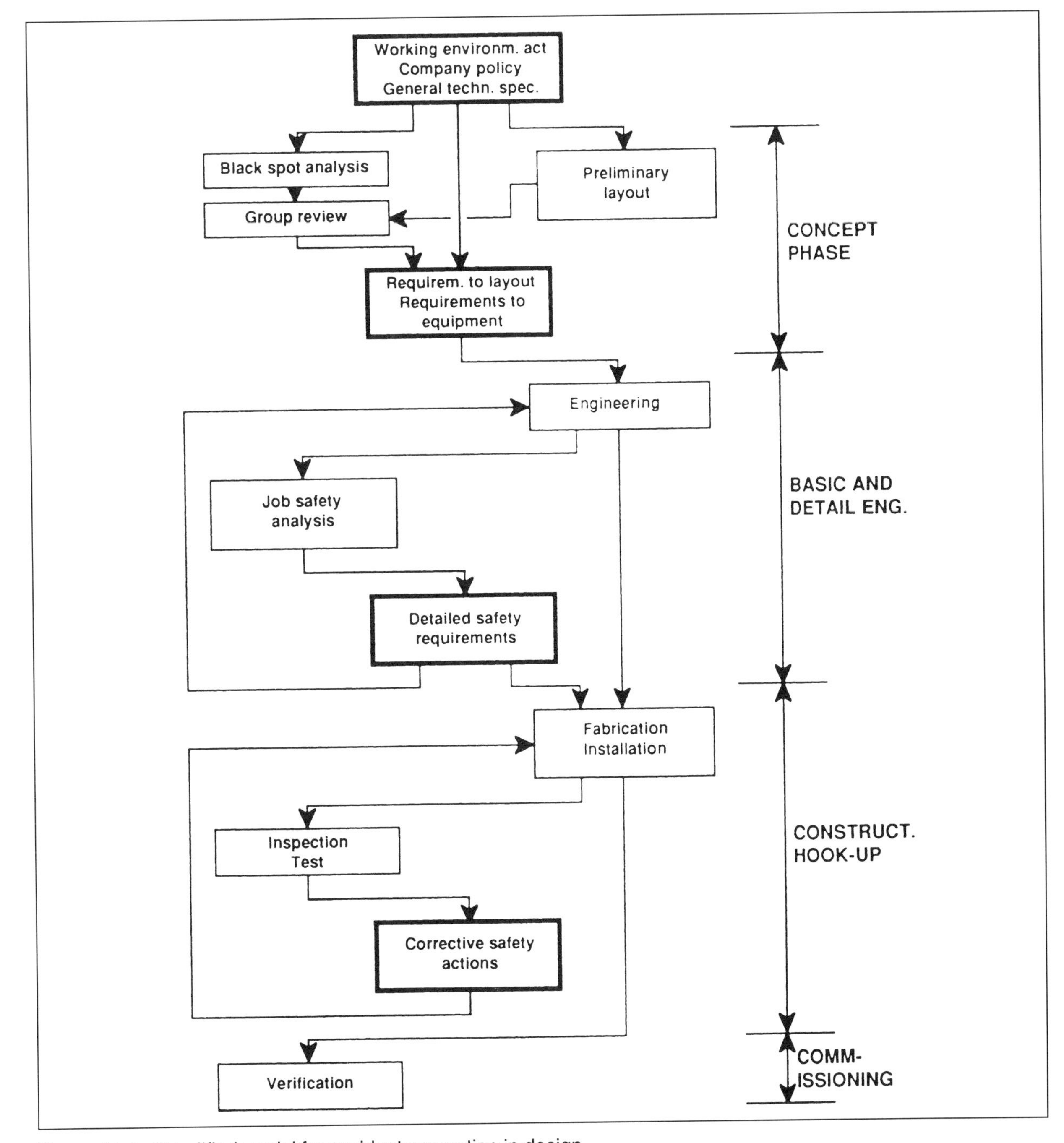

Figure 11-4. Simplified model for accident prevention in design.
Source: Kjellén, U. Safety control in design: experiences from an offshore project. *Journal of Occupational Accidents,* Vol. 12, June 1990, p. 58.

concerns can provide valuable information to designers. They can then work to eliminate or minimize these problems.

The information that designers gather from these sources enables them to ask fundamental questions about how to design safety into the machine, equipment, or process. This multidisciplinary approach is being used in the petroleum industry, for example, in the design of offshore drilling and layout equipment (Kjellén, 1990; Figure 11-4). The use of past experience, multidisciplinary evaluations, and group reviews has enabled firms to resolve design problems more quickly and to address safety concerns within budget and schedule constraints.

Working in cooperation with all groups concerned, designers should strive for safe designs to eliminate hazards both in terms of the equipment, machinery, or processes and in terms of the possible actions or interactions of people with these items and processes. Only when elimination, substitution, or engineering controls are not feasible should there be a reliance on barriers, warning systems, training, and personal protective equipment.

In general there is a need for more reliability, more inherent safety features, and more interactive technology that allows the machine to communicate with workers. For a simple example, when a copying machine jams repeatedly, workers may begin to disregard safety cautions and open up the machine to remove the jam without shutting off the power, putting themselves at risk of electrical shock or hand injury to mechanical movement. The designers might ask themselves such questions as:

- Should we redesign the copying mechanism to prevent or reduce the number of jams? Do we know where jams occur and how often?
- Should we design the interior of the machine that is accessible to operators to be free of nip points and thermal/electrical contact?
- Should we put a switch on the doors that automatically shuts off the power when the doors are opened?
- Should we make the machine self-diagnostic so it can tell workers what to do each time a jam occurs?
- Should we design a machine that can unjam itself?
- Should we design a machine that has several of these features?
- Should we continue to blame poor alignment of paper in the feed tray or poor paper quality?

Each of these questions carries its own implications in terms of resources, costs, complexity, and reliability. The place that safety concerns have in the corporate culture is likely to dictate which option is adopted. If safety is largely a matter of lip service, then even though the designer may be instructed to consider safety concerns, he or she will be evaluated on strictly operating performance results. Does the machine run faster, reduce defects, produce more, cost less? The example above can be transferred to the factory floor where employees make decisions daily regarding how they will cope with manufacturing upsets that require their intervention. Often the options are not jet black or snow white but seductive shades of gray! Design can either inhibit or contribute to safety associated with the operation, servicing, or maintenance of the process. The right corporate culture is critical to the role of design in enhancing the control of hazardous energy.

Formula for the Future

In the face of increasing complexity in the workplace, manufacturers and employers must move beyond mere problem solving into creating new methods for the future that incorporate safety in design. Fred A. Manuele, in his book *On the Practice of Safety*, recommends the following model for a policy requiring that safety issues be addressed from the outset of the design process. The model is reproduced below (Manuele, 1993, pp. 61–65):

> **Process Design and Equipment Review**
>
> **PURPOSE**
>
> To provide operations, engineering, and design personnel with guidelines and methods to foresee, evaluate, and control hazards related to occupational safety and health, and the environment when considering new or redesigned equipment and process systems.
>
> **SCOPE AND DEFINITIONS**
>
> This guideline is applicable to all processes, systems, manufacturing equipment, and test fixtures regardless of size or materials used.
>
> These conditions will be necessary for an exemption from design review:
>
> - No hazardous materials are used (as defined by 29 *CFR* 1910.1200);
> - Operating voltage of equipment is <15 volts and the equipment will be used in nonhazardous atmospheres and dry locations.
> - No hazards are present that could cause injury to personnel (for example, overexertion, repetitive motion, error-prone situations, falls, crushing, lacerations, dismemberment, projectiles, visual injury, etc.).
> - Pressures in vessels or equipment are less than 2 psi.
> - No hazardous wastes as defined by 40 *CFR* 26 and 262 are generated.
> - No radioactive materials or sealed source devices are used.
>
> *If other exemptions are desired, they are to be cleared by the safety, health, and environmental professional.*

PHASE I—PRE-CAPITAL REVIEW

This review is to be completed prior to submission of a project request or a request for equipment purchase, in accord with the outlined capital levels. Pre-capital reviews are crucial for planning facilities needs, such as appropriateness of location, power supply, plumbing, exhaust ventilation. Process and project feasibility are determined through this review. A complete "What if" hazard analysis is to accompany that request. Non-capital projects should also be reviewed utilizing these procedures, but a formal "What if" hazard analysis is not required.

PHASE II—INSTALLATION REVIEW

This review requires a considerably more detailed hazards and failure analysis relative to equipment design, production systems, and operating procedures. Detailed information is documented, including equipment operating procedures, a work methods review giving emphasis to ergonomics, control systems, warning and alarm systems, et cetera. A "What if" system of hazard analysis may be used and documented. Other methods of hazards analysis will be applied if the hazards identified cannot be properly evaluated through the "What if" system.

The Project Manager shall be responsible for the establishment of a Hazard Review Committee and for managing its functions.

HAZARD REVIEW COMMITTEE

This committee will conduct all phases of design review for equipment and processes. In addition to the Project Manager, members will include the safety, health, and environmental professional, the facilities engineer, the design engineer, the manufacturing engineer, and others (financial, purchasing) as needed. For particular needs, outside consultants for the equipment design or hazard analysis may be recommended by the safety, health, and environment professional.

WHAT IF" HAZARD ANALYSIS

This method of hazard assessment utilizes a series of questions focused on equipment, processes, materials, and operator capabilities and limitations, including possible operator failures, to determine that the system is designed to a level of acceptable risk. Users of the "What if" method would be identifying the possibility of unwanted energy release or unwanted release of hazardous materials, deriving from the characteristics of facilities, equipment, and materials and from the actions or inactions of people.

For some hazards, a "What if" check list will be inadequate and other hazard analysis methods may be used.

RESPONSIBILITIES

Project Manager

The Project Manager will be responsible for all phases of the design review, from initiation to completion. That includes initiation of the design review, forming the design review committee, compiling and maintaining the required information, distribution of documents, setting meeting schedules and agendas, and preparation of the final design review report. Also, the Project Manager will be responsible for coordination and communication with all outside design, engineering, and hazard analysis consultants.

Department Manager

Department Managers will see that design reviews are completed for capital expenditure or equipment purchase approvals, and previous to placing equipment or processes in operation, as required under "Installation Review."

Signatures of Department Managers shall not be placed on asset documents until they are certain that all design reviews have been properly completed, and that their findings are addressed.

Design Engineer

Whether an employee or a contractor, the design engineer shall provide to the Project Manager and to the Hazard Review Committee documentation, including:

- detailed equipment design drawings
- equipment installation, operation, preventive maintenance, test instructions, etc.
- details of and documentation for codes and design specifications
- requirements and information needed to establish regulatory permitting and/or registrations.

For all of the foregoing, information shall clearly establish that the required consideration has been given to safety, health, and environmental matters.

Safety, Health, and Environmental Professional

Serving as a Hazard Review Committee member, the safety, health, and environmental professional will assist in identifying and evaluating hazards in the design process and provide counsel as to their avoidance, elimination, mitigation, or control. Special training programs for the review committee may be recommended by the safety, health, and environmental professional. Also, consultants may be recommended who would complete hazards analyses, other than for the "What if" system.

ADMINISTRATIVE PROCEDURES
In this section, the administrative procedures would be set forth such as the amount of time prior to submission of a capital expenditure or equipment purchase request to be allowed the Hazard Review Committee for its work, information distribution requirements, assuring that the dates for Installation Review meetings are planned in advance, assuring that findings of hazards analyses are addressed, and resolving differences of opinion of Hazard Review Committee members.

This model can serve as a starting point for companies and suppliers who understand the importance of designing safety into machines, equipment, and processes. Unless safety issues are addressed from the beginning, it is doubtful whether industry will be able to improve safety records significantly. This fact will become more true as the pace of technological change continues to accelerate in the coming years.

MANAGING CHANGE

How to manage change has become one of the key organizational issues of the decade and an essential element in total quality management. In *The Third Wave*, Alvin Toffler (1980, p. 10) has described the magnitude of the turmoil and opportunities that are now confronting people and organizations worldwide:

> Humanity faces a quantum leap forward. . . . Until now the human race has undergone two great waves of change. . . . The First Wave—the agricultural revolution—took thousands of years to play itself out. The Second Wave—the rise of industrial civilization—took a mere three hundred years. . . .
>
> The Third Wave brings with it a genuinely new way of life based on diversified, renewable energy sources; on methods of production that make most factory assembly lines obsolete; on new, non-nuclear families; on a novel institution that might called the "electronic cottage;" and on radically changed schools and corporations of the future.

Management, government, and other leaders are struggling to make the transition from the Industrial Age to Post–Industrial Age. As they develop these "new corporations of the future," they also must ensure the health and safety of their workers, often in the face of new, rapidly changing technologies. How well they achieve this goal depends on how well they understand the nature of change and how well they learn to manage the many transitions along the way. Organizations must strive to build a culture in which unsafe behavior, equipment, and processes are not tolerated (Petersen, 1993).

The Dynamics of Change

When Heraclitus, a philosopher of ancient Greece, stated, "Nothing is constant but change," he could not have foreseen the rate of change taking place in today's world. In the early 1970s, Alvin Toffler's (1970) ground-breaking book, *Future Shock*, described the cumulative effects of such rapid change. At first, people and institutions adjust to the initial disruptions and new ways of doing things. As the rate and variety of change increase, however, people and institutions become less able to anticipate and respond flexibly. Eventually their coping mechanisms become totally overwhelmed and their resistance to change becomes active hostility. When the inevitable forces of change encounter this resistance, the clash often precipitates personal and institutional crises. Thus, the need to learn how to manage the transitions that change brings is critical to companies' survival. Today's "learning corporations" must make anticipating and responding to change part of their strategic policies (Bridges, 1991).

Change has more than one component. It can be *linear* (or directional) or *exponential* (Johnson, 1980). Linear change refers to changes that occur along a predictable, slowly rising curve. They are not likely to change back to the former state. Steady improvements in established technology, standard equipment, or routine work processes are examples of linear change.

Exponential change means that changes interact to compound the effects on the environment. For example, the explosive growth in industries such as telecommunications, advanced materials, and biotechnology; changes in the makeup of the work force; dramatic shakeouts in many industries; introduction of new energy sources; and the restructuring of the nature of work all interact to create a world where people feel overwhelmed and where safety measures are often compromised or given low priority.

Unfortunately, the human capacity to deal with the rapid rate of change has not kept pace. It is primarily through the aid of technology (computers, telecommunications, interactive video training, etc.) that human performance is improving. Problem solving and hazard analysis technology and tools are being developed to help management and employees respond to changes either from within the company or from the outside.

Paradigm for Disaster

Too many organizations, instead of creating a climate for change and learning, set up various paradigms for disaster by introducing change in a manner that maximizes risk. These approaches often have disastrous effects not only on the transition process but also on safety activities. Some of the more common paradigms for disaster include the following:

- Change is introduced as an accomplished fact, and workers are simply forced to deal with it. This is almost a guarantee for a poor performance. Not only are workers forced to learn new systems, skills, or processes without preparation, but they are distracted by their resentment, fear, and anxiety. Change may be forced through quickly but at a high cost in worker safety, goodwill, and productivity.
- Change is introduced in a haphazard, contradictory manner. Maintenance hears one story, shipping hears another, and manufacturing another. Rumors fly and productivity and safety drop. Conflicting messages about the type and nature of change throw operations into confusion. During the transition, training is provided to some workers but not to others, and safety is often an afterthought, usually attended to after someone has been injured or machinery has been damaged.
- Management pays lip service to involving workers but largely ignores their feedback and suggestions. This approach may be worse than excluding workers altogether; at least in that case management is consistent in ignoring their employees. In the former case, management is saying one thing but doing another. There are few quicker ways to destroy trust that giving mixed messages. Workers, in retaliation, may actually sabotage the changes.
- Changes introduced, although making business sense, are incompatible with the capacity of employees to perform the work. For example, management installs a faster conveyor system that moves products along the line at 30 units per minute. However, even the best employees can perform their tasks at only 20 units per minute. Within a short time, several things are likely to happen: Assembly defects will rise, complaints from customers will increase, and employees may be more susceptible to stress-related injuries and illnesses trying to keep up with the faster pace.
- Changes introduced result in employees working in a manner that puts them at more risk. For example, one company installed a new pallet-feeding mechanism that was to automate a task formerly done by workers manually. Unfortunately, the feed unit kept jamming. The station operators were then required to use various techniques such as prying the material out with a piece of pipe, kicking the pallet with their feet, and climbing into a risky position to use their hands to dislodge the wedged pallet. The company now had a semiautomated task, and several workers experienced injuries. Both productivity and safety were compromised while the debugging process proceeded.

These paradigms for disaster are all too likely to occur in many organizations unless the corporate culture is such that management has learned how to manage change. They must do so through planning safety into design and engineering, involving workers in the process, developing a plan to implement and monitor change, and following up to ensure a successful transition.

Managing Change

Many of the problems associated with change result from people's fear of the unknown and management's lack of knowledge about how to handle the transitions from the old to the new. Although there have been many methods proposed to help organizations manage transitions, they boil down to a few time-worn principles (Bridges, 1991):

1. Develop a strategy for change that involves all workers and includes a time for ending, a neutral zone of partial transition to the new way, and the new beginning.
2. Acknowledge and plan for the sense of loss, grieving, and resistance people will experience when they must let go of the old systems.
3. Be aware that a "neutral zone" occurs in which the old ways are discontinued but the new systems have not yet been fully implemented. This time is likely to be chaotic and difficult, but it is a normal and inevitable part of any change. It is during this time that management must communicate over and over the vision of the new system or organization that workers are helping to bring into existence. At this point, safety issues can be part of that vision.
4. Launch the new system after people in the organization have been properly prepared and briefed. Management can reinforce acceptance of the new beginning by adhering to four general rules:
   - Be consistent: As far as possible, tell workers the same story about what is going to change, how and why it will change, who will be affected, and how it will be introduced. Conflicting messages can destroy trust and cooperation.
   - Ensure some type of quick success. Try to ensure that some part of the initial changes are implemented successfully right away. The success may be as simple as installing new lighting or creating a new message-routing system. If people have small successes to build on, the larger changes tend to go more smoothly.
   - Symbolize the new order. Hold a companywide meeting or some other public form of recognition of the new system, identity, process, or organization. This is a good time to recognize the efforts of all concerned in making the change possible.
   - Celebrate the success. This should be done soon

after the new order is in place and running fairly smoothly.

This model of "ending, neutral zone, beginning" can be used to manage the nonstop change that is hitting many organizations. Regardless of the type of changes that a company or institution may face, industry experts are clear on one fact: A new world of work requires new rules and new survival techniques (Gilbreath, 1991). Safety professionals, like other managers and staff, must learn to anticipate and respond to change instead of merely reacting, as discussed in the following section.

Effects of Change

With regard to safety, change can produce a host of problems as side effects of either planned or unplanned alterations. Workers no sooner get used to the new method of stacking raw materials when their supervisor arrives to tell them they will be given new equipment for hauling multiple pallets or that their manual control presses are being replaced by computerized versions. Even though change is unsettling for most of us, this scenario is being repeated in thousands of workplaces every day. Change is pervasive and has exponential effects over time, particularly in conjunction with energy. In fact, it may be said that when an incident occurs on a familiar or new system, it is because something has changed. This truism has value in four common management situations (Johnson, 1980):

- The cause of a safety problem is obscure. Analytic methods that take into account something in the system has changed can identify the true cause more quickly.
- A planned change, intended to improve conditions, has unwanted and unrealized side effects. The effects of change in this instance were inadequately analyzed when the change was in the planning stages.
- An unplanned change is not identified. Changes can provoke more changes that go unrecognized. For example, an off specification metal supply may cause more press jamming and lead frustrated workers to circumvent de-energizing procedures to keep up their production performance.
- An unplanned change is identified but its significance is not recognized. The supervisors may realize that the metal is off specification but fail to appreciate its impact on press lockout practice for machine jams. As a result, the change is not analyzed for effects, nor are the necessary counter-changes made.

Change-based analysis. Certain methods can help organizations conduct effective analyses of incidents to determine how and what type of change contributed to the problem. However, the more pressing need may be for preventive, before-the-fact analyses to decrease operating and safety problems due to change. Organizations need to:

- Determine the significance of the changes causing trouble, starting with top management actions and statements. Middle managers and supervisors can be trained to understand the potential for changes to produce problems.
- Develop an analytic format for analysis of change and the potential problems that may arise. Figure 11-5 presents a sample potential problem analysis worksheet for anticipating the effects change may have on safety concerns. This worksheet can be used at the beginning of any new project and expanded as the project continues. This low-cost approach to change analysis is unusually effective in focusing attention on the causes of future problems and on the necessary countermeasures to ensure safer operations.

The principle behind this type of analysis is that whenever hardware, processes, or people are changed, the changes should be reviewed up and down the system. In this way, it is possible to evaluate what impact the changes are likely to have on more remote sites and tasks. For example, if a change is made in a conveyor system, how will that affect the activities leading up to and away from the adjusted section? How might it affect shipping or other more remotely positioned tasks?

Problems also occur *after* change is introduced or an unplanned change occurs. When such an incident happens, there is a six-step MORT safety system process to analyze (Figure 11-6):

1. Evaluate the incident situation.
2. Evaluate comparable incident-free situations.
3. Compare the incident situation with the incident-free situation.
4. Set down *all* known differences no matter how unrelated they may seem.
5. Analyze the differences for their effect on producing the incident, for example, a change in written instructions on how to operate a machine, a change in color coding, or a change in materials.
6. Integrate this information into the incident investigation report.

Sensitivity to change. One of the best preventive mechanisms for anticipating the negative effects of change is to train middle managers and supervisors to be sensitive to changes in their operations and to report them. The sensitivity is a key component in a good manager, who may have an almost intuitive feel for when something has or is about to change. In some cases it may be a difference in the way a machine sounds or in the way workers are behaving that signals an impending problem.

CHANGE-BASED POTENTIAL PROBLEM ANALYSIS WORKSHEET

Specify Problem ____________________

Factors	Present	Prior Comparable	Differences Distinctions	Affecting Changes	Counter Changes

Figure 11-5. The change-based potential problem analysis worksheet shows a preventive counterchange column added to the problem worksheet. Specify the changes in a project as compared with recent conditions or comparable projects.
Source: Johnson WG. *MORT Safety Assurance Systems.* New York: Marcel Dekker, 1980, p. 65. Reprinted by courtesy of Marcel Dekker, Inc.

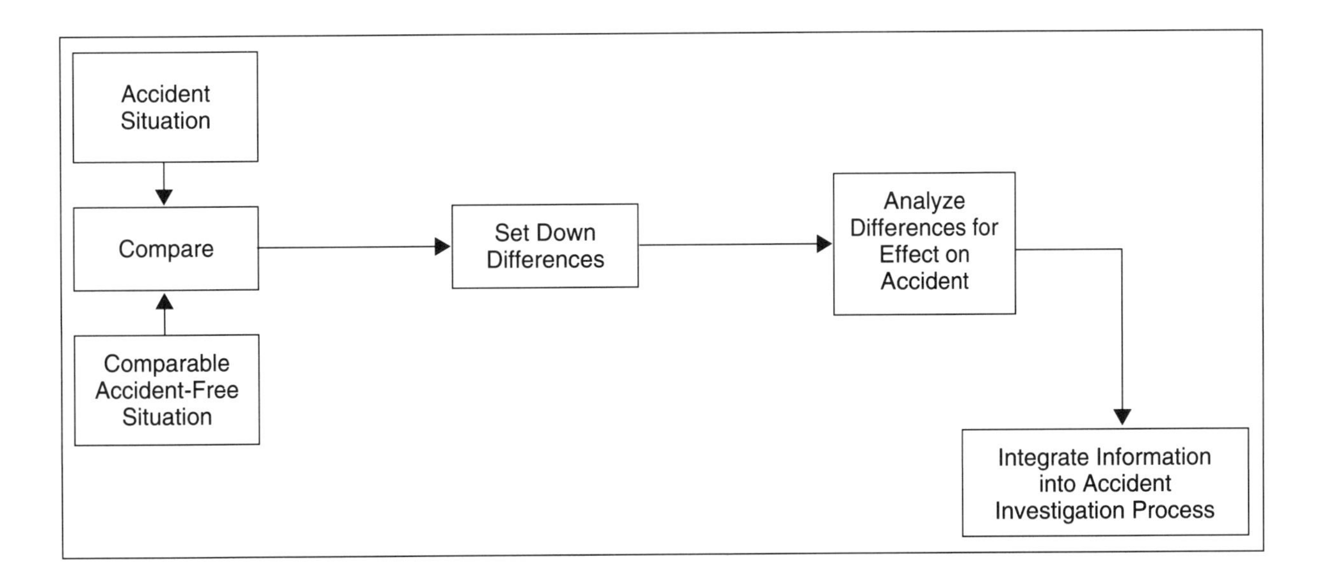

Figure 11-6. The six steps in change analysis.
Source: Johnson WG. *MORT Safety Assurance Systems.* New York: Marcel Dekker, 1980, p. 59. Reprinted by courtesy of Marcel Dekker, Inc.

Organizations may establish safety committees who help train management and workers in this skill, which includes the effort to distinguish between insignificant and significant changes.

Many managers claim they and their workers do not need training in this area because accidents have sensitized them. Becoming sensitive to change after several incidents have occurred, however, is not an effective or low-cost method of training.

Types of changes and counterchanges. Johnson (1980), in his book on MORT safety systems, describes the types of changes and counterchanges that managers and workers in an organization should be aware of. These include the following:

1. Planned versus unplanned change
   a. Planned: Require scaled hazard analysis process (HAP) and affirmative safety action
   b. Unplanned: First, detect by monitoring. When detected, make immediate correction when necessary and require scaled HAP
2. Actual versus potential or possible change
   a. Actual change is detected by reports and observations
   b. Potential or possible change requires analysis
3. Time: Deterioration of a process over time, interaction with other changes
4. Technological: The new projects and processes, particularly near technological boundaries
5. Personal: The many variables that affect performance
6. Sociological: Closely related to personal changes but of broader significance
7. Organizational: Shifts in unit responsibilities may leave interface gaps, particularly when hazard analysis was ill-defined but done by custom by some people
8. Operational: Changes in procedures without safety review
9. Macro versus micro change
   a. Macro: Overall organization data, for example, new employees, transfers, and other operating data suggesting need for preventive counterchanges
   b. Micro: Particular events. A useful subdivision of micro-events might be early detection and counteraction, for example, a plan to promote a supervisor and then to promote his or her assistant. What does this change imply?

When problems or incidents are caused by one or more of these types of change, counterchanges or measures should be taken. These include the following:

- Modify color, shape, sound, odor, motion, meaning, and light.
- Rearrange sequence, pace, components, schedule, and pattern.
- Reverse order and direction.
- Reduce, omit, shorten, split, and condense.
- Substitute ingredients, power, process, and approach.
- Combine units, assortments, ensembles, and blends.

Resistance and Adaptability

Perhaps no element is more important than the human factor in managing change. Although new machinery may be installed, it must be operated by the same people who handled the old machinery. In addition, machines are becoming more complex and "smart," whereas workers are less able physically and mentally to keep up with their speed and functions. Finally, even though the work force may be more educated, it is also more diverse with a multitude of languages, cultures, disabilities, and skill levels found in the same work environment. This diversity complicates communication, jeopardizes training, and raises the risks on the job.

The question is, Can an organization introduce change in a way that reduces the risks people encounter? In the midst of this process, how do they make learners out of their workers and improve human behavior?

There are several critical components that can help reduce resistance and increase acceptance of changes. At a minimum, management must do the following:

- Clarify and communicate the reason(s) for the change. Why is it being done? What makes it necessary? What was wrong or inadequate about the old way? People need to know the purpose of the change rather than simply being ordered to accept it.
- Explain how the change will work. What will change? What is the expected outcome? How will workers' jobs change? How long will it take? What will the impact be on safety procedures and lockout/tagout rules?
- Develop a plan to implement the change through all phases: initial design, installation, follow-up and monitoring. Management must know at each stage how workers are adapting to or resisting change. Are training sessions held addressing the fears and complaints employees have, as well as teaching new skills? After the systems are in place, are people going back to the old way of doing things? Are they refusing to use new equipment, new safety mechanisms, etc.? Are their complaints normal "griping," or are they more significant complaints about defects or flaws in the new systems, machinery, or procedures?
- Involve line managers, supervisors, and workers in the process as early as possible. Everyone concerned must have some ownership of the change process to reduce resistance. Have employees and line managers been polled regarding their suggestions or concerns for changes in the design or engineering stages? Do workers feel included in the change process or the victims of it? To be successful and safe, the new machinery, processes, etc. must reflect the way people actually work and the shortcuts or unusual actions they are likely to take.

The importance of these four factors cannot be overstated in terms of ensuring a relatively smooth transition from one system to another. Otherwise, companies risk an unnecessarily high incidence of adverse events during and shortly after the transition.

Information Systems: Curse or Cure?

Safety information systems are designed to supply the necessary information in usable form to those who need it, where and when they need it. During times of change, however, these systems can either help or hinder the transition. When an information system is well integrated and provides accurate, timely information about the number and type of incidents, relevant causes, safety measures taken, and etc., the data can be used to guide future safety efforts and improve design.

In many organizations, however, the use of information technology in terms of electronic data bases, interconnected systems, and currently available software generally is fair to poor (Benjamin & Levinson, 1993). Too often safety information, instead of being readily available from one source, is scattered in many subsystems (Johnson, 1980). For example, although accident circumstance data are kept, data regarding the causes of accidents may be subjective and of questionable value. Reports may be difficult to retrieve from computer systems, and basic records of inspection, monitoring, and audit systems are not usually well organized. The system may be partly manual and partly computerized, which makes retrieving information in a timely manner difficult if not impossible. In addition, the system may focus on gathering quantitative data (number of hours of machine downtime, number of dollars spent on workers' compensation, number of incidents reported) and neglect qualitative data (analysis of causes, workers' input on safety issues, preventive measures).

In terms of lockout/tagout issues, companies may have an overload of information in areas such as production downtime or quality assurance figures and a lack of information in other areas, for example:

- *why* machines experience downtime (cause specific)
- what specifically workers must do when equipment needs to be serviced for various reasons
- historical data regarding past interventions to service a process and the safety problems encountered by workers.

Thus, when an organization is about to introduce a new technology, new equipment, and new processes, too often it must rely on safety information systems that provide an incomplete picture of actual working conditions and the risks that workers face on the job.

Likewise, safety information systems that train workers in lockout/tagout procedures can be equally deficient. Where they exist, these systems typically consist of crude, printed materials. Only a few organizations are using electronic media to educate and train workers in lockout/tagout safety. Given the fact that today's work force is likely to be multilingual and highly varied in educational background, it is risky at best for an organization to rely solely on the printed word to communicate with workers. In addition, the system must be updated constantly to reflect current conditions and standards. It is easy for new printed materials to become confused with the old or for workers to miss key points where they must change a previous procedure or behavior.

Thus, part of managing change in any organization should be a thorough evaluation and overhaul of the information technology and safety information systems in place. This includes training, inspection, auditing, incident reporting, analysis, feedback, and other safety-related activities. The information systems should take advantage of the sophisticated electronic media available to even the smallest firms. They should interface with other manufacturing and process systems that exist in the organization. Companies also can create a systematic process of upgrading information technology as it proves of value to a company's systems. This process would be part of the organization's continuous improvement process.

In this manner, safety information systems can become a clearinghouse mechanism for reporting and storing data on safety-related issues, gathering feedback on lockout/tagout tasks where flaws were detected, problems encountered, innovative ideas generated, new techniques conceived, etc. This information could then be used to develop training programs delivered through interactive media such as video/computer programs, graphics, and animated presentations. Other data on incidents could be used in new designs of equipment or processes, with graphic displays illustrating safety problems and design flaws of current machines or processes. Such an information system, illustrated in Figure 11-7, would permit much more rapid adjustments to change and would represent actual conditions on the job. Data bases tend to make knowledge available for all parties, equalizing access to information for workers from management to operators (Benjamin & Levinson, 1993). Employees would be involved in direct feedback on their jobs and would tend to have more investment in the safety procedures developed as a result.

CONTINUOUS IMPROVEMENT

Steady improvement in safety areas, particularly energy isolation, requires what Joel Barker (1991) in *Future Edge* calls the three keys: anticipation, innovation, and excellence. Companies must not simply adapt to change but begin to change themselves to meet the challenges of the twenty-first century and the new competitive environment. Safety must be part of that reengineering effort, or unacceptable incidence rates

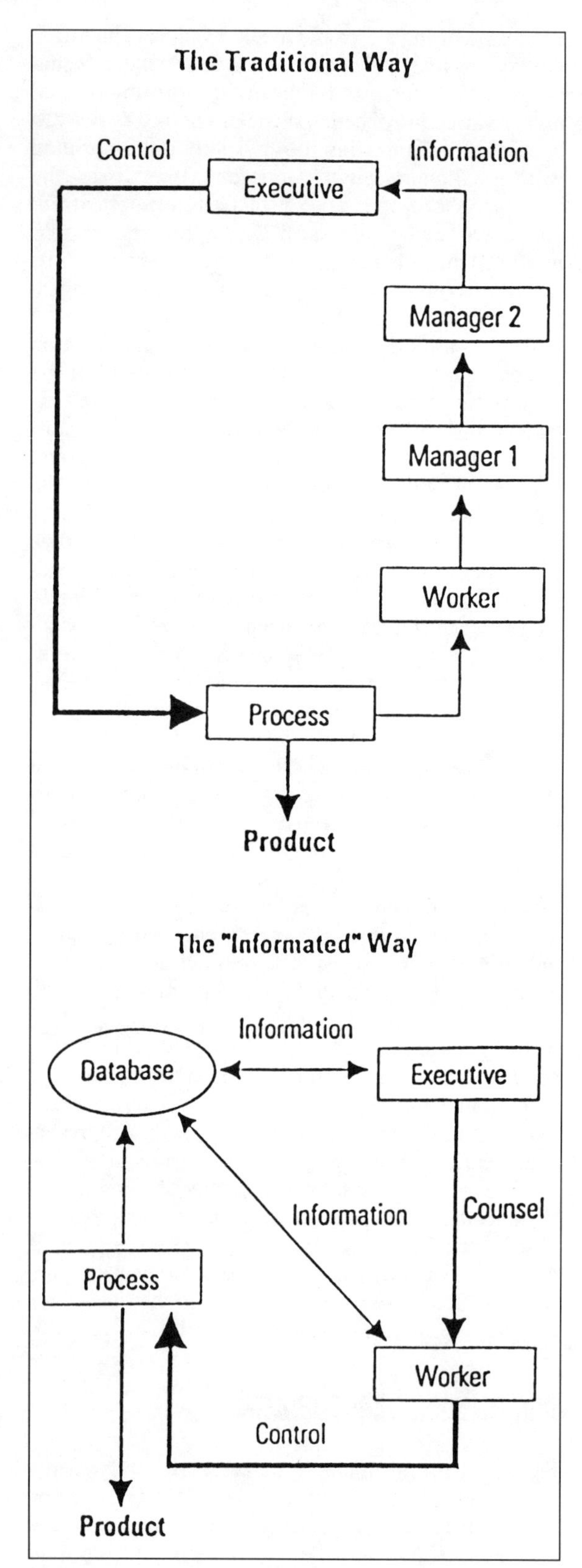

Figure 11-7. Control systems: old and new.
Source: Benjamin B., Morton MS. *Personal Computers and Intelligent Systems: Information Processing,* Vogt F, editor, Vol 3. Amsterdam: Elsevier, 1992, p. 141.

will inevitably be part of a company's liabilities. This section will explore methods organizations can adopt to raise hazardous-energy-control efforts to new levels.

Current Attitudes Toward Safety Improvement

For the most part, organizations tend to address hazardous energy control in a fragmented way. Although they analyze and document the operational performance of a machine or process, they tend not to think about the complex working environment in which that machine or process is used. Issues such as ergonomics, how operators and maintenance workers interact with the machine or process, environmental conditions, energy isolation in abnormal or unusual circumstances, and interaction of the machine or process with other systems are largely ignored until an incident happens.

Yet many safety professionals believe that industry will not make much progress on energy-isolation efforts unless organizations adopt more progressive methods. They need to take a look at their operations in terms of four areas (see Figure 6-1):

- *Human/machine interaction.* How is the machine really used versus how the designer or engineer envisioned its use? Machines and workers do not operate in isolation. Organizations need to conduct more studies documenting how their operators really work and how the machines perform under actual working conditions (normal and abnormal).
- *Environment.* This includes noise, temperature, lighting, spatial arrangement, and weather, as well as local community conditions (homogenous or mixed population, range of skills available, languages spoken, etc.). For example, a company may expect their machines and employees to work the same in their Georgia plant, where it is hot and humid throughout the summer, as they do in the Oregon plant, where it is cooler. The inside temperature of the plant in Georgia may regularly reach 120 F (49 C), whereas the plant in Oregon rarely goes above 80 F (28 C). Fatigue, discomfort, and irritability are likely to exist among workers in the Georgia plant, and workers may be inclined to take shortcuts around safety procedures to minimize the time spent working under demanding or arduous conditions. Organizations need to understand how workers are likely to behave under different environmental conditions and what impact their behavior may have on safety.
- *Management practices and procedures.* Management policies for lockout/tagout procedures are generally written for normal conditions. However, when repairs or maintenance must be done under abnormal conditions, management may have an unstated expectation that workers will do whatever is necessary and still work safely. Unfortunately, if a worker can free a stuck valve only by sliding under a machine and bracing his feet against the frame-

work, even a careful worker can be at high risk. If management is not receptive to employee complaints or suggestions about specific work procedures, real improvement is unlikely.

- *Design and engineering.* Management needs to consult more with the operators, maintenance workers, and supervisors when either designing new machines, equipment, or processes or changing the old ones. Too often management may be threatened by the idea of including workers in the design and engineering stage, fearing they will delay the work or complicate it unnecessarily. Generally, however, the opposite is true. Few people know a machine or process better than the people who work on it or take care of it. They can provide invaluable feedback when it comes time to upgrade the machine or process or to design a new one that can reduce worker risks and increase the safety margin.

In brief, organizations need to pay attention to MEP, procedures, and human resources when integrating safety into their operations. They need to consider all of these factors early in the process of design/engineering and change. They also need to analyze what aspects of their old ways of doing business can be circumvented by workers. For example, if it takes more time to lock out a machine than it does to repair it, maintenance workers are likely to skip the lockout procedure and simply fix the machine with the primary power still on. Perhaps the new machine needs a streamlined lockout procedure or a simpler way to isolate power for maintenance and repairs.

Because change is so much a feature of today's business world, management knows when it is coming and should be able to establish routine methods for evaluating a broader range of factors that need to be considered. This means documenting not only machine performance but also environmental conditions, employee work methods, the number and type of abnormal circumstances requiring energy isolation, etc. If a major renovation or modification is due, the organization has a chance to be truly innovative and creative in using analysis and improvement methods that involve a broader range of employees. If the renovations or changes are more modest, the organization still has a chance to alter beneficially their "business as usual" methods.

More organizations are beginning to realize that to make real progress on energy-isolation efforts, they are going to have to involve workers much more in the process than has been true in the past. Those companies that have already made this move are seeing positive results in terms of higher productivity and a better safety record.

Employee Empowerment

The Japanese made employee involvement famous through the use of "quality circles" (Schmidt & Finnigan, 1992). In these circles, management listened while the workers talked about their work problems, suggestions, and observations. These comments were then incorporated into company policies, design changes, work methods, traffic flow, and dozens of other applications. Management ran their businesses more effectively, and workers had a stronger commitment to and investment in their jobs.

The need to stay competitive has led more and more U.S. companies to take a look at various involvement initiatives that have evolved into employee empowerment, or EE. In addition, today's workers are better educated and want more individual respect and recognition for their efforts. Factors that make EE work include the following (Kaiser, 1984):

- Employees have unique sources of expertise about their jobs.
- This expertise is not being tapped effectively.
- Employees want to and can contribute.
- Employees want to have more say about what affects them.
- Better management decisions are made with employee input.

In this area, corporate culture is a crucial factor. The organizational barriers to EE may be long standing and should not be underestimated. They include the following:

- lack of management commitment to the concept
- management resistance to giving workers any real voice
- inadequate planning and preparation
- unrealistic or inappropriate objectives and time frame
- insufficient union involvement in the process
- desire to have a quick fix to problems
- inadequate management and employee training in the basics of EE
- lack of attention to the planned and unplanned consequences of EE programs and meetings.

Chapter 6, The Process Approach, discusses the importance of EE in the success of hazardous-energy-control systems, illustrating the concept in the employee power pyramid in Figure 6-10. As employees move up the pyramid from abstention to participation and empowerment, they simultaneously increase their level of commitment to their jobs, their work teams, and their organizations.

Dr. Donald Millar (1993, pp. 100–101), former U.S. Assistant Surgeon General and Director of NIOSH, commented on the total quality movement and employee empowerment:

> Respect for people, empowering each individual, continuous improvement, the quest for knowledge and balance, and teamwork—all of these principles are essential to the

quality philosophy. . . . Enlightened management and empowered workers are our best hope for achieving safe and healthful working conditions.

Companies who wish to institute EE and make the fullest use of employees' knowledge and expertise need to know what is required to make employee empowerment successful. Leadership is critical, and top management must be the role models for the rest of the company. Basic requirements for an EE program are as follows (American National Can, 1989):

- *Long-term perspective and effort.* The program may take some time to work effectively.
- *Management and union commitment.* The two must work in partnership to ensure that managers and workers have the same understanding and level of involvement:
  - attitudes and behaviors in line with stated goals
  - willingness to support those involved
  - adequate resources supplied, include budget, time, facilities, and staff
  - systematic evaluation and flexibility to modify and adjust the program
  - realistic expectations; success will require extensive efforts, results will take time, and problems will arise
  - continual maintenance.

In safety terms, the general results of EE can be a routine method of obtaining worker input regarding machine or process operations and problems. Workers can also help management anticipate potential safety problems, take part in developing innovative solutions, and contribute to the company mission of achieving quality or excellence.

Team Building

One form that employee empowerment can take is the establishment of employee teams. The teams can be made up of manufacturing, technical, engineering, or other professional and hourly employees. The objectives of these teams are (General Motors, 1990):

- to increase opportunities for employees to use problem-solving skills
- to create a satisfying and productive work environment
- to improve communication between supervisors/ other managers and employees
- to develop team spirit
- to encourage stronger concerns for safety and quality
- to develop a better understanding of job requirements and safety issues
- to improve employees' self-confidence and leadership abilities.

The teams are generally headed by a coordinator and either hold regular meetings to discuss problems and focus on particular objectives or may be called together for specific projects. If the company is changing from manual to computerized manufacturing systems, for example, or redesigning work flow, an employee-management team may be called together to work on the change. This approach ensures that all viewpoints are heard and that all employees' expertise and knowledge are included in the process.

The commitment and resources needed to build teams within an organization require that top management strongly support the effort and be dedicated to making it work.

Kaizen (Japan)

Another approach to achieve continuous improvement has been developed by Japanese firms through the philosophy of *kaizen.* In Japanese, the term kaizen means "good change" and refers to making small, incremental improvements on a day-by-day basis rather than by introducing major upgrades or technological changes (Peters, 1987).

These small improvements are made in processes and products to increase their quality and effectiveness. The focus is on producing successive generations of better, more efficient, and less costly goods and services. To do so, organizations examine each step in the work process—raw materials, manufacturing, finishing, ordering, shipping, customer service, etc.—for areas where the work flow can be improved, even in small ways. Techniques such as Just-in-Time Inventory or Time-Based Continuous Improvement Process (American National Can, 1989) have helped to streamline manufacturing processes dramatically in many instances. For example, some companies using this approach have been able to reduce their pollution emissions to zero by the relentless pursuit of small improvements over weeks, months, and even years (McInereny & White, 1993).

This approach can be applied to safety in terms of monitoring work and safety equipment and procedures for time-based improvements. For example, a company may set a goal for a 50% reduction in incidents within two years and conduct a detailed examination of its safety training, on-the-job education, and procedures and policies. By making continuous improvements in each phase of its safety program, the organization can gradually achieve its goal. As in other changes, employee involvement is critical, because workers are in the front lines of work procedures and processes and must deal with equipment and machines on a daily basis.

Management may also use the same elements of time-based improvements used to reduce setup time in their efforts to improve safety. For example:

- Establish corrective action teams or safety action committees that involve workers in identifying and correcting safety problems.

- Devise measurements that record actions taken and their results.
- Videotape new procedures or problem areas to demonstrate where safety problems lie and videotape the corrected conditions.
- Question how and why things are done as they are, particularly those things that everyone takes for granted. Why does a worker have to change belts on a die grinder several times in one week? Why do all the sanding machines have to be in one area? When do the machine breakdowns occur the most often and why? Such questions keep everyone thinking about continuous improvements throughout the days, weeks, and years.
- Devise a method to make improvements quickly and efficiently.
- Find ways to eliminate unnecessary, repetitious, risky, or inefficient procedures to make a machine or process more efficient and safer.

Continuous improvement can be a major tool in achieving a new level of safety and in building a safety orientation into the corporate culture of the 1990s and beyond.

FORWARD DIRECTIONS

The future environment in which employees' supervisors, managers, and safety/health/environmental (SHE) professionals will work is likely to be highly automated with sophisticated interconnected, computer-controlled equipment producing goods and running most of a business's systems. This trend will continue well into the twenty-first century.

This fact means there will probably be fewer employees and fewer managers in the work force in many organizations, and their functions will be considerably different from the types of jobs created by the industrial revolution (Gilbreath, 1991). As companies downsize and reengineer themselves, layers of management are disappearing, along with many traditional blue-collar jobs. In all likelihood, these jobs will never return. In other companies where automation may not be so pervasive, the work force still is likely to change dramatically. There will be more women and more minorities, creating a multicultural, multilingual work force that will present special safety challenges to management.

Challenges of the New Workplace

The worker and manager of the future are likely to find their jobs more alike than in the past. In the Industrial Age, workers were hired to perform manual, repetitive jobs that rarely varied. Managers were hired basically to make sure the workers did their jobs and that company production and quality goals were met. Workers and managers came to be regarded almost as natural enemies, and a great deal of management time and energy was expended on how to make employees work harder, faster, and better (Toffler, 1980). Employees often worked equally as hard to find ways around the more restrictive, stifling management rules. Unfortunately, this attitude at times extended to safety matters (Petersen, 1993).

In the Post–Industrial Age, however, workers and managers may perform functions that blend their activities. With new technology, a great many repetitive manual jobs are now automated. The few workers who are left do far less manual labor and far more monitoring of advanced systems. As a result, their education level must be higher, they must have more sophisticated training, and they must have more authority, autonomy, and responsibility for managing their areas. The sum total of their experience, training, and background will be higher than that of workers in the past.

Managers, on the other hand, are likely to do fewer of the traditional management activities of motivation and control. Workers and systems will be virtually self-managing, and computerized systems will handle more of the training and recordkeeping functions. Management will spend more time coordinating systems and resolving problems or glitches in those systems (for example, in problems of raw material supply).

For the whole organization, the old authoritarian style will decline and interactive, collective involvement will rise (Peters, 1988). The demand for quality and performance will be so high that all workers will need to become actively involved in the operations of the organization. Employees will in all likelihood be members of various teams that meet regularly to discuss problems, suggestions, and changes in company operations. Management, for example, may ask line workers to talk with a customer to find out what is needed and how those needs can be met. Management will provide specialized assistance to workers, giving information and feedback and monitoring the health of the system. Managers will spend more time looking at how to enhance the system than at how to get it to work (Bridges, 1991). In general, organizations of the future will use workers and their minds rather than workers and their backs.

Generally this challenging new environment bodes well for lockout/tagout. There will be better equipment and better people to carry out the procedures. However, the future is not without risks of its own (Petersen, 1993):

- First, machine and process design for safety will be a critical element. Because so many of the processes and equipment will be automated, there must be built-in safeguards and procedures for isolating energy. For example, a worker may be able to isolate a machine by remote control or through computer code or computer-controlled sequence. In some instances, "smart" machines will inform the worker what is wrong, diagnose the problem, and self-correct or take steps to isolate energy so the problem can be fixed. They may also generate and store their own electronic library of faults, interruptions, etc. for future analysis. Systems will need to have feedback capabilities to tell workers when a procedure has been successful or when a problem remains. New designs will also tend to minimize risk by eliminating poorly placed valves, switches, and controls and by ensuring the safer conduction of energy.

- Second, with fewer people on the floor of a plant, the lockout/tagout responsibilities will be placed more squarely on each worker's shoulders. In situations that do not exactly fit established lockout/tagout procedures, employees will have to decide on their own what is the correct method using appropriate principles. They will have to resist the temptation to take shortcuts and should have safe behavior regularly reinforced.
- The life span of factories, machinery, and equipment will be shorter, requiring workers to continually update their knowledge of safety procedures. Global competition will keep the rate of change at a high level for some time to come, which means rapid turnovers in technology and even whole industries. Workers of the future will not talk about one career in a lifetime but perhaps of two or three careers.
- The new multicultural work force will need higher levels of skills if their companies are to remain competitive, which means greater use of computer-assisted instruction and training. Programs will offer training in a variety of formats and languages, along with skills tests that can be checked by supervisors.
- Electronic communications will also ensure that different operations are connected to one another, giving workers in one area instant, real-time information on processes in other areas. This technique will facilitate problem solving and coordination of activities during times of transition, maintenance, or repair.

Increasing the effectiveness of the hazardous-energy-control process should be a priority in every manufacturing organization. A "six I's strategy" is proposed as the foundation for elevating present preventive activities to a state of excellence. It is premised on the existence or development of an organizational culture that values safety.

Six I's Strategy

The six I's strategy—involvement, inquiry, innovation, intervention, intensity, and improvement—is designed as an action program to take a company beyond mere safety compliance. Practically speaking, there is no way to regulate these factors into company policy; they have more to do with the culture and character of an organization's management. However, these six factors can be measured if management desires or monitored through surveys that reveal workers' perceptions of how the six I's are being implemented and the impact they are having on the workplace. Basically the six I's strategy is a test of a company's commitment to deeds, not words.

Involvement. Involvement in this context is more than simply participating in a safety program but a real ability to influence beneficial change. This is a planned rather than spontaneous activity and employs vertical and horizontal organizational engagement. For example, management may deliberately assign safety tasks or projects to groups that are usually overlooked in the general safety effort: accounting, quality control, purchasing, etc.

Inquiry. This activity involves the investigation of near misses, infractions, machinery limitations, incidents, and tasks where zero energy is not practicable. The organization also conducts research into other company efforts, new devices and training materials/technology, incidence experience, attitude sampling, and safety auditing. The basic philosophy behind this factor of the six I's is that what you don't know *can* hurt you. Any suspected safety problem areas should be thoroughly inspected on a routine basis. The objective is to minimize or eliminate surprises through proactive initiatives.

Innovation. "Imagineering" rather than simple engineering means comparing the current reality with what can be. To achieve results, the organization can foster and encourage the generation of ideas for safety programs that will inspire a quantum leap in safety rather than merely incremental changes. The organization must be willing to invest the time required to make real innovation happen. Hazard control creativity can flourish only in a climate where it is acceptable to challenge the status quo. An important aspect of this factor is for management to promote and acknowledge workers' contributions.

Intervention. The emphasis is on action—if something is wrong . . . act to correct it! Management can take the lead to allow employees the freedom to self-correct their mistakes. To make this method work, management must adopt a nonpunitive style in which mistakes are acknowledged and used as learning experiences. There must be clear definitions for what are right or correct procedures or actions and what are not. It is well understood that uncorrected unsafe behavior serves to negatively condition employees over time and sets the stage for future mishaps. The behavioral improvement process (BIP) is based on these principles and works well given the right management attitude and administration.

Intensity. This factor has to do with how intensely an organization wants to achieve improved levels of safety performance and the consistency of that desire over time. In this regard, a little fanaticism goes a long way. Management must take the lead and show by example that the safety issue is a top priority not just for a week or month but for the long haul. Such consistency of effort helps to build credibility among the work force and can plant the roots of safety orientation deep in an organization's culture.

Improvement. Improvement must be seen not as a goal or an event but as an ongoing, continuous process that everyone is responsible for developing. In today's business environment, no system can exist for even a few months without being changed either through upgrades or technological adjustments. As a result, to build in a system of continuous improvement, an organization should work toward the following:

- enhancements that come from worker suggestions, ideas, or innovations, that is, "worker-driven" improvements
- connection of improvements with "quality processes," so workers see a direct relationship between safety and quality
- specific measures of progress, for example, fewer failures, incidents, flaws, deficiencies, and complaints
- third-party assessment of improvements over time, that is, where the organization was in the past versus where it is now in terms of safety and quality.

These six factors will prove valuable in making the lockout/tagout process a more effective part of an organization's drive to raise their safety efforts to the next level in the coming post–industrial age.

SUMMARY

- After more than 50 years of efforts to control hazardous-energy release, workers are still being seriously injured. Government agencies, employers, workers, and associations need to ask what more can be done beyond mere compliance to reach the next level in safety. Prevention of "energy-release incidents" will require an unprecedented commitment by manager and employer.
- Three critical areas have been neglected and need to be addressed regarding existing lockout/tagout programs: designing for safety, managing change, and continuous improvement of process and system. The post–industrial revolution is likely to bring profound changes in work life and an increasingly complex technological environment in which to manage safety.
- For improvements in safety to become a reality, an organization's corporate culture must place safety high on the list of priorities. In such companies, management must require accountability in SHE affairs; create a strong SHE organization and staff; establish effective information systems; use safety in design; establish strong safety training programs; ensure employee involvement; and effectively design, implement, and update SHE programs.
- Design-in safety procedures are one of the most effective and economical ways to prevent or minimize anticipated hazards when new or updated equipment is to be installed. Concurrent engineering can be used to develop a systematic method for designers to anticipate, evaluate, and eliminate or control hazards and risks before design and engineering work begins.
- Management today must address safety issues within a climate of continual change. The rapid development of new technologies and sources of power mean equipment and machines in the workplace will be constantly replaced. Human capacities have not kept pace with these developments. Companies must develop methods and information systems for analyzing how this continual change contributes to safety problems on the job and how to involve workers in addressing these problems.
- Continuous improvement in safety areas, particularly energy isolation, requires anticipation, innovation, and excellence. Safety must be part of companies reengineering efforts or unacceptable incidence rates will be part of a company's liabilities. Organizations need to look at safety in human/machine interactions, environment, management practices and procedures, and design and engineering.
- EE will be a significant factor in helping companies to improve their safety record and stay competitive in the post–industrial era. Employees and employee teams can provide expertise and knowledge that can help managers make better decisions. Kaizen, or incremental improvements, is another approach to achieve continuous improvement in safety.
- The future workplace environment in which supervisors, managers, and SHE professionals work is likely to be highly automated with sophisticated interconnected, computer-controlled equipment producing goods and running systems. As a result, lockout/tagout issues may be more difficult to address and require a highly trained work force to implement.
- The main issues to be addressed involve increasing (1) machine and process design for safety, (2) workers' knowledge and skill in lockout/tagout procedures, (3) management of rapid turnover in technology and systems, (4) multicultural training, and (5) more sophisticated and effective information and communication systems.
- The six I's strategy—involvement, inquiry, innovation, intervention, intensity, and improvement—is designed as an action program to take a company beyond mere safety compliance. It is premised, however, on the existence or development of an organizational culture that values safety. Increasing the effectiveness of hazardous energy control should be a top priority in every organization in this post–industrial age.

REFERENCES

American National Can. *Employee Participation Teams*. Chicago: American National Can, 1989.

Barker JA. *Future Edge: Discovering the New Rules of Success*. New York: Morrow, 1992.

Benjamin RI, Levinson E. A framework for managing IT-enabled change. *Sloan Management Review* (Summer 1993):23–33.

Bowles J, Hammond J. *Beyond Quality: New Standards of Total Performance that Can Change the Future of Corporate America.* New York: Berkley Books, 1991.

Bridges W. *Managing Transitions: Making the Most of Change*. Reading, MA: Addison-Wesley, 1991.

Bruss LR, Roos HT. Operations, readiness, and culture: Don't reengineer without considering them. *Inform* (April, 1992):57–64.

Fabrysky WJ, Mize JH. *Systems Analysis and Design for Safety*. Englewood Cliffs, NJ: Prentice-Hall, 1976.

Federal Register, February 24, 1992, pp. 6356–6417.

General Motors Corporation. *Comparison of Traditional Safety Approach vs. Design In.* Internal memorandum. Detroit: General Motors Corporation, October 26, 1990.

General Motors Corporation. *Design-in*. #4: Future Vision booklet. Detroit: General Motors Corporation, 1992.

General Motors Corporation. *Pro(ACT) Process*. Detroit: General Motors Corporation, 1992.

Gilbreath RD. *Save Yourself! Six Pathways to Achievement in the Age of Change*. New York: McGraw-Hill, 1991.

Hall RK. *It's Back to Basics with the Time-Based Continuous Improvement Process*. Chicago: American National Can, August 1989, pp. 1–3.

Hall RK. *Setup Reduction: the Building Block for Time-Based Improvements*. Chicago: American National Can, December 1989, pp. 1–4.

Johnson WG. *MORT Safety Assurance Systems*. New York: Marcel Dekker, 1980.

Kaiser Aluminum & Chemical Corporation. *Employee Involvement*. Training Series, Kaiser Aluminum & Chemical Corporation, January 23, 1984.

Kjellén U. Safety control in design: Experiences from an offshore project. *Journal of Occupational Accidents* 12 (June 1990):49–61.

Manuele FA. *On the Practice of Safety*. New York: Van Nostrand Reinhold, 1993.

McInerney F, White S. *Beating Japan*. New York: Dutton, 1993.

Millar D. Commentary. *Occupational Safety & Health* 23, no. 9 (Sept. 1993):100–101.

Naibett J, Aburdene P. *Megatrends 2000: Ten New Directions for the 1990's*. New York: Avon Books, 1990.

OSHA's Voluntary Protection Programs. *Federal Register*, July 12, 1988, p. 53.

Peters T. *Thriving on Chaos: Handbook for a Management Revolution.* New York: Knopf, 1987.

Petersen D. *The Challenge of Change: Creating a New Safety Culture, Implementation Guide*. Video series and sofware. Safety Training Systems, 1993.

Process safety management of highly hazardous chemicals; final rule. *Federal Register*, Vol. 57, No. 36, February 24, 1993, pp. 6356–6417.

Rose J. Engineering design safety. *Occupational Safety & Health* 18, no. 1 (Jan. 1988):10–14.

Schmidt WH, Finnigan JP. *The Race Without a Finish Line*. San Francisco: Jossey-Bass Publishers, 1992.

Toffler A. *Future Shock*. New York: Bantam Books, 1970.

Toffler A. *The Third Wave*. New York: Bantam Books, 1980.

Appendix 1

The Control of Hazardous Energy (Lockout/Tagout), 29 *CFR* 1910.147

1910.147—THE CONTROL OF HAZARDOUS ENERGY (LOCKOUT/TAGOUT)

(a) Scope, application and purpose.

(1) Scope.

(i) This standard covers the servicing and maintenance of machines and equipment in which the unexpected energization or start up of the machines or equipment, or release of stored energy could cause injury to employees. This standard establishes minimum performance requirements for the control of such hazardous energy.

(ii) This standard does not cover the following:

(a) Construction, agriculture and maritime employment;

(b) Installations under the exclusive control of electric utilities for the purpose of power generation, transmission and distribution, including related equipment for communication or metering; and

(c) Exposure to electrical hazards from work on, near, or with conductors or equipment in electric utilization installations, which is covered by Subpart S of this part; and

(d) Oil and gas well drilling and servicing.

(2) Application.

(i) This standard applies to the control of energy during servicing and/or maintenance of machines and equipment.

(ii) Normal production operations are not covered by this standard (See Subpart 0 of this Part). Servicing and/or maintenance which takes place during normal production operations is covered by this standard only if:

(a) An employee is required to removed or bypass a guard or other safety device; or

(b) An employee is required to place any part of his or her body into an area on a machine or piece of equipment where work is actually performed upon the material being processed (point of operation) or where an associated danger zone exists during a machine operating cycle.

NOTE: Exception to paragraph (a)(2)(ii): Minor tool changes and adjustments, and other minor servicing activities, which take place during normal production operations, are not covered by this standard if they are routine, repetitive, and integral to the use of the equipment for production, provided that the work is performed using alternative measures which provide effective protection (see Subpart 0 of this Part).

(iii) This standard does not apply to the following.

(a) Work on cord and plug connected electric equipment for which exposure to the hazards of unexpected energization or start up of the equipment is controlled by the unplugging of the equipment from the energy source and by the plug being under the exclusive control of the employee performing the servicing or maintenance.

(b) Hot top operations involving transmission and distribution systems for substances such as gas, steam, water or petroleum products when they are performed on pressurized pipelines, provided that the employer demonstrates that (1) continuity of service is essential; (2) shutdown of the system is impractical; and (3) documented procedures are followed, and special equipment is used which will provide proven effective protection for employees.

(3) Purpose.

(i) This section requires employers to establish a program and utilize procedures for affixing appropriate lockout devices or tagout devices to energy isolating devices, and to otherwise disable machines or equipment to prevent unexpected energization, start-up or release of stored energy in order to prevent injury to employees.

(ii) When other standards in this part require the use of lockout or tagout, they shall be used and supplemented by the procedural and training requirements of this section.

(b) Definitions applicable to this section.

Affected employee. An employee whose job requires him/her to operate or use a machine or equipment on which servicing or maintenance is being performed under lockout or tagout, or whose job requires him/her to work in an area in which such servicing or maintenance is being performed.

Authorized employee. A person who locks or implements a tagout system procedure on machines or equipment to perform the servicing or maintenance on that machine or equipment. An authorized employee and an affected employee may be the same person when the affected employee's duties also include performing maintenance or service on a machine or equipment which must be locked or a tagout system implemented.

"Capable of being lockout out." An energy isolating device will be considered to be capable of being locked out either if it is designed with a hasp or other attachment or integral part to which, or through which, a lock can be affixed, or if it has a locking mechanism built into it. Other energy isolating devices will also be considered to be capable of being locked out, if lockout can be achieved without the need to dismantle, rebuild, or replace the energy isolating device or permanently alter its energy control capability.

Energized. Connected to an energy source or containing residual or stored energy.

Energy isolating device. A mechanical device that physically prevents the transmission or release or energy, including but not limited to the following: A manually operated electrical circuit breaker; a disconnect switch; a manually operated switch by which the conductors of a circuit can be disconnected from all ungrounded supply conductors and, in addition, no pole can be operated independently; a slide gate; a slip blind; a line valve; a block; and any similar device used to block or isolate energy. The term does not include a push button, selector switch, and other control circuit type devices.

Energy source. Any source of electrical, mechanical, hydraulic, pneumatic, chemical, thermal, or other energy.

Hot tap. A procedure used in the repair, maintenance and services activities which involves welding on a piece of equipment (pipelines, vessels or tanks) under pressure, in order to install connections or appurtenances. It is commonly used to replace or add sections of pipeline without the interruption of service for air, gas, water, steam, and petrochemical distribution systems.

Lockout. The placement of a lockout device on an energy isolating device, in accordance with an established procedure, ensuring that the energy isolating device and the equipment being controlled cannot be operated until the lockout device is removed.

Lockout device. A device that utilizes a positive means such as a lock, either key or combination type, to hold an energy isolating device in the safe position and prevent the energizing of a machine or equipment.

Normal production operations. The utilization of a machine or equipment to perform its intended production function.

Servicing and/or maintenance. Workplace activities such as constructing, installing, setting up, adjusting, inspecting, modifying, and maintaining and/or servicing machines or equipment. These activities include lubrication, cleaning or unjamming of machines or equipment and making adjustments or tool changes, where the employee may be exposed to the unexpected energization or startup of the equipment or release of hazardous energy.

Setting up. Any work performed to prepare a

machine or equipment to perform its normal production operation.

Tagout. The placement of a tagout device on an energy isolating device, in accordance with an established procedure, to indicate that the energy isolating device and the equipment being controlled may not be operated until the tagout device is removed.

Tagout device. A prominent warning device, such as a tag and a means of attachment, which can be securely fastened to an energy isolating device in accordance with an established procedure, to indicate that the energy isolating device and the equipment being controlled may not be operated until the tagout device is removed.

(c) General.

(1) Energy control program. The employer shall establish a program consisting of an energy control procedure and employee training to ensure that before any employee performs any servicing or maintenance on a machine or equipment where the unexpected energizing, start up or release of stored energy could occur and cause injury, the machine or equipment shall be isolated, and rendered inoperative, in accordance with paragraph (c)(4) of this section.

(2) Lockout/tagout.

(i) If an energy isolating device is not capable of being locked out, the employer's energy control program under paragraph (c)(1) of this section shall utilize a tagout system.

(ii) If an energy isolating device is capable of being locked out, the employer's energy control program under paragraph (c)(1) of this section shall utilize lockout, unless the employer can demonstrate that the utilization of a tagout system will provide full employee protection as set forth in paragraph (c)(3) of this section.

(iii) After October 31, 1989, whenever major replacement, repair, renovation or modification of machines or equipment is performed, and whenever new machines or equipment are installed, energy isolating devices for such machines or equipment shall be designed to accept a lockout device.

(3) Full employee protection.

(i) When a tagout device is used on an energy isolating device which is capable of being locked out, the tagout device shall be attached at the same location that the lockout device would have been attached, and the employer shall demonstrate that the tagout program will provide a level of safety equivalent to that obtained by using a lockout program.

(ii) In demonstrating that a level of safety is achieved in the tagout program which is equivalent to the level of safety obtained by using a lockout program, the employer shall demonstrate full compliance with all tagout-related provisions of this standard together with such additional elements as are necessary to provide the equivalent safety available from the use of a lockout device. Additional means to be considered as part of the demonstration of full employee protection shall include the implementation of additional safety measures such as the removal of an isolating circuit element, blocking of a controlling switch, opening of an extra disconnecting device, or the removal of a valve handle to reduce the likelihood of inadvertent energization.

(4) Energy control procedure.

(i) Procedures shall be developed, documented and utilized for the control of potentially hazardous energy when employees are engaged in the activities covered by this section.

NOTE: Exception: The employer need not document the required procedure for a particular machine or equipment, when all of the following elements exist: (1) The machine or equipment has no potential for stored or residual energy or reaccumulation of stored energy after shut down which could endanger employees; (2) the machine or equipment has a single energy source which can be readily identified and isolated; (3) the isolation and locking out of that energy source will completely deenergize and deactivate the machine or equipment; (4) the machine or equipment is isolated from that energy source and locked out during servicing or maintenance; (5) a single lockout device will achieve a locked-out condition; (6) the lockout device is under the exclusive control of the authorized employee performing the servicing or maintenance; (7) the servicing or maintenance does not create hazards for other employees; and (8) the employer, in utilizing this exception, has had no accidents involving the unexpected activation or reenergization of the machine or equipment during servicing or maintenance.

(ii) The procedures shall clearly and specifically outline the scope, purpose, authorization, rules, and techniques to be utilized for the control of hazardous energy, and the means to enforce compliance including, but not limited to, the following:

(a) A specific statement of the intended use of the procedure;

(b) Specific procedural steps for shutting down, isolating, blocking and securing machines or equipment to control hazardous energy;

(c) Specific procedural steps for the placement, removal and transfer of lockout devices or tagout devices and the responsibility for them; and

(d) Specific requirements for testing a machine or equipment to determine and verify the effectiveness of lockout devices, tagout devices, and other energy control measures.

(5) Protective materials and hardware.

(i) Locks, tags, chains, wedges, key blocks, adapter pins, self-locking fasteners, or other hardware shall be provided by the employer for isolating, securing or blocking of machines or equipment from energy sources.

(ii) Lockout devices and tagout devices shall be singularly identified; shall be the only devices(s) used for controlling energy; shall not be used for other purposes; and shall meet the following requirements:

(a) Durable.

(1) Lockout and tagout devices shall be capable of withstanding the environment to which they are exposed for the maximum period of time that exposure is expected.

(2) Tagout devices shall be constructed and printed so that exposure to weather conditions or wet and damp locations will not cause the tag to deteriorate or the message on the tag to become illegible.

(3) Tags shall not deteriorate when used in corrosive environments such as areas where acid and alkali chemicals are handled and stored.

(b) Standardized. Lockout and tagout devices shall be standardized within the facility in at least one of the following criteria: Color; shape; or size; and additionally, in the case of tagout devices, print and format shall be standardized.

(c) Substantial.

(1) Lockout devices. Lockout devices shall be substantial enough to prevent removal without the use of excessive force or unusual techniques, such as with the use of bolt cutters or other metal cutting tools.

(2) Tagout devices. Tagout devices, including and their means of attachment, shall be substantial enough to prevent inadvertent or accidental removal. Tagout device attachment means shall be of a non-reusable type, attachable by hand, self-locking, and non-releasable with a minimum unlocking strength of no less than 50 pounds and having the general design and basic characteristics of being at least equivalent to a one-piece, all-environment-tolerant nylon cable tie.

(d) Identifiable. Lockout devices and tagout devices shall indicate the identity of the employee applying the device(s).

(iii) Tagout devices shall warn against hazardous conditions if the machine or equipment is energized and shall include a legend such as the following: **Do Not Start, Do Not Open, Do Not Close, Do Not Energize, Do Not Operate.**

(6) Periodic inspection.

(i) The employer shall conduct a periodic inspection of the energy control procedure at least annually to ensure that the procedure and the requirements of this standard are being followed.

(a) The periodic inspection shall be performed by an authorized employee other than the ones(s) utilizing the energy control procedure being inspected.

(b) The periodic inspection shall be designed to correct any deviations or inadequacies observed.

(c) Where lockout is used for energy control, the periodic inspection shall include a review, between the inspector and each authorized employee, of that employee's responsibilities under the energy control procedure being inspected.

(d) Where tagout is used for energy control, the periodic inspection shall include a review, between the inspector and each authorized and affected employee, of that employee's responsibilities under the energy control procedure being inspected, and the elements set forth in paragraph (c)(7)(ii) of this section.

(ii) The employer shall certify that the periodic inspections have been performed. The certification shall identify the machine or equipment on which the energy control procedure was being utilized, the date of the inspection, the employees included in the inspection, and the person performing the inspection.

(7) Training and communication.

(i) The employer shall provide training to ensure that the purpose and function of the energy control program are understood by employees and that the knowledge and skills required for the safe application, usage, and removal of energy controls are required by employees. The training shall include the following:

(a) Each authorized employee shall receive training in the recognition of applicable hazardous energy sources, the type and magnitude of the energy available in the workplace, and the methods and means necessary for energy isolation and control.

(b) Each affected employee shall be instructed in the purpose and use of the energy control procedure.

(c) All other employees whose work operations are or may be in an area where energy control procedures may be utilized, shall be instructed about the procedure, and about the prohibition relating to attempts to restart or reenergize machines or equipment which are locked out or tagged out.

(ii) When tagout systems are used, employees shall also be trained in the following limitations of tags:

(a) Tags are essentially warning devices affixed to energy isolating devices, and do not provide the physical restraint on those devices that is provided by a lock.

(b) When a tag is attached to an energy isolating means, it is not to be removed without authorization of the authorized person responsible for it, and it is never to be bypassed, ignored, or otherwise defeated.

(c) Tags must be legible and understandable by all authorized employees, affected employees, and all other employees whose work operations are or may be in the area, in order to be effective.

(d) Tags and their means of attachment must be made of materials which will withstand the environmental conditions encountered in the workplace.

(e) Tags may evoke a false sense of security, and their meaning needs to be understood as part of the overall energy control program.

(f) Tags must be securely attached to energy isolating devices so that they cannot be inadvertently or accidentally detached during use.

(iii) Employee retraining.

(a) Retraining shall be provided for all authorized and affected employees whenever there is a change in their job assignments, a change in their job assignments, a change in machines, equipment or processes that present a new hazard, or when their is a change in the energy control procedures.

(b) Additional retraining shall also be conducted whenever a periodic inspection under paragraph (c)(6) of this section reveals, or whenever the employer has reason to believe, that there are deviations from or inadequacies in the employee's knowledge or use of the energy control procedures.

(c) The retraining shall reestablish employee proficiency and introduce new or revised control methods and procedures, as necessary.

(iv) The employer shall certify that employee training has been accomplished and is being kept up to date. The certification shall contain each employee's name and dates of training.

(8) Energy isolation. Implementation of lockout or the tagout system shall be performed only by authorized employees.

(9) Notification of employees. Affected employees shall be notified by the employer or authorized employee of the application and removal of lockout devices or tagout devices. Notification shall be given before the controls are applied, and after they are removed from the machine or equipment.

(d) Application of control. The established procedure for the application of energy control (implementation of lockout or tagout system procedures) shall cover the following elements and actions and shall be done in the following sequence:

(1) Preparation for shutdown. Before an authorized or affected employee turns off a machine or equipment, the authorized employee shall have knowledge of the type and magnitude of the energy, the hazards of the energy to be controlled, and the method or means to control the energy.

(2) Machine or equipment shutdown. The machine or equipment shall be turned off or shut down using the procedures required by this standard. An orderly shutdown must be utilized to avoid any additional or increased hazard(s) to employees as a result of equipment deenergization.

(3) Machine or equipment isolation. All energy isolating devices that are needed to control the energy to the machine or equipment shall be physically located and operated in such a manner as to isolate the machine or equipment from the energy source(s).

(4) Lockout or tagout device application.

(i) Lockout or tagout devices shall be affixed to each energy isolating device by authorized employees.

(ii) Lockout devices, where used, shall be affixed in a manner to that will hold the energy isolating devices in a "safe" or "off" position.

(iii) Tagout decices, where used, shall be affixed in such a manner as will clearly indicate that the operation or movement of energy isolating devices from the "safe" or "off" position is prohibited.

(a) Where tagout devices are used with energy isolating devices designed with the capability of being locked, the tag attachment shall be fastened at the same point at which the lock would have been attached.

(b) Where a tag cannot be affixed directly to the energy isolating device, the tag shall be located as close as safely possible to the device, in a position that will be immediately obvious to anyone attempting to operate the device.

(5) Stored energy.

(i) Following the application of lockout or tagout devices to energy isolating devices, all potentially hazardous stored or residual energy shall be relieved, disconnected, restrained, and otherwise rendered safe.

(ii) If there is a possibility of reaccumulation of stored energy to a hazardous level, verification of isolation shall be continued until the possibility of such accumulation no longer exists.

(6) Verification of isolation. Prior to starting work on machines or equipment that have been locked out or tagged out, the authorized employee shall verify that isolation and deenergization of the machine or equipment have been accomplished.

(e) Release from lockout or tagout. Before lockout or tagout devices are removed and energy is restored to the machine or equipment, procedures shall be followed and actions taken by the authorized employee(s) to ensure the following:

(1) The machine or equipment. The work area shall be inspected to ensure that nonessential items have been removed and to ensure that machine or equipment components are operationally intact.

(2) Employees.

(i) The work area shall be checked to ensure that all employees have been safely positioned or removed.

(ii) Before lockout or tagout devices are removed and before machines or equipment are energized, affected employees shall be notified that the lockout or tagout devices have been removed.

(3) Lockout or tagout devices removal. Each lockout or tagout device shall be removed from each energy isolating device by the employee who applied the device. **Exception to paragraph (e)(3):** When the authorized employee who applied the lockout or tagout device is not available to remove it, that device may be removed under the direction of the employer, provided that specific procedures and training for such removal have been developed, documented and incorporated into the employer's energy control program. The employer shall demonstrate that the specific procedure provides equivalent safety to the removal of the device by the authorized employee who applied it. The specific procedure shall include at least the following elements:

(i) Verification by the employer that the authorized employee who applied the device is not at the facility;

(ii) Making all reasonable efforts to contact the authorized employee to inform him/her that his/her lockout or tagout device has been removed; and

(iii) Ensuring that the authorized employee has this knowledge before he/she resumes work at that facility.

(f) Additional requirements.

(1) Testing or positioning of machines, equipment or components thereof. In situations in which lockout or tagout devices must be temporarily removed from the energy isolating device and the machine or equipment energized to test or position the machine, equipment or component thereof, the following sequence of actions shall be followed:

(i) Clear the machine or equipment of tools and materials in accordance with paragraph (e)(1) of this section;

(ii) Remove employees from the machine or equipment area in accordance with paragraph (e)(2) of this section;

(iii) Remove the lockout or tagout devices as specified in paragraph (e)(3) of this section;

(iv) Energize and proceed with testing or positioning;

(v) Deenergize all systems and reapply energy control measures in accordance with paragraph (d) of this section to continue the servicing and/or maintenance.

(2) Outside personnel (contractors, etc.).

(i) Whenever outside servicing personnel are to be engaged in activities covered by the scope and application of this standard, the on-site employer and the outside employer shall inform each other of their respective lockout or tagout procedures.

(ii) The on-side employer shall ensure that his/her personnel understand and comply with restrictions and prohibitions of the outside employer's energy control procedures.

(3) Group lockout or tagout.

(i) When servicing and/or maintenance is performed by a crew, craft, department or other group, they shall utilize a procedure which affords the employees a level of protection equivalent to that provided by the implementation of a personal lockout or tagout device.

(ii) Group lockout or tagout devices shall be used in accordance with the procedures required by paragraph (c)(4) of this section including, but not necessarily limited to, the following specific requirements:

(a) Primary responsibility is vested in an authorized employee for a set number of employees working under the protection of a group lockout or tagout device (such as an operations lock);

(b) Provision for the authorized employee to ascertain the exposure status of individual group members with regard to the lockout or tagout of the machine or equipment and

(c) When more than one crew, craft, department, etc. is involved, assignment of overall job-associated lockout or tagout control responsibility to an authorized employee designated to coordinate affected work forces and ensure continuity of protection; and

(d) Each authorized employee shall affix a personal lockout or tagout device to the group lockout device, group lockbox, or comparable mechanism when he or she begins work, and shall remove those devices when he or she stops working on the machine or equipment being serviced or maintained.

(4) Shift or personnel changes. Specific procedures shall be utilized during shift or personnel changes to ensure the continuity of lockout or tagout protection, including provision for the orderly transfer of lockout or tagout devices between off-going and oncoming employees, to minimize exposure to hazards from the unexpected energization, start-up of the machine or equipment, or release of stored energy.

"(The information collection requirements contained in this section are approved by the Office of Management and Budget (OMB) and listed under OMB control number 1218-0150.)"

[FR Doc. 89–24467 Filed 10-16-89; 8:45 am]

NOTE: The following Appendix to § 1910.147 services as a non-mandatory guideline to assist employers and employees in complying with the requirements of this section, as well as to provide other helpful information. Nothing in the Appendix adds to or detracts from any of the requirements of this section.

APPENDIX A—TYPICAL MINIMAL LOCKOUT OR TAGOUT SYSTEM PROCEDURES

General

Lockout is the preferred method of isolating machines or equipment from energy sources. To assist employers in developing a procedure which meets the requirements of the standard, however, the following simple procedure is provided for use in both lockout or tagout programs. This procedure may be used when there are limited number or types of machines or equipment or there is a single power source. For more complex systems, a more comprehensive procedure will need to be developed, documented, and utilized.

Lockout (or Tagout) Procedure for (Name of Company).

Purpose

This procedure establishes the minimum requirements for the lockout or tagout of energy isolating devices. It shall be used to ensure that the machine or equipment are isolated from all potentially hazardous energy, and locked out or tagged out before employees perform any servicing or maintenance activities where the unexpected energization, start-up or release of stored energy could cause injury (Type(s) and Magnitude(s) of Energy and Hazards).

Responsibility

Appropriate employees shall be instructed in the safety significance of the lockout (or tagout) procedure (Name(s)/Job Title(s) of employees authorized to lockout or tagout). Each new or transferred affected employee and other employees whose work operations are or may be in the area shall be instructed in the purpose and use of the lockout or tagout procedure (Name(s)/Job Title(s) of affected employees and how to notify).

Preparation for Lockout or Tagout

Make a survey to locate and identify all isolating devices to be certain which switch(s), valve(s) or other energy isolating devices apply to the equipment to be locked or tagged out. More than one energy source (electrical, mechanical, or others) may be involved. (Type(s) and Location(s) of energy isolating means).

Sequence of Lockout or Tagout System Procedure

(1) Notify all affected employees that a lockout or tagout system is going to be utilized and the reason therefore. The authorized employee shall know the type and magnitude of energy that the machine or equipment utilizes and shall understand the hazards thereof.

(2) If the machine or equipment is operating, shut it down by the normal stopping procedure (depress stop button, open toggle switch, etc.)

(3) Operate the switch, valve, or other energy isolating device(s) so that the equipment is isolated from its energy source(s). Stored energy (such as that in springs, elevated machine members, rotating flywheels, hydraulic systems, and air, gas, steam, or water pressure, etc.) must be dissipated or restrained by methods such as repositioning, blocking, bleeding down, etc. (Type(s) of Stored Energy-methods to dissipate or restrain).

(4) Lockout and/or tagout the energy isolating devices with assigned individual lock(s) or tag(s) (Method(s) Selected; i.e., locks tags, additional safety measures, etc.)

(5) After ensuring that no personnel are exposed, and as a check on having disconnected the energy sources, operate the push button or other normal operating controls to make certain the equipment will not operate (Type (s) of Equipment checked to ensure disconnections).

CAUTION: Return operating control(s) to "neutral" or "off" position after the test.

(6) The equipment is now locked out or tagged out.

Restoring Machines or Equipment to Normal Production Operations

(1) After the servicing and/or maintenance is complete and equipment is ready for normal production operations, check the area around the machines or equipment to ensure that no one is exposed.

(2) After all tools have been removed from the machine or equipment, guards have been reinstalled and employees are in the clear, remove all lockout or tagout devices. Operate the energy isolating devices to restore energy to the machine or equipment.

Procedure Involving More Than One Person

In the preceding steps, if more than one individual is required to lockout or tagout equipment, each shall place his/her own personal lockout device or tagout device on the energy isolating device(s). When an energy isolating device cannot accept multiple locks or tags, a multiple lockout or tagout device (hasp) may be used. If lockout is used, a single lock may be used to lockout the machine or equipment with the key being placed in a lockout box or cabinet which allows the use of multiple locks to secure it. Each employee will then use his/her own lock to secure the box or cabinet. As each person no longer needs to maintain his or her lockout protection, that person will remove his/her lock from the box or cabinet (Name(s)/Job Title(s) of employees authorized for group lockout or tagout).

Basic Rules for Using Lockout or Tagout System Procedure

All equipment shall be locked out or tagged out to protect against accidental or inadvertent operation when such operation could cause injury to personnel. Do not attempt to operate any switch, valve, or other energy isolating device where it is locked or a tagged out.

Appendix 2

Guidelines for Controlling Hazardous Energy During Maintenance and Servicing, NIOSH Pub. No. 83–125

Examples of Alternative Methods of Isolating or Blocking Energy and Securing the Point(s) of Control (NIOSH Publication No. 83-125)

Energy Type		Method of Isolating or Blocking Energy		Method of Securing the Point of Control	Remarks
Mechanical motion Rotation Translation Linear Oscillation	1.	Remove segments of operating mechanical linkages such as dismantling push rods, removing belts, and removing flywheels.	(1)	Tag the linkages and place them in a locked cabinet away from the machine. or	
			(2)	Attach warning tags where the linkages were removed and restrict access to trained personnel. or	
			(3)	Post a person to protect against unauthorized reinstallation of the linkages.	
	2.	Use blocking devices such as wood or metal blocks.	(1)	Chain and lock in point of control or use metal pins driven or welded in place. or	
			(2)	Attach warning tags on the blocking devices and restrict access into the area to trained personnel. or	
			(3)	Post a person to protect against unauthorized removal of the blocking devices.	
	3.	Remove power or energy from the driving mechanism, such as: a. Disconnect main electrical source. b. Close hydraulic or pneumatic valves, bleed.	(1)	a. Padlock in the OFF position. b. Disconnect pneumatic and hydraulic lines and tag. or	
			(2)	Attach warning tags at control points and restrict access to trained personnel. or	
			(3)	Post a person to protect against unauthorized reconnection of the energy sources.	
Electrical	1.	Place the main electrical disconnect switch in the OFF position.	(1)	Secure by a padlock, a clip and padlock, or a bar and padlock. or	Check for alternate sources of power.
			(2)	Attach a warning tag and restrict access into the area to trained personnel. or	
			(3)	Post a person to protect against unauthorized actuation of the switch.	
	2.	Remove segments of electrical circuit, such as printed circuit modules.	(1)	Tag the module and place in a locked cabinet away from the control center and tag the control center door. or	
			(2)	Attach a warning tag at the module location and restrict access to trained personnel. or	

Energy Type		Method of Isolating or Blocking Energy		Method of Securing the Point of Control	Remarks
			(3)	Have a person remain at the control center to protect from unauthorized installation of a spare or replacement module.	
Thermal (steam)	1.	Close valves and maintain an open bleed.	(1)	Chain and padlock valve or use blind flanges or slip blinds. or	Allow time for residual heat to dissipate.
			(2)	Attach warning tags to the valves and restrict access to the area to trained personnel. or	
			(3)	Station a person at the valve locations to protect against unauthorized or inadvertent opening of valves.	
Potential (pressure)	1.	Close valves and maintain open vent to relieve.	(1)	Secure, block, blind flange, slip bind, or valve with locking device. or	
			(2)	Attach warning tags and restrict access to trained personnel. or	
			(3)	Station a person at the valves to protect against unauthorized actuation.	
Potential (gravity)	1.	Block in place by using metal or wood blocks under the mechanism or pin the linkages in a position where gravity will not cause the mechanism to inadvertently fall.	(1)	Secure block, or pin with a locking device. or	Energy could be dissipated by lowering to a point where gravity could no longer cause inadvertent falling.
			(2)	Attaching warning tags to blocks, linkages, and pins and restrict access to trained personnel. or	
			(3)	Station a person at the mechanism to prevent unauthorized removal of blocks and pins and reinstallation of linkages.	
Potential (spring)	1.	Block in a safe position by pinning or clamping the device, eliminating the potential of unrestricted and undesired travel.	(1)	Secure pin or clamp in place with a locking device. or	Spring energy could be dissipated by release or dismantling of the mechanism.
			(2)	Attach warning tags to the pins and clamps and restrict release or access to trained personnel. or	
			(3)	Station a person at the control point to protect against pin or clamp removal and unauthorized activation of the spring mechanisms.	

Appendix 3

Diagram for Controlling Hazardous Energy During Maintenance and Servicing NIOSH Pub. No. 83–125

Legend

Rectangle Symbol
Identifies an event that results from the combination or exclusion of activities or events.

And Gate
Describes an operation whereby co-existence of all inputs is required to produce an output event.

Or Gate
Defines a situation whereby an output event will occur if one of the inputs exists.

Decision Gate
Exclusive Or Gate which functions as an Or Gate but provides a footnoted (▷) list of decision criteria that are not self-evident.

Diamond
Describes an event that is considered basic in a given logic sequence. Event is not developed further because development is obvious.

Decision Criteria

1 ▷ Implementation of safeguards to control hazards with energy present may be chosen instead of hazardous energy elimination by devices or techniques, when it can be demonstrated that hazards are controlled with energy present by:
1. Identifying all hazardous energy sources and hazardous residual energy, and
2. Documenting a procedure for and demonstrating that the procedure will control hazards resulting from each hazardous energy identified.

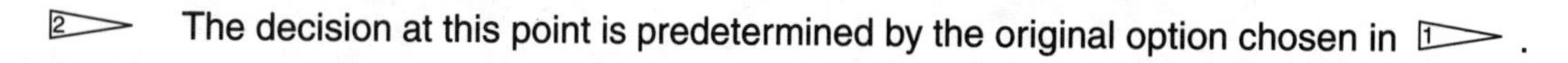

2 ▷ The decision at this point is predetermined by the original option chosen in 1 ▷ .

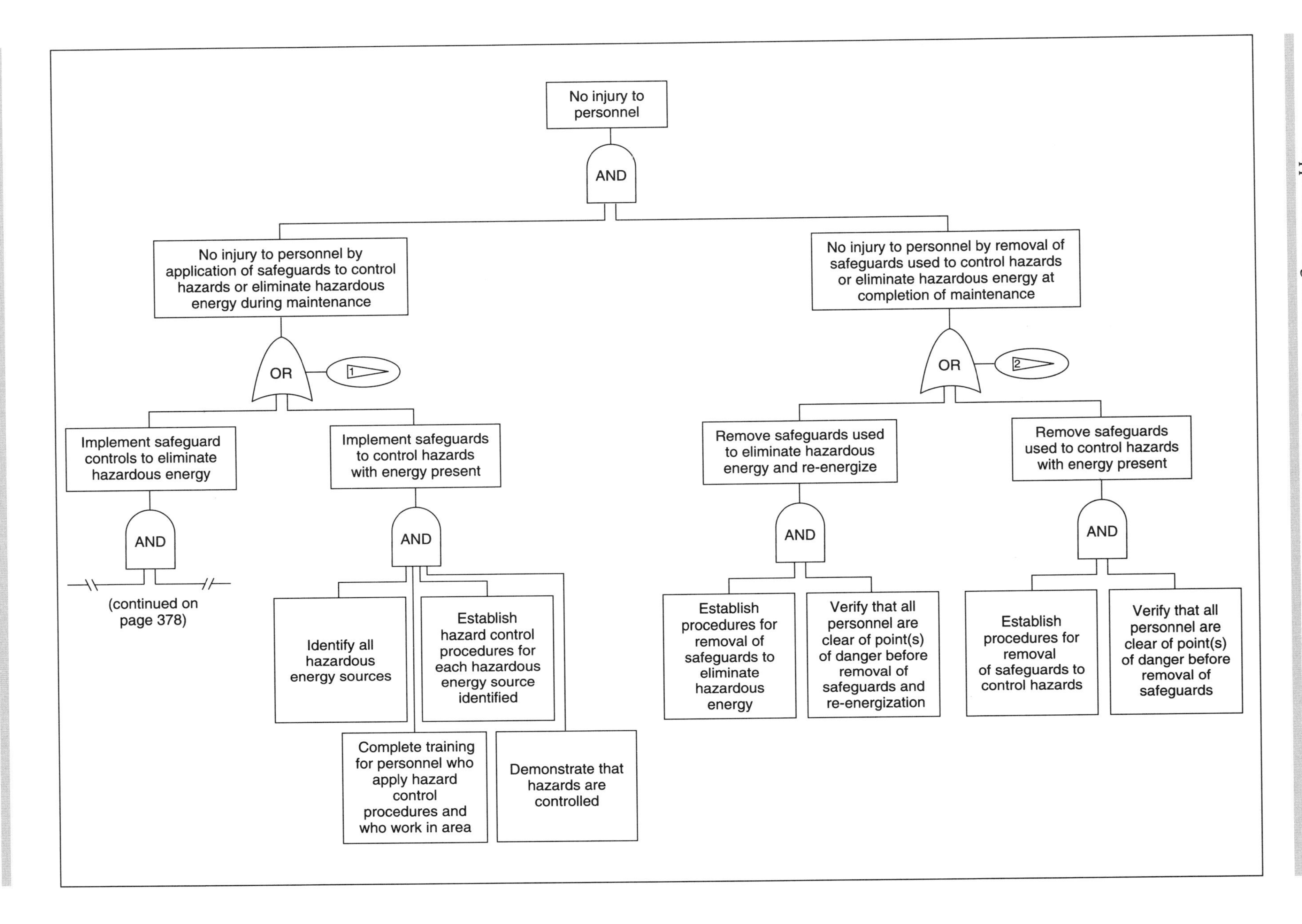
No injury to personnel
AND
No injury to personnel by application of safeguards to control hazards or eliminate hazardous energy during maintenance
OR
1
Implement safeguard controls to eliminate hazardous energy
AND
(continued on page 378)
Implement safeguards to control hazards with energy present
AND
Identify all hazardous energy sources
Establish hazard control procedures for each hazardous energy source identified
Complete training for personnel who apply hazard control procedures and who work in area
Demonstrate that hazards are controlled
No injury to personnel by removal of safeguards used to control hazards or eliminate hazardous energy at completion of maintenance
OR
2
Remove safeguards used to eliminate hazardous energy and re-energize
AND
Establish procedures for removal of safeguards to eliminate hazardous energy
Verify that all personnel are clear of point(s) of danger before removal of safeguards and re-energization
Remove safeguards used to control hazards with energy present
AND
Establish procedures for removal of safeguards to control hazards
Verify that all personnel are clear of point(s) of danger before removal of safeguards

(continued from page 377)

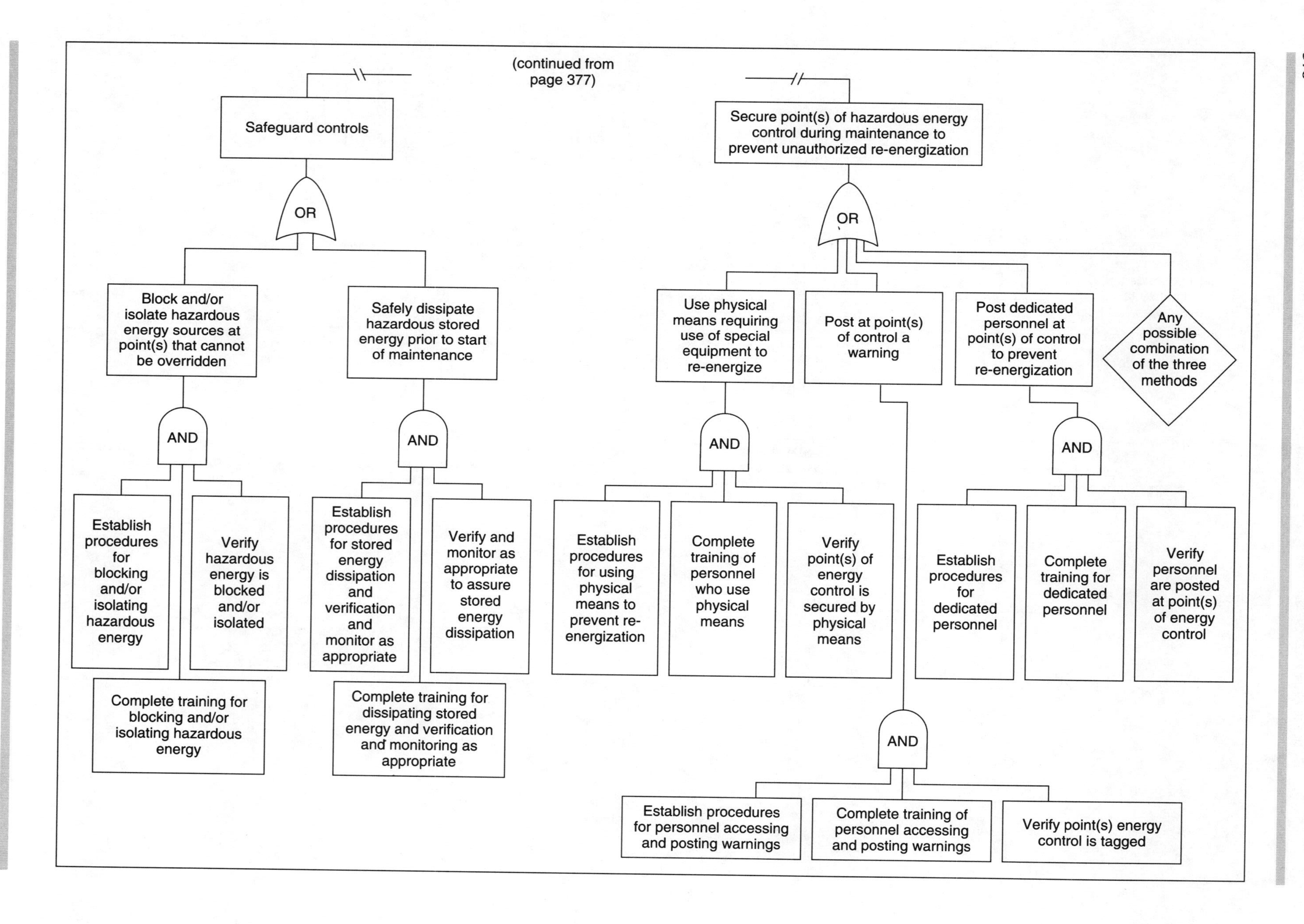

Appendix 4

Sample Lockout/Tagout Policy and Procedure

Title: Lockout/Tagout Process

No.:

Date:

I. Purpose

To establish a control system and utilize procedures to prevent the unexpected release or transmission of equipment/process energy.

II. Objectives

A. Prevent inadvertent operation or energization of the equipment/process in order to protect personnel.

B. Establish methods for achieving zero energy state.

C. Comply with applicable regulatory standards.

III. Scope

A. This policy applies to activities such as, but not limited to: erecting, installing, constructing, repairing, adjusting, inspecting, cleaning, operating or maintaining the equipment/process.

B. This policy applies to energy sources such as, but not limited to: electrical, mechanical, hydraulic, pneumatic, chemical, radiation, thermal, compressed air, energy stored in springs, and potential energy from suspended parts (gravity).

C. International facilities will comply with the substance of this policy or the prevailing national requirements whichever is more stringent.

D. Installation/design requirement. After January 1, 1990, whenever replacement or major repair, renovation or modification of a machine or equipment is performed, and whenever new machines or equipment are installed, energy isolating devices shall be designed to accept a lockout device (29 *CFR* 1910.147).

IV. Definitions

A. *Energy isolating device*. A mechanical device that physically prevents the transmission or release of energy, including but not limited to the following: a manually operated electrical circuit breaker; a disconnect switch; a manually operated switch by which the conductors of a circuit can be disconnected from all ungrounded supply conductors and, in addition, no pole can be operated independently; a line valve; a block; and any similar device used to block or isolate energy. The term does not include pushbutton, selector switch, and other control circuit type devices.

B. *Lockout.* The placement of a lockout device on an energy isolating device, in accordance with an established procedure, ensuring that the energy isolating device and the equipment being controlled cannot be operated until the lockout device is removed.

C. *Lockout device.* A device that utilizes a positive means such as a lock, to hold an energy isolating device in a safe position and prevent the energizing of a machine or equipment.

D. *"Capable of being locked out."* An energy isolating device that is designed with a hasp or other means of attachment to which, or through which a lock can be affixed, or if it has a locking mechanism built into it. Other energy isolating devices will also be considered to be capable of being locked out, if lock out can be achieved without the need to dismantle, rebuild, or replace the energy isolating device or permanently alter its energy control capability.

E. *Tagout.* The placement of a tagout device on an energy isolating device, in accordance with established procedure, to indicate that the energy isolating device and equipment being controlled may not be operated until the tagout device is removed.

F. *Tagout device.* A prominent warning device such as a tag and means of attachment which can be securely fastened to an energy isolating device in accordance with an established procedure, to indicate that the energy isolating device and the equipment being controlled may not be operated until the tagout device is removed.

G. *Caution tag.* A warning device such as a tag and means of attachment used to warn employees of an existing or potential hazard. Its legend cautions personnel of the hazard(s) and identifies the applier.

H. *Affected employee.* An employee whose job requires him/her to operate or use a machine or equipment on which servicing or maintenance is being performed under lockout or tagout.

I. *Authorized employee.* A person who locks out or tags out machines or equipment to perform the servicing or maintenance on that machine.

J. *Other employee.* An employee whose job requires him/her to work in an area in which machine/equipment servicing or maintaining is being performed.

K. *Servicing/maintenance.* Work place activities such as constructing, installing, setting up, adjusting, inspecting, modifying, and maintaining and/or servicing machines or equipment. These activities include lubrication, cleaning or unjamming of machines or equipment and making adjustments to tool changes where the employee may be exposed to the unexpected energization or start up of the equipment or release of hazardous energy.

V. Procedure

A. *Lockout/tagout system.* Each facility shall develop a written lockout/tagout policy, which incorporates the following elements:

1. Principles

a. *All* personnel (hourly and salary) shall comply with the provision of the lockout/tagout system. Supervision must enforce the use of personal locks/tags to ensure protection when personnel performing tasks where exposure to unexpected energization may occur.

b. The locks/tags shall be standardized throughout the facility and the only authorized method used of the lockout/tagout of energy sources. Locks and employee tags shall not be used for any purpose other than personal protection.

c. Individual locks/tags shall be applied and removed by each person exposed to the unexpected release of energy, other than in those special situations where specific facility procedures have been developed.

d. Where equipment is lockable, use of a lock is required by all exposed personnel.

e. Where equipment is not lockable, tagout application or special lockout/tagout procedures shall be utilized.

f. When locks are used in the lockout/tagout application, they shall always be accompanied by tags.

(1) Locks used for personnel protection shall be accompanied by employee tags.

(2) Locks used to protect against

hazards shall be accompanied by caution tags.

g. Energy isolating devices shall be clearly labeled or identified to indicate their function unless located and arranged so their purpose is evident. Such identification is necessary to reduce possible errors in applying the lockout/tagout.

h. The lockout/tagout of electrical energy sources shall occur at the circuit disconnect switch. (Note: Facilities shall identify any situations where the circuit cannot be positively interrupted and develop procedures providing equivalent protection. Feasibility of effective circuit isolation shall be considered in future engineering improvements.)

i. The use of electrical control circuitry to accomplish lockout-tagout is normally prohibited since it does not offer *positive* personnel protection. Examples:
   (1) Electrical shorts. (Water in lines and some types of dust can create a path to complete the control circuit.)
   (2) Vibration or switch component failure.
   (3) Remote or interlocked switches not affected by control circuitry.

2. Protective appliances
   a. Locks. Shall be purchased specifically for lockout applications. They shall be of such design and durability that removal by other than abnormal means would require excessive force or unusual techniques. In addition, they shall possess individual keying/combination capability.
   b. Tags. Appliances which are used to provide warning or information.
      (1) *Employee tag* (mandatory). Used only for personnel protection; clearly distinguishable from caution tags and shall include a legend such as DO NOT START; DO NOT OPERATE, or a similar directive that informs employees working in the area not to start up the equipment.
      (2) *Caution tag* (mandatory). Provides a warning of hazards. It does not indicate that the applier is currently exposed to the unexpected release or transmission of energy. The use of a caution tag is provided to preserve the integrity of the employee tag.
   c. Lockout fixture. An appliance which accommodates one or more locks to secure an energy isolating device.
   d. Additional protecting appliances. Some exposures may require additional protective techniques or mechanical safeguards.

3. Application and exposure survey
   a. Each facility shall conduct an application survey to determine if the equipment/process can be safety isolated.
      (1) The survey should determine if energy isolating devices are available, adequate, and practically located for positive protection.
      (2) A plan shall be developed to correct the surveyed deficiencies or provide interim alternative protection in order to make the lockout/tagout system effective.
   b. Each facility shall conduct an exposure survey to determine what tasks are being done, i.e., cleaning rolls, removing jams, etc., with equipment energized. Each situation shall be evaluated to determine if the task can be accomplished with the power off or alternatively what method must be used to reduce employee risk.

4. Responsibilities
   a. Management is responsible for the development, implementation and administration of an effective lockout/tagout system.
   b. All employees are responsible for complying with the provisions of the facility lockout/tagout system.
   c. Affected employees shall be aware of lockout/tagout procedures used to guard against unexpected startups.
   d. Only authorized individuals shall operate energy isolating devices and place locks/tags on controls to prevent unexpected startups.
   e. Other employees who work in the

area where lockout/tagout procedures are used shall be instructed about their purpose and prohibited from attempting to restart machines or equipment which are locked or tagged out.

B. System utilization
   1. Preparation for lockout/tagout
      a. All personnel affected by the intended lockout/tagout shall be notified by the supervisor or authorized employee before commencing any work.
      b. A method shall be established to permit access to the equipment/process. This method should involve acknowledgement and release by the individual(s) responsible for the equipment/process.
      c. A prejob plan shall be developed to insure appropriate lockout/tagout when the equipment/process complexity or nature and scope of work warrants (i.e., job objectives and involved equipment/process; estimated job duration; crafts involved; type, number, and location of energy isolating devices, start-up provisions, etc.).
   2. Application of lockout/tagout
      a. Use appropriate equipment/process shutdown procedure(s) to deactivate operating controls or return them to the neutral mode.
      b. All involved energy isolating devices shall be operated/positioned in such a manner as to isolate the equipment/process from the energy source(s).
      c. Locks and employee tags shall be applied to each energy isolating device by authorized individuals.
         (1) Lockout fixtures and locks shall be attached in such a manner as to hold the energy isolating device(s) in a safe position.
         (2) Employee tags shall be completed by the applier and attached to the energy isolating device(s)
      d. After lockout/tagout application and prior to commencement of work, one or more of the following actions shall be taken:
         (1) Operate the equipment/process controls (push buttons, switches, etc.) to verify that energy isolation has been accomplished. Controls must be deactivated or returned to the neutral mode after test.
         (2) Check the equipment/process by use of test instruments and/or visual inspection to verify that energy isolation has been accomplished.
      f. The equipment/process shall be examined to detect any residual energy. If detected, action must be taken to relieve or restrain the stored energy.
   3. Release from lockout/tagout
      a. Each lock/tag shall be removed by the authorized individual who applied it prior to leaving the job.
         (1) A procedure shall be developed to deal with instances where employees have left the jobsite without clearing their personal lock/tag.
      b. The individual responsible for the equipment/process (affected employee) shall be notified when the work is complete and the overall lockout/tagout has been cleared.
         (1) Before equipment/process energization, visual inspection of the work area should be made to insure that all personnel are in the clear and that all nonessential items have been removed and components are operationally intact.
   4. Specific procedures
      Each facility will develop specific energy isolation procedures for major machines/equipment/process components/utilities, etc. A "Lockout/Tagout Checklist for Energy Isolation or Job Safety Analysis" is suitable for this requirement.

C. Special lockout/tagout situations
   1. Lockout/tagout interruption (energized testing)
      In situations where the energy isolating device(s) is locked/tagged and there is a need for testing or positioning of the equipment/process, the following sequence shall apply:

a. Clear equipment/process of tools and materials.
b. Clear personnel.
   c. Clear the energy isolating device(s) of locks/tags according to established procedure.
   d. Proceed with test.
   e. De-energize and relock/tag energy isolating device(s) to continue the work.
   f. Operate controls, etc., to verify energy isolation.

2. Exposure of noncompany personnel
   a. Company and outside employers (contractors, etc.) shall inform each other of their respective lockout/tagout procedures.
   b. Each facility shall ensure that its employees understand and comply with the requirements of the outside employer's or mutually agreed upon energy control procedures.
3. Multiple personnel protection
   For major process/equipment overhaul, rebuilds, etc., which require crew, craft, department or other group lockout/tagout, a system is required that affords employees a level of protection equivalent to that provided by personal lockout/tagout.
4. High voltage work
   Special written procedures shall be developed to describe the lockout/tagout measures necessary when employees are required to work on high voltage circuits or equipment (above 600 volts).
5. Shift change
   Facilities shall develop specific written procedures to accommodate those situations where it is necessary to continue the current lockout/tagout of the equipment/process into subsequent shifts.

VI. Exception
Unique requirements for equipment/process service such as jogging, threading coil/stock, etc. may necessitate employee activity under energized conditions. Each such task must be evaluated to provide safeguarding techniques to protect employees from equipment/process exposures (see V.A.3.b.).

VII. Education and Training
A. Training shall be provided prior to assignment to ensure that the purpose and function of the plant lockout/tagout program is understood by employees and that the knowledge and skills required for the safe application, use, and removal of energy controls are acquired. The training shall include the essential elements of 1910.147 and the following:
   1. Each affected employee shall be instructed in the purpose and use of the energy control procedure.
   2. Each authorized employee shall receive training in the recognition of applicable hazardous energy sources, the type and magnitude of the energy available in the work place, the methods and means necessary for energy isolation and control, and the means of verification of control.
   3. Other employees whose work operations are or may be in an area where energy control procedures may be utilized shall be instructed about the procedure and about the prohibition relating to attempts to restart or reenergize machines or equipment which are locked out or tagged out.
B. Retraining shall be provided annually to reestablish employee proficiency with control methods and procedures.
   1. Retraining also shall be provided for all affected and authorize employees whenever there is a change in job assignments, a change in machines, equipment or processes that present a new hazard or when there is a change in the energy control procedures or revision of control methods.
   2. Additional retraining shall also be conducted whenever periodic audits (see VIII, Management Controls) reveal or whenever supervisory observations give reason to believe that there are deviations from or inadequacies in the employees knowledge or use of energy control procedures.
C. Plant documentation shall certify that employee training has been accomplished and is being kept up-to-date. The certification shall contain each employees name, clock number and dates of training.

VIII. Management Controls
A. Each facility shall develop and document a formal compliance audit of the lockout/tagout energy control procedure *semiannually* to ensure that employees are knowledgeable

and utilize the designated procedures. The documentation shall identify the machine or equipment on which the energy control procedure was being utilized, the date of the inspection, the employees included in the inspection and the person performing the inspection.

1. The semiannual audits shall be performed by an authorized management employee.
2. The amount of lockout/tagout auditing should adequately represent the size of the plant and number of authorized employees.
3. The audits shall be designed to correct any deviations or inadequacies observed.
4. Where lockout is used for energy control, the audit shall include a review between the inspector, and each authorized employee of that employee's responsibilities under the energy control procedure being audited.
5. Where tagout is used for energy control, the audit shall include a review, between the inspector and each authorized and affected employee, of that employees responsibilities under the energy control procedure being audited to ensure that employees understand the limitations of a tagout system and their purpose in the energy control program.
6. Where tagout procedures are used, other employees whose work operations are or may be in the area shall be contacted by the inspector to ensure that they are aware of and understand the purpose of the procedures.

NOTE: If compliance with any element of this procedure is not possible, equivalent protection must be provided by an alternate system approved by the Safety/Health Department.

Appendix 5

Case Studies

The following energy-release-incident case studies have been taken from a variety of sources.

All are actual incidents with certain source, personal data, etc., removed to protect the privacy and confidentiality of the individuals/organizations involved. They represent a cross section of incidents with a variety of causal factors that represent proximate, contributory, and root causes. Most of the cases do not reveal *all* of the causal factors involved in the incident's occurrence. In some instances, the investigations apparently did not proceed sufficiently to discover the less obvious or remote causal events. In others, they were not available.

The case studies can be useful in a number of ways: (1) to increase awareness of employees/managers, (2) to sharpen investigative skills, (3) to identify deficiencies in existing hazardous energy control systems/procedures, and (4) to identify hazard potentials not previously anticipated.

Cases 1, 3, 4-11, 13-14, and 16-20 were prepared by Edward V. Grund. Cases 2, 12, and 15 were adapted from the State of Florida, Department of Labor. Finally, cases 21-38 were adapted from MSHA's "Fatalgrams." (See 30 *CFR* 56-57 for MSHA regulations.

CASE 1: CLEANING ENERGIZED MACHINERY

The employee reached over a barrier guard to clean the rubber leveler rolls on the coil line with the unit operating. While cleaning the rolls, the rag and the glove he was wearing became caught and pulled the right hand into the rolls crushing the thumb, index finger, and middle finger.

Investigation revealed that the barrier guarding was in place, but the employee was tall (6 ft 4 in. [1.9 m]) and was able to reach over the guard to clean the roll.

Recommendations

The barrier guard was replaced with a tighter fitting point of operation guard to preclude access to the roll nip points. Employees were instructed to de-energize the machine when cleaning and lockout the power. The existing procedure was inadequate in describing the requirement for lockout during this cleaning procedure. The employee believed it was easier and quicker to clean the rolls while they were turning. A study is being conducted to determine what could be done in a "jog mode" to facilitate roll cleaning yet ensuring employee safety. The incident was reviewed with all crews, stressing the hazard and the required controls.

CASE 2: INADEQUATE SAFEGUARDS

"Go ahead!" shouted a tile pourer for a roofing tile manufacturing company, and his co-worker closed a cement mixing machine wall mounted activating switch. Within seconds the pourer had been caught and mutilated by the mixer's turning blade.

The two workers had been assigned the task of cleaning the mixer. The date was Monday, October 9, 1978, and the clock read 3:15 p.m.

The work had started at 2:30 p.m. with the two men using a shovel, scraper, hammer, and so on, until they were midway through the job. Access to the mixer blades was gained by standing on an elevated platform installed at the front of the machine.

The victim was standing on the platform, both hands holding a scraper. The mixer control switch was open, but not locked out. The co-worker was standing 25 ft (7.6 m) away at the switch waiting for the victim's instruction to close it. Upon hearing the go ahead signal, he closed the switch, and almost immediately heard the victim call out a single word, "Oh!" and saw the latter's body disappear into the mixer.

Re-opening the switch, the co-worker ran the 25-ft (7.6-m) distance to the mixer, and saw the victim lying at the bottom of the machine, his body badly mangled by the blades.

Help arrived in the form of other employees, and in five minutes an emergency squad was on the scene with three other such units. The victim was removed from the mixer and taken to the hospital where he was pronounced dead on arrival.

Recommendations

- OSHA Standard for General Industry states in Section 29 *CFR* 1910.212(a)(1): One or more methods of machine guarding shall be provided to protect the operator and other employees in the machine area from hazards such as those created by point of operation, ingoing nip points, rotating parts, flying chips, and sparks. Examples of guarding methods are barrier guards, two-handed tripping devices, electronic safety devices, etc.
- Further, 29 *CFR* 1910.212 (a)(4) states: Revolving drums, barrels, and containers shall be guarded by an enclosure which is interlocked with the drive mechanism, so that the barrel, drum, or container cannot revolve unless the guard enclosure is in place.
- Whenever maintenance work is being performed on cement mixers or other machines the master switch controlling the equipment should be locked out by positive means such as a padlock the key to which should be in the sole possession of the employee doing the work. If two employees are working at the job, each should have his/her own lock and key. Using this lockout method, the probability of the worker(s) being trapped by machinery that might be inadvertently activated by other persons closing the unit's operating switch will be greatly reduced.
- Employers should make certain that all their employees engaged in operating or performing maintenance on machinery are trained in the safe performance of their work.

CASE 3: TROUBLESHOOTING DANGERS

The plant electrician had removed a guard to view timing cams on a printing unit. The removed guard covered both the cams and planetary gears on the unit. The electrician was sitting on a platform as he was observing the equipment run. In attempting to stand up, he grabbed the printer near the planetary gears for leverage. The gear caught his fingers, crushed his index and ring fingers, and amputated the tip of his middle finger on his left hand. It was reported that as the hand cycled through the gears, he did not pull his hand out.

Recommendations

The incident was reviewed with maintenance employees who were reinstructed on moving machinery hazards. The plant also modified the existing guard so that it would be possible to view the timing cams without removing the planetary gear guard. No troubleshooting procedure existed which defined what alternative protective measures were to be taken when guards were removed and equipment was required to be under power.

CASE 4: EQUIPMENT MALFUNCTION

The electrician was servicing a 440-V power supply for glass furnaces. After locating a blown 100-amp fuse in the electrical switch box, he removed the fuse and replaced it with a new one. He closed the electrical box with his left hand and pushed the lever to the ON position. A short in the circuit caused the fuse to arc and flash. The force of the arc blew the box open, exposing the electrician to the arc, causing burns to both hands and wrists and to the left side of his head, ear, and face. His eyes were not injured by the flash because he looked away when the circuit was energized.

Recommendations

The plant investigation showed that the latches on the electrical box were defective and could not secure the door when it was closed. The incident was reviewed with all electricians emphasizing the need to properly position oneself before activating circuits and checking to see that cabinets-boxes can be properly closed before restoring energy. All electrical boxes in the plant were also audited to ensure that the doors on electrical panels and boxes are working properly so that a similar injury would not occur. All electricians were also reinstructed in how to position themselves in relation to the electrical box before activating the energy circuit.

CASE 5: CONTROL CIRCUITRY ASSUMPTION

A maintainer trainee suffered a fracture of the right index finger and thumb when another employee jogged the bodymaker while he was checking the ram for scrap. The incident indicates the need for proper procedures to isolate energy sources as required by company policy and regulatory standards on lockout/tagout procedures.

The investigation showed that bodymaker unit 42 was down while a maintainer was working on the trimmer. During this time, the adjacent bodymaker unit 41 stopped and the maintainer trainee, who was standing in the area with the supervisor, opened the unit to see if there was any scrap in the domer. When the maintainer finished working on the trimmer, he started unit 42. Then he noticed the red light on unit 41, and instinctively jogged the machine without observing if it was clear to do so. The activation of the equipment caught the trainee's right hand between the ram carriage and the redraw carriage.

Recommendations

The incident was reviewed with front end mechanics, operators, and supervisors on all crews. The plant is also installing E-stop controls, which will require resetting the stop controls before the equipment can be activated at the control console.

CASE 6: QUESTIONABLE WORK PRACTICE

A break had occurred in the stream or web of foil being fed through rollers in a laminating operation. The assistant laminator operator was attempting to reduce the resulting foil loss.

Foil is initially drawn into the laminator from a roll. It is then advanced along a series of idlers to two glue applicator rollers. From there, the foil moves around a "combiner" roller, where it is joined by a stream or web of paper coming from the opposite direction (Figure A5-1).

In this incident, the foil stream broke between the glue impression roller and the combiner roller. The foil began to wind around the glue roller. To prevent further foil accumulating around the roller, the employee broke the stream between the foil roll and the first idler by striking his hand and wrist through the foil, as is customary. The distance between the foil roll and the first idler was one inch. After his hand broke

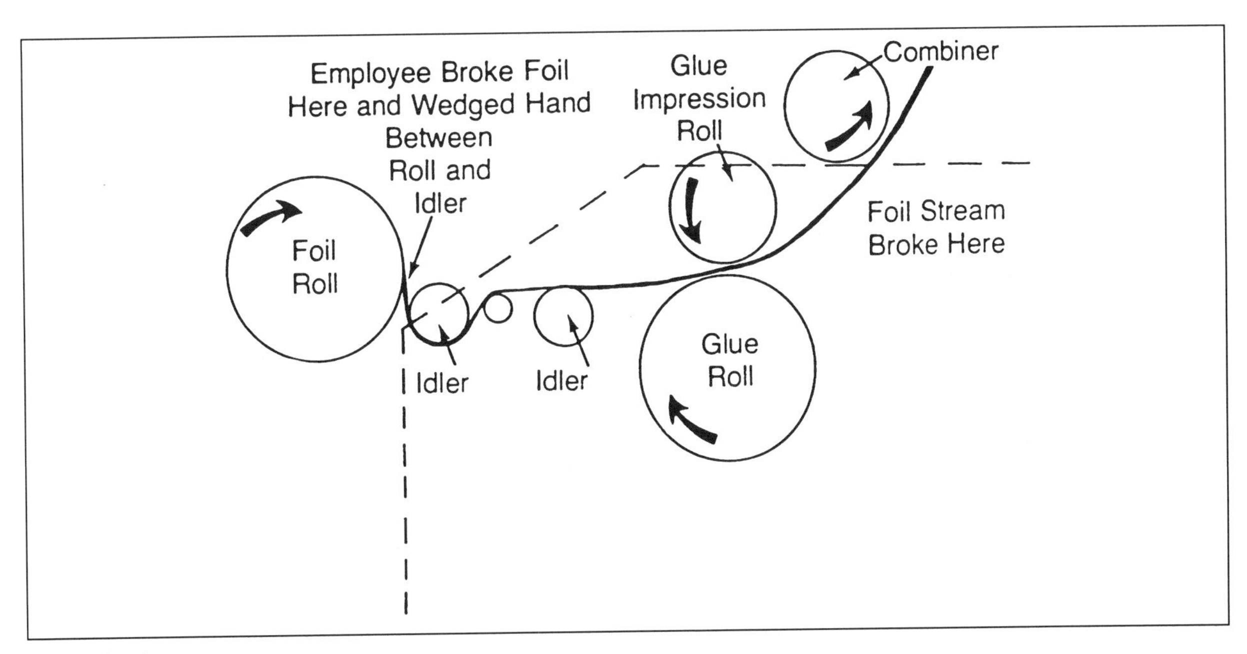

Figure A5-1.

through the foil, it continued downward into the one-inch gap. The moving roll and idler wedged it downward between them. He sustained a hairline fracture to his right wrist.

Recommendations

- A two- to three-inch space will be maintained between the foil roll and idler.
- A mechanical means to stop the foil stream once it has been broken will be investigated.
- Review of the customary practice of breaking the foil stream with the hand will take place to determine if other methods are available without exposure to roll rotation.

CASE 7: IMPROPER CLEANING METHOD

An employee used a piece of wood to clear the sawdust from around the blade of a cross-out table saw. The still revolving blade grabbed and pulled the wood and the employee's hand into the blade (Figure A5-2). Several tendons were severed; after reconstructive surgery, the employee has 90% use of the hand. No procedure existed for saw clean-up activity.

Recommendations

- Clear disks were installed as guards on either side of the blade so the operator's hand will strike the guard, allowing sufficient time to release his/her grasp and prevent his/her hand from contacting the blade.
- A procedure requiring opening the wall mounted electrical disconnect after saw turn-off was prepared.
- Proper cleaning tools were defined in the procedure.

CASE 8: CONTRACTOR CONFUSION

Two welders were at a fish processing plant doing general welding and cutting operations. They were to cut 3 in. x 3 in. (8 cm x 8 cm) squares on a 16 in. (41 cm) auger (screw) and had been in the trough about five minutes when the incident occurred. At the time of the incident, two other workers were repairing leaks on a dryer and it was necessary to rotate the dryer one-half turn to get to the leaks on the bottom. One of the men went to the control panel board and since it was not witnessed, it is assumed that he started the auger instead of the dryer. The two welders in the auger trough were caught in the activated auger resulting in death to one man and serious injuries to the second. The man who went to the control panel board could neither read nor write. The superintendent for the plant stated that the switches (a total of 27) in the control panel board were not locked out nor were they tagged. The screw (auger) installation was new and the cutting of the auger blade was the last work to be accomplished on the new installation.

The employer was cited for violation of 29 *CFR* 1926.555(a)(7): Failure to provide lockout or otherwise, render inoperable or tagged out with a DO NOT OPERATE on the screw conveyor electrical switch box controlling a 16 in. (41 cm) screw in trough 18 in. (46 cm) in depth.

CASE 9: FATAL MISTAKE

The deceased had been doing clean-up work in the area for seven months. The equipment that he was cleaning at the time of the incident was a conveyor system, which fed meat through and past a 18-in. (46-cm) circular meat saw mounted on one side. Op-

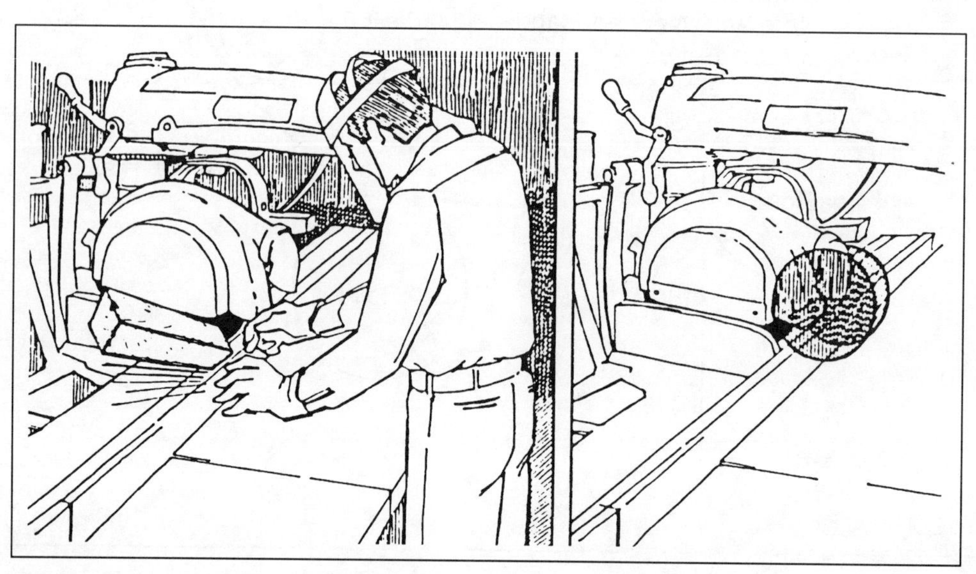

Figure A5-2.

posite the saw were twin switches for starting the saw and conveyor system. The conveyor at the saw is approximately 45 in. (114 cm) off the floor and 29 in. (74 cm) wide at this point. The two switches located on the side opposite the saw are about 44 in. (112 cm) above the top of the conveyor. The guard is removed from the blade when cleaning the machine. The deceased was found in a semi-sitting position with his T-shirt wrapped around the saw shaft and a deep cut through the back and beltline. Neither the saw or conveyor was operating when the body was found. It was determined that the saw switch was on but the saw motor had tripped off the circuit breaker by means of an overload protection. The weight of the deceased's body caused the saw, which had been operating, to overload. It was concluded that the deceased, rather than walk around to the switch side, climbed up on the braces, hit the wrong switches (he wanted the conveyor switch) while his T-shirt rested on the exposed 18-in. (46-cm) blade and was immediately pulled into the saw. There were facilities for locking out the saw switch but they were not used.

CASE 10: INADEQUATE SAFEGUARDS

The deceased's job was to fill an open auger with spilled fertilizer materials that would eventually become a saleable mixture. The procedure followed was for the foreman to turn the auger off while the deceased filled it and then the deceased would return to the foreman to tell him to turn it back on. No one was to go near the auger while it was on. This procedure had been followed for several years but the company had no written procedures for this operation. The controls for the auger consist of one green and one red button located 40 ft (12 m) from the auger so the person standing at the control buttons cannot see someone standing in the narrow hall where the auger is located. On the day of the incident, the deceased had called (later, the foreman said the deceased had come to him) to start the auger. The foreman then called back to the deceased to stay away from the auger and turned it on. While talking with a representative of a scales company, the two men heard the deceased yell "Turn it." The foreman did so and the two men ran to the auger and found the deceased completely in it except for portions of his legs. The deceased was wearing oversized coveralls before the incident. The narrow hall had several piles of fertilizer material in it, some of which had been "caked" into semi-hard texture. It is possible that the deceased's oversized clothing may have become tangled in the auger or he may have slipped on the piles of fertilizer. His shovel was found outside the hallway.

The employer was cited with 29 *CFR* 1910.212(a)(1): One or more methods of machine guarding were not provided to protect the operator(s) against hazards created by nip points and rotating parts.

CASE 11: CREW TASKS—INCREASED POTENTIAL

Three employees were assigned to the winder area to clean up the area in preparation for a reel of paper. All three employees were familiar with the area and had knowledge of the equipment and how to clean the area. Two employees (including the deceased) were using air hoses to clean the winder and the third was in the scale area cleaning equipment. During normal clean-up operations, the winder cradle or gate is raised to the UP position. This allows the crew to clean under the winder. While the cleaning-up activity was being completed, the dry end operator was becoming concerned about the paper going into the pulp in the pit. He then instructed an operator to have the "showers" turned off so that the consistency in the pit could be maintained but instead he (operator) proceeded to the winder control panel in another area and lowered the winder cradle. There was a scream and the deceased was found under the winder cradle crushed from above the knee to the lower chest cavity region. The winder cradle was raised and the deceased removed. It was surmised that the deceased had gone to the front side of the winder and saw scrap material under the winder and drive rolls. He then laid down on his stomach and went under the equipment.

The employer was cited with the following incident related standard: 29 *CFR* 1910.261(b)(4); Devices such as padlocks were not provided for locking out the source of power before any maintenance, inspection, cleaning, adjusting, etc., that require entrance into or close contact with machinery etc.

CASE 12: NEW EMPLOYEE RISKS

A worker, employed for only four weeks by a mining company, was crushed to death by the paddles of a log washer's rotating shaft when a co-worker accidentally switched on the equipment with the victim inside.

Although the victim was performing routine maintenance on the equipment, investigation of the incident showed that the man was not given proper instructions concerning the safety procedures to be followed before entering the log washer.

Some instructions were given by the plant operator and foreman; however, lockout procedures were not discussed. The machine's switches were labeled but the main power, which was located in the main control room elsewhere on the premises, was not shut off. Gang locking devices and padlocks were available in the main control room but were not used.

The incident happened while other employees of the mining operation were servicing another piece of equipment. A worker entered the control room to activate a second machine and inadvertently pushed the switch that activated the log washer. Another worker managed to escape to safety when the shaft started to move.

Recommendations

- MSHA regulations require that a newly hired worker must have at least 24 hours of training before being assigned to work duties. The training program must include an introduction to the work environment, hazard recognition, and health and safety aspects of the new worker's job.
- MSHA regulations also requires that electrically powered equipment must be de-energized before mechanical work is done on such equipment. Power switches shall be locked out or other measures taken to prevent the equipment from being energized without the knowledge of the individuals working on it. Suitable warning notices shall be posted at the power switches and signed by the individuals who are doing the work.
- Lockout procedures must be developed, instituted, and strictly adhered to, to prevent repetition of similar incidents.

CASE 13: KNOWING IS NOT ENOUGH

A general foreman in a refractories plant had been assisting the safety professional develop the facility lockout/tagout procedures. Pictures were taken illustrating the proper way to de-energize equipment for use in future training sessions. Approximately six months later, a brick press operator notified the same general foreman that the mix material was not feeding properly to his press. The general foreman and the operator ascended the stairs to the upper feed level after pushing the press STOP button. They removed the screw auger cover to determine the problem.

The general foreman saw a plug obstructing the feed flow to the press charging chute. He reached in to remove the plug without returning to the ground level to lockout the electrical breaker. At that moment, shift change was in process. The on-coming operator noticed the idle press and assumed the operator had already gone home. He hit the START button, which caused the auger to rotate, taking the general foreman's arm off the bicep.

The general foreman immediately applied direct pressure to the wound and hurried to the plant manager's office. He told the manager to call for a nearby ambulance and that he was sorry that he really had fouled-up. In the investigation, it was determined that another employee some years earlier had lost an arm and several fingers in a press feed gearing mechanism under similar circumstances. The employee had been performing routine maintenance (i.e., oiling/lubricating) without de-energizing the press and locking or tagging.

CASE 14: HIGH-PRESSURE HOSE

A laborer for a contractor, with approximately five years experience, was using a cold-water-high-pressure (2,500 psig) washer to clean the walls of a hot mill roll stand pit. While cleaning, he stood on the roll screw support which was in the DOWN position. The trailing high pressure hose moved in front of the screw support proximity switch which activated the air cylinders. The activation caused the screw support to raise, trapping his left foot between the support arm and screw. The high-pressure hose contained a metal lining that caused the switch to function.

Recommendations

Investigation revealed that the contractor failed to lockout the air cylinder to the screw support before commencing work. The laborer was unaware of the potential associated with the equipment in the immediate work area. A specific safe job procedure was not available and deficiencies in the contractor safety standards were identified. Re-instruction regarding maintenance shutdown safety was initiated. Lockout/tagout procedures for shutdowns were reviewed and standard practices established.

CASE 15: BLOCKING FAILURE

An iron worker was working for a street paving contractor. During the early part of his first day on the job, he was assigned the task of steam cleaning a motorized scraper haulage bed. He stood on the ground and steamed while positioned between the bed body and the upraised apron. The apron (2,500 lb [72 kg]) fell crushing him against the cutting edge of the bed. The apron hydraulic cylinder pressure released unexpectedly allowing the apron to fall (Figure A5-3). No attempt had been made to block or secure the apron from movement. There was no indication that the iron worker had been given any instruction regarding this potential hazard or what specific cleaning practice to follow. OSHA Construction Standard 29 *CFR* 1926.600(a)(3)(i) states: Bulldozer and scraper blades and similar equipment shall either be fully lowered or blocked when being repaired or not in use.

CASE 16: SUPERVISORY NEGLIGENCE

The plant maintenance manager, the electrical supervisor and an hourly electrician were investigating a pallet misfeed problem on the palletizer. The electrician pushed the STOP button to shut down the equipment and climbed the outside frame of the pallet loading area. Access was possible because the side panel, which guarded the unit, was missing.

At the time of the incident, the electrician was bal-

Figure A5-3.

ancing his right foot on a 2 ft x 2 ft (0.6 m x 0.6 m) rib section, 8 ft (2.4 m) above the ground. As he focused his efforts on locating the problem, the machine cycled. The hoist support bar of the pallet infeed mechanism acting as a shear, chopped through his right safety shoe and amputated the right big toe. The pallet feed mechanism cycled because the electrician was attempting to adjust the sensors and pneumatic actuators.

Recommendations

Investigation disclosed that a series of acts and conditions contributed to the incident: failure to lock-out equipment so it could not restart; uncontrolled system of work—three people were involved, but each was unaware of the other's actions; side panel guard was missing, allowing access; and climbing on equipment and standing on a rib section bar that had minimal clearance for the hoist support bar. Corrective action addressing these failures was initiated.

CASE 17: PROCESS POWER-ON

The coil coater lead mechanic was preparing a second coil for splicing to a first coil, which was being run on the unit. In attaching a filler strip to the leading edge of the second coil that was laying on an enclosed table above the running coil, the filler strip slipped from the employee's hands and fell onto the moving coil below. When the employee instinctively reached for the strip to prevent it from moving further down the line, his hand was pinned between the moving coil and a nonenergized roll. This action resulted in a crushing injury, which partially denuded the left hand and fractured four bones. Alert responses from fellow crew members in shutting down the operation and freeing the employee prevented a more serious injury.

The investigation showed that when the first coil is being unwound, the tension brings the strip against the roll and creates a pinch point hazard. The roll functions as an idler roller only when the second coil is being rewound.

The condition (roll arrangement) existed for 20 years without a similar incident or detection as a hazard. The lead mechanic had 19 years of service with only one reportable injury during his employment.

Recommendations

A guard was installed which prevents access to the pinch point and an emergency STOP control was placed at the work station. Management is also reviewing the entire line for pinch points and a lower risk method of preparing for the coil changeover. Energy isolation is not feasible since power is needed to make the splice. However, safer alternatives that further minimize exposure may be possible.

CASE 18: WORKER COORDINATION

A production mechanic, who was assigned responsibility for operating two end presses, suffered an amputation at the first digits of the middle and ring fingers when he placed his right hand into the die area while the press was being jogged.

One assigned end press was experiencing a sporadic operational problem of dropped tabs, and a production maintenance mechanic was troubleshooting the problem. After observing the most recent jam, the maintenance mechanic engaged the press at the main control panel and began to use the two hand jog buttons to turnover the press and remove the jam.

It was reported that the maintenance mechanic had checked the safety mirror prior to jogging the unit. Although he saw the injured employee standing at the discharge belt, he never expected him to place his hand into the unit since they had previously removed the jams by jogging the unit until the damaged end cleared the die and was on the discharge belt.

Recommendations

The investigation showed that the injured production mechanic was primarily responsible for the incident for the following reasons:

- Failure to communicate with fellow employee that he was going to reach into the unit to remove the tab.
- Failure to de-energize equipment if hands were going to be placed into unit.
- Failure to use die blocks when putting hands into press.
- Failure to follow safety instructions. Earlier on the shift, the employee was given safety talk on use of die blocks.

The press discharge area is not guarded and the injured employee had to reach almost 2 ft (0.6 m) into the unit where the injury occurred. The plant is evaluating the need for additional guarding.

All departments of the plant were shut down to communicate the incident to all employees. Plant specific safety procedures were also reviewed regarding requirements before placing hands into machinery exposure points.

CASE 19: INEXPERIENCED ELECTRICIAN

An electrician working the midnight shift with one month's plant service was fatally electrocuted while attempting to run wiring to a 110-V lighting fixture. The employee was working at ceiling level approximately 18 ft (5.5 m) from the floor. A machine operator looked up and saw the deceased draped over some piping and tangled in the chain holding the fluorescent light. He had been running wire from the light fixture to a junction box.

The investigation revealed no exposed wiring, but the circuit was energized. There was no necessity for the circuit to be hot for the task that was being performed. A pair of uninsulated side cutters was found on the floor nearby. The plastic nut covering the hot wires at the junction box had a small sharp cut in it. It is believed that the employee was trying to remove the plastic nut by pulling and twisting with the side cutters. He apparently made contact with the live circuit and was well grounded on the surrounding piping and steelwork.

The employee had failed to de-energize the power circuit and follow the facility lockout/tagout procedure. The employee's prior electrical experience was less than five years and associated with residential construction. Nothing was found to indicate the deceased had any real meaningful exposure to or training in lockout/tagout concepts or procedures. However, it is believed that he understood the dangers and safe practices necessary for working on wiring circuits. No evidence was found that any special safety indoctrination was provided the employee during his first 30 days at the facility.

CASE 20: JUDGMENT ERROR

A maintenance mechanic with 11 years of service was fatally injured while working on the night shift.

At approximately 2:30 a.m. on Saturday morning, the furnace operator observed that the batch bin was overflowing, an indication that the "bindicator" (an electronic control) was not functioning. This malfunction caused the auger, which moves the batch material to the bin, to continue to run after the proper furnace level had been reached. The furnace operator verified this by going to the platform where the auger was located. The steel floor plate covering the auger was in place at that time.

The furnace operator advised another operator of the problem and that he would get maintenance. He returned with the deceased, a mechanic, and assisted in cleaning a device that controls the glass level in the furnace. This did not correct the problem and the auger continued to run.

The mechanic then climbed the stairs to the batch platform, leaving the furnace operator below. After a period of approximately one minute, the furnace operator heard a scream. The furnace operator looked up, and because he did not see the mechanic, he immediately hit the STOP button shutting down the auger. He quickly ran up the stairs to the platform and found the deceased had slipped or fallen into the auger and that his right leg had been pulled in up to the pelvis. Death was immediate due to loss of blood and shock.

Apparently, the deceased had removed the auger inspection cover to troubleshoot the equipment. Means were available to isolate the auger (control circuit and electrical disconnect) during the inspection. No negative process impact would have been experienced during a short shutdown period for the needed maintenance. No communication occurred between the mechanic and furnace operator regarding what he intended to do while on the upper platform.

CASE 21: SERVICING MOVING MACHINERY

A truckdriver was killed when he was drawn into a conveyor belt-head pulley pinch while applying belt dressing (Figure A5-4). The victim had 11 years experience in his regular truck driving job and 42 days as a crusher operator in the job where he was filling in at the time of the incident.

The victim was applying belt dressing to the 30-in. (76-cm) diameter unguarded head pulley using a small squeeze can, and working from a platform around the head pulley. The belt was 42-in. (107-cm) wide. The incident was not witnessed as the other workers were busy, but just after completing another task, the regular crusher operator discovered the victim caught in the pinch point.

Recommendations

Guarding for in-running nip points is required and belt dressing must not be applied manually with belts in motion. Equipment shall be de-energized before service work of this nature is performed.

CASE 22: CONTROL CIRCUIT CALAMITY

A plant laborer was fatally injured when a log washer was inadvertently energized while he was working inside (Figure A5-5). The victim had a total of four weeks mining experience.

The control switch for the log washer was on a multiswitch console inside the main plant control house. This console contained the push-button control switches for all equipment in the plant which included the log washer, screens and feeder that were shut down for scheduled maintenance. The plant operator wanting to "bump" the pan feeder into a better position for the scheduled maintenance, pushed the wrong switch, and started the log washer.

Recommendations

Employees should receive proper training regarding lockout/tagout requirements with emphasis on personal responsibility for safeguarding themselves. Electrical energy supply shall be isolated and locked out or other measures taken which prevents machinery movement/operation.

CASE 23: SAFETY BLOCK FAILS

A dozer operator was fatally injured by a slab of limestone that fell and crushed him against the front of a dozer (Figure A5-6). The victim and a co-worker were replacing cutting plates on the dozer's blade. A larger limestone boulder was positioned under the raised blade as a safety block. The victim was working in the confined area between the dozer blade and the radiator, cutting away the rusted nuts and bolts so that so that the old cutting plates could be removed. A slab weighing about 2,000 lb (907 kg) separated from the limestone safety block and fell toward the victim. The slab pinned the victim's head against the radiator nose guard of the dozer before he could escape.

Recommendations

Using unprepared natural materials for safety purposes (props, blocks, chocks, etc.) increases risk to exposed personnel. Integrity of the material may be impossible to determine under field conditions. Use designed/prepared materials for restraints for gravitational forces and avoid application of only a single device.

Figure A5-4.

Figure A5-5.

CASE 24: INEXPERIENCED SUPERVISOR

A pit foreman, with 10½ months of experience in his job and about a year of previous mining experience, died when a dump truck body fell on him as he worked on the drive shaft beneath it (Figure A5-7).

The drive shaft had a mechanical problem to be corrected, and the victim and his companion had decided to remove it from the truck entirely to facilitate bringing the truck into the shop. They raised the body in order to carry out their plan, but failed to block it. The victim leaned over the truck frame, removing bolts, while his companion supported part of the mechanism from the ground beneath the truck. As they were doing this, the body came down, pinning and crushing the victim.

The foreman relied on the trucks hydraulic system to restrain the dump body while the intended work was performed. He compounded his error by placing himself in the primary danger zone. These actions were in direct conflict with the requirement to block any piece of mobile equipment in a raised position.

CASE 25: ENERGIZED ERROR

A first-class electrician was fatally injured when he came in contact with an energized 13,800-V phase leg of a delta system, receiving approximately 7,976-V phase to ground (Figure A5-8). The victim had a total of 5 years and 6 months mining experience, 4 years and 6 months of this time as a first class electrician, all at this mine.

The victim and his working partner were replacing damaged and broken frame ground wires at a substation which consisted of a power pole, two banks of transformers, switchgear, and wiring all enclosed within a cyclone fence. The elements involved in this incident were the primary side of the 13,800/480-V transformer and the uninsulated phase leg jumper connecting the two outside transformers. One end of the ground wire had been disconnected and while the victim was working to disconnect the other end, the ground wire came in contact with the uninsulated phase leg.

The employees were knowingly performing the

Figure A5-6.

work under "live/hot" conditions without conforming to appropriate practices. Hot-line tools, protective equipment, and insulating gear was not used nor was the electrical power circuit de-energized.

CASE 26: DANGER ZONE

A crusher operator with one year of mining experience was fatally injured when he was pulled around a shaft after his clothing became entangled in the unguarded sprocket attached to the shaft (Figure A5-9).

The victim had crawled into a confined area under an inclined conveyor and in close proximity to the unguarded sprocket to make adjustments to the crusher while the machinery was energized.

A number of failures and deficiencies were involved in the incident, which may suggest that there was a basic problem with the management safety system. A repair/adjustment task had been undertaken without de-energizing the equipment. Elements of the lockout/tagout procedure (i.e. employee training, maintenance shutdown authority/scheduling, notification, etc.) may be ineffective. The unguarded sprocket apparently was believed to be guarded by its position although access was possible. Employees failed to react to the combination of exposures presented when sizing up the work to be done.

CASE 27: TROUBLE WITH TROUBLESHOOTING

A crushing plant foreman was fatally injured when he was caught between an unguarded conveyor belt and the plant structural steel framework (Figure A5-10).

The victim had signaled the plant operator to start a screen feed belt conveyor while he observed the screen for a possible source of oversize stone. He became entangled between the conveyor and the plant framework. The conveyor was not guarded nor equipped with an emergency stop device.

Recommendations

The foreman failed to maintain a safe distance while observing the performance of the lift type conveyor belt. He apparently attempted to investigate some condition by using his hand while the belt was in motion. These actions were unwarranted under operating conditions. Interlocked access guards with viewports may have prevented the fatal behavior.

Figure A5-7.

CASE 28: HIGH-VOLTAGE HAZARDS

A laborer with 18 years mining experience was fatally injured when he contacted an energized 34,500-V conductor (Figure A5-11). The victim and a co-worker were cleaning the top of a transformer installation and an oil circuit breaker. Neither installation had been de-energized or locked out prior to commencement of work and both miners were exposed to the hazard.

The victim contacted the energized line, was severely burned and fell approximately 14 ft (4 m). His death occurred five days later.

Recommendations

Power circuits shall be de-energized before work is done on such circuits unless hot-line tools are used. Suitable warning signs shall be posted by the individuals who are to do the work. Switches shall be locked out or other measures taken which shall prevent the power circuits from being energized without the knowledge of the individuals working on them. Such locks, signs, or preventive devices shall be removed only by the person who installed them or by authorized personnel.

CASE 29: CO-WORKER CONFUSION

A repairman-welder, with no recorded previous mining experience, was killed while trying to clean a tail pulley prior to making repairs (Figure A5-12).

The guard had been removed to facilitate the work to be done. Prearranged procedures between the victim and his partner for a momentary conveyor start-up failed to include some means for an audible warning signal prior to throwing the start-up switch. Vision from the start-up switch to the tail pulley area was obscured by a large motor and the approaching darkness. The victim unwarned of the start-up became caught between the conveyor belt and the tail pulley then dragged to the opening beneath the tail pulley and dropped 35 ft (10.7 m) to the ground below.

Recommendations

- Pulleys of conveyors shall not be cleaned manually while the conveyor is in motion.
- When the entire length of the conveyor is not visible from the starting switch, a positive audible or visible warning system shall be installed and operated to warn persons that the conveyor will be started.

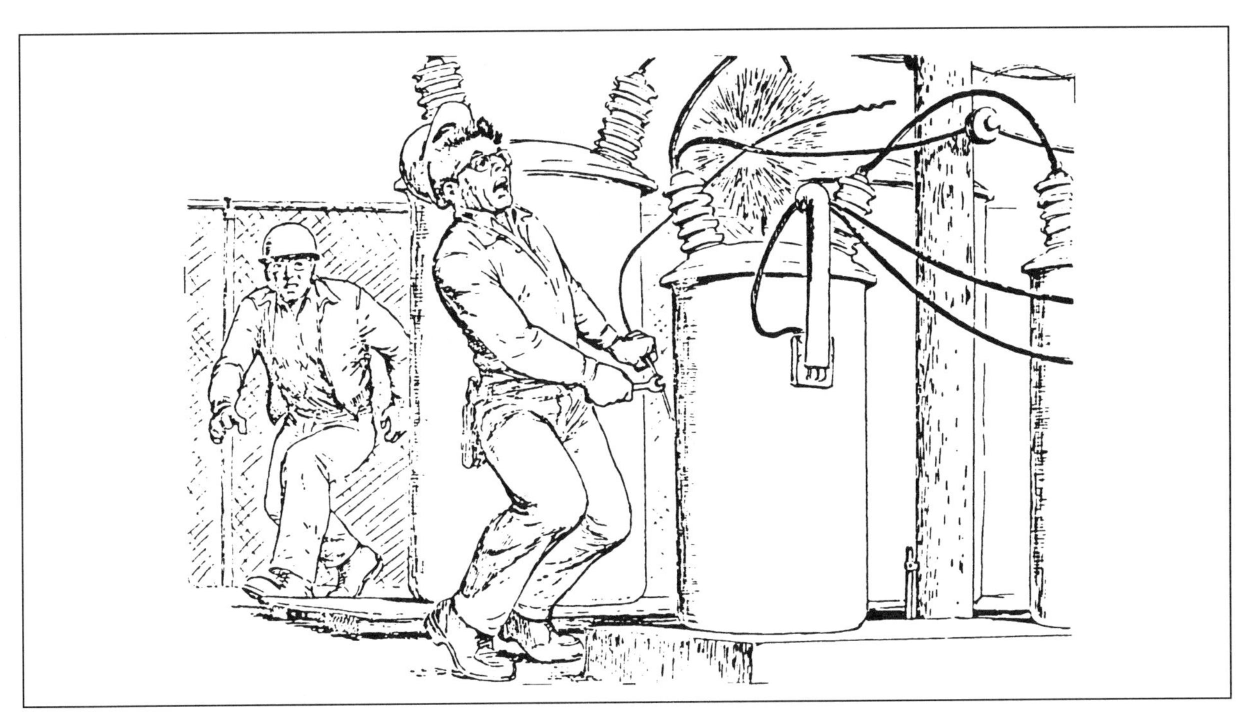

Figure A5-8.

Figure A5-9.

- Illumination sufficient to provide safe working conditions shall be provided in and on all surface structures, paths, walkways, stairways, switch panels, loading and dumping sites, and working areas.
- Electrically powered equipment shall be de-energized before mechanical work is done on such equipment. Power switches shall be locked out or other measures taken which shall prevent the equipment from being energized without the knowledge of the individual working on it.

Figure A5-10.

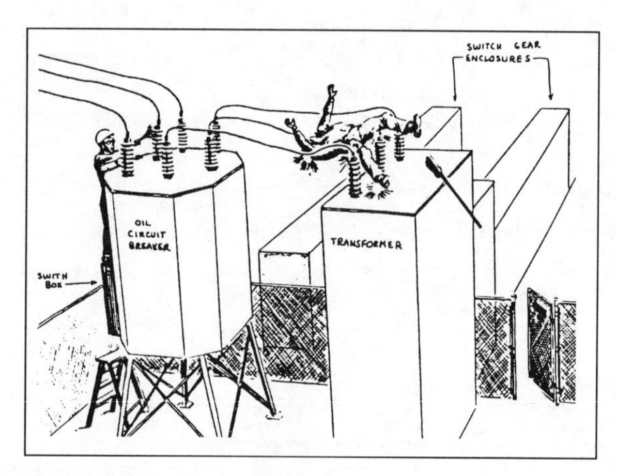

Figure A5-11.

CASE 30: CONTROL CIRCUIT FOLLY

A load-out operator with eight months mining experience, four months on this job, was fatally injured when the bucket elevator where he was installing a new bucket was energized by his co-worker.

The victim and his co-worker had removed the access cover and had spotted the elevator belt in a missing bucket location when the bucket replacement was interrupted to load a customer's truck. The plant operator inadvertently started the elevator not knowing his co-worker had resumed the installation of the bucket. The elevator was not de-energized, locked, or tagged while maintenance was being performed.

Recommendations

- Electrical powered equipment shall be de-energized before mechanical work is done on such equipment. Power switches shall be locked out or other measures taken which shall prevent the equipment from being energized without the knowledge of the individuals working on it. Suitable warning notices shall be posted at the power switch and signed by the individuals who are to do the work. Such locks or preventive devices shall be removed only by the persons who installed them or by authorized personnel.
- New employees shall be indoctrinated in safety rules and safe work procedures.

CASE 31: MULTIPLE FEED

An electrician, with 10 months experience in mining and 10 years as an electrician, was electrocuted while attempting to position himself on the steel framework of a substation. He contacted an energized 13,200-V power line (Figure A5-13).

The victim with three others, two of them also electricians, was preparing to clean the insulators on the substation. The installation received power from two sources at 13,200 V in order to continue operation in

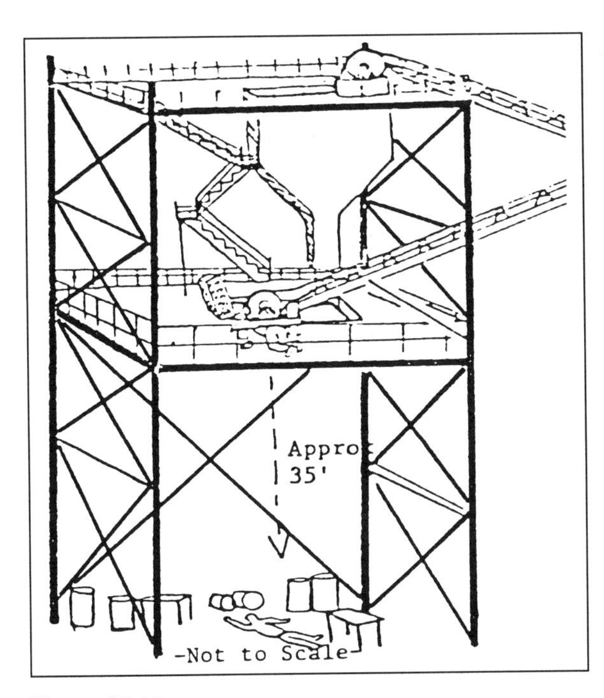

Figure A5-12.

the event of a power failure. Because of this, there were several points to disconnect in order to shut off power completely for work on the substation. In this case, something was missed and the victim contacted 13,200 V when he undoubtedly believed it had been disconnected.

Recommendations

- Power circuits shall be de-energized before work is done on such circuits unless hot-line tools are used.
- A procedural routine should be developed for disconnecting the power to the transformers, which should be reduced to written instructions to be followed whenever substation maintenance is required.

CASE 32: SHOCK-TRIGGERED FALL

The owner-operator of a sand and gravel pit was fatally injured in a fall from a ladder after sustaining electrical shock while attempting a repair to wires supplying power to his operation without shutting off the power. The victim had 25 years of mining experience, the last 10 in his current position.

Power was supplied to the operation in a three phase system at 12,500 V, and reduced through a pole-mounted transformer to 480 V, which was carried to a second pole only 10½ ft (3.2 m) away and then entered conduit and passed to a fused disconnect. The victim was working with the 480-V wiring on the secondary pole trying to find the two bare wires at the 480-V potential. He was paralyzed by the shock and could not release himself. No disconnect point was

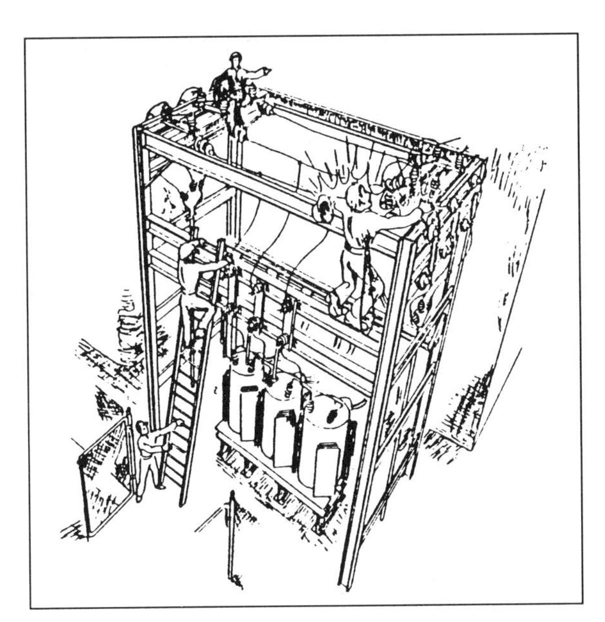

Figure A5-13.

conveniently located, so his son, who was tending the ladder, shook it until the victim fell. He broke his neck in the resulting 15 ft (4.6 m) fall, which was the immediate cause of death.

Recommendations

Power circuits shall be de-energized before work is done on them unless hot-line tools are used. Switches shall be locked out, and the locks removed only by the person who installed them.

CASE 33: LOW-VOLTAGE DANGER

A laborer was electrocuted when his upper back contacted a 40-watt florescent tube end.

The victim was astride a screw conveyor he was helping to install, which was suspended on metal brackets about 30 in. (76 cm) from the ceiling. He had been employed by the construction contractor for 4½ months and had worked 2 days on the mining property.

Recommendations

Where new installations are installed in existing shops or plants, the location where the installation is planned to be should be examined by the designer, supervisor, and the worker involved. Each of these individuals should be aware of any potential hazard they may encounter as their work progresses.

CASE 34: NO WHITE-COLLAR IMMUNITY

A plant engineer with more than 13 years mining experience was fatally injured when he fell from a plant superstructure onto a moving conveyor belt and was pinned beneath a steel discharge chute (Figure A5-14).

The victim had enlisted the aid of a front-end-loader operator to run the plant control room while he positioned himself on the plant superstructure to observe a problem area. He signaled the loader operator to start up the belts from a position where only his forearm was visible to the control room. The control room operator responded to the signal by starting two successive belts. The plant engineer could not be seen from the control room at this time. He apparently fell onto one of the belts while they were in motion. When the control room operator observed a problem on the other belt, he de-energized both belts and assumed that the plant engineer would return to the control room. His failure to return prompted a search, and he was found pinned between the bottom of a hopper and the surface of the belt onto which he had fallen.

Figure A5-14.

Recommendations

- Safe access should be provided to work areas where observations must be made. The victim had climbed the superstructure of a conveyor belt and a shaker screen.
- Operators should be certain, by signal or other means, that all persons are clear before starting equipment.

Figure A5-15.

CASE 35: FAILURE TO BLOCK

A lead maintenance man, with 23 years experience in that job, was fatally injured in a fall from the top of a rotary kiln when it rotated unexpectedly (Figure A5-15).

The victim had just finished welding a wedge under the kiln rotating ring. He fell, striking a parked forklift before falling to the floor, a total height of about 19 ft (5.8 m). The kiln had not been blocked to prevent movement and the victim had failed to use a safety belt and line.

Recommendations

- Repairs or maintenance shall not be performed on machinery until the power is off and the machinery is blocked against motion, except where machinery motion is necessary to make adjustments.
- Safety belts and lines shall be worn when men work where there is danger of falling.

CASE 36: GRAVITATIONAL FORCE

A maintenance mechanic, with three years mining experience (all on this job), was fatally injured when a shuttle car he was working under rolled off of support blocks and pinned him beneath the car (Figure A5-16).

The victim had driven the shuttle car up on the support blocks to gain access to the underside of the car so he could repair a malfunctioning conveyor system.

Recommendations

Repairs or maintenance shall not be performed on machinery until the power is off and the machinery is blocked against motion, except where machinery motion is necessary to make adjustments.

CASE 37: CLEANING—POWER ON

A crusher mechanic with about six months of known mining experience was killed when he was drawn into the pinch point of a conveyor belt breakover pulley (Figure A5-17).

The victim was attempting to clean the breakover pulley by using a steel rod as a scraper while the conveyor was in motion. Contrary to instructions, the victim started up the conveyor apparently to facilitate the cleaning of the breakover pulley. The guard at this location extended sufficiently to prevent a person from "accidentally" reaching behind the guard and becoming caught, but not far enough to prevent reaching behind with an implement to extend access to the rotating pulley.

Recommendations

- Pulleys of conveyors shall not be cleaned manually while the conveyor is in motion.
- Electrically powered equipment shall be de-energized before mechanical work is done on such

Figure A5-16.

Figure A5-17.

equipment. Power switches shall be locked out or other measures taken which shall prevent the equipment from being energized without the knowledge of the individuals working on it. Suitable warning notices shall be posted at the power switch and signed by the individuals who are to do the work. Such locks or preventive devices shall be removed only by the person who installed them or by authorized personnel.

CASE 38: ENERGIZED TRAP

An electrician with eight years mining experience was electrocuted while troubleshooting an electrical malfunction in an open pit power shovel (Figure A5-18).

The shovel was powered by a 4160-V, three-phase system. The victim and his co-worker, the shovel oiler, had determined that the suspect circuits were a set of 480-V AC contacts which were housed in the same oil switch container as the DC coil for the 4,160-V switch contacts. The 4,160-V AC power entered the

Figure A5-18.

contact compartment of the oil switch through bushings which were located close to the suspect contacts.

The victim apparently lost his balance while kneeling into position to work and contacted two 4,160-V bushings, allowing the current to travel from hand to hand across his body. The shovel oiler received voltage across his body to ground when he pulled the victim from the contacts. This caused the ground relay to trip, but he was not injured since his contact was limited to phase to ground with resistance limiting voltage. The proper operation of the ground fault interrupter circuit probably prevented a second fatality in this incident.

Recommendations

Power circuits in quarry shovels shall be de-energized before work is performed on or near the circuits.

Appendix 6

Glossary

This glossary is provided to assist in understanding terms related to the control of hazardous energy. The glossary is divided into three sections: regulatory definitions, consensus definitions, and general definitions. The division reflects the following differences in definition that may exist:

- *Regulatory:* Definitions that have the force and effect of laws, regulations, or governmental standards
- *Consensus:* Definitions arrived at by consensus as a result of standards-developing activities of organizations such as the American National Standards Institute, the American Society of Testing and Materials, and the British Standards Institution
- *General:* Definitions obtained from various sources that represent common usage, professional practice, and technical and scientific perspectives.

In some cases, definitions in each category are available with different wording or possibly different literal interpretation. This is not intended to create confusion but to expand the reader's appreciation for the existence of varying definitions of common terms. A broader understanding of a term may be possible when several definitions are compared.

From a regulatory standpoint, definitions will impact what, how, and when something must be done to achieve compliance. An "authorized person" according to the U.S. OSHA standards definition may be quite different from an authorized person in a dictionary or consensus definition and may also carry with it specific legal consequences.

The following sources were used to create this glossary:

- Regulatory
  - 29 *CFR* 1910.146, *Permit-Required Confined Spaces*
  - 29 *CFR* 1910.147, *The Control of Hazardous Energy (Lockout/Tagout)*
  - 29 *CFR* 1910.211, Subpart O, *Machinery and Machine Guarding*
  - 29 *CFR* 1910.399, Subpart S, *Electrical*
- Consensus
  - *Safety Requirements for the Lock Out/Tag Out of Energy Sources,* ANSI Z244.1-1992. American Standards Institute, 1992
  - *National Electrical Code,* NFPA 70. National Fire Protection Association, 1993
  - *Safety of Machinery,* BS 5304-1988. British Standards Institution
  - *Safety Requirements for Sand Preparation, Molding, and Coremaking in the Sand Foundry Industry,* ANSI Z241.1-1989. American National Standards Institute, 1989
  - *Safety Requirements for Confined Spaces,* ANSI Z117.1-1989. American National Standards Institute, 1989

- General
  - *Accident Prevention Manual for Business & Industry: Engineering & Technology,* 10th edition. National Safety Council, 1992. All definitions in the General Section are from this manual.
  - An additional source of general definitions is the American Society of Safety Engineers, *The Dictionary of Terms Used in the Safety Profession,* 3d ed., Des Plaines, IL: 1988.

Definitions are reprinted form the sources given in parentheses after the definition. Acknowledgment is given to the following:

- Definitions are reproduced with permission from American National Standard *Safety Requirements for the Lock Out/Tag Out of Energy Sources*, ANSI Z244.1-1982, copyright 1982 by the American National Standards Institute. Copies of this standard may be purchased form the American National Standards Institute, 11 West 42nd Street, New York, NY 10036.
- The definition of Zero Mechanical State (ZMS) is printed with permission from the American Foundrymen's Society, Inc.
- Definitions are reprinted with permission from NFPA 70-1993, the *National Electric Code®*, Copyright © 1992, National Fire Protection Association, Quincy, MA 02269. This reprinted material is not the complete and official position of the National Fire Protection Association on the referenced subject, which is represented only by the standard in its entirety.
- Extracts from BS 5304: 1988 are reproduced with the permission of the British Standards Institution. Complete copies can be obtained through national standards bodies.
- Definitions are reproduced with permission from American National Standard *Safety Requirements for Sand Preparation, Molding, and Coremaking in the Sand Foundry Industry,* ANSI Z241.1-1989, copyright © 1989 by the American National Standards Institute. Copies of this standard may be purchased form the American National Standards Institute, 11 West 42nd Street, New York, NY 10036.
- Definitions from ANSI Z117.1 are reprinted with permission from the American Society of Safety Engineers.

REGULATORY DEFINITIONS

Accessible (as applied to equipment). Admitting close approach; not guarded by locked doors, elevation, or other effective means (*see* Readily accessible). (29 *CFR* 1910.399)

Accessible (as applied to wiring methods). Capable of being removed or exposed without damaging the building structure or finish, or not permanently closed in by the structure or finish of the building (*see* Exposed). (29 *CFR* 1910.399)

Affected employee. An employee whose job requires him/her to operate or use a machine or equipment on which servicing or maintenance is being performed under lockout or tagout, or whose job requires him/her to work in an area in which such servicing or maintenance is being performed. (29 *CFR* 1910.147)

Ampacity. Current-carrying capacity of electric conductors expressed in amperes. (29 *CFR* 1910.399)

Authorized employee. A person who locks or implements a tagout system procedure on machines or equipment to perform the servicing or maintenance on that machine or equipment. An authorized employee and an affected employee may be the same person when the affected employee's duties also include performing maintenance or service on a machine or equipment which must be locked or a tagout system implemented. (29 *CFR* 1910.147)

Authorized person. One to whom the authority and responsibility to perform a specific assignment has been given by the employer. (29 *CFR* 1910.211)

Belts. All power transmission belts, such as flat belts, round belts, V-belts, etc., unless otherwise specified.

Bite. The nip point between any two inrunning rolls. (29 *CFR* 1910.211)

Blanking or blinding. The absolute closure of a pipe, line, or duct by the fastening of a solid plate (such as a spectacle blind or a skillet blind) that completely covers the bore and that is capable of withstanding the maximum pressure of the pipe, line, or duct with no leakage beyond the plate. (29 *CFR* 1910.146)

Bonding. The permanent joining of metallic parts to form an electrically conductive path which will assure electrical continuity and the capacity to conduct safely any current likely to be imposed. (29 *CFR* 1910.399)

Branch circuit. The circuit conductors between the final overcurrent device protecting the circuit and the outlet(s). (29 *CFR* 1910.399)

"Capable of being locked out." An energy isolating device will be considered to be capable of being locked out either if it is designed with a hasp or other attachment or integral part to which, or through which, a lock can be affixed, or if it has a locking mechanism built into it. Other energy isolating devices will also be considered to be capable of being locked out, if lockout can be achieved without the need to dismantle, rebuild, or replace the energy isolating device or permanently alter its energy control capability. (29 *CFR* 1910.147)

Circuit breaker.

600 volts nominal, or less. A device designed to open and close a circuit by nonautomatic means and to open the circuit automatically on a predetermined overcurrent without injury to itself when properly applied within its rating.

Over 600 volts, nominal. A switching device capable of making, carrying, and breaking currents under normal circuit conditions, and also making, carrying for a specified time, and breaking currents under specified abnormal circuit conditions, such as those of short circuit. (29 *CFR* 1910.399)

Competent person. One who is capable of identifying existing and predictable hazards in the surroundings or working conditions that are unsanitary, hazardous, or dangerous to employees, and who has the authorization to take prompt corrective measures to eliminate them. (U.S. OSHA 1926.32)

Conductor.

Bare. A conductor having no covering or electrical insulation whatsoever.

Covered. A conductor encased within material of composition or thickness that is not recognized as electrical insulation.

Insulated. A conductor encased within material of composition and thickness that is recognized as electrical insulation. (29 *CFR* 1910.399)

Control system. Sensors, manual input and mode selection elements, interlocking and decision-making circuitry, and output elements to the press-operating mechanism. (29 *CFR* 1910.211)

Controller. A device or group of devices that serves to govern, in some predetermined manner, the electrical power delivered to the apparatus to which it is connected. (29 *CFR* 1910.399)

Cutout (over 600 volts, nominal). An assembly of a fuse support with either a fuseholder, fuse carrier, or disconnecting blade. The fuseholder or fuse carrier may include a conducting element (fuse link), or may act as the disconnecting blade by the inclusion of a nonfusible member. (29 *CFR* 1910.399)

Die. The tooling used in a press for cutting or forming material. An upper and a lower die make a complete set. (29 *CFR* 1910.211)

Die setter. An individual who places or removes dies in or from mechanical power presses, and who, as a part of his/her duties, makes the necessary adjustments to cause the tooling to function properly and safely. (29 *CFR* 1910.211)

Disconnecting means. A device, or group of devices, or other means by which the conductors of a circuit can be disconnected from their source of supply. (29 *CFR* 1910.399)

Disconnecting (or isolating) switch (over 600 volts, nominal). A mechanical switching device used for isolating a circuit or equipment from a source of power. (29 *CFR* 1910.399)

Double block and bleed. The closure of a line, duct, or pipe by closing and locking or tagging two in-line valves and by opening and locking or tagging a drain or vent valve in the line between the two closed valves. (29 *CFR* 1910.146)

Energized. Connected to an energy source or containing residual or stored energy. (29 *CFR* 1910.147)

Energy isolating device. A mechanical device that physically prevents the transmission or release of energy, including but not limited to the following: a manually operated electrical circuit breaker; a disconnect switch; a manually operated switch by which the conductors of a circuit can be disconnected from all ungrounded supply conductors and, in addition, no pole can be operated independently; a slide gate; a slip blind; a line valve; a block; and any similar device used to block or isolate energy. The term does not include a push button, selector switch, and other control circuit type devices. (29 *CFR* 1910.147)

Energy source. Any source of electrical, mechanical, hydraulic, pneumatic, chemical, thermal, or other energy. (29 *CFR* 1910.147)

Equipment. A general term including material, fittings, devices, appliances, fixtures, apparatus, and the like, used as a part of, or in connection with, an electrical installation. (29 *CFR* 1910.399)

Exposed (as applied to live parts). Capable of being inadvertently touched or approached nearer than a safe distance by a person. It is applied to parts not suitably guarded, isolated, or insulated (*see* Accessible). (29 *CFR* 1910.399)

Exposed (as applied to wiring methods). On or attached to the surface or behind panels designed to allow access (*see* Accessible [as applied to wiring methods]). (29 *CFR* 1910.399)

Exposed to contact. The location of an object is such that a person is likely to come into contact with it and be injured. (29 *CFR* 1910.211)

Externally operable. Capable of being operated without exposing the operator to contact with live parts. (29 *CFR* 1910.399)

Feeder. All circuit conductors between the service equipment, or the generator switchboard of an isolated plant, and the final branch-circuit overcurrent device. (29 *CFR* 1910.399)

Fuse (over 600 volts, nominal). An overcurrent protective device with a circuit-opening fusible part that is heated and severed by the passage of overcurrent through it. A fuse comprises all the parts that form a unit capable of performing the prescribed functions. It may or may not be the complete device necessary to connect it into an electrical circuit. (29 *CFR* 1910.399)

Ground. A conducting connection, whether intentional or accidental, between an electrical circuit or equipment and the earth, or to some conducting body that serves in place of the earth. (29 *CFR* 1910.399)

Grounded. Connected to earth or to some conducting body that serves in place of the earth. (29 *CFR* 1910.399)

Ground-fault circuit-interrupter. A device whose function is to interrupt the electrical circuit to the load when a fault current to ground exceeds some predetermined value that is less than that required to operate the overcurrent protective device of the supply circuit. (29 *CFR* 1910.399)

Guarded. Covered, shielded, fenced, enclosed, or otherwise protected by means of suitable covers, casings, barriers, rails, screens, mats, or platforms to remove the likelihood of approach to a point of danger or contact by persons or objects. (29 *CFR* 1910.399)

Hot tap. A procedure used in the repair, maintenance, and services activities that involves welding on a piece of equipment (pipelines, vessels, or tanks) under pressure, in order to install connections or appurtenances. It is commonly used to replace or add sections of pipeline without the interruption of service for air, gas, water, steam, and petrochemical distribution systems. (29 *CFR* 1910.147)

Hot work permit. The employer's written authorization to perform operations (for example, riveting, welding, cutting, burning, and heating) capable of providing a source of ignition. (29 *CFR* 1910.146)

Inch. An intermittent motion imparted to the slide (on machines using part-revolution clutches) by momentary operation of the INCH operating means. Operation of the INCH operating means engages the driving clutch so that a small portion of one stroke or indefinite stroking can occur, depending upon the length of time the INCH operating means is held operated. INCH is a function used by the die setter for setup of dies and tooling, but is not intended for use during production operations by the operator. (29 *CFR* 1910.211)

Inerting. The displacement of the atmosphere in a permit space by a noncombustible gas (such as nitrogen) to such an extent that the resulting atmosphere is noncombustible. (29 *CFR* 1910.146)

Interrupter switch (over 600 volts, nominal). A switch capable of making, carrying, and interrupting specified currents. (29 *CFR* 1910.399)

Isolated. Not readily accessible to persons unless special means for access are used. (29 *CFR* 1910.399)

Isolation. The process by which a permit space is removed from service and completely protected against the release of energy and material into the space by such means as: blanking or blinding; misaligning or removing sections of lines, pipes, or ducts; a double block and bleed system; lockout or tagout of all sources of energy; or blocking or disconnecting all mechanical linkages. (29 *CFR* 1910.146)

Job-lock. A device used to ensure the continuity of energy isolation during a multishift operation. It is placed upon a lockbox. A key to the job-lock is controlled by each assigned primary authorized employee from each shift.

Job-tag with tab. A special tag for tagout of energy isolating devices during group lockout/tagout procedures. The tab of the tag is removed for insertion into the lock-box. The company procedure would require that the tagout job-tag cannot be removed until the tab is rejoined to it.

Jog. An intermittent motion imparted to the slide by momentary operation of the drive motor, after the clutch is engaged with the flywheel at rest. (29 *CFR* 1910.211)

Line breaking. The intentional opening of a pipe, line, or duct that is or has been carrying flammable, corrosive, or toxic material, an inert gas, or any fluid at a volume, pressure, or temperature capable of causing injury. (29 *CFR* 1910.146)

Lockout. The placement of a lockout device on an energy-isolating device, in accordance with an established procedure, ensuring that the energy-isolating device and the equipment being controlled cannot be operated until the lockout device is removed. (29 *CFR* 1910.147)

Lockout device. A device that utilizes a positive means such as a lock, either key or combination type, to hold an energy isolating device in the safe position and prevent the energizing of a machine or equipment. (29 *CFR* 1910.147)

Master lockbox. The lockbox into which all keys and tabs from the lockout or tagout devices securing the machine or equipment are inserted and which would be secured by a "job-lock" during multishift operations.

Master tag. A document used as an administrative control and accountability device. This device is normally controlled by the operations department personnel and is a personal tagout device if each employee personally signs on and signs off on it and if the tag clearly identifies each authorized employee who is being protected by it.

May. If a discretionary right, privilege, or power is conferred, the word "may" is used. If a right, privilege, or power is abridged or if an obligation to abstain from acting is imposed, the word "may" is used with a restrictive "no," "not," or "only" (e.g., no employer may . . .; an employer may not . . .; only qualified persons may . . . .) (29 *CFR* 1910.399)

Normal production operations. The utilization of a machine or equipment to perform its intended production function. (29 *CFR* 1910.147)

Oil (filled) cutout (over 600 volts, nominal). A cutout in which all or part of the fuse support and its fuse link or disconnecting blade are mounted in oil with complete immersion of the contacts and the fusible portion of the conducting element (fuse link), so that arc interruption by severing of the fuse link or by opening of the contacts will occur under oil. (29 *CFR* 1910.399)

Operator's station. The complete complement of controls used by or available to an operator on a given operation for stroking the press. (29 *CFR* 1910.211)

Outlet. A point on the wiring system at which current is taken to supply utilization equipment. (29 *CFR* 1910.399)

Overcurrent. Any current in excess of the rated current of equipment or the ampacity of a conductor. It may result from overload (see definition), short circuit, or ground fault. A current in excess of rating may be accommodated by certain equipment and conductors for a given set of conditions. Hence the rules for overcurrent protection are specific for particular situations. (29 *CFR* 1910.399)

Overload. Operation of equipment in excess of normal, full load rating, or of a conductor in excess of rated ampacity which, when it persists for a sufficient length of time, would cause damage or dangerous overheating. A fault, such as a short circuit or ground fault, is not an overload (*see* Overcurrent). (29 *CFR* 1910.399)

Panelboard. A single panel or group of panel units designed for assembly in the form of a single panel; including buses, automatic overcurrent devices, and with or without switches for the control of light, heat, or power circuits; designed to be placed in a cabinet or cutout box placed in or against a wall or partition and accessible only from the front (*see* Switchboard). (29 *CFR* 1910.399)

Pinch point. Any point other than the point of operation at which it is possible for a part of the body to be caught between the moving parts of a press or auxiliary equipment, or between moving and stationary parts of a press or auxiliary equipment or between the material and moving part or parts of the press or auxiliary equipment. (29 *CFR* 1910.211)

Point of operation. That point at which cutting, shaping, or forming is accomplished upon the stock and shall include such other points as may offer a hazard to the operator in inserting or manipulating the stock in the operation of the machine. . . . The area of the press where material is actually positioned and work is being performed during any process such as shearing, punching, forming, or assembling. (29 *CFR* 1910.211)

Press. A mechanically powered machine that shears, punches, forms, or assembles metal or other material by means of cutting, shaping, or combination dies attached to slides. A press consists of a stationary bed or anvil, and a slide (or slides) having a controlled reciprocating motion toward and away from the bed surface, the slide being guided in a definite path by the frame of the press. (29 *CFR* 1910.211)

Primary authorized employee. The authorized employee who exercises overall responsibility for adherence to the company lockout/tagout procedure (*see* Appendix 1, 29 *CFR* 1910.147[f][3][ii][A]).

Prime movers. Steam, gas, oil, and air engines, motors, steam and hydraulic turbines, and other equipment used as a source of power. (29 *CFR* 1910.211)

Principal authorized employee. An authorized employee who oversees or leads a group of servicing/maintenance workers (e.g., plumbers, carpenters, electricians, metal workers, mechanics).

Qualified person. One familiar with the construction and operation of the equipment and hazards involved.

NOTE 1: Whether an employee is considered to be a "qualified person" will depend upon various circumstances in the workplace. It is possible and, in fact, likely for an individual to be considered "qualified" with regard to certain equipment in the workplace, but "unqualified" as to other equipment (*see* 29 *CFR* 1910.332[b][3] for training requirements that specifically apply to qualified persons).

NOTE 2: An employee who is undergoing on-the-job training and who, in the course of such training, has demonstrated an ability to perform duties safely at his or her level of training and who is under the direct supervision of a qualified person is considered to be a qualified person for the performance of those duties. (29 *CFR* 1910.399)

Readily accessible. Capable of being reached quickly for operation, renewal, or inspections, without requiring those to whom ready access is requisite to climb over or remove obstacles or to resort to portable ladders, chairs, etc. (*see* Accessible). (29 *CFR* 1910.399)

Receptacle. A receptacle is a contact device installed at the outlet for the connection of a single attachment plug. A single receptacle is a single contact device with no other contact device on the same yoke. A multiple receptacle is a single device containing two or more receptacles. (29 *CFR* 1910.399)

Remote-control circuit. Any electric circuit that controls any other circuit through a relay or an equivalent device. (29 *CFR* 1910.399)

Repeat. An unintended or unexpected successive stroke of the press resulting from a malfunction. (29 *CFR* 1910.211)

Safety block. A prop that, when inserted between the upper and lower dies or between the bolster plate and the face of the slide, prevents the slide from falling of its own deadweight. (29 *CFR* 1910.211)

Safety system. The integrated total system, including the pertinent elements of the press, the controls, the safeguarding and any required supplemental safeguarding, and their interfaces with the operator, and the environment, designed, constructed, and arranged to operate together as a unit, such that a single failure or single operating error will not cause injury to personnel due to point-of-operation hazards. (29 *CFR* 1910.211)

Satellite lockbox. A secondary lockbox or lockboxes to which each authorized employee affixes his/her personal lock or tag.

Service. The conductors and equipment for delivering energy from the electricity supply system to the wiring system of the premises served. (29 *CFR* 1910.399)

Servicing and/or maintenance. Workplace activities such as constructing, installing, setting up, adjusting, inspecting, modifying, and maintaining and/or servicing machines or equipment. These activities include lubrication, cleaning or unjamming of machines or equipment, and making adjustments or tool changes, where the employee may be exposed to the unexpected energization or start up of the equipment or release of hazardous energy. (29 *CFR* 1910.147)

Sheaves. Grooved pulleys, and shall be so classified unless used as fly-wheels. (29 *CFR* 1910.211)

Stroking selector. The part of the clutch/brake control that determines the type of stroking when the operating means is actuated. The stroking selector generally includes positions for OFF (CLUTCH CONTROL), INCH, SINGLE STROKE, and CONTINUOUS (when CONTINUOUS is furnished). (29 *CFR* 1910.211)

Switchboard. A large single panel, frame, or assembly of panels that have switches, buses, instruments, overcurrent and other protecting devices mounted on the face or back or both. Switchboards are generally accessible from the rear as well as from the front and are not intended to be installed in cabinets (*see* Panelboard). (29 *CFR* 1910.399)

Switches.

General-use switch. A switch intended for use in general distribution and branch circuits. It is rated in amperes, and it is capable of interrupting its rated current at its rated voltage.

General-use snap switch. A form of general-use switch so constructed that it can be installed in flush device boxes or on outlet box covers, or otherwise used in conjunction with wiring systems recognized by this subpart.

Isolating switch. A switch intended for isolating an electric circuit from the source of power. It has no interrupting rating, and it is intended to be operated only after the circuit has been opened by some other means.

Motor-circuit switch. A switch, rated in horsepower, capable of interrupting the maximum operating overload current of a motor of the same horsepower rating as the switch at the rated voltage. (29 *CFR* 1910.399)

Switching devices (over 600 volts, nominal). Devices designed to close and/or open one or more electric circuits. Included in this category are circuit breakers, cutouts, disconnecting (or isolating) switches, disconnecting means, interrupter switches, and oil (filled) cutouts. (29 *CFR* 1910.399)

Tagout. The placement of a tagout device on an energy-isolating device, in accordance with an established procedure, to indicate that the energy-isolating device and the equipment being controlled may not be operated until the tagout device is removed. (29 *CFR* 1910.147)

Tagout device. A prominent warning device, such as a tag and a means of attachment, which can be securely fastened to an energy-isolating device in accordance with an established procedure, to indicate that the energy-isolating device and the equipment being controlled may not be operated until the tagout device is removed. (29 *CFR* 1910.147)

Trip or tripping. Activation of the clutch to "run" the press. (29 *CFR* 1910.211)

Utilization equipment. Equipment which utilizes electric energy for mechanical, chemical, heating, lighting, or similar useful purpose. (29 *CFR* 1910.399)

Voltage, nominal. A nominal value assigned to a circuit or system for the purpose of conveniently designating its voltage class (as 120/240, 480Y/277, 600, etc.). The actual voltage at which a circuit operates can vary from the nominal within a range that permits satisfactory operation of equipment. (29 *CFR* 1910.399)

Work permit. A control document which authorizes specific tasks and procedures to be accomplished.

CONSENSUS DEFINITIONS

Affected employee. A person whose job includes activities such as erecting, installing, constructing, repairing, adjusting, inspecting, operating, or maintaining the equipment/process. (ANSI Z244.1)

Ampacity. The current in amperes that a conductor can carry continuously under the conditions of use without exceeding its temperature rating. (NFPA 70)

Attachment plug (plug cap) (cap). A device that, by insertion in a receptacle, establishes connection between the conductors of the attached flexible cord and the conductors connected permanently to the receptacle. (NFPA 70)

Attainment of Zero Mechanical State (ZMS)© Condition. The procedure and sequence of lockoff provided by the manufacturer to attain ZMS© shall be followed in its entirety. (ANSI Z241.1)

Authorized individual. A knowledgeable individual to whom the authority and responsibility to perform a specific assignment has been given by an employer. (ANSI Z244.1)

Automatic. A machine in automatic control mode is one wherein each function in the machine cycle is automatically performed and sequenced, including load, unload, and repeat cycle. (ANSI Z241.1) Self-acting, operating by its own mechanism when actuated by some impersonal influence, as for example, a change in current strength, pressure, temperature, or mechanical configuration (*see* Nonautomatic). (NFPA 70)

Blinding/blanking. Inserting a solid barrier across the open end of a pipe leading into or out of the confined space, and securing the barrier in such a way to prevent leakage of material into the confined space. (ANSI Z117.1)

Branch circuit. The circuit conductors between the final overcurrent device protecting the circuit and the outlet(s). (NFPA 70)

Circuit breaker. A device designed to open and close a circuit by nonautomatic means and to open the circuit automatically on a predetermined overcurrent without damage to itself when properly applied within its rating. (NFPA 70)

Consensus standards. Standards for which consensus must be reached by those having substantial concern with its scope and provisions. Consensus standards as used in this standard refer to those standards developed under the auspices of an approved standards-writing organization. (ANSI Z241.1)

Controller. A device or group of devices that serves to govern, in some predetermined manner, the electric power delivered to the apparatus to which it is connected. (NFPA 70)

Disconnecting means. A device, or group of devices, or other means by which the conductors of a circuit can be disconnected from their source of supply. (NFPA 70)

Double block and bleed. A method used to isolate a confined space from a line, duct, or pipe by physically closing two in-line valves on a piping system and opening a "vented-to-atmosphere" valve between them. (ANSI Z117.1)

Energy isolating device. A physical device that prevents the transmission or release of energy, including, but not limited to, the following: a manually operated electrical circuit breaker, a disconnect switch, a manually operated switch, a slide gate, a slip blind, a line valve, blocks, and similar devices with a visible indication of the position of the device. (Push buttons, selector switches, and other control-circuit type devices are not energy-isolating devices). (ANSI Z244.1)

Energy source. Any electrical, mechanical, hydraulic, pneumatic, chemical, nuclear, thermal, or other energy source that could cause injury to personnel. (ANSI Z244.1)

Exposed (as applied to live parts). Capable of being inadvertently touched or approached nearer than a safe distance by a person. It is applied to parts not suitably guarded, isolated, or installed (*see* Accessible). (NFPA 70)

Exposed (as applied to wiring methods). On or attached to the surface or behind panels designed to allow access (*see* Accessible [as applied to wiring methods]). (NFPA 70)

Externally operable. Capable of being operated without exposing the operator to contact with live parts. (NFPA 70)

Failure to danger. Any failure of the machinery, its associated safeguards, control circuits or its power supply that leaves the machinery in an unsafe condition. (BSI)

Failure to safety. Any failure of the machinery, its associated safeguards, control circuits or its power supply that leaves the machinery in a safe condition.

> NOTE: Nothing can be guaranteed always to fail to safety. (BSI)

Feeder. All circuit conductors between the service equipment or the source of a separately derived system and the final branch-circuit overcurrent device. (NFPA 70)

Fuse. An overcurrent protective device with a circuit-opening fusible part that is heated and severed by the passage of overcurrent through it. (NFPA 70)

Ground. A conducting connection, whether intentional or accidental, between an electrical circuit or equipment and the earth, or to some conducting body that serves in place of the earth. (NFPA 70).

Grounded. Connected to earth or to some conducting body that serves in place of the earth. (NFPA 70)

Ground-fault circuit-interrupter. A device intended for the protection of personnel that functions to de-energize a circuit or portion thereof within an established period of time when a current to ground exceeds some predetermined value that is less than that required to operate the overcurrent protective device of the supply circuit. (NFPA 70)

Guard. A physical barrier that prevents or reduces access to a danger point or area. (BSI)

Guarded. Covered, shielded, fenced, enclosed, or otherwise protected by means of suitable covers, casings, barriers, rails, screens, mats, or platforms to remove the likelihood of approach or contact by persons or objects to a point of danger. (NFPA 70) Shielded, fenced, enclosed, or otherwise protected by means of suitable enclosure, covers, casing, shield guards, trough guards, barrier guard, railing guards, or guarded by location, or other protective devices, so as to reduce the possible risk of personnel injury from accidental contact or approach, or in the case of spill guards so as to reduce possibility of personnel injury from material being spilled into the area protected. Where it is impossible or physically impractical to guard the hazard, or where the guard in itself creates a hazard, the potential hazard shall be marked prominently to warn of its existence. (ANSI Z241.1)

Guarded by location or position. In order to be guarded by location or position according to height above a walkway, platform, or workspace, any moving part shall be at last 8 ft (2.46 m) above same. However, pinch pins of all descriptions and moving projections shall not be guarded by location unless they are a minimum of 9 ft (2.74 m) above the pertinent floor.

When moving parts are guarded by their remoteness from the floor, platform, walkway, or other working level, or by their location with reference to frame, foundation, or structure when located so as to reduce the possibility of accidental contact by persons or objects. Remoteness from regular or frequent presence of public or employed personnel may, in reasonable circumstances, constitute guarding by location.

When moving parts are remote from floors, platforms, walkways, other working levels or by their location with reference to frames, foundations, or structures which minimize the probability of accidental contact by personnel, they shall be considered to be guarded by position or location. (ANSI Z241.1)

Hazard evaluation. A process to assess the severity of known, or real, or potential hazards or all three, at or in the confined space. (ANSI Z117.1)

Hostage control. A type of control in which the physical act of operating the initiator prevents operator exposure to the motion or response produced by the initiator. (ANSI Z241.1)

Hot work. Work within a confined space that produces arcs, sparks, flames, heat, or other sources of ignition. (ANSI Z117.1)

In sight from (within sight from, within sight). Where this Code specifies that one equipment shall be "in sight from," within sight from," or "within sight," etc., of another equipment, one of the equipments specified is to be visible and not more than 50 ft (15.24 m) distant from the other. (NFPA 70)

Inch initiator. A hostage control which causes machine motion in single or repeated small increments only when controlled by manual pressure. It is intended for use in set up or maintenance, but not in normal operation. (ANSI Z241.1)

Initiator. A device that causes an action of controls or power actuator. (ANSI Z241.1)

Integrity. The ability of devices, systems, and procedures to perform their function without failure or defeat. (BSI)

Interlock. A safety device that interconnects a guard with the control system or the power system of the machinery. (BSI) A device in a system which when actuated permits or prevents the operation of one or more components in the system. (ANSI Z241.1)

Interlocked barrier guard. A barrier attached to the machine and interlocked with the machine power or control so that the machine cycle cannot be initiated with the operating controls unless the guard, or the hinged or movable sections, effectively encloses the hazardous zone. (ANSI Z241.1)

Isolated. Not readily accessible to persons unless special means for access are used. (NFPA 70)

Isolation. A process of physically interrupting, or disconnecting, or both, pipes, lines, and energy sources from the confined space. (ANSI Z117.1)

Knowledgeable individual. One who knows the effect of operating the controls or equipment. (ANSI Z244.1)

Lockout device. A device that utilizes a lock and key to hold an energy-isolating device in the safe position for the purpose of protecting personnel. (ANSI Z244.1)

Lockout/tagout. The placement of a lock/tag on the energy-isolating device in accordance with an established procedure, indicating that the energy isolating device shall not be operated until removal of the lock/tag in accordance with an established procedure. (The term "lockout/tagout" allows the use of a lockout device, a tagout device, or a combination of both.) (ANSI Z244.1) The placement of a lock/tag on the energy-isolating device in accordance with an established procedure, indicating that the energy-isolating device shall not be operated until removal of the lock/tag in accordance with an established procedure. (ANSI Z117.1)

Manual. A machine in manual control mode is one wherein each machine function in the machine cycle and load cycle is manually initiated and controlled. (ANSI Z241.1)

Nip zone. A point or zone where a portion of the body may be caught hold of and squeezed between two surfaces, edges, or points. (ANSI Z241.1)

Nonautomatic. Action requiring personal intervention for its control (*see* Automatic). (NFPA 70)

Operator's work zone(s). The operator's work zone(s) of equipment is that area in which the operator's presence is required, while operating in the intended manner. (ANSI Z241.1)

Outlet. A point on the wiring system at which current is taken to supply utilization equipment. (NFPA 70)

Overcurrent. Any current in excess of the rated current of equipment or the ampacity of a conductor. It may result from overload, short circuit, or ground fault (*see* Overload). (NFPA 70)

Overload. Operations of equipment in excess of normal, full-load rating, or of a conductor in excess of rated ampacity that, when it persists for a sufficient length of time, would cause damage or dangerous overheating. A fault, such as a short circuit or ground fault, is not an overload (*see* Overcurrent). (NFPA 70)

Panelboard. A single panel or group of panel units designed for assembly in the form of a single panel; including buses, automatic overcurrent devices, and equipped with or without switches for the control of light, heat, or power circuits; designed to be placed in a cabinet or cutout box placed in or against a wall or partition and accessible only from the front (*see* Switchboard). (NFPA 70)

Pinch zone. Any point other than the point of operation at which it is possible for a part of the body to be caught between the moving parts of a machine or mechanism, or between moving and stationary parts of a machine or mechanism. (ANSI Z241.1)

Power locked off. The state in which the device that turns power off is locked in the OFF position with the padlock of every individual who is working on the machine. (ANSI Z241.1)

Power off. The state in which power cannot flow to the equipment. (ANSI Z241.1)

Protection—primary. Protective means that cannot be disconnected by malfunction of a device or intentional bypassing of it. Prevention of machine movement by direct control of the power source or direct prevention by mechanical stops.

Protection—secondary. A device or structure that can be bypassed by foreseeable means or may malfunction resulting in defeat of the protection sought. All ordinary control devices are of this nature, since they do not prevent application of power directly, but do so only through their action in controlling a device that does prevent such power application. (ANSI Z241.1)

Protective device. A means whereby personnel access to a hazardous zone or area is denied through the use of means other than a guard. (ANSI Z117.1)

Qualified person. A person who by reason of training, education and experience is knowledgeable in the operation to be performed and is competent to judge the hazards involved. (ANSI Z117.1) A person determined by the employer to have the training and/or experience to operate and/or maintain the equipment involved. (ANSI Z241.1) One familiar with the construction and operation of the equipment and the hazards involved. (NFPA 70)

Receptacle. A contact device installed at the outlet for the connection of a single attachment plug. (NFPA 70)

Receptacle outlet. An outlet where one or more receptacles are installed. (NFPA 70)

Remote-control circuit. Any elecrical circuit that controls any other circuit through a relay or an equivalent device.

Safe working practice. A safe system of work, i.e., a method of working that eliminates or reduces the risk of injury. (BSI)

Safeguard. A guard or device designed to protect persons from danger. (BSI)

Safety device. A device other than a guard that eliminates or reduces danger. (BSI)

Semiautomatic. A machine in semiautomatic control mode is one wherein at least one machine function in the cycle is automatically performed and sequenced, but which requires the operator to initiate at least one function manually. (ANSI Z241.1)

Separately derived system. A premises wiring system whose power is derived from generator, transformer, or converter windings and that has no direct electrical connection, including a solidly connected grounded circuit conductor, to supply conductors originating in another system. (NFPA 70)

Service equipment. The necessary equipment, usually consisting of a circuit breaker or switch and fuses, and their accessories, located near the point of entrance of supply conductors to a building or other structure, or an otherwise defined area, and intended to constitute the main control and means of cutoff of the supply. (NFPA 70)

Shall. The word "shall" denotes a mandatory requirement. (ANZI Z244.1) Denotes a mandatory requirement. (ANSI Z117.1)

Should. A recommendation that is sound safety and health practice; it does not denote a mandatory requirement. (ANSI Z117.1) Denotes an advisory recommendation. (ANSI Z244.1)

Slash (/). A slash (/) denotes "and/or" and indicates that two words or expressions, such as lockout/tagout or equipment/process, are to be taken together or individually. (ANSI Z244.1)

Stop block—manual. A restraining device that will prevent hazardous machine movement. (ANSI Z241.1)

Stop block—mechanical. A mechanical member incorporated into and acting as a part of a machine or piece of equipment for the purpose of stopping, holding, or restricting movement of a member(s). (ANSI Z241.1)

Switchboard. A large, single panel, frame, or assembly of panels on which are mounted, on the face or back, or both, switches, overcurrent and other protective devices, buses, and usually instruments. Switchboards are generally accessible from the rear as well as from the front and are not intended to be installed in cabinets (*see* Panelboard). (NFPA 70)

Switches.

Bypass isolation switch. A bypass isolation switch is a manually operated device used in conjunction with a transfer switch to provide a means of directly connecting load conductors to a power source, and of disconnecting the transfer switch.

General-use snap switch. A form of general-use switch so constructed that it can be installed in flush device boxes or on outlet box covers, or otherwise used in conjunction with wiring systems recognized by this Code.

General-use switch. A switch intended for use in general distribution and branch circuits. It is rated in amperes, and it is capable of interrupting its rated current at its rated voltage.

Isolating switch. A switch intended for isolating an electric circuit from the source of power. It has no interrupting rating, and it is intended to be operated only after the circuit has been opened by some other means.

Motor-circuit switch. A switch, rated in horsepower, capable of interrupting the maximum operating overload current of a motor of the same horsepower rating as the switch at the rated voltage.

Transfer switch. A transfer switch is a device for transferring one or more load conductor connections from one power source to another. (NFPA 70)

Switching device. A device designed to close, open, or both, one or more electrical circuits. (NFPA 70)

Switching devices (> 600 volts).

Circuit breaker. A switching device capable of making, carrying, and breaking currents under normal circuit conditions, and also making, carrying for a specified time, and breaking currents under specified abnormal circuit conditions, such as those of short circuit.

Cutout. An assembly of a fuse support with either a fuseholder, fuse carrier, or disconnecting blade. The fuseholder or fuse carrier may include a conducting element (fuse link), or may act as the disconnecting blade by the inclusion of a nonfusible member.

Disconnecting (or isolating) switch (disconnector, isolator). A mechanical switching device used for isolating a circuit or equipment from a source of power.

Disconnecting means. A device, group of devices, or other means whereby the conductors of a circuit can be disconected from their source of supply.

Interrupter switch. A switch capable of making, carrying, and interrupting specified currents.

Oil cutout (oil-filled cutout). A cutout in which all or part of the fuse support and its fuse link or disconnecting blade is mounted in oil with complete immersion of the contacts and the fusible portion of the conducting element (fuse link), so that arc interruption by severing of the fuse link or by opening of the contacts will occur under oil.

Oil switch. An oil switch is a switch having contacts that operate under oil (or askarel or other suitable liquid).

Regulator bypass switch. A specific device or combination of devices designed to bypass a regulator. (NFPA 70)

Tagout device. A prominent warning device that is capable of being securely attached and that, for the purpose of protecting personnel, forbids the operation of an energy-isolating device and identifies the applier or authority who has control of the procedure. (ANSI Z244.1)

Two-hand maintained initiators. A type of control in which the operator causes a motion by manually operating an initiator concurrently with each hand, the motion stopping or reversing upon deactuation of either or both initiators. (ANSI Z241.1)

Two-hand momentary initiators. A type of control in which the operator causes a motion by manually operating an initiator concurrently with each hand, the motion continuing to completion whether the initiators continue to be held actuated or not. (ANSI Z241.1)

Utilization equipment. Equipment that utilizes electric energy for electronic, electromechanical, chemical, heating, lighting, or similar purposes. (NFPA 70)

Voltage, nominal. A nominal value assigned to a circuit or system for the purpose of conveniently designating its voltage class (e.g., 120/240, 480Y/277, 600). (NFPA 70)

Zero Mechanical State (ZMS)©. That mechanical state of a machine in which:

(1) Every power source that can produce a machine member movement has been locked off.

(2) Pressurized fluid (air, oil, or other) power lockoffs (shut-off valves), if used, will block pressure from the power source and will reduce pressure on the machine side port of that valve by venting to atmosphere or draining to tank.

(3) All accumulators and air surge tanks are reduced to atmospheric pressure or are treated as power sources to be locked off, as stated in paragraphs (1) and (2) of this section.

(4) The mechanical potential energy of all portions of the machine is at its lowest practical value; i.e. opening of pipe(s), tubing, hose(s), or actuation of any valve(s) will not produce a movement which could cause injury.

(5) Pressurized fluid (air, oil, or other) trapped in the machine lines, cylinders, or other components is not capable of producing a machine motion upon actuation of any valve(s).

(6) The kinetic energy of the machine members is at its lowest practical value.

(7) Loose or freely movable machine members are secured against accidental movement.

(8) A workpiece or material supported, retained, or controlled by the machine shall be considered as part of the machine if the workpiece or material can move or can cause machine movement.

(9) Part of the Zero Mechanical State (ZMS©) procedure shall be actuation of the start initiator after lockout. (ANSI Z241.1)

GENERAL DEFINITIONS

Accident. An unplanned event, not neccessarily injurious or damaging to property, interrupting the activity in process.

Accident causes. Hazards and those factors that, individually or in combination, directly cause accidents or accident potential.

Atomic energy. Energy released in nuclear reactions. The energy is released when a neutron splits an atom's nucleus into smaller pieces (fission) or when two nuclei are joined together under millions of degrees of heat (fusion).

Barrier guard. Protection for operators and other individuals from hazard points on machinery and equipment.

Adjustable barrier guard. An enclosure attached to the frame of the machinery or equipment with front and side sections that can be adjusted.

Fixed barrier guard. A point-of-operation enclosure attached to the machine or equipment.

Gate or movable barrier guard. A device designed to enclose the point of operation completely before the clutch can be engaged.

Interlocked barrier guard. An enclosure attached to the machinery or equipment frame and interlocked with the power switch so that the operating cycle cannot be started unless the guard is in its proper position.

Casual factor (of an accident). A combination of simultaneous or sequential circumstances directly or indirectly contributing to an accident. Modified to identify several kinds of causes such as direct, early, mediate, proximate, distal, etc.

CFR. *Code of Federal Regulations.* A collection of the regulations that have been promulgated under U.S. law.

Circuit. A complete path over which electrical current may flow.

Circuit breaker. A device that automatically interrupts the flow of an electrical current when the current becomes excessive.

Codes. Rules and standards that have been adopted by a government agency as mandatory regulations having the force and effect of law. Also used to describe a body of standards.

Consensus standard. A standard developed through a consensus process or general opinion among representatives of various interested or affected organizations and individuals.

Energy-control program. A program consisting of an energy-control procedure and employee training to ensure that a machine or equipment is isolated or inoperative before servicing or maintenance, thus protecting the employee from unexpected machine start-up or energizing.

Energy-isolating device. A mechanical device that prevents the release or transmission of energy. Some examples of energy-isolating devices include: a manually operated circuit breaker, a disconnect switch, a line valve, a block, and other similar devices. The following are not energy-isolating devices: push buttons, selector switches, and other circuit-control devices.

Engineering controls. Methods of controlling employee exposures by modifying the source or reducing the quantity of hazards.

Force. That which changes the state of rest or motion in matter.

Imminent danger. An impending or threatening dangerous situation that could be expected to cause death or serious injury to persons in the immediate future unless corrective measures ar taken.

Interlock. A device that interacts with another device or mechanism to govern succeeding operations. For example, an interlocked machine guard will prevent the machine from operating unless the guard is in its proper place. An interlock on an elevator door will prevent the car from moving unless the door is properly closed.

Job safety analysis. A method for studying a job in order to (a) identify hazards or potential accidents associated with each step or task and (b) develop solutions that will eliminate, nullify, or prevent such hazards or accidents. Sometimes called Job Hazard Analysis.

Kinetic energy. Energy due to motion (*see* Work).

Lockout/tagout. A program or procedure that prevents injury by eliminating unintentional operation or release of stored energy within machinery or processes during set-up, start-up, or maintenance repairs (*see* Energy-control program).

MORT. Management Oversight and Risk Tree.

Negligence. The lack of reasonable conduct or care, characterized by "accidental" or "thoughtlessness," that a prudent person would ordinarily exhibit. There need not be a legal duty.

OSHA. The U.S. Occupational Safety and Health Administration of the Department of Labor; federal agency with safety and health regulatory and enforcement authorities for general U.S. industry and business.

Potential energy. Energy due to position of one body with respect to another or to the relative parts of the same body.

Power. Time rate at which work is done; units are the watt (one joule per second) and the horsepower

(33,000 foot-pounds per minute). One horsepower equals 746 watts.

Pressure. Force applied to, or distributed over a surface; measured as force per unit area.

Preventive maintenance. The systematic actions performed to maintain equipment in normal working condition and prevent failure.

Safeguarding. The term used to cover all methods of keeping employees away from points of operation.

Work. When a force acts against resistance to produce motion in a body, the force is said to do work. Work is measured by the product of the force acting and the distance moved through against the resistance. The units of measurement are the erg (the joule is 1×10^7 ergs) and the foot-pound.

Zero mechanical energy (ZME). An old term indicating a piece of equipment without any source of power that could harm someone.

Index

A

F

G

I

N

P

T

U